Biomimetic Organic Synthesis

Edited by Erwan Poupon and Bastien Nay

Related Titles

Nicolaou, K. C., Chen, J. S.

Classics in Total Synthesis III

Further Targets, Strategies, Methods

2011

ISBN: 978-3-527-32958-8

Dewick, P. M.

Medicinal Natural Products

A Biosynthetic Approach

Third Edition

2009

ISBN: 978-0-470-74168-9

Dalko, P. I. (ed.)

Enantioselective Organocatalysis

Reactions and Experimental Procedures

2007

ISBN: 978-3-527-31522-2

Breslow, R. (ed.)

Artificial Enzymes

2005

ISBN: 978-3-527-31165-1

Berkessel, A., Gröger, H.

Asymmetric Organocatalysis

From Biomimetic Concepts to Applications in Asymmetric Synthesis

2005

ISBN: 978-3-527-30517-9

Nicolaou, K. C., Snyder, S. A.

Classics in Total Synthesis II

More Targets, Strategies, Methods

2003

ISBN: 978-3-527-30684-8

Biomimetic Organic Synthesis

Volume 1
Alkaloids

Edited by Erwan Poupon and Bastien Nay

WILEY-VCH Verlag GmbH & Co. KGaA

The Editors

Prof. Dr. Erwan Poupon
Université Paris-Sud
Faculté du Pharmacie
5, rue Jean-Baptiste Clément
92260 Châtenay-Malabry
France

Dr. Bastien Nay
Museum National d'Histoire
Naturelle, CNRS
57, rue Cuvier
75005 Paris
France

Library of Congress Card No.: applied for

British Library Cataloguing-in-Publication Data
A catalogue record for this book is available from the British Library.

Bibliographic information published by the Deutsche Nationalbibliothek
The Deutsche Nationalbibliothek lists this publication in the Deutsche Nationalbibliografie; detailed bibliographic data are available on the Internet at <http://dnb.d-nb.de>.

Composition Laserwords Private Ltd., Chennai
Printing and Binding Strauss GmbH, Mörlenbach
Cover Design Schulz Grafik-Design, Fußgönheim

Printed in the Federal Republic of Germany
Printed on acid-free paper

ISBN: 978-3-527-32580-1
ePDF ISBN: 978-3-527-63477-4
ePub ISBN: 978-3-527-63476-7
Mobi ISBN: 978-3-527-63478-1

Foreword

The beauty and diversity of the biochemical pathways developed by Nature to produce complex molecules is a good source of inspiration for chemists who want to guided in their synthetic approach by biomimetic strategies. The first biomimetic syntheses were reported at the beginning of the 20th century, with the famous examples of Collie's and Robinson's related to the synthesis of phenolics (orcinol) and alkaloids (tropinone). Since then, the number of reported biomimetic syntheses, especially in the last twenty years, has increased, demonstrating the power of these approaches in contemporary organic and bioorganic chemistry. Biomimetic strategies allow the construction of complex natural products in a minimum of steps which is in accordance with the "atom economy" principle of green chemistry and, in addition, simple reagents can be used to access the targets. Furthermore, the bioorganic consequences of such successful syntheses allow the comprehension of the biosynthetic origin of natural compounds and these processes can produce sufficient quantities of pure products to achieve biological investigations.

The biomimetic synthesis field came to maturity thanks to interconnexions between biosynthetic studies and organic synthesis, especially in the total synthesis of complex molecules. Biomimetic syntheses could even be considered as the latest stage of biosynthetic studies, confirming or invalidating the intimate steps leading to natural product skeletons. For example, the Johnson's polycyclization of squalene precursors is one of the most impressive achievements in this field. This is still organic synthesis as the reactions are taking place in the chemist's flask under chemically controlled experimental conditions, while biosynthetic steps can involve enzymatic catalysis, at least to a certain extent. However, concerning complex biochemical transformations, the exact role of enzymes has not always been clear, and has even been questionned by synthetic chemists.

The two book volumes "*Biomimetic Organic Synthesis*" fill the gap in the organic chemistry literature on complex natural products. These books gather 25 chapters from outstanding authors, not only dealing with the most important families of natural products (alkaloids, terpenoids, polyketides, polyphenols...), but also with biologically inspired reactions and concepts which are truly taking part in biomimetic processes. By assembling these books, the editors ***E. Poupon and B. Nay*** succeeded in gathering specialists in complex natural product chemistry for the benefit of the synthetic chemist community. With an educational effort in discussions and schemes, and in comparing both the biosynthetic routes and the biomimetic achievements, the demonstration of the power of the biomimetic strategies will become obvious to the readers in both research and teaching areas. These books will be a great source of inspiration for organic chemists and will ensure the continued development in this exciting field.

ESPCI-ParisTech Paris, France *Janine Cossy*

Contents to Volume 1

Contents to Volume 2

Preface

When we decided to start this project, at the end of 2008, we were perfectly aware that the amount of work to provide on it, the Biomimetic Organic Synthesis saga, would be very important. In fact, we were far from reality since the field not only concerns the huge universe of natural product chemistry, but also tends to embrace many fields beyond. We tried to design this book according to natural product chemistry principles, mainly by compound classes, and hope that few of them slipped our notice. Hopefully, the contributors who were asked to write a chapter in their respective field have welcomed this project with a great enthusiasm and worked hard to finish their chapter on time. Our editing adventure is now ending and we want now to warmly thank all of them for their outstanding contribution to this lengthy book. We also want to pay tribute to Professor François Tillequin, so happy with natural product chemistry, who recently passed away. Special thanks are also due to the staff of Wiley-VCH especially to Dr Gudrun Walter and Lesley Belfit for excellent collaboration.

Biomimetic synthesis is the construction of natural products by chemical means using Nature's hypothetical or established strategies, *i.e.* starting from synthetic mimicry of Nature's biosynthetic precursors, ideally by way of biologically compatible reactions. In theory, this principle can be applied to all natural product classes, from the simplest to the most complex compounds. Yet the activation methods in the laboratory can be far from Nature's enzymatic environment, and the biomimetic step can then be more difficult than expected at first glance. The way may therefore be tricky, even for a skilled chemist. We hope this book will delight readers by materializing most of organic synthesis concepts built from biochemical (biosynthetic) inspirations. Fortunately, readers may find solutions to synthetic problems or, at least, find a new way to improve their knowledge, as we did.

Enjoy reading.

March 2011

Erwan Poupon
Université Paris-Sud, Châtenay-Malabry, France
Bastien Nay
Muséum National d'Histoire Naturelle, Paris, France

List of Contributors

Ali Al-Mourabit
Centre de Recherche de
Gif-sur-Yvette
Institut de Chimie des
Substances Naturelles
UPR 2301 CNRS
Avenue de la Terrasse
91198 Gif-sur-Yvette
France

Jérôme Appenzeller
Centre de Recherche de
Gif-sur-Yvette
Institut de Chimie des
Substances Naturelles
UPR 2301 CNRS
Avenue de la Terrasse
91198 Gif-sur-Yvette
France

Hans-Dieter Arndt
Technische Universität
Dortmund
Fakultät Chemie
Otto-Hahn-Strasse 6
44221 Dortmund
Germany

and

Max-Planck Institut für
Molekulare Physiologie
Otto-Hahn-Strasse 11
44227 Dortmund
Germany

Marco A. Ciufolini
The University of
British Columbia
Department of Chemistry
2036 Main Mall
Vancouver
British Columbia V6T 1Z1
Canada

Romain Duval
Institut de Recherche pour le
Dévelopement
UMR 152
Faculté des Sciences
Pharmaceutiques
118 Route de Narbonne
31062 Toulouse
France

Tanja Gaich
Leibniz Universität Hannover
Institute of Organic Chemistry
Schneiderberg 1
30167 Hannover
Germany

Edmond Gravel
CEA, iBiTecS
Service de Chimie
Bioorganique et de Marquage
91191 Gif-sur-Yvette
France

Huan Liang
The University of
British Columbia
Department of Chemistry
2036 Main Mall
Vancouver
British Columbia V6T 1Z1
Canada

Roman Lichtenecker
Technische Universität
Dortmund
Fakultät Chemie
Otto-Hahn-Strasse 6
44221 Dortmund
Germany

and

Max-Planck Institut für
Molekulare Physiologie
Otto-Hahn-Strasse 11
44227 Dortmund
Germany

Patrick Loos
Technische Universität
Dortmund
Fakultät Chemie
Otto-Hahn-Strasse 6
44221 Dortmund
Germany

and

Max-Planck Institut für
Molekulare Physiologie
Otto-Hahn-Strasse 11
44227 Dortmund
Germany

Sylvie Michel
Université Paris Descartes
Faculté de Pharmacie
Laboratoire de Pharmacognosie
U.M.R.-C.N.R.S. n° 8638
4 Avenue de l'Observatoire
75006 Paris
France

Lech-Gustav Milroy
Technische Universität
Dortmund
Fakultät Chemie
Otto-Hahn-Strasse 6
44221 Dortmund
Germany

and

Max-Planck Institut für
Molekulare Physiologie
Otto-Hahn-Strasse 11
44227 Dortmund
Germany

Johann Mulzer
University of Vienna
Institute of Organic Chemistry
Währinger Strasse 38
1090 Vienna
Austria

Erwan Poupon
Université Paris-Sud 11
Faculté de Pharmacie
5 rue Jean-Baptiste Clément
92260 Châtenay-Malabry
France

Stefano Rizzo
Technische Universität
Dortmund
Fakultät Chemie
Otto-Hahn-Strasse 6
44221 Dortmund
Germany

and

Max-Planck Institut für
Molekulare Physiologie
Otto-Hahn-Strasse 11
44227 Dortmund
Germany

Rim Salame
Université Paris-Sud 11
Faculté de Pharmacie
5 rue Jean-Baptiste Clément
92260 Châtenay-Malabry
France

François Tillequin
Université Paris Descartes
Faculté de Pharmacie
Laboratoire de Pharmacognosie
U.M.R.-C.N.R.S. n° 8638
4 Avenue de l'Observatoire
75006 Paris
France

Timothy R. Welch
Colorado State University
Department of Chemistry
Fort Collins, CO 80523-1872
USA

Robert M. Williams
Colorado State University
Department of Chemistry
Fort Collins, CO 80523-1872
USA

Lok-Hang Yan
Université Paris-Sud 11
Faculté de Pharmacie
5 rue Jean-Baptiste Clément
92260 Châtenay-Malabry
France

Biomimetic Organic Synthesis: an Introduction

Bastien Nay and Erwan Poupon

Nature always makes the best of possible things
Aristotle

1
General remarks

"Biomimetic", "biomimicry" and "biologically inspired" are terms that can be used whenever Nature symphonic processes inspire human creation. This will encompass science, arts, architecture and so on. In this book, we will focus our attention on organic chemistry. To assemble these two volumes, we were spoiled for choice. A selection of topics was made to give a wide perspective on biomimetic synthesis, especially when dedicated to natural product chemistry and total synthesis which will constitute the major part and a guiding principle all along this book. We are of course conscious that entire fields are left aside such as material chemistry or supramolecular chemistry. Yet we wish that this book will convince most readers that applying biomimetic strategies to organic synthesis can provide a shortcut toward efficiency, beauty and originality, in a wide scope of fields (Figure 1).

2
Natural products as a vital lead

For many decades, the question why living organisms of all kingdoms produce secondary metabolites ("natural products") has been the subject of many debates. As soon as the first structures were determined, chemists also started thinking about the possible origin of the molecules [1].

Natural products are at the center of chemical ecology and have been forged in the crucible of Darwinian evolution. Many theories have tried to explain the incredible diversity of natural substances, including appealing views concluding that the living organisms that may be selected by evolution are the ones that favor chemical diversity [2], which may be the product of biochemical combinatorial processes. It is needless to remind here the importance of secondary metabolites

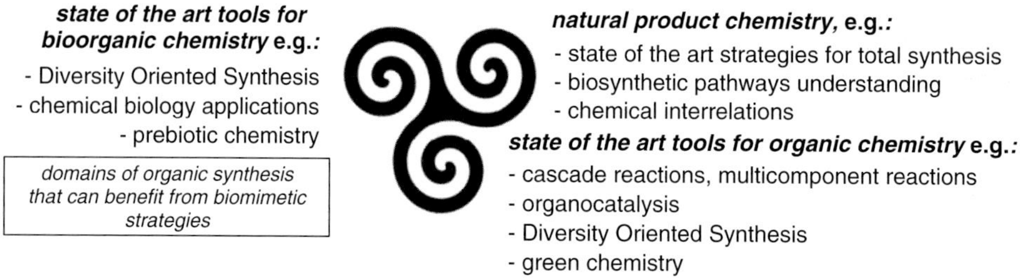

Figure 1 Tentacular influence of biomimetic strategies.

for Humanity notably as a source of drug candidates, pharmaceuticals, flavours, fragrances, food supplements. This aspect has been widely covered over the years.

Back to the biological functions, activities of natural substances may be explain because they interact with and modulate almost all type of biological targets including proteins (enzymes, receptors, and cytoskeleton), membranes, or nucleic acids. Here again, important notions such as the conservation of protein domains in living organisms or the selection of privileged scaffolds have been discussed and should not be ignored by chemists interested in natural substances [3].

3 Biomimetic synthesis

Biomimetic synthesis is the construction of natural products by chemical means using Nature's hypothetical or established strategy. It therefore stands in close relation with biosynthetic studies. Engaged in biomimetic strategies, the chemists

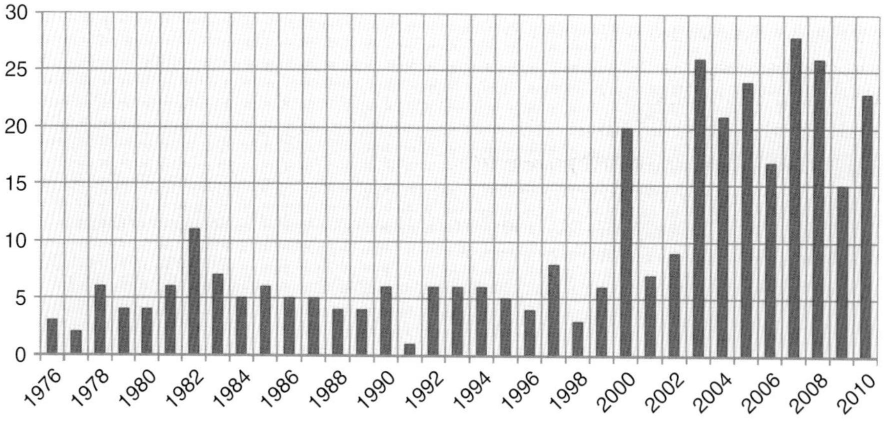

Figure 2 Analysis of bibliographical search in SciFinder with the terms "biomimetic total synthesis" from 1960 to 2010, leading to 339 references (30/10/2010).

will face pragmatic issues for planning their synthesis but will undoubtedly wonder about the exact role of enzymes in nature's way to construct sometimes highly complex structures. From highly evolved biosynthetic pathways involving enzymes with very high selectivity to less evolved routes or less specific enzymic catalysis, secondary metabolism pathways embrace a wide range of chemical efficiency. Biomimetic strategies will often, usually unintentionally, point out this aspect.

An increasing number of total syntheses have been termed "biomimetic" or "biosynthetically inspired" and so on, especially during the last decade. Basically a quick search in *SciFinder*, using the term "biomimetic total synthesis" over the period 1960–2010, afforded 339 occurrences, beginning in 1976. As illustrated in Figure 2, the last ten years have shown an increasing number of publications in this field.

Different situations and "degree of biomimicry" can then be instinctively distinguished when closely analyzing the final total synthesis including:

- a total synthesis featuring a biomimetic crucial step after a multistep total synthesis of the natural product precursor;
- a total synthesis featuring a biomimetic cascade reaction from more or less simple precursors.

Many examples in both situations will be found in this book. Simple parameters can at first sight help defining the relevance of the biomimetic step especially in terms of complexity generation. These include among others: the number of new carbon-carbon bonds and cycles formed, the number of changes in hybridization state of carbon atoms, the global oxidation state and also the stereochemical changes. The chemical reactions borrowed from Nature tool box for building carbon-carbon bonds that will emerge will particularly stand out century-old reactions such as aldolization, Claisen condensation, Mannich reaction and Diels-Alder and other cycloadditions. Situations where self-assembly relies merely on inherent reactivity of the precursors are probably situations that will be most likely mimicked successfully in the laboratory [4]. Beautiful examples will be presented in this book. Since simplicity should be the hallmark of total syntheses approaching the perfect or ideal total synthesis [5], the use of biomimetic strategies can advantageously bring solutions to intricate synthetic problems [6]. Let us finally add that by many aspects we will not debate here, biomimetic strategies may fulfill the criteria of "green chemistry" and "atom economy" when exploiting for example multicomponent strategies [7].

4
On the organization of the book, in close relation with secondary metabolism biochemistry

The chief purpose of this book is not to give a full coverage of the main biosynthetic pathways of secondary metabolites. Yet a particular care has been brought by authors in providing basic key elements of biosynthesis in the different chapters, to make them as comprehensive as possible to readers. If more has to be known about biosynthetic elements, we suggest referring to excellent books that have already covered the subject [8].

4.1 Alkaloids

An alkaloid is a cyclic organic compound containing nitrogen in a negative oxidation state which is of limited distribution among living organisms. This is a modern definition for a heterogeneous class of natural substances given by S. W. Pelletier in the first volume of the series of famous periodical books *Alkaloids* [9].

In our *Biomimetic Organic Synthesis*, all chapters related to alkaloids have been gathered in the first volume of this edition. Many classifications were proposed for this class of compounds. They could be based on the biogenesis, structure, biological origin, spectroscopic properties or also biological properties. The great lack of general principles towards a unified classification is obvious and the borderline between alkaloids *sensu stricto* and other natural nitrogen-containing secondary metabolites (such as peptides or nucleosidic compounds) is often unclear. A classification based on the nitrogen source of the alkaloid will guide

Chapter 2.1 by R.Salame, L.-Y. Yan and E. Poupon

Biomimetic Synthesis of Ornithine/Arginine and Lysine-Derived Alkaloids: Selected Examples

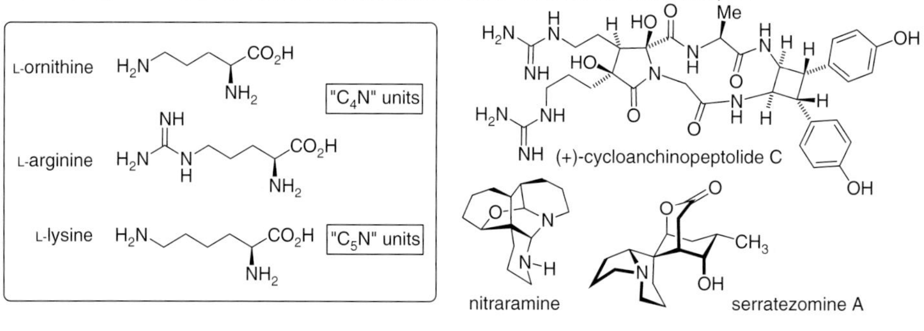

Chapter 2.2 by H. Liang and M. A. Ciufolini

Biomimetic synthesis of alkaloids derived from tyrosine: the case of FR-901483 and TAN 1251 compounds

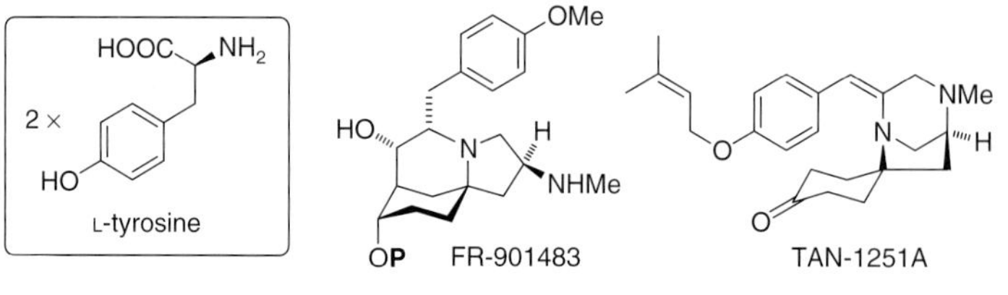

Scheme 1 Alkaloids derived from ornithine/arginine, lysine and tyrosine.

our choice of topics tackled in this book. This has the advantage of linking the biosynthetic origin (and thereby the biomimetic approach) and the chemical structure of the secondary metabolites. Accordingly, chapters will be devoted to alkaloids primarily deriving from ornithine, arginine, lysine, tyrosine (Scheme 1) and of course tryptophan (which highly diverse chemistry will be envisaged in three chapters, Scheme 2). Particularly, we thought that the important class of indolomonoterpenic alkaloids, despite largely discussed along the years, deserved an overview chapter putting forward crucial ideas and challenges when approaching their chemistry. A large array of natural substances isolated from microorganisms displays a diketopiperazine ring system in more or less rearranged form. Because of constant efforts towards the comprehension of their biosynthesis and their total synthesis, a chapter is dedicated to these alkaloids. In fact, they are probably among the secondary metabolites that have largely benefited from biomimetic strategies

Chapter 2.3 by S. Michel and F. Tillequin

Biomimetic synthesis of alkaloids derived from tryptophan: indolemonoterpene alkaloids

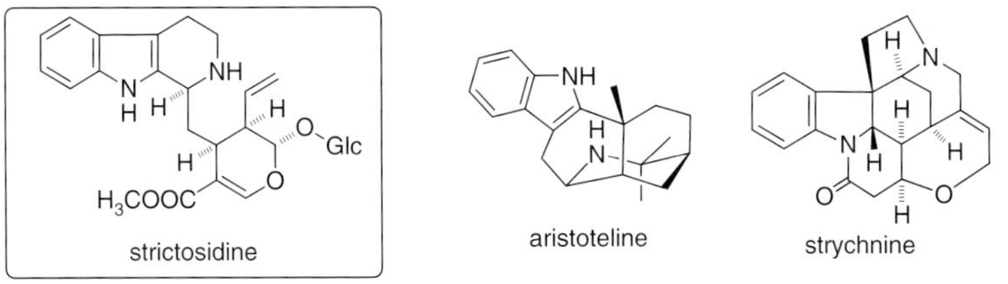

Chapter 2.4 by T. R. Welch and R. M. Williams

Biomimetic Synthesis of Alkaloids Derived from Tryptophan: Dioxopiperazine Alkaloids

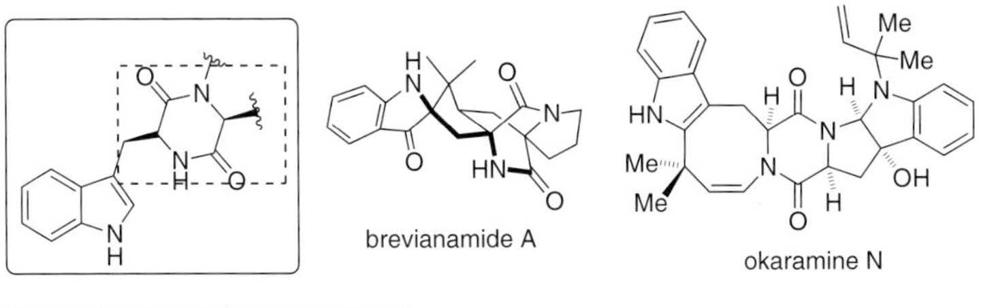

Chapter 2.5 by T. Gaich and J. Mulzer

Biomimetic synthesis of alkaloids with a modified indole nucleus

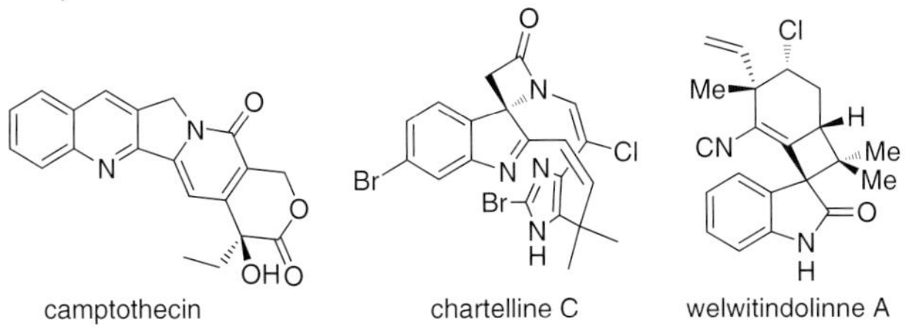

Scheme 2 Alkaloids derived from tryptophan.

Chapter 2.6 by R. Duval and E. Poupon

Biomimetic synthesis of manzamine alkaloids

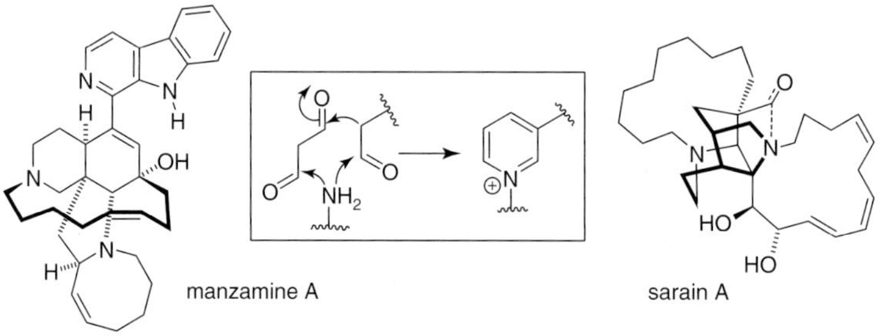

Chapter 2.7 by J. Appenzeller and A. Al Mourabit

Biomimetic synthesis of marine pyrrole-2-aminoimidazole and guanidium alkaloids

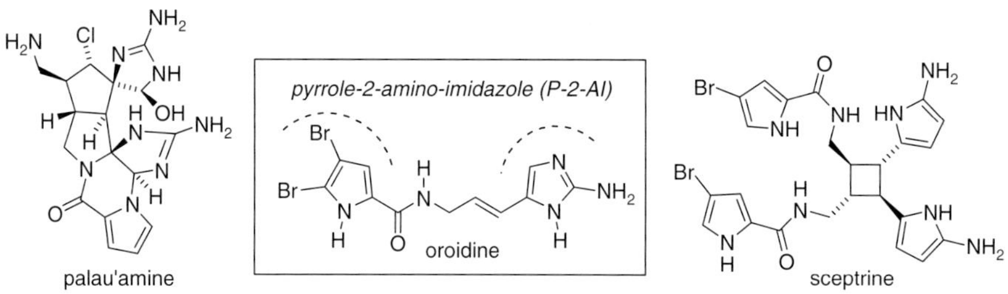

Scheme 3 Focus on two classes of complex marine alkaloids.

with undeniable success. Also of special interest in biomimetic chemistry, several alkaloids are derived in nature by profound modifications of the indole nucleus itself giving rise to secondary metabolites for which the biosynthetic origin is not obvious at first glance. The examples of quinine and camptothecin were among the first structures where such phenomena were suspected. Up to now, such biomimetic syntheses imply an initial oxidation step of the indole nucleus, which selected examples are disclosed in a proper chapter.

Two chapters will also cover the biomimetic synthesis of two classes of important marine alkaloids: the manzamine type alkaloids and the pyrrole-2-aminoimidazole alkaloids (Scheme 3). Alkaloids encompass also secondary metabolites obviously deriving from terpenes/steroids or polyketides, they will be considered as well in an individualized chapter (Scheme 4). Despite more related to polyketides in terms of biosynthetic machinery (see below), peptides alkaloids will be covered by two chapters in this section (Scheme 5).

4.2
Terpenes and terpenoids

Terpenes and terpenoids will be covered by chapters of the second volume of *Biomimetic Organic Synthesis*. They are made by terpene cyclases which catalyze

Chapter 2.8 by E. Gravel

Biomimetic synthesis of alkaloids with a non amino-acid origin

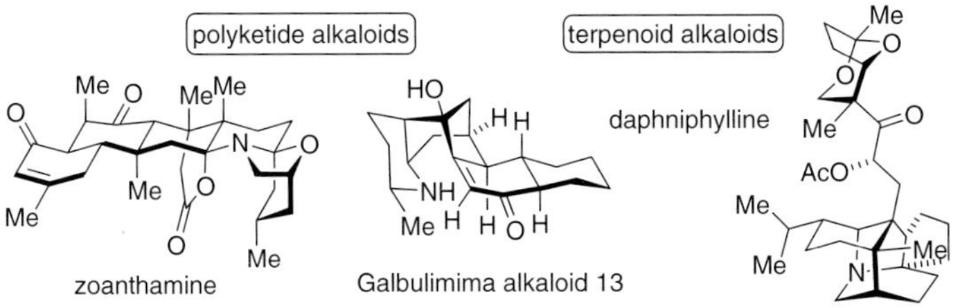

Scheme 4 Polyketide and terpenoid alkaloids.

Chapter 2.9 by H.-D. Arndt, R. Lichtenecker, P. Loos and L.-G. Milroy

Biomimetic synthesis of azole- and aryl-peptide alkaloids

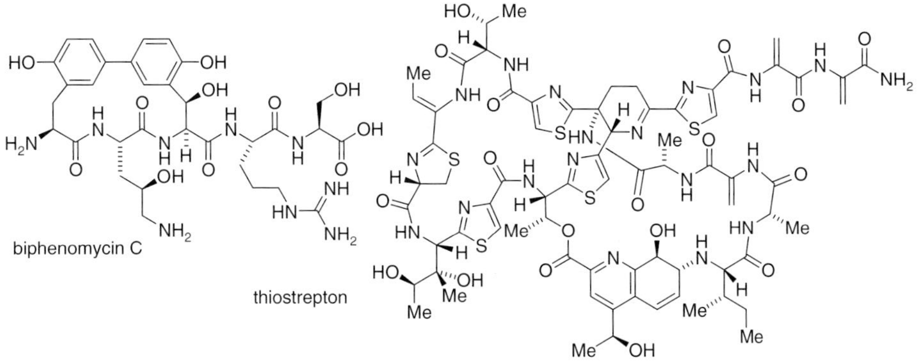

Chapter 2.10 by H.-D. Arndt, L.-G. Milroy, S. Rizzo

Biomimetic synthesis of indole-oxidized and complex peptide alkaloids

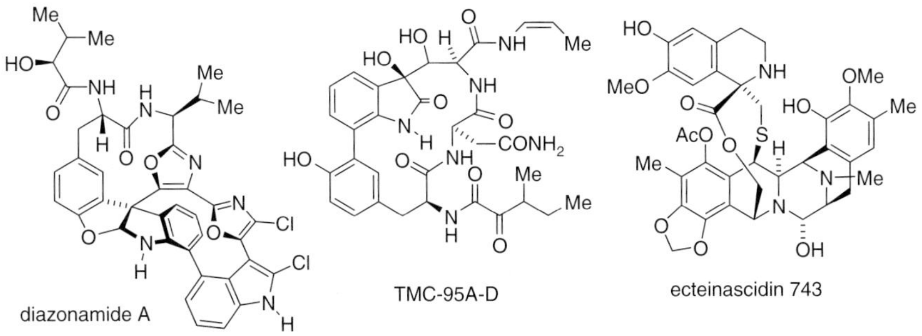

Scheme 5 Complex peptide alkaloids.

highly efficient reactions at the origin of such a rich chemistry. The cationic cascade aspect of terpene biosynthesis from oligomers of activated forms of isoprene is very appealing for many biomimetic endeavors. Current aspects of terpene biosynthesis include the interesting notion of accuracy of terpene cyclases, an issue that is closely

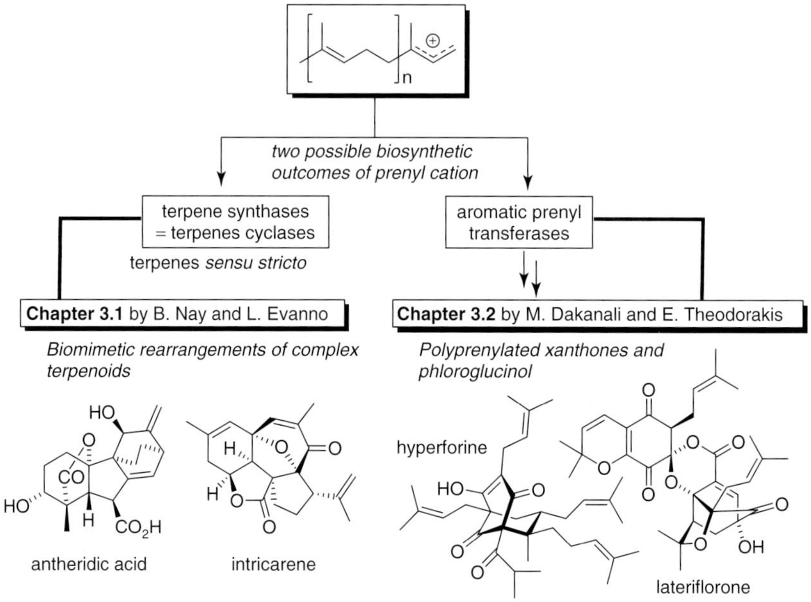

Scheme 6 Biomimetic synthesis of terpenes and terpenoids.

related to the quest for selectivity in biomimetic synthesis of such compounds. Leading review articles have already been published elsewhere about cationic cascade cyclizations [10]. A chapter will focus on the post-polycyclization events, dealing with biomimetic rearrangements of already complex terpene structures. Polyprenylated secondary metabolites resulting primarily from the transfer of prenyl units to aromatic rings by aromatic prenyl transferases, and sometimes followed by rearrangements, will be covered by another chapter in the second volume (Scheme 6). Other aspects of terpene alkaloids have been developed in the first volume of *Biomimetic Organic Synthesis* (especially, the reader can refer to chapters 2.3 and 2.8).

4.3
Polyketides

Manipulations of polyketide gene clusters have contributed to a revolution in the comprehension of polyketides (PK), and also of non-ribosomal peptides (NRP), biosynthesis. The genome sequencing of numerous PK and NRP producing microorganisms has revealed a large number of cryptic metabolites mostly unknown. Current challenges include the discovery of such natural compounds by allowing the expression of the corresponding genes ("turn them on") and the programming/reprogramming of fungal PKS. Not to be forgotten is the implication of the PK pathways in aromagenesis in nature *via* the biosynthesis of phenols. We therefore

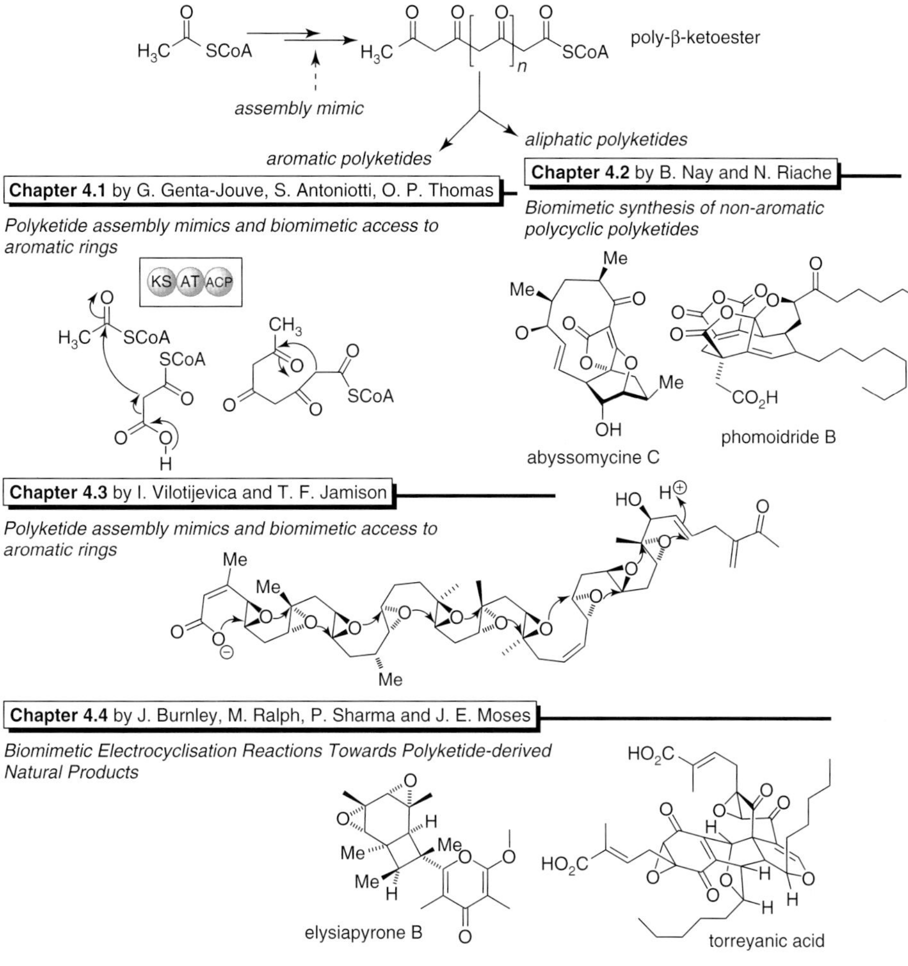

Scheme 7 Biomimetic synthesis of polyketides.

decided to ask for a contribution on biomimetic mimics of the fundamental steps of PK assembly and phenol ring formation. Turning our attention to more complex structures, beautiful examples of biomimetic synthesis of complex non aromatic polycyclic-PKs will be presented. The two following chapters will deal with two specific classes of natural substances characterized by their seminal mechanism of formation: *i.e.* polyepoxide ring opening and electrocyclization (Scheme 7).

4.4
Polyphenolic compounds

Another important biosynthetic route to aromatic rings in nature is provided by the shikimate/chorismate pathway. Simple phenolic acids enter the biosynthesis

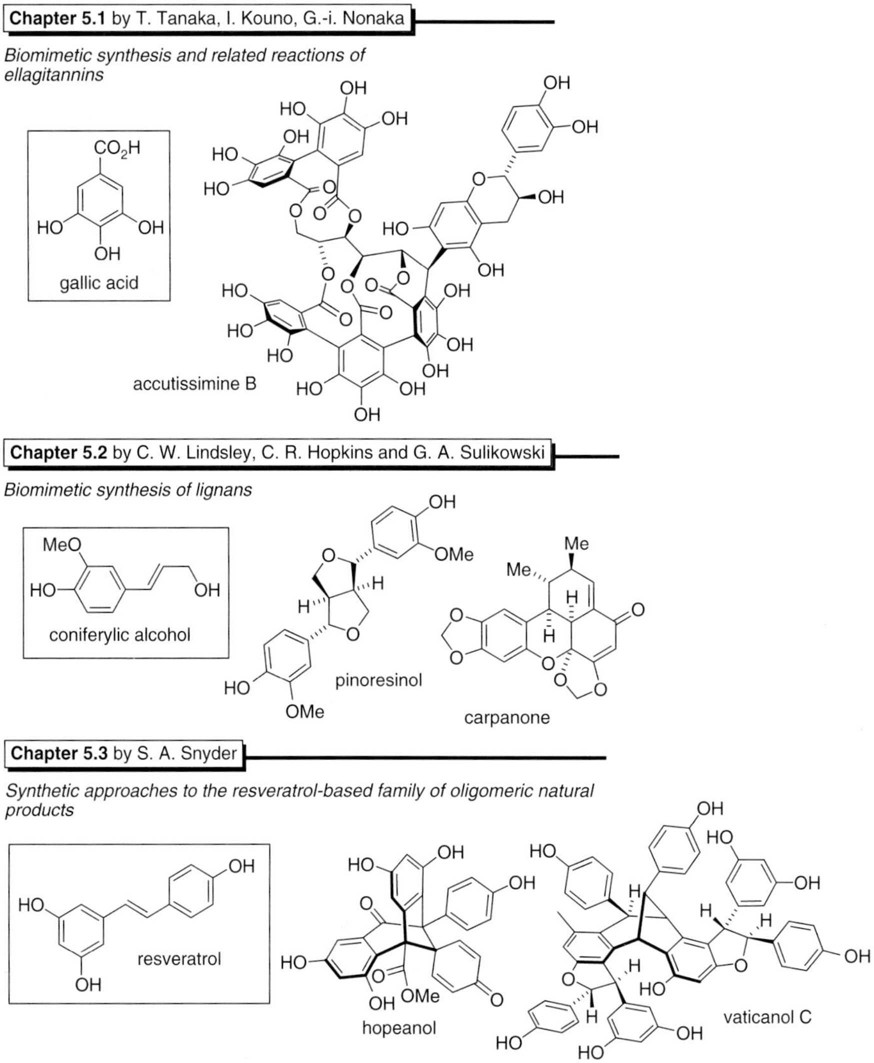

Scheme 8 Biomimetic synthesis of polyphenolic natural substances.

of sometimes highly complex ellagitannins, a class of hydrolysable tannins widely studied for their health benefits (Scheme 8). Among phenylpropanoids natural substances directly derived from chorismate are the lignans that are discussed in the following chapter. Typical extended phenylpropanoids include compounds such as flavonoids and stilbenes. The chemistry of flavonoids has been widely studied and reviewed over the years [11]. This is not the case for natural substances deriving from resveratrol which hold center stage in the last few years because of the growing importance of resveratrol itself in human health, and because of new developments in the total synthesis of this very interesting class of polycyclic molecules. For these last three classes of molecules (ellagitannins, lignans, resveratrol derived), radical phenolic couplings plays a center role as the main source of carbon-carbon bonds.

4.5
Frontiers in biomimetic synthesis

At the cross-roads of methodology and total synthesis, a few topics will show how nature observation, especially enzymic mechanisms, can lead to new discoveries in organic chemistry (Scheme 9). A discussion on the engaging issue of occurrence of the Diels-Alder reaction in nature will be conducted in a chapter. The exponential impact of organocatalysis in organic chemistry will be illustrated by the challenging problem of transfer hydrogenations in a bio-inspired manner. Once again a plethora of review articles and books deals with the other aspects of organocatalysis [12]. Finally, by many aspects, biomimetic organic chemistry may be closely linked

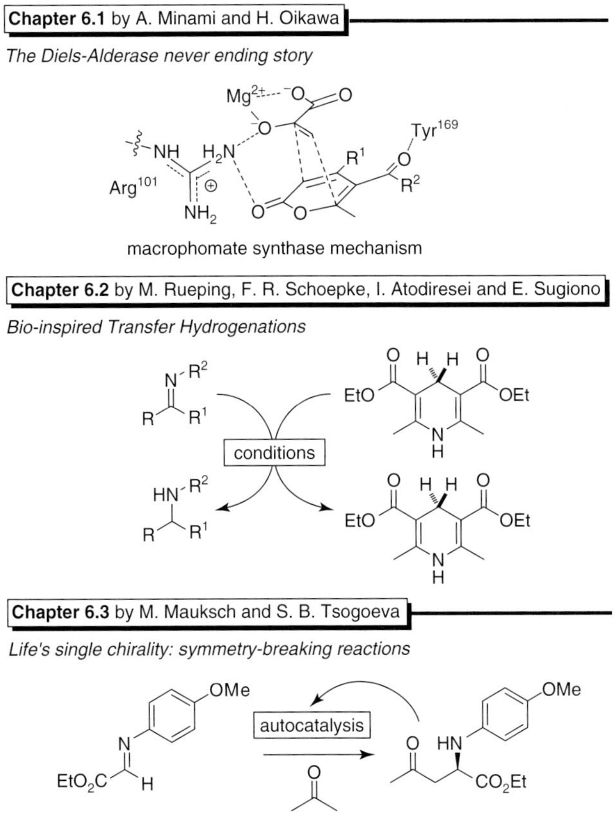

Chapter 6.1 by A. Minami and H. Oikawa
The Diels-Alderase never ending story
Arg101
Tyr169
macrophomate synthase mechanism
Chapter 6.2 by M. Rueping, F. R. Schoepke, I. Atodiresei and E. Sugiono
Bio-inspired Transfer Hydrogenations
conditions
Chapter 6.3 by M. Mauksch and S. B. Tsogoeva
Life's single chirality: symmetry-breaking reactions
autocatalysis

Scheme 9 Frontiers in biomimetic organic synthesis.

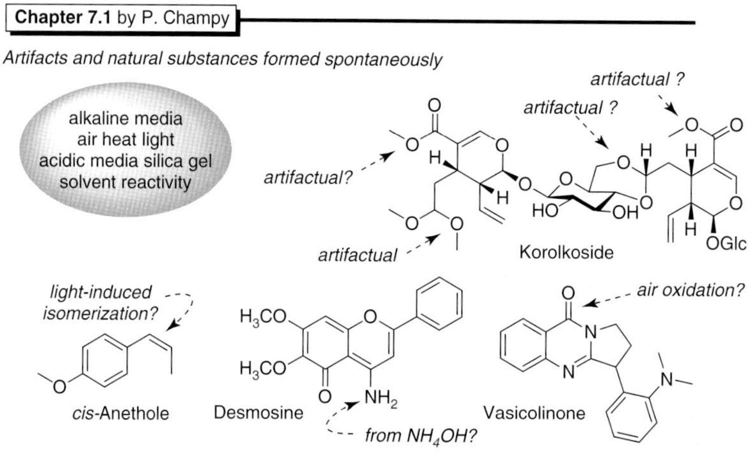

Scheme 10 Artifacts as a matter of debate for the conclusion.

to prebiotic chemistry. Key-words such as spontaneous evolution, molecular and supramolecular self-organization of organic molecules can indeed refer to both domains. A chapter will be devoted to the emergence of life single chirality on earth in a manner, once again, understandable to a broad readership.

Eventually, we thought that a chapter about artifacts in natural product chemistry might provide the matter of debate for an open conclusion, just to spin out the discussion (Scheme 10). May the readers enjoy their trip in the fascinating science of *Biomimetic Organic Synthesis.*

References

1. See, this article of great interest: Thomas, R. (2004) *Nat. Prod. Rep.*, **21**, 224–248.
2. See among others: (a) Firn, R.D. and Jones, C.G. (2009) *J. Exp. Bot.*, **60**, 719–726 and references cited therein; (b) Jenke-Kodama, H. and Dittmann, E. (2009) *Phytochemistry*, **70**, 1858–1866.
3. See among others: (a) Breinbauer, R., Vetter, I.R., and Waldmann, H. (2002) *Angew. Chem. Int. Ed.*, **41**, 2878–2890; (b) Bon, R.S. and Waldmann H. (2010) *Acc. Chem. Res.*, **43**, 1103–1114 and references cited therein; (c) Dobson, C.M. (2004) *Nature*, **432**, 824–828 and references cited therein; (d) Welsch, M.E., Snyder, S.A., and Stockwell, B.R. (2010) *Curr. Opin. Chem. Biol.*, **14**, 347–361.
4. (a) Gravel, E. and Poupon, E. (2008) *Eur. J. Org. Chem.*, 27–42; (b) E.J. Sorensen (2003) *Bioorg. Med. Chem.*, **11**, 3225–3228.
5. (a) Wender, P.A., Handy, S.T., and Wright, D.L. (1997) *Chemistry & Industry*, 765; (b) Wender, P.A. and Miller, B.L. (2009) *Nature*, **460**, 197–20; (c) Gaich, T. and Baran, P.S. (2010) *J. Org. Chem.*, **75**, 4657–4673.
6. Among other review articles, interesting thoughts and historical perspectives are discussed in: (a) Scholz, U. and Winterfeldt, E. (2000) *Nat. Prod. Rep.*, **17**, 349–366; (b) de la Torre, M.C. and Sierra, M.A. (2004) *Angew. Chem. Int. Ed.*, **43**, 160–181; (c) Heathcock, C.H. (1996) *Proc. Natl. Acad. Sci. USA*, **93**, 14323–14327.
7. Touré, B.B. and Hall, D.G. (2009) *Chem. Rev.*, **109**, 4439–4486.

8. (a) Dewick, P.M. (2009) *Medicinal natural products: a biosynthetic approach*, 3rd Edition, Wiley, Chichester (UK); (b) Bruneton, J. (2009) *Pharmacognosie, phytochimie et plantes médicinales*, 4th Edition, Tec et Doc, Paris; (c) see also the book series: Barton, D., Nakanishi, K., Meth-Cohn, O. (Eds) (1999) *Comprehensive Natural Products Chemistry, 1–9*, Elsevier Science Ltd, Oxford; (d) Mander, L. and Liu, H.-W. (Eds) (2010) *Comprehensive Natural Products Chemistry II, 1–10*, Elsevier Science Ltd, Oxford; (d) See also the monthly issues of *Nat. Prod. Rep.*
9. Pelletier, S.W. (1983) The nature and definition of an alkaloid in *Alkaloids: Chemical and Biological Perspectives*, Vol. 1 (ed. Pelletier, S.W.) Wiley-Interscience, New York, pp. 1–32.
10. For example, see the following early and late reviews: (a) Johnson, W.S. (1976) *Bioorg. Chem.*, **5**, 51–98; (b) Yoder, R.A. and Johnston, J.N. (2005) *Chem. Rev.*, **105**, 4730–4756.
11. Andersen, Ø. M. and Markham, K.R. (Eds) (2006) *Flavonoids: chemistry, biochemistry, and applications*, CRC Taylor and Francis, Boca Raton.
12. (a) Berkessel, A. and Groger, H. (2005) *Asymmetric organocatalysis: from Biomimetic Concepts To Applications In Asymmetric Synthesis*, Wiley-VCH, Weinheim; (b) Reetz, M.T., List, B., Jaroch, S., and Weinmann, H. (eds) (2008) *Organocatalysis*, Springer Verlag, Berlin; (c) with specific applications in total synthesis, see for example: Marquéz-López, E., Herrera, R.P., and Christmann, M. (2010) *Nat. Prod. Rep.*, **27**, 1138–1167.

Part I
Biomimetic Total Synthesis of Alkaloids

Biomimetic Organic Synthesis, First Edition. Edited by Erwan Poupon and Bastien Nay.

1 Biomimetic Synthesis of Ornithine/Arginine and Lysine-Derived Alkaloids: Selected Examples

Erwan Poupon, Rim Salame, and Lok-Hang Yan

1.1 Ornithine/Arginine and Lysine: Metabolism Overview[1)]

1.1.1 Introduction: Three Important Basic Amino Acids

- L-Ornithine (L-**1**) (L-Orn, Figure 1.1) is a non-proteinogenic amino acid produced from L-glutamic acid (**4**) in plants and from L-arginine (**2**) in animals. L-Ornithine plays a central role in the urea cycle in terrestrial vertebrates [1].
- L-Arginine: With its guanidine residue, L-Arginine (L-**2**) (L-Arg, R) is a highly basic amino acid. It is encoded by DNA and is the direct precursor of L-ornithine (L-**1**), urea, and also nitric oxide. It is also be encountered in some natural products (see below) [1, 2].
- L-Lysine (L-**3**) (L-Lys, K): is the only amino acid to have two different biosynthetic pathways. One is the aspartate (**5**) pathway present in bacteria, plants, and algae. The other starts from α-ketoglutarate (**6**) and is present in fungi [3, 4]. Lysine is an essential amino acid for humans.

Scheme 1.1 reflects some of the biochemical relations between L-ornithine (L-**1**)/L-arginine (L-**2**) and L-lysine (L-**3**). It is of course not the aim of this chapter to provide further details concerning their respective biosynthesis.[2)] Only important metabolic intermediates, helpful for a better comprehension of the following sections, have been stressed.

1) These three amino acids (Figure 1.1) share common chemical reactivity and are implicated in more or less similar biosynthetic pathways. We thus decided, in an effort to establish useful comparisons, to garner lysine and arginine/ornithine derived natural substances in a single chapter.
2) The interested reader is referred to classical biochemistry textbooks.

Biomimetic Organic Synthesis, First Edition. Edited by Erwan Poupon and Bastien Nay.

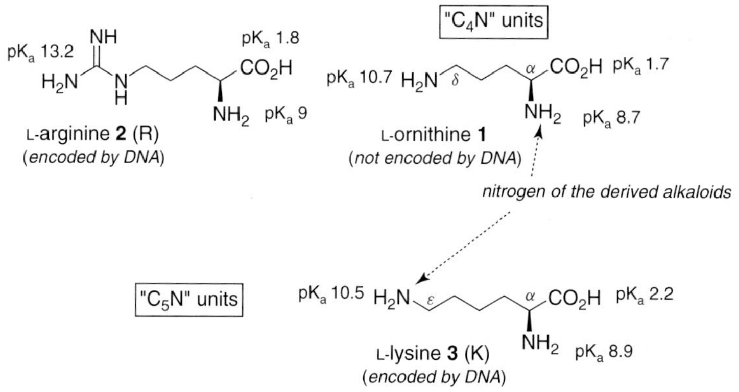

Figure 1.1 Structure of the amino acids.

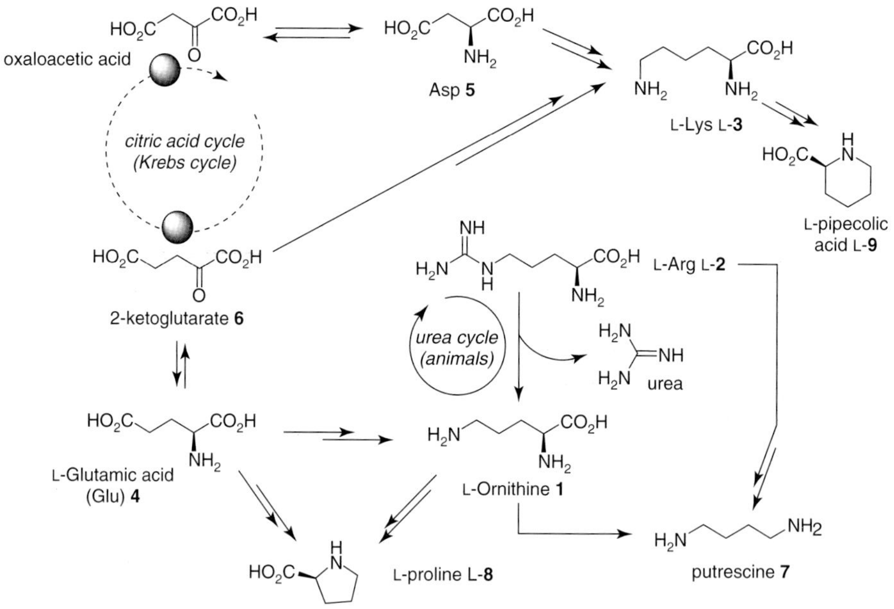

Scheme 1.1 Place of the three amino acids in primary metabolism.

1.1.2
From Primary Metabolism to Alkaloid Biosynthesis

The parallel between L-ornithine (L-**1**) and L-lysine (L-**3**) metabolism concerning their catabolism/biotransformation and subsequent chemical reactivity to form alkaloids is obvious (even if the incorporated nitrogen atom is different, that is, incorporation of the α-amino group for ornithine and ε-amino group for lysine). Mainly, both amino acids will be able to undergo decarboxylation to the corresponding diamine [putrescine **7** (C_4) and cadaverine **10** (C_5), respectively] and then oxidative deamination into aminoaldehydes (see below). Thereby, rather stable amino acids are turned into highly reactive units suitable for natural organic chemistry.

1.1.2.1 L-Ornithine Entry into Secondary Metabolism[3)]

The diamine putrescine (**7**) can be formed directly from the decarboxylation of L-ornithine (**1**) (Scheme 1.2); it can also be derived from L-arginine (**2**) [6] after decarboxylation and transformation of the guanidine functional group. Putrescine (**7**) is then mono-*N*-methylated[4)] by putrescine *N*-methyltransferase (PMT). This reaction is the first purely "secondary metabolite" step and **11** is the first specific metabolite towards alkaloids. *N*-Methylputrescine (**11**) may then be oxidatively deaminated by diamine oxidase to 4-methylaminobutanal (**12**), which generates the *N*-methyl-Δ^1-pyrrolinium cation **13**, a cornerstone electrophilic intermediate and a central precursor of numerous alkaloids belonging to the pyrrolidine or tropane groups when the reaction with an appropriate nucleophile occurs.

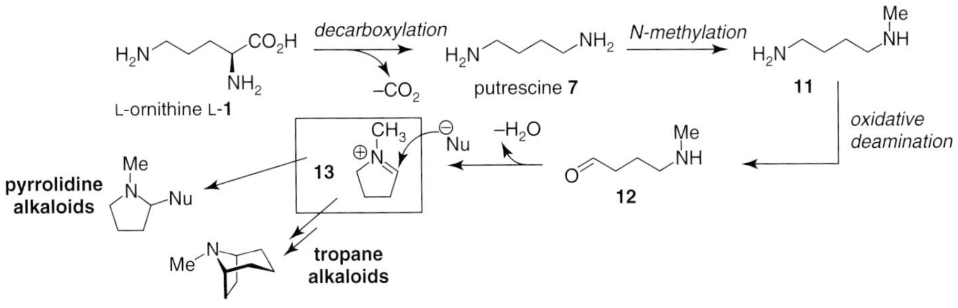

Scheme 1.2 Key elements of ornithine metabolism towards alkaloids.

1.1.2.2 L-Lysine Entry into Secondary Metabolism[5)]

Decarboxylation of L-lysine (L-**3**) into the diamine cadaverine (**10**) (Scheme 1.3) followed by oxidative deamination leads to aminopentanal **14**, which can cyclize into tetrahydropyridine **15**.[6)] This latter is most likely the universal intermediate

3) The fundamental first steps of ornithine (or lysine) catabolism towards alkaloids now constitute a classic in biosynthesis textbooks. See, among others, Reference [5].

4) This important step has been widely studied. PMT is closely related to spermidine synthase. See Reference [7] for a valuable review.

5) The interested reader is referred to classical biochemistry textbooks.

6) Synonymous terms: 2,3,4,5-tetrahydropyridine = Δ^1-piperideine (imine form); 1,2,3,4-tetrahydropyridine = Δ^2-piperideine (enamine form).

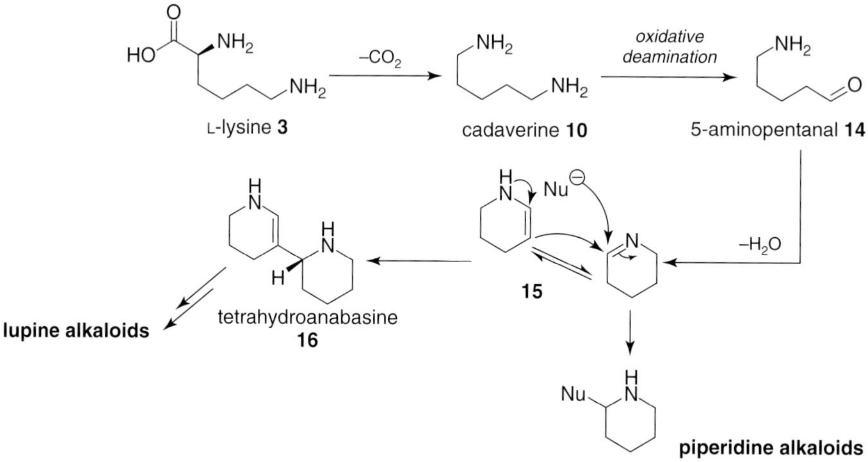

Scheme 1.3 Key elements of lysine metabolism toward alkaloids.

to lysine-derived piperidine alkaloids. This imine is too unstable and reactive to be isolated as such from plant material. Nucleophilic addition reactions at the imine function with suitable nucleophiles help stabilize **15** and are at the origin of various piperidine alkaloids. Dimerization into tetrahydroanabasine (**16**) is an important alternative in the lysine metabolism (the enamine form of **15** being the nucleophile). This latter reaction is spontaneous at physiological pH (*vide infra*) though stereospecific coupling involves an appropriate enzymic intervention in plants [8]. Tetrahydroanabasine (**16**) (which is probably also quite unstable as such *in vivo*) can then undergo various transformations and is at the origin of a class of alkaloids known as "*lupine alkaloids*" (Section 1.4.2).

1.1.3 Closely Related Amino Acids

- L-Proline: (L-**8**) (L-Pro, P, Scheme 1.1) is one of the 20 amino acids of the genetic code but the only one with a secondary amine function. It is biosynthesized from L-glutamic acid [9]. L-Proline recently increased in importance with the successful development of organocatalysis.
- L-Pipecolic acid: Unlike L-proline (L-**8**), L-pipecolic acid (L-**9**) (Scheme 1.1) is a non-proteogenic amino acid. It derives from L-lysine and is at the origin of several classes of secondary metabolites. Biosynthetic and biomimetic aspects of pipecolic acid are discussed below.[7]

7) Other close amino acids exist; of particular interest are piperazic acids (structure shown here). Their biosynthesis from lysine is discussed in Chapter 10 (Indole-Oxidized and Complex Peptide Alkaloids); see also Reference [10].

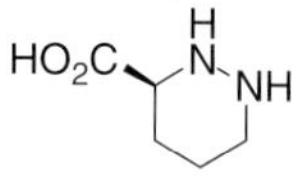

1.1.4
The Case of Polyamine Alkaloids

Cases where a C_4N_2 building block is incorporated [i.e., the polyamine putrescine (**7**)] include "polyamine alkaloids" (Scheme 1.4). Comprehensive review articles, especially by M. Hesse and colleagues, have appeared detailing the massive amount of work done in the field of these secondary metabolites during the last 20 years [11, 12]. Essentially, six basic backbone components, namely, putrescine (**7**), spermidines (**17, 18**), homospermidines (**19, 20**), spermine (**21**), and homospermine (**22**), participate in the skeleton of polyamine alkaloids. Figure 1.2 gives examples of cyclic polyamine alkaloids [piriferine (**23**), celacinnine (**24**), aphelandrine (**25**), and lipogrammistin A (**26**)], organized according to the classification of M. Hesse and colleagues (see Reference [11]). Interestingly, cadaverine units are very rarely present in such cyclic molecules. Despite interesting biomimetic syntheses [13], polyamine alkaloids will not be covered in this chapter (except for the biosynthesis of pyrrolidine alkaloids; see Section 1.3.1).

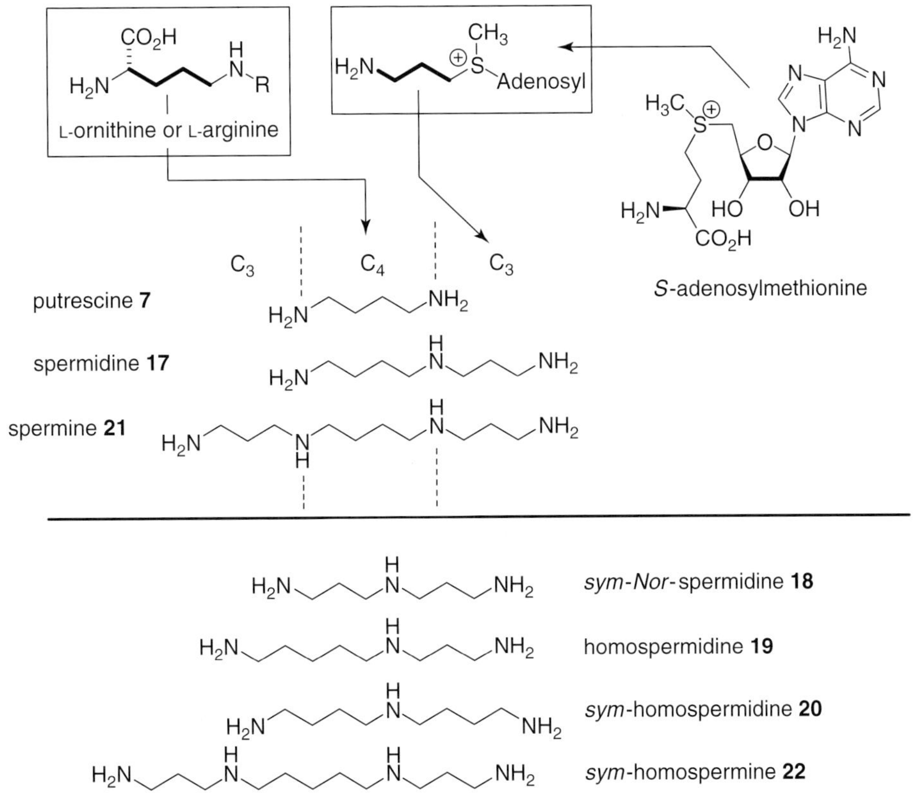

Scheme 1.4 Main polyamine backbones encountered in polyamine alkaloids.

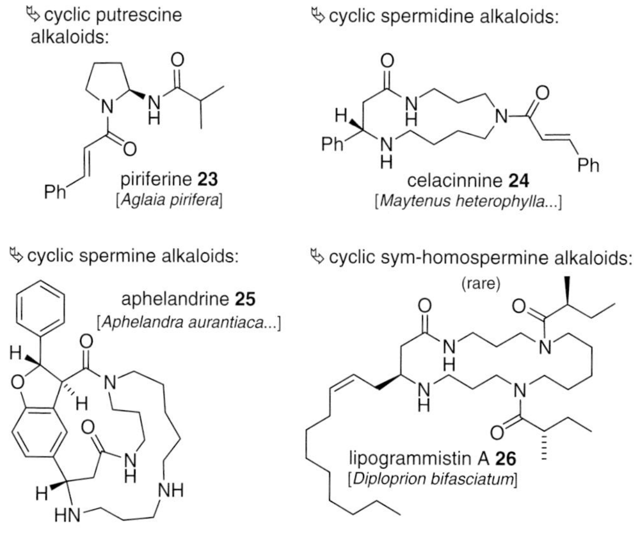

Figure 1.2 Representative polyamine alkaloids.

1.1.5 Biomimetic Synthesis of Alkaloids

Ionic iminium/enamine reactions are central to the construction of the title alkaloids, especially the Mannich reaction. This is probably one reason why biomimetic strategies have been particularly efficient in this class of secondary metabolites. We will, of course, approach in this chapter only some of the multifaceted aspects of the biomimetic chemistry of L-lysine or L-ornithine/arginine derived secondary metabolites. We will select topics and examples that are not covered (or not with the scrutiny we think opportune) in other review articles. This is especially the case when dealing with the manipulation of small reactive C_4 and C_5 units derived from the three amino acids. Some selected examples are presented in the following sections and organized as follows:

- biomimetic syntheses from L-lysine and L-ornithine/L-arginine or C_4, C_5 reactive units presumably derived from the amino acids;
- selected examples of biomimetic syntheses of more complex structures.[8]

8) For the implementation of highly complex metabolisms within marine sponges and involving, according to biosynthetic proposals, L-lysine, L-arginine, and L-proline see Chapter 7 (Biomimetic Synthesis of Marine Pyrrole-2-aminoimidazole and Guanidinium Alkaloids).

1.2
Biomimetically Related Chemistry of Ornithine- and Lysine-Derived Reactive Units

1.2.1
Ornithine-Derived Reactive Units

1.2.1.1 Biomimetic Behavior of 4-Aminobutyraldehyde

The Christophersen group studied the evolution of aqueous solutions of 4-aminobutyraldehyde (**27**) [prepared from aminobutyraldehyde dimethyl acetal (**28**), Scheme 1.5] by ^{1}H NMR over a wide pH range (1–12) [14]. Entropic factors explain the rapid formation of cyclic imine **29**, which can trimerize into **30**. When in aqueous solution, different species are in equilibrium and are depicted in Scheme 1.5. Along with aminobutanal **27**, two neutral (pyrrolidine **29** and trimer **30**), and four protonated entities (**31**–**34**) were detected and tracked as a function of pH. Around physiological pH, the four protonated species predominate. Owing to the rapid emergence of acid-catalyzed aldol condensation products, the authors cautiously avoided concentrated solution. As we will detail in a coming section, no dimeric structure such as **35** has been characterized from the mixtures.

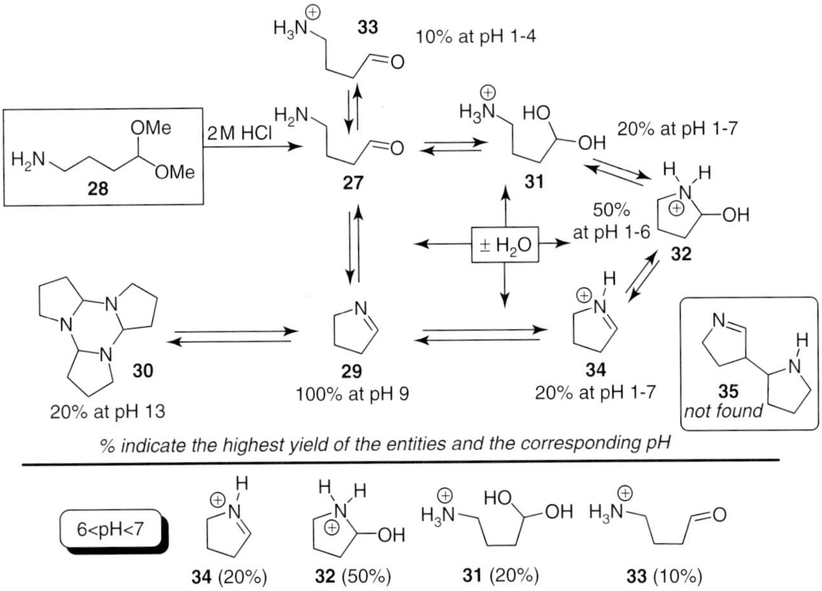

Scheme 1.5 Biomimetic reactions from 4-aminobutyraldehyde (**27**).

With substituted nitrogen atom, for example, with a biosynthetically relevant aminobutyl side chain (Scheme 1.6), the formation of the pyrrolidinium ring is nearly exclusive, with **36** predominant among several other entities [15].

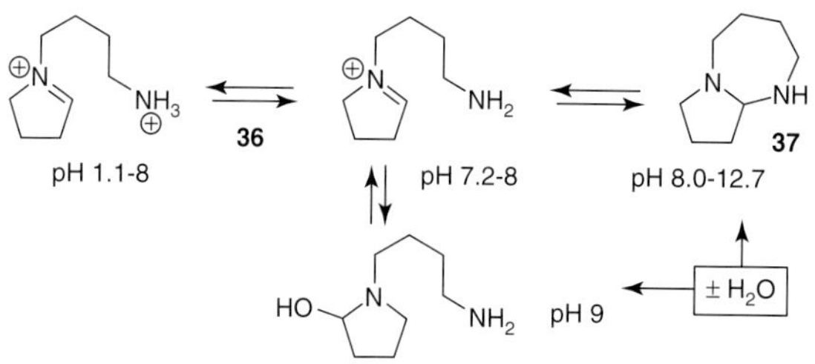

Scheme 1.6 Biomimetic behavior of a substituted pyrrolidinium ion.

Notably, aminal **37** is a bicyclic compound that is formed spontaneously at basic pH.[9)]

1.2.1.2 Dimerization

Whereas dimerization of six-membered ring enamines is a major outcome (as we will discuss in detail when dealing with lysine-derived units), the dimerization of the corresponding five-membered ring systems such as **29** is by far less described in the literature. The chemistry of these molecules is, correspondingly, simpler than that of lysine, probably because of a lesser propensity to react as an enamine (Scheme 1.7).

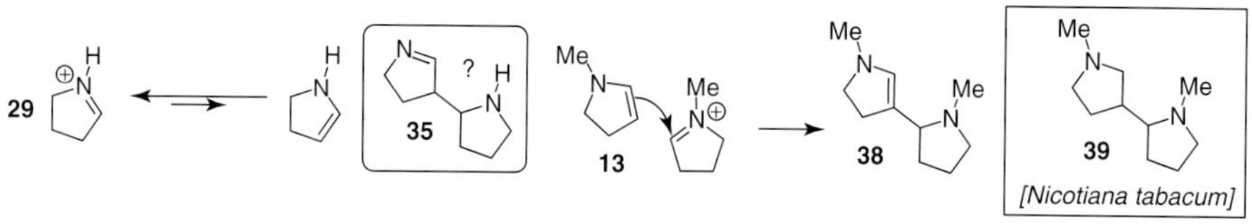

Scheme 1.7 Dimerization in the pyrrolidine series.

The dimer type **35** has never been described, although it has been postulated in some biosynthesis (Section 1.3.3). In the case of *N*-methyl substituted **13**,[10)] dimerization has been observed and dimer **38** characterized. Notably, the corresponding reduced dimer **39** has been detected as a mixture of diastereomers from the root system of *Nicotiana tabacum*, as well as monomeric **13** [18]. In fact, at physiological pH (around 7.2 in a growing tobacco plant), the coexistence of imine and enamine forms of **13** should provide the opportunity for more or less spontaneous condensations.

In the laboratory (Scheme 1.8), starting from *N*-methyl-4-aminobutanal diethyl acetal (**40**) in acidic conditions, deprotection affords **12**, which cyclizes into biosynthetic intermediate **13**, which in turn can dimerize into **38**. The product of a retro-Michael reaction, **41**, has also been fully characterized; it was usually observed as an impurity in the course of the synthesis of monomers **13** [19]. Alternative

9) Diazabicyclononane **37** was isolated earlier from *in vitro* chemical or enzymic conversion of spermidine with pea seedling diamine oxidase (PSDO); see Reference [16].

10) Is monomeric **29** too reactive and unstable? *N*-Monomethylputrescine (**11**) is in fact an early and central precursor of many alkaloids as the first specific metabolite *en route* to alkaloids such as nicotine or tropanes via *N*-methylpyrrolinium (**13**) (which could therefore be more stable than **15**); see References [5, 17].

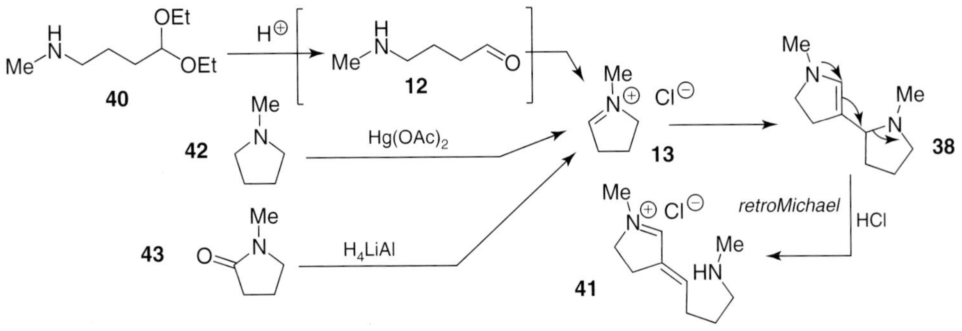

Scheme 1.8 Biomimetic dimerization in the pyrrolidine series.

procedures toward **13** consist of the oxidation (e.g., with mercuric diacetate [20]) of *N*-methylpyrrolidine (**42**) or the reduction of lactam **43** with aluminum hydrides [21].

1.2.2 Lysine-Derived Reactive Units

1.2.2.1 Oxidative Degradation of Free L-Lysine

Despite a seemingly simple pathway, mimicking the fundamental L-lysine (and L-ornithine) catabolism pathways in the laboratory is far from trivial. Only a few publications report on the direct oxidation of L-lysine. In 1966, B. Franck and colleagues oxidized L-lysine with alkaline NaOCl in water (Scheme 1.9) [22]. Cyclic oxidation products were identified and compared with authentic samples. Scheme 1.9 highlights the possible pathways toward the isolated compounds; it may be assumed that the cyclic oxidation products arise from L-lysine by one (intermediates **46**, **47**), two (intermediate **48**) or three oxidation steps (**15**, **16**, **44**, **45**). Despite an incomplete conversion of L-lysine and the complex mixtures obtained, this simple reaction is of great interest as a totally biomimetic reaction that mimics (i) the decarboxylation/oxidative deamination steps and (ii) the spontaneous evolution of the resulting reactive species (**15**, dimer **16**, lactam **44**).

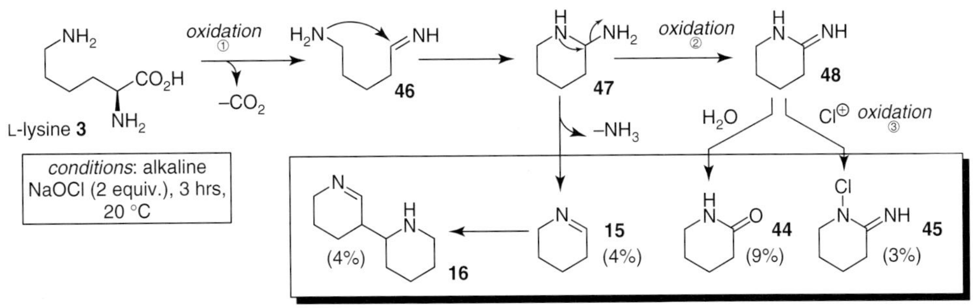

Scheme 1.9 Cyclic oxidative products of L-lysine.

1.2.2.2 Clemens Schöpf's Heritage: 50 Years of Endocyclic Enamines and Tetrahydroanabasine Chemistry

Numerous compounds resulting from the self-condensations of endocyclic enamines, which are closely related to metabolic pathways, have been described, especially in the pioneering work of Schöpf, starting from the late 1930s.

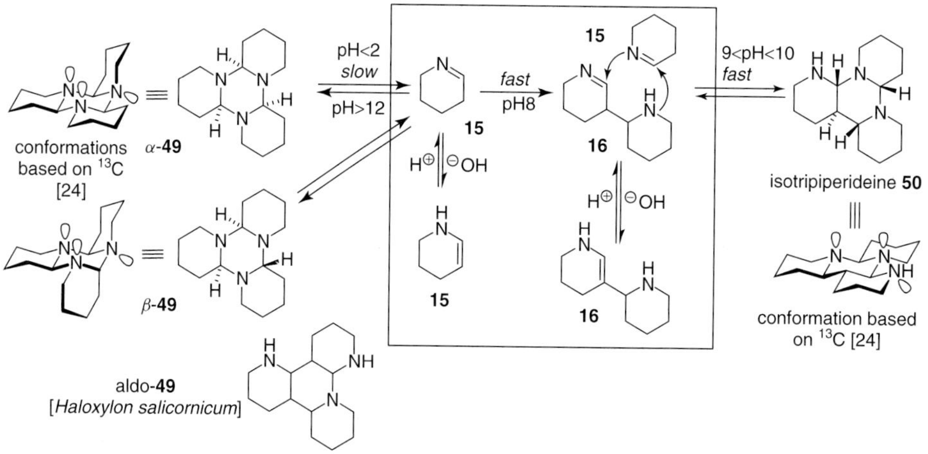

Scheme 1.10 Evolution of tetrahydropyridine.

The simplest compound, that is, Δ^1-piperideine (**15**), does not exist in monomeric forms but, instead, trimeric assemblies of types α- and β-**49** and **50** have been isolated (Scheme 1.10).[11] Structures **16**, **49**, and **50** were characterized by Schöpf in 1948 [23] and the configurations and conformations studied by the Kessler group in 1977 by ^{13}C NMR [24]. At neutral or slightly basic pH (~8), and as in nature, monomers **15** dimerize into tetrahydroanabasine (**16**) (which can be isolated as a dihydrobromide crystalline salt in the laboratory [25]). As early as 1956, Schöpf and colleagues clearly demonstrated the importance of pH on the kinetic and yield of conversion of **15** into **16**. Studies were conducted with pioneering and clear-sighted biosynthetic considerations (*zellmöglichen Bedingungen*, see Scheme 1.11) [26]. Tetrahydroanabasine also reacts in solution at pH 9 with imine **15** and gives in almost quantitative yield trimer **50**, also called isotripiperideine [23]. Consequently, trimer **50** is, in turn, in equilibrium with tetrahydroanabasine **16** and free Δ^1-piperideine **15** [27] and can, therefore, be considered as a stable, protected form of tetrahydroanabasine with interesting synthetic potential.

Aldotripiperideine (aldo-**49**) is another trimer that results from a rearrangement of α-tripiperideine in acidic conditions or in basic conditions at pH 9.2 at 100 °C. Interestingly, aldo-**49** was isolated from *Haloxylon salicornicum* as a natural substance [28].

11) Besides, they constitute a suitable way to store piperideine as a crystalline solid on a 100 g scale.

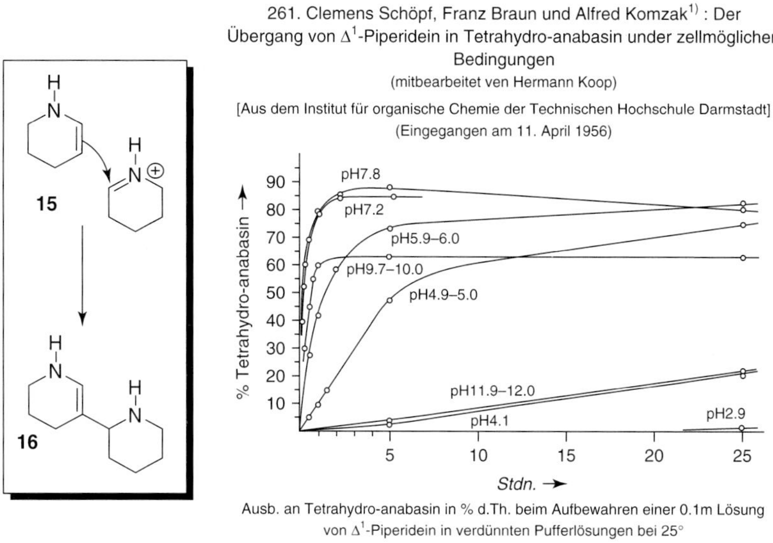

Scheme 1.11 Schöpf's pioneering works.

1.2.2.3 Spontaneous Formation of Alkaloid Skeletons from Glutaraldehyde

Glutaraldehyde (**51**) is a well-known crosslinking agent in biochemistry or histology and is used as a biocide.[12] It can also be advantageously seen as a convenient surrogate of lysine by considering a hypothetical oxidative deamination on aminopentanal **14**.[13]

Simple reactions (Scheme 1.12) were recently disclosed, highlighting an impressive propensity of **51** to mimic several lysine metabolism elements [29]. Whereas **51** was known to polymerize[14] according to different mechanisms and kinetics (Scheme 1.12) depending, for example, on the pH, very few studies previously described compounds resulting from self-condensations into small molecules. Products formed during the treatment of **51** in an aqueous solution at pH 8.5 and 60 °C were investigated. A double homoaldolization followed by crotonization of one of the aldol adducts can easily explain the formation of bicyclic **52**. This compound was previously described but no information was available concerning its stereochemistry [30]. Oxidation of **52** with Dess–Martin periodinane permitted crystallization of the major diastereomer **53**. Compound **52** is, interestingly, related to a biosynthetic intermediate postulated in the course of the biosynthesis of

12) The numerous applications of glutaraldehyde have been extensively reviewed: see Reference [8] in Reference [29].

13) Of course, one needs to keep in mind that in plants reactive functional groups such as aldehydes will most likely be masked in a transitory way; they are, in this chapter, presented in their reactive form for simplicity.

14) Glutaraldehyde is quite stable as an aqueous solution, where it exists as several hydrates with the slow formation of oligomers and polymers (Scheme 1.12). It undergoes rapid polymerization in neat conditions with a catalytic amount of water.

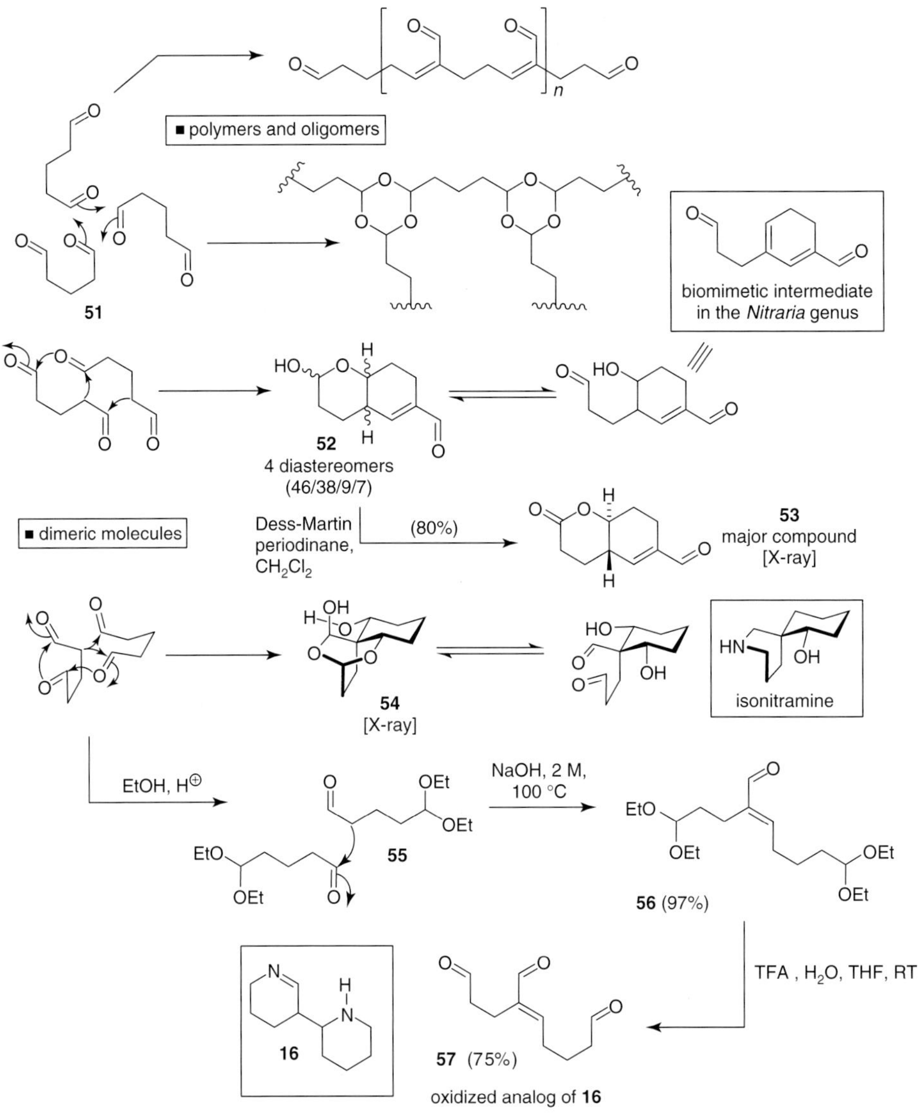

Scheme 1.12 Various condensations of glutaraldehyde.

Nitraria alkaloids (Section 1.4.3). From the same mixture, a crystalline compound **54** that displayed a tricyclic structure with a spiranic quaternary carbon and contiguous acetal and hemiacetal functions was isolated. This intriguing molecule has a striking analogy with known simple spiroalkaloids such as nitramine also isolated from different species of the *Nitraria* genus and its plausible mechanism of formation totally parallels the postulated biosynthesis of such natural substances [31]. Diethoxypentanal **55**, which is easily available from monoprotection of **51**, has

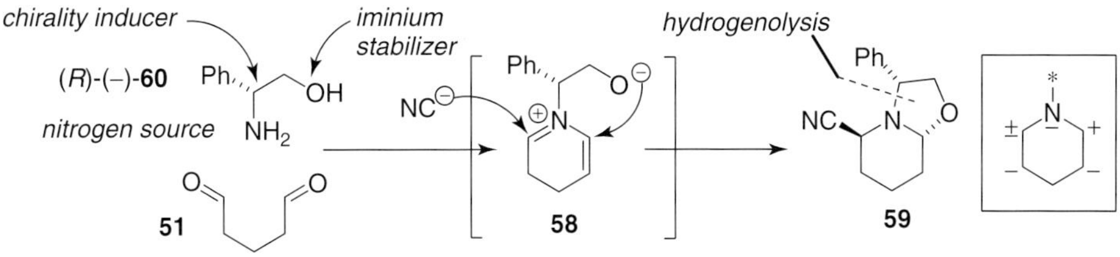

Scheme 1.13 Cyanophenyloxazolopiperidine: a convenient building block.

been treated in a boiling sodium hydroxide solution to furnish quantitatively compound **56** by aldolization/crotonization as a single *(E)*-stereoisomer. Deprotection in acidic conditions gave **57**, an interesting oxidized analog of tetrahydroanabasine.

Glutaraldehyde (**51**) has been widely used for the synthesis of various heterocycles through the formation of dihydropyridine-type intermediates. The most powerful applications are probably the so-called "CN(*R,S*)"[15] method (Scheme 1.13) with the use of the chiral non-racemic *N*-cyanomethyloxazolidine ring system integrated in a piperidine structure as an ideal way to stabilize dihydropyridine **58** into compound **59** [prepared in a single step from glutaraldehyde **51**, *(R)*-(−)-phenylglycinol (**60**) and a source of cyanide ions in water]. This strategy, developed by the Husson and Royer groups, permitted the total synthesis of many alkaloids in a diastereoselective manner. Some of them are closely related to biomimetic strategies. This strategy has been extensively reviewed over the years [32]. As, interestingly, naturally occurring piperidine alkaloids bearing α-side chains are either in the *(R)* or *(S)* configuration, the possibility of using building blocks such as **59** to modulate the stereochemistry at α or α' positions constituted a real breakthrough in piperidine total synthesis.

1.2.3
Biomimetic Access to Pipecolic Acids

1.2.3.1 Pipecolic Acids: Biosynthesis and Importance

L-Pipecolic acid (L-**9**) was first identified in 1952 as a constituent of leguminous plants [33]. It is now recognized as a universal lysine-derived entity present in plants, animals, and microorganisms [34]. In natural substances, L-**9** is a key element in molecules as diverse as swainsonine (**61**) or castanospermine (**62**) (small indolizidine alkaloids known for their glycosidase-inhibiting properties, Scheme 1.14) or FK 506/tacrolimus (**63**) (a polyketide/non-ribosomal peptide hybrid clinically approved as an immunosuppressant). Over the years, many studies have sought to establish the biosynthetic routes to L-**9**. Different metabolic pathways (with different proposed mechanisms [35]) are involved in the formation of L-**9** (Scheme 1.15). Chemically speaking, these basic routes are distinguishable at the loss of the amino group of lysine **3**. The reality of immediate precursors, that is, piperideine carboxylic acids **64** (known as *P2C*) and **65** (*P6C*), has been

15) Named after the French Centre National de la Recherche Scientifique (CNRS).

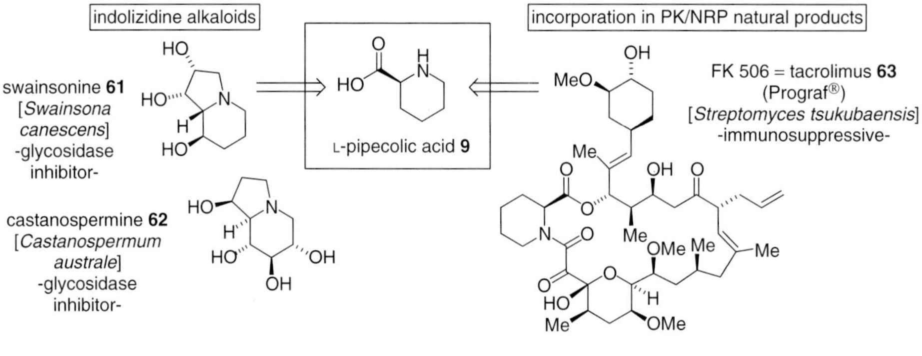

Scheme 1.14 Selected example of pipecolic acid derived alkaloids and pipecolic acid containing secondary metabolites.

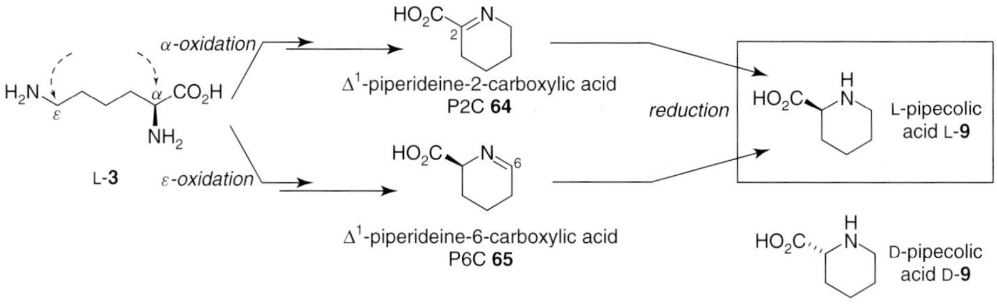

Scheme 1.15 Biosynthesis of pipecolic acid.

demonstrated by many feeding experiments and will not be further developed in the present chapter as many research and review articles are available [33]. The reverse pathways converting L-pipecolic acid (L-**9**) into (P6C **65**) [36] or lysine (**3**) [37] are also known. D-Pipecolic acid (D-**9**) was also reported and derives from D-lysine (whereas L-**9** can be biosynthesized from both L- and D-lysine (**3**) [38]) and was found as a constituent of a few natural substances.

1.2.3.2 Biomimetic Access to Pipecolic Acids

The chemical synthesis of pipecolic acid has been a subject of great interest. Powerful methods of asymmetric piperidine synthesis have been developed toward this aim [39] and have been reviewed [40]. We only outline, in this section, reactions directly involving L-lysine (**3**) (or protected L-**3**) to access **9** in a somewhat biomimetic way. In the 1970s, a first conversion of L-lysine (**3**) into optically active pipecolic acids was disclosed by Yamada and colleagues (Scheme 1.16) [41]. Sodium nitrite–hydrochloric acid was used as a deaminating agent of L-lysine, followed by barium or sodium hydroxide treatment to afford D-pipecolic acid (D-**9**) with more than 90% optical purity and satisfactory overall yield. Net retention of configuration was explained by the formation of lactonic intermediate **66** followed by halogeno-acid **67**. On the other hand, L-lysine could be converted directly into natural L-pipecolic acid starting from ε-tosyl-l-lysine (**68**) but in very low yield (~1%)

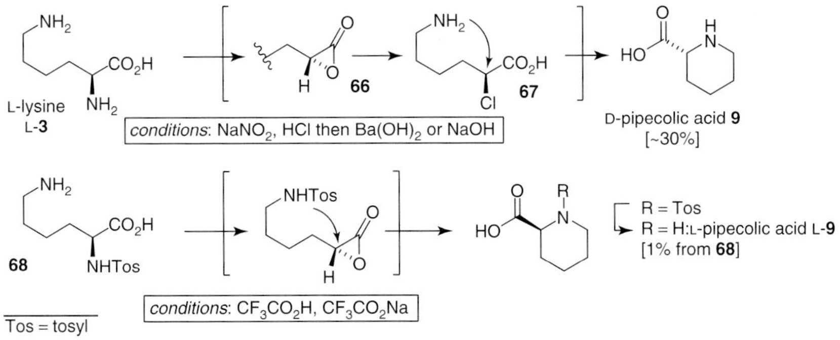

Scheme 1.16 Yamada's biomimetic access to pipecolic acids.

in strong acidic conditions. This work by Yamada and colleagues is not "truly" biomimetic, but is worth mentioning as it converts in a single step the acyclic skeleton of L-**3** into the piperidine ring of **9**.

In 2004, the fully protected L-lysine **69** was converted into aminal **70** by selective oxidation of the side chain using a $Mn(OAc)_2$/peracetic acid system by the Rossen group (Scheme 1.17) [42]. No epimerization occurred under the buffered oxidation conditions. Treatment under mild acidic conditions gave enamide **71**, which could be reduced to L-pipecolate **72**. This study constitutes one of the rare examples of biomimetic conversion of a side chain amino group into an aldehyde oxidation state (obtained as the cyclic *N,N*-acetal **70**).

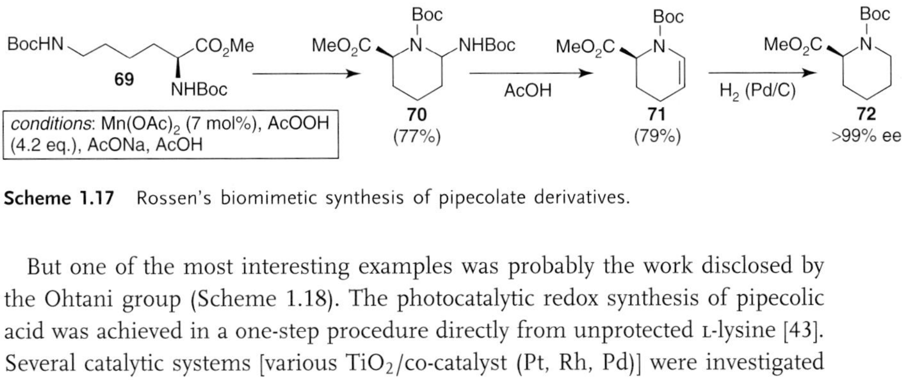

Scheme 1.17 Rossen's biomimetic synthesis of pipecolate derivatives.

But one of the most interesting examples was probably the work disclosed by the Ohtani group (Scheme 1.18). The photocatalytic redox synthesis of pipecolic acid was achieved in a one-step procedure directly from unprotected L-lysine [43]. Several catalytic systems [various TiO_2/co-catalyst (Pt, Rh, Pd)] were investigated to define the best conditions in terms of selectivity (oxidation of ε-amino versus α-amino – which influences the final optical purity of pipecolic acid), yields, and rates. The mechanism was proved to proceed via (i) oxidation of L-lysine with positive holes, leading to P2C (**64**) and P6C (**65**), depending on the oxidized nitrogen and (ii) reduction of the imines with electron, with both steps taking place at the surface of the catalyst. Titanium oxides were shown to predominantly oxidize the ε-amino group, permitting enantiomeric excess up to 90%. The ins and outs of the multiple combinations of catalysts/co-catalysts were studied; the interested reader is referred to the original article for details. With the recourse to catalysis

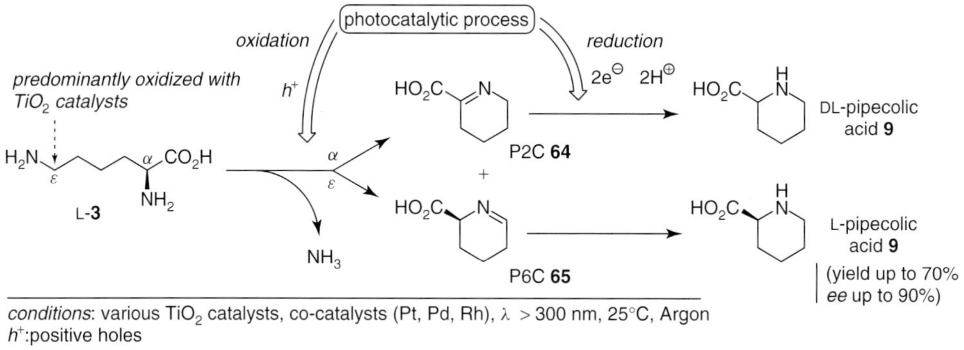

Scheme 1.18 Totally biomimetic synthesis of pipecolic acid by photocatalysis.

and the release of ammonia as the only by-product of the reaction, this synthesis is indubitably a green chemistry process and a beautiful achievement.[16)]

1.3
Biomimetic Synthesis of Alkaloids Derived from Ornithine and Arginine

1.3.1
Biomimetic Access to the Pyrrolizidine Ring

The pyrrolizidine nucleus consists of two fused pyrrolidine cycles; similarities between both biosyntheses may therefore be expected. These alkaloids are biosynthesized from homospermidine, which comes from putrescine (**7**) (Scheme 1.19). Oxidative deamination and subsequent formation of a first five-membered cycle (**74**) through dehydration is followed by an intramolecular Mannich reaction exploiting the enolization capacity of the remaining aldehyde of **75**. Biosyntheses of such alkaloids have been reviewed, as well as their structure elucidation, chemistry, and pharmacology [44].

The direct conversion of an acyclic precursor into a pyrrolizidine-type bicyclic structure was accomplished by the Marson group in 2000 (Scheme 1.20) [45]. Treatment of compound **76** in acidic conditions permitted the deprotection of the masked aldehyde, formation of the Δ^1-pyrrolidine, and subsequent formation of bicyclic **77** by an aza-Prins type cyclization. Such a cationic cyclization (radical cyclizations to pyrrolizidine are known) is closely related to biosynthetic pathways and was the first example of a non-enzymic synthesis of the pyrrolizidine ring from an acyclic precursor. The relative configuration of **77**, which was the sole diastereomer isolated, was ascertained by X-ray crystallography.

16) The term "*biomimetic*" does not appear in this article and no references to biosynthesis are made.

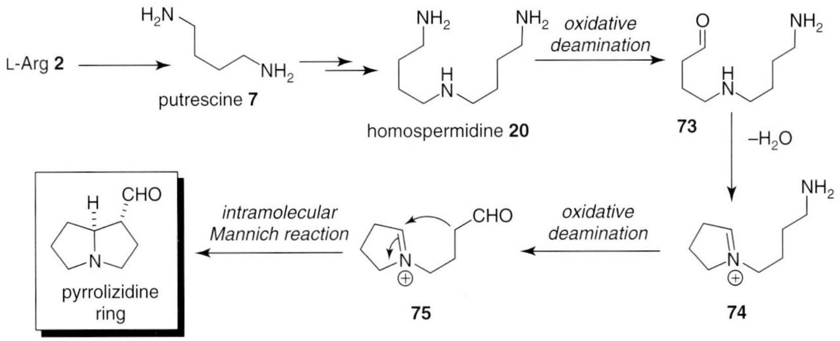

Scheme 1.19 Pyrrolizidine ring biosynthesis.

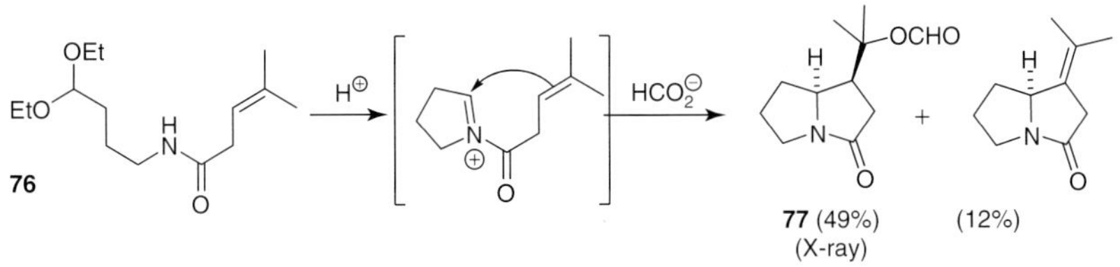

Scheme 1.20 Biomimetic access to the pyrrolizidine ring.

1.3.2
Biomimetic Syntheses of *Elaeocarpus* Alkaloids

Homologous to intermediate **75** derived from homospermidine and implicated in the biosynthesis of indolizidine alkaloids, intermediate **78** (Scheme 1.21) has been postulated to be at the origin of interesting natural substances. Biomimetically speaking, aldehydic intermediate **78** was prepared in the laboratory by the Gribble group in the 1980s (Scheme 1.22) [46]. Bis-acetal **79** was prepared in three steps from chloroacetal **80** in 47% overall yield. It constitutes a stable protected equivalent of intermediate **81**, which under acidic conditions can be deprotected to give *in situ* **78**. In fact, non-isolated **81** when buffered at pH 5.5 presumably generates pyrrolinium aldehyde **78**. This latter was reduced to pyrrolidine **82** or more interestingly trapped by various nucleophiles in a biomimetic manner, thus giving a very interesting unified access to a group of alkaloids isolated from different species of *Elaeocarpus* and *Peripentadiena* (commonly known as *Elaeocarpus alkaloids* [47], see examples on Scheme 1.21, **83–86**). Most of them share a common indolizidine backbone functionalized at positions 7 and 8. Onaka, in the early 1970s, suggested intermediate **78** as the universal precursor of these natural substances [48]. The isolation some years after of alkaloids such as peripentadenine (**86**) [49] ascertained a spermidine (**17**) metabolism pathway. Although numerous syntheses of *Elaeocarpus* alkaloids have been reported [47], we deliberately delineate herein syntheses based on the *in situ* generation of intermediate **78**.

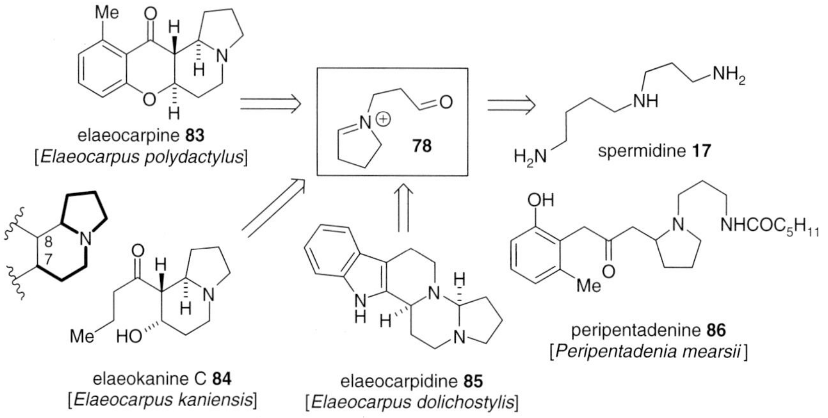

Scheme 1.21 *Elaeocarpus* alkaloids.

Trapping of intermediate **78** with tryptamine gave a straightforward isohypsic synthesis of (±)-elaeocarpidine (**85**) in a stereoselective cascade of reaction from acetal **79** by just adjusting the pH of the solutions (Scheme 1.22) [46]. The elaeocarpidine aminal function was then reduced with sodium cyanoborohydride to give tarennine (**87**). In turn, trapping of intermediate **78** with β-keto ester **88**, thus engaged in a tandem Mannich/aldol condensation, gave **89** as a mixture of two major diastereomers (axial and equatorial hydroxyl at C7) in 62% overall yield. Compound **89** appeared to be a common intermediate for the synthesis of both (±)-elaeokanine A (**90**) and C (**84**). Decarboalkoxylation in strong acidic conditions was accompanied with dehydration at position 7 and gave rise to (±)-elaeokanine A (**90**) in 91% yield. Milder conditions were necessary to carry out the synthesis of (±)-elaeokanine C (**84**). The conditions were carefully studied by the authors, who finally chose catalytic transfer hydrogenation with ammonium formate and palladium in methanol to afford the desired reaction. Although some degree of selectivity was expected, the outcome of the reaction clearly showed a preference for the wrong isomer, with a predominance of (±)-7-epi-elaeokanine C (**91**) over (±)-elaeokanine C (**84**) despite a good overall yield of 60% from bis-acetal **79**. This experiment and its outcome in terms of stereochemistry can be rationalized when considering transition state **92** in which an equatorial hydroxyl is to be expected. This supposition was compared to a similar case studied by Tufariello and Ali [50] a few years before with the intramolecular kinetic aldol reaction of a ketone instead of a β-ketoester. In the latter, a stereocontrol, totally in favor of the axial configuration, could have been governed by transition state **93**. Returning to a biosynthesis hypothesis and taking into account these findings, one can suggest that should such a pathway occur in Nature the decarboxylation step has to be prior to the aldol reaction.

Starting from readily accessible acetal **79**, it is worth noting how simple starting materials (tryptamine, ketoesters) and reaction conditions (aqueous solutions at

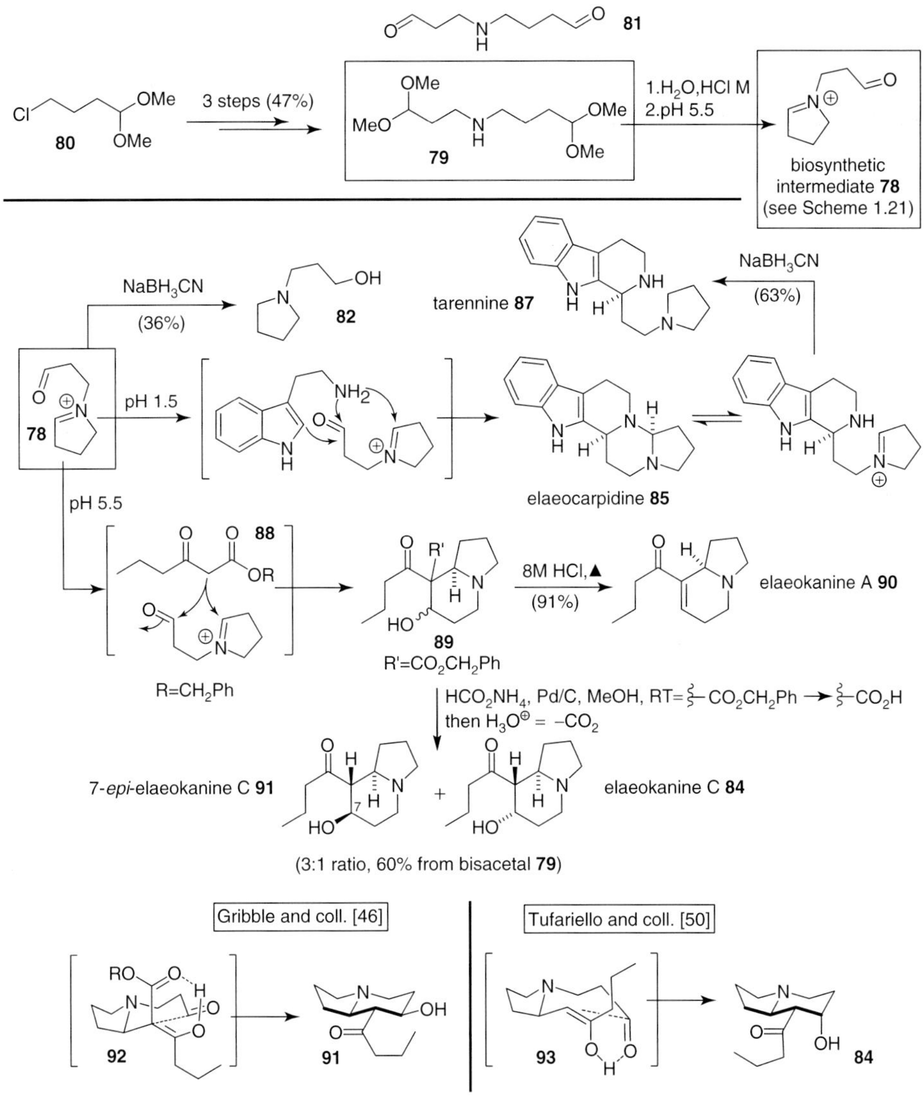

Scheme 1.22 *Elaeocarpus* alkaloids: biomimetic synthesis.

various pH) enabled the design of a divergent pathway to *Elaeocarpus* alkaloids following, and thereby reinforcing, previously proposed biosynthetic hypotheses.

Concomitant with Gribble's work, the Hortmann group used α-cyanopyrrolidine **94** as a surrogate of **79** (Scheme 1.23) [51]. It was prepared by oxidative cyanation of the corresponding pyrrolidine **95** with chlorine dioxide (as an alternative to classical mercury acetate oxidation or modified Polonovski reaction). Compound **94** was used in a total synthesis of (±)-elaeocarpidine (**85**). Lactam **96** was prepared by Lévy and colleagues for the synthesis of **85** [52]. It was engaged in a reductive

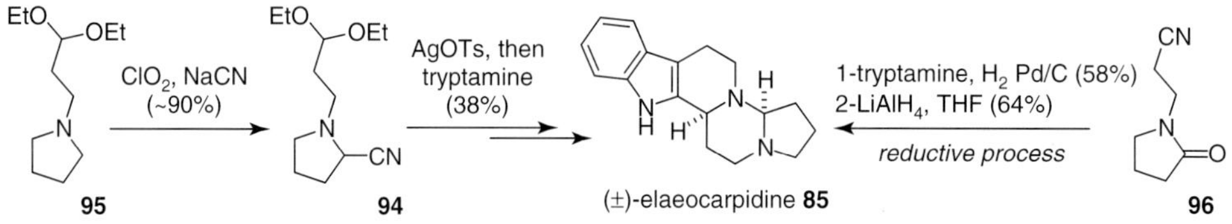

Scheme 1.23 *Elaeocarpus* alkaloids: alternative biomimetic synthesis.

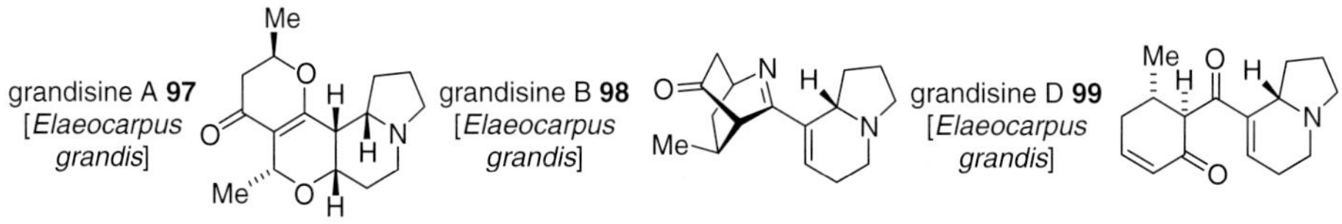

Figure 1.3 Other *Elaeocarpus* alkaloids.

Pictet–Spengler reaction under catalytic hydrogenation conditions. The recourse to trivalent functional groups (lactam of **96**) where divalent ones (imine) are needed places this former approach at the borderline of biomimetic synthesis.

The *Elaeocarpus* alkaloid group gained constant interest with the discovery of new structures (see examples **97–99** in Figure 1.3) along with interesting biological properties (such as selective γ-opioid receptor affinity for grandisine-type alkaloids) [53]. These complex structures also stimulated state of the art total syntheses [54].

1.3.3 Biomimetic Synthesis of Fissoldhimine

Fissoldhimine (**100**) was isolated in 1994 by Sankawa *et al.* [55] from fresh stems of *Fissistigma oldhamii* (Annonaceae), a shrub mainly found in Southern China and Taiwan (Scheme 1.24). Its structure was unambiguously confirmed by X-ray analysis. Since *n*-butanol was used to extract this basic molecule from an alkaline solution, it was suggested that fissoldhimine was an artifact resulting from the aminoacetalization of compound **101** (which may therefore be the "true" natural product) with a molecule of *n*-butanal presumably present in *n*-butanol.[17] The authors proposed a biosynthetic pathway (Scheme 1.25) to fissoldhimine (**100**/**101**) from two molecules of cyclic enamine **29** that came from L-ornithine via a dimerization (see above). R. A. Batey and colleagues revisited the hypotheses in 2007 when disclosing the first biomimetic investigations toward fissoldhimine [56].

17) Defined as spontaneously formed natural products, artifacts are closely related to biomimetic chemistry (see Chapter 24 of this book).

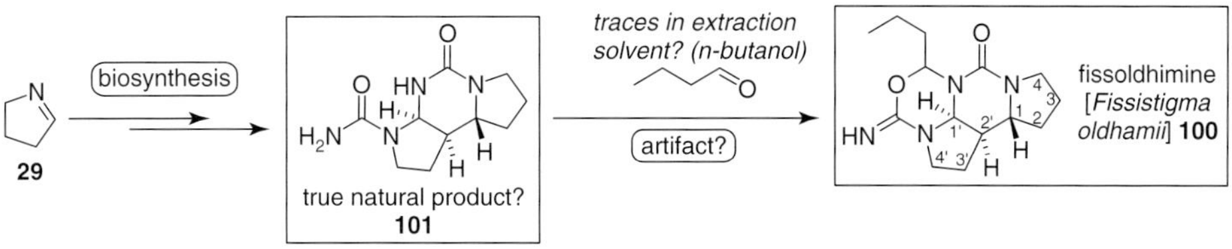

Scheme 1.24 Fissoldhimine: possible structures.

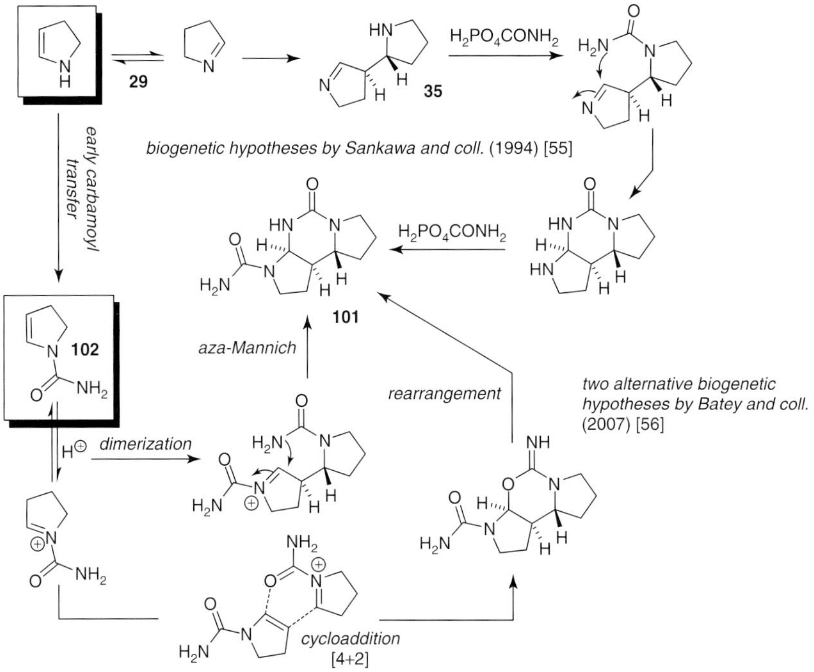

Scheme 1.25 Biosynthetic hypotheses.

Benzyl- or *para*-methoxybenzyl-protected urea **103** (Scheme 1.26) was used as biomimetic equivalent of biosynthetic intermediate **102** postulated in Scheme 1.25. The best conditions, the authors found, for the formation of the desired *exo*-dimer **104b** were the use of trifluoroacetic anhydride in THF (*endo*/*exo*: 2 : 1 ratio) but separation of the two diastereomers was impossible. In addition, final deprotection of the urea nitrogens was also problematic (removal of the benzyl group was unsuccessful and DDQ (2,3-dichloro-5,6-dicyano-1,4-benzoquinone)-mediated deprotection of PMB groups resulted in the deprotection of only one of the nitrogens). Despite obtaining an *endo* relative stereochemistry that contrasts with the *exo* stereochemistry of natural fissoldhimine **100**/**101**, this example is interesting as it opens the way to discussions concerning the biosynthesis and the implication of enzymes, particularly since fissoldhimine was shown, by X-ray, to be a racemate in nature! Another point is that an early carbamoyl transfer is likely, in direct

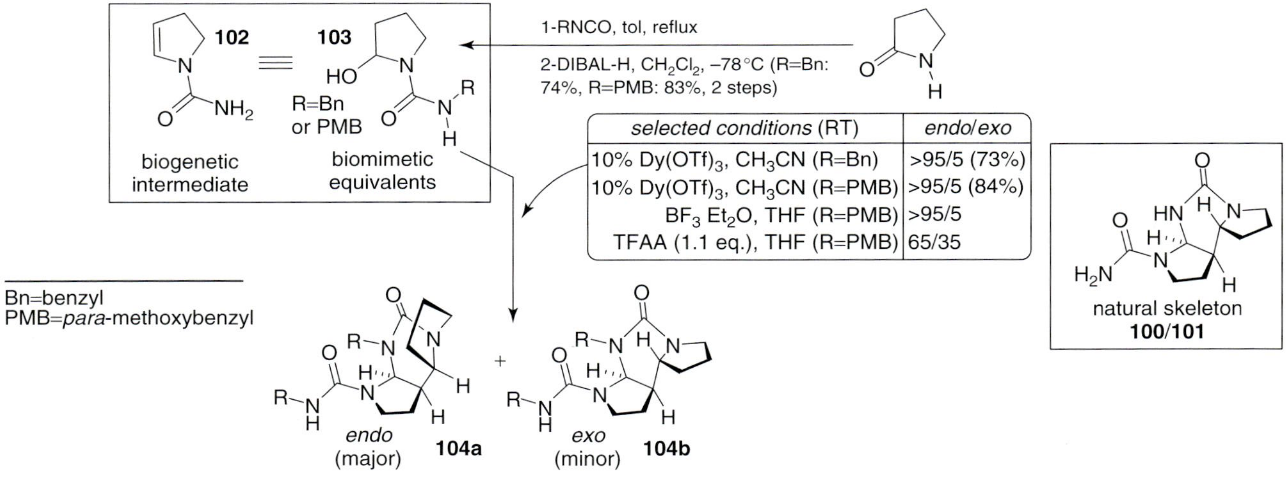

Scheme 1.26 Biogenetically inspired heterodimerization toward fissoldhimine.

correlation with the critical difficulty of dimerization of imine **29** as already evoked (see above, Scheme 1.5).

1.3.4
Biomimetic Synthesis of Ficuseptine, Juliprosine, and Juliprosopine

As we will now discuss, imine **29** was successfully engaged by the Snider group [57] in a biomimetic synthesis of two dihydroindolizinium alkaloids: ficuseptine (**105**) [58] and juliprosine (**106**) [59] (Scheme 1.27). Biosynthetic considerations led to the postulation of a Chichibabin pyridine synthesis type reaction as a cornerstone convergent step for both **105** and **106**. Mixing aldehyde **107** (which is presumably a tyrosine metabolite in Nature) with imine **29**, generated *in situ* from **108**, in acetic acid provided ficuseptine (**105**) in a single step in 52% yield (Scheme 1.28).

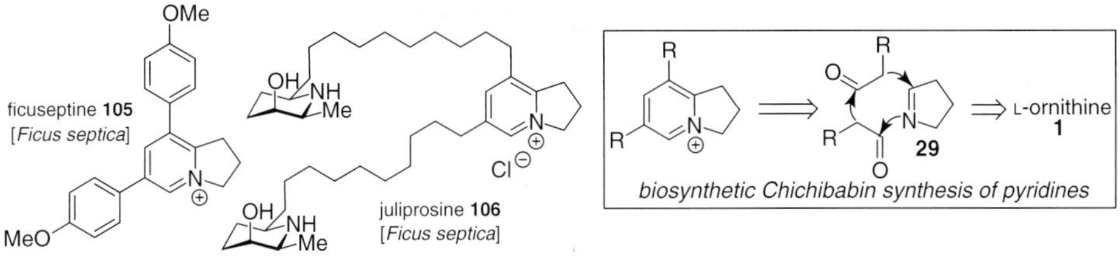

Scheme 1.27 Ficuseptine (**105**) and juliprosine structures (**106**), and plausible biogenesis.

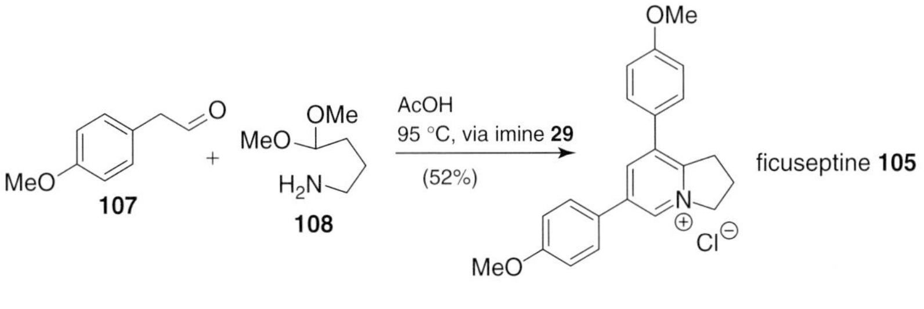

Scheme 1.28 Ficuseptine: biomimetic synthesis.

This strategy appeared to be as effective (Scheme 1.29) when dealing with more complex juliprosine (**106**) in terms of side chains engaged in the pyridine formation (probably of polyketide origin in Nature). Aldehyde **109** was prepared in six steps from bromopyridinol **110** and reacted with preformed imine **29** at room temperature in acetic acid to give Troc-**106**, from which juliprosine **106** was liberated by protecting group removal. Juliprosopine (**111**) [60], a reduced tetrahydropyridine counterpart of juliprosine was then targeted by the authors as a final destination. Exposure of Troc-**106** to sodium borohydride gave a separable 1 : 1 mixture of diastereomers. Prior to this last step, model studies concluded that the natural stereochemistry of **111** should be trans. Therefore, the trans isomer was deprotected to afford juliprosopine (**111**), for which the stereochemistry was in consequence clarified 25 years after its isolation. It is striking in terms of

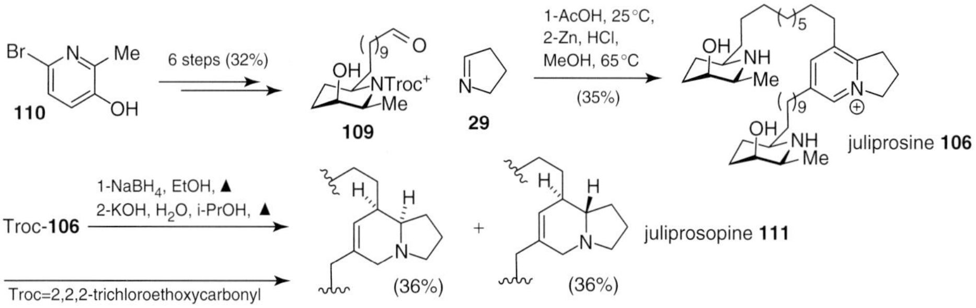

Scheme 1.29 Juliprosine (**106**) and juliprosopine (**111**) biomimetic synthesis.

evolutionary convergence that a pivotal biosynthetic (and synthetic!) reaction can give rise to different natural substances in different plants. These examples once again raise questions about the intervention of enzymes in such biosyntheses.

1.3.5 Biomimetic Synthesis of Arginine-Containing Alkaloids: Anchinopeptolides and Eusynstyelamide A

1.3.5.1 Natural Products Overview

Original dihydroxybutyrolactams were isolated from marine ascidian *Eusynstyela latericius* collected in the Australian Great Reef and named eusynstyelamides A–C (**112–114**, Figure 1.4) by the Tapiolas group in 2009 [61]. They are in fact closely related to other natural substances, namely, anchinopeptolides A–D (**115–118**) and surprisingly cycloanchinopeptolide C (**119**) (with its central core consisting in a tricyclic 5-12-4 ring system) isolated earlier from a marine sponge [62]. Biosynthetically, they might all arise from dimerization of monomeric modified α-ketoacid-containing peptides (dipeptides or tripeptides in the case of eusynstyelamides and anchinopeptolides, respectively) by an aldol reaction (formation of the carbon–carbon bond) and an addition to the amide (Figure 1.4, box). Biomimetic syntheses of these two classes of alkaloids have been performed by the Snider group.

1.3.5.2 Biomimetic Synthesis

(±)-Anchinopeptolide D (118) The aldol dimerization of protected modified tripeptide **120** (synthesis not described herein) was undertaken (Scheme 1.30) [63]. Treatment of **120** with potassium hydroxide in MeOH afforded a mixture of three isomeric compounds (Boc-**118**, Boc-**121**, **122**), two of which were deprotected to afford **118** (anchinopeptolide D) and **121** (epi-anchinopeptolide D). Equilibration studies clearly demonstrated epimerization at the hemiaminal center: heating pure Boc-**118** or Boc-**121** for 1 h resulted in a 2 : 1 mixture of Boc-**118** and Boc-**121**. Equilibration of the aldol stereocenter occurred more slowly (from pure Boc-**118**

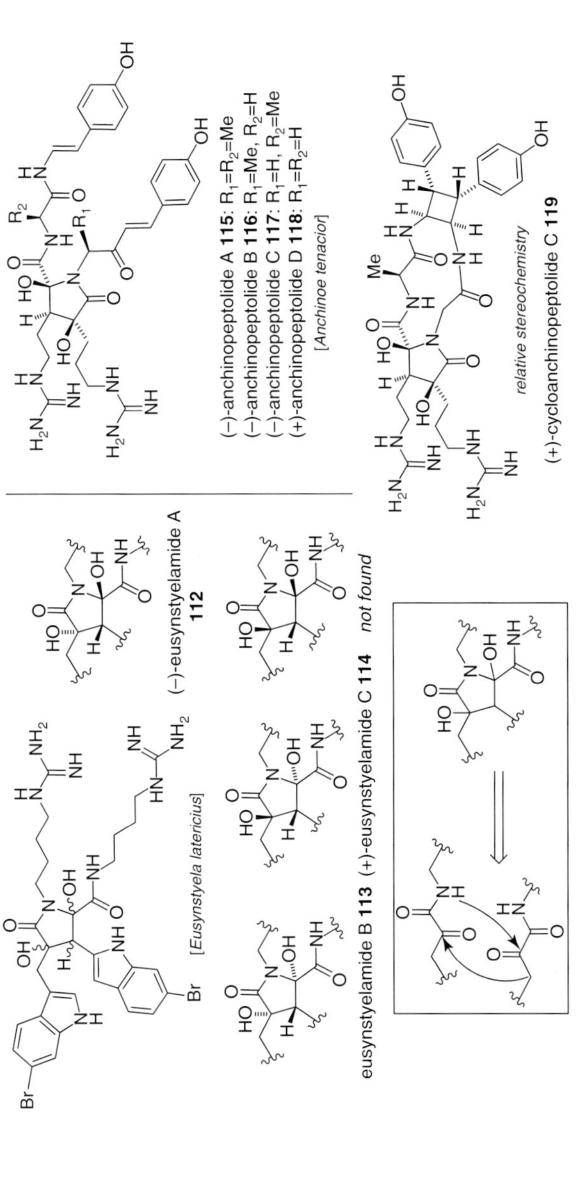

Figure 1.4 Eusynstyelamides and anchinopeptolides.

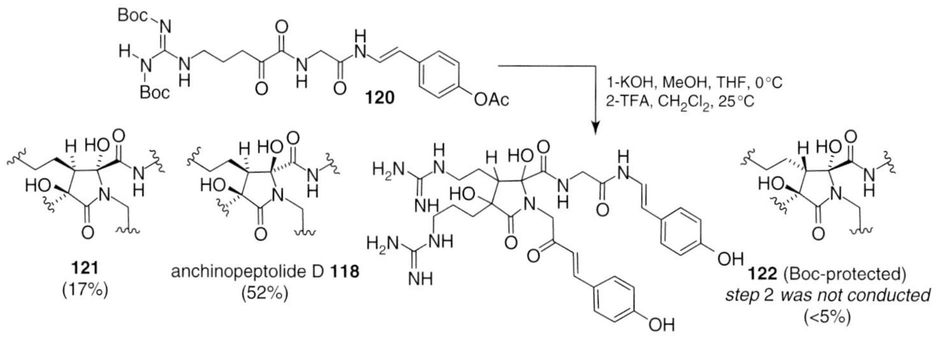

Scheme 1.30 Biomimetic synthesis of anchinopeptolides.

or Boc-**121**, 17 h of heating in KOH THF–MeOH provided a 6 : 3 : 1 mixture of Boc-**118**, Boc-**121**, and **122**).

"(±)-Cycloanchinopeptolide D" (123) A head to head [2 + 2] cycloaddition between the two double bonds of the plausibly tyrosine-derived part of anchinopeptolide C (**117**) may explain the formation of the cyclobutane ring of cycloanchinopeptolide C (**119**). To confirm this hypothesis, access to analog **123** (unnatural "cycloanchinopeptolide D") was targeted (Scheme 1.31). Careful choice of the photochemical conditions, especially the solvent, permitted the obtaining of **123** avoiding, thereby, the isomerization of the double bonds. Indeed, in water, irradiation of a 0.005 M solution of **118** at 350 nm afforded cycloanchinopeptolide D (**123**) in 48% yield. In water, a hydrophobic effect may cause the nonpolar chain to pack closely and favor the [2 + 2] cycloaddition *versus* the *trans*/*cis* isomerization.

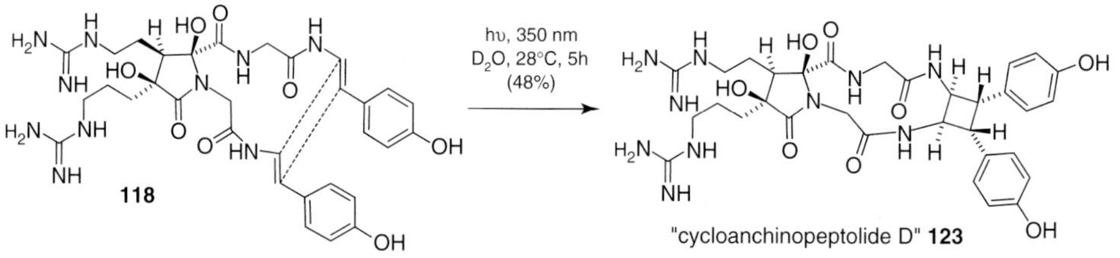

Scheme 1.31 Biomimetic synthesis of cycloanchinopeptolide D.

(±)-Eusynstyelamide A (112) Adopting a similar philosophy, access to **112** was possible (Scheme 1.32) [64] from unstable dipeptide **124** (prepared in four steps), which could undergo butyrolactam core formation under basic conditions. The obtaining of Boc-protected-**112** as a major compound (37%) was accompanied by small amounts of Boc-**113** (<4%). Deprotection of Boc-**112** afforded eusynstyelamide A (**112**). At this stage, chemical and reactivity observations stimulated interesting thoughts. As previously mentioned, anchinopeptolide D (**118**) was shown to equilibrate into a mixture of epimeric compounds. The question of

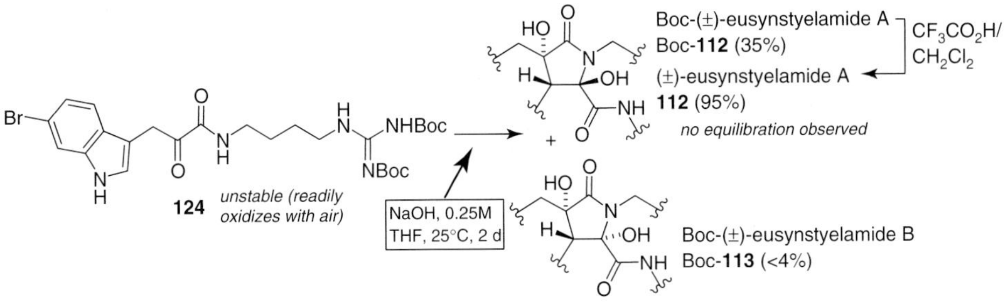

Scheme 1.32 Biomimetic synthesis of eusynstyelamide A.

whether eusynstyelamides B (**113**) and C (**114**) are natural products or artifacts spontaneously generated during extraction was logically posed. In fact, when the isolation of eusynstyelamides was conducted, some samples, carefully processed, of the sponge contained only eusynstyelamide A (**112**) whereas some others contained the two isomeric eusynstyelamides A (**112**) and B (**113**) (1 : 1 ratio) or the three isomeric compounds [eusynstyelamides A–C (**112–114**), 2 : 6 : 1 ratio]. Additionally, no evidence of equilibration was observed by the authors as confirmed by Snider and colleagues with synthetic eusynstyelamide A (**112**). Indeed, synthetic **112** did not equilibrate under various conditions (acidic, storage in CD_3OD) and decomposed in basic conditions. The formation of Boc-eusynstyelamide A (Boc-**112**) as the major isomer from monomeric **124** is consistent with the isolation of eusynstyelamide A (**112**) as a unique compound from some sample of raw material. In the laboratory, the aldol reaction gave rise to a major isomer with high selectivity according to the proposed mechanism (Scheme 1.33) and may be correlated with the fact that it is the main relative stereochemistry observed in the two series of alkaloids. The influence of the side chains when comparing the eusynstyelamide and anchinopeptolide series is probably prominent in controlling both the equilibration propensity and the stereochemical outcome of the aminal formation.

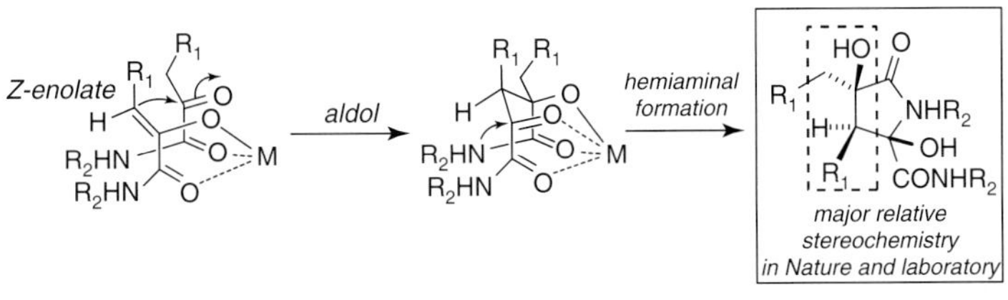

Scheme 1.33 Dimerization: a plausible mechanism.

1.3.6
A Century of Tropinone Chemistry

The biomimetic synthesis of tropinone (**125**) by Robinson in 1917 [65] is probably among the most renowned synthesis of organic chemistry. More than many other, it has been analyzed, cited and is usually presented as the first biomimetic total synthesis and as a pioneering multicomponent reaction (Scheme 1.34). We will therefore not develop further this textbook classic. The reaction was revisited by Schöpf in the 1930s with an in-depth study of the reaction conditions [66] and has been adapted to other target molecules many times since then. Questions interfacing biosynthesis and chemistry arose continuously concerning tropane alkaloids; they were compiled and analyzed by O'Hagan and Humphrey in a review article [67]. The now well-established biosynthesis of the tropinone (**125**) skeleton is depicted in Scheme 1.34.

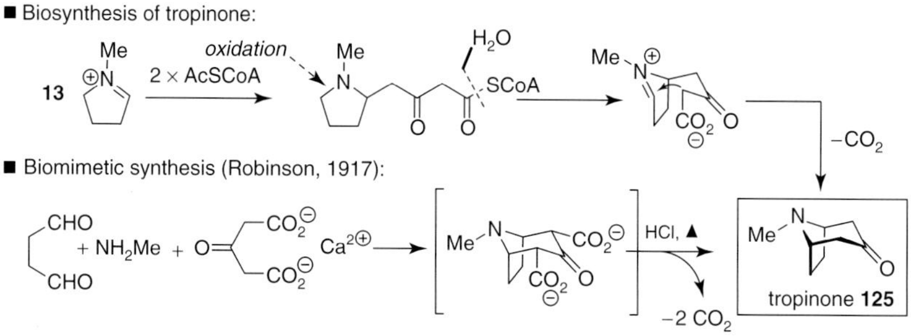

Scheme 1.34 Tropinone biosynthesis and landmark biomimetic synthesis.

Nearly one century later, this reaction still stimulates interesting discoveries. By way of an example, a solid-phase version was developed as an interesting means of generating tropane analogs in a combinatorial way [68]. The use of siloxy(allyl)silane was also reported in place of acetonedicarboxylates (avoiding thereby the double thermal decarboxylation step) and afforded tropinone in good yield [69]. Asymmetric syntheses of substituted tropinones, some using biomimetically related intramolecular Mannich reactions have also been disclosed [70]. The Mannich reaction and the Robinson–Schöpf condensation are absolute musts in the chemical toolbox for biomimetic synthesis of alkaloids [71].[18]

1.4 Biomimetic Synthesis of Alkaloids Derived from Lysine

1.4.1 Alkaloids Derived from Lysine: To What Extent?

A list (if not a classification *sensu stricto*) according to the number of carbons that have been incorporated via the C_5N units originating from L-Lys can be drawn

18) "Time can never diminish the exquisite beauty of Robinson's total synthesis of tropinone. The simplicity of this remarkable synthesis, the splendor of the cascade and the stunning analysis that led to its conception will always stand out, marking this endeavor as a true classic in total synthesis, and inspiring new generations of chemists" [72].

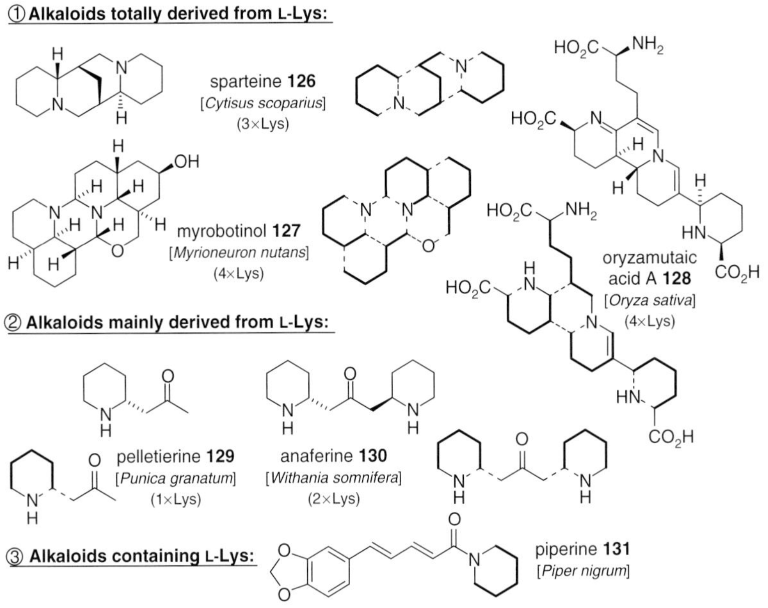

Figure 1.5 Lysine-derived alkaloids: selected examples.

up – some examples are given in Figure 1.5. One may then distinguish, more or less artificially, between three categories, that is, alkaloids: (i) exclusively derived from L-Lys[19] [e.g., sparteine (**126**), myrobotinol (**127**),[20] oryzamutaic acid A (**128**)[21]), (ii) mainly derived from L-Lys [e.g., pelletierine (**129**) and anaferine (**130**)], and (iii) containing L-Lys [e.g., piperine (**131**)]. For each of the two first groups, examples of biomimetic syntheses will be discussed in this section.

1.4.2
Lupine Alkaloids

1.4.2.1 Overview and Biosynthesis Key Steps

The "lupine alkaloids" are an important and biosynthetically homogeneous group of natural substances found especially, but not only, in the *Lupinus* genus (Fabaceae). The vast majority of them are structurally characterized by a quinolizidine rings and they are classified in ten or so main subgroups defined by the central skeleton. Lupinine (**132**) (the simplest alkaloid of this group), sparteine (**126**),

19) Every single carbon atom of the alkaloids comes from L-Lys. It is, at this point, striking to see how complex structures can arise from a single amino acid.
20) See below for comments on this intriguing new family of alkaloids.
21) Oryzamutaic acid A (128) is one of the rare alkaloids that contain four molecules of L-lysine [22]. It is a novel yellow pigment isolated from the endosperm of an *Oryza sativa* (rice) mutant. Its rather elaborate architecture presumably incorporates several types of units derived from L-lysine, including lysine itself. It undoubtedly constitutes an appealing target for biomimetic synthesis! [73].

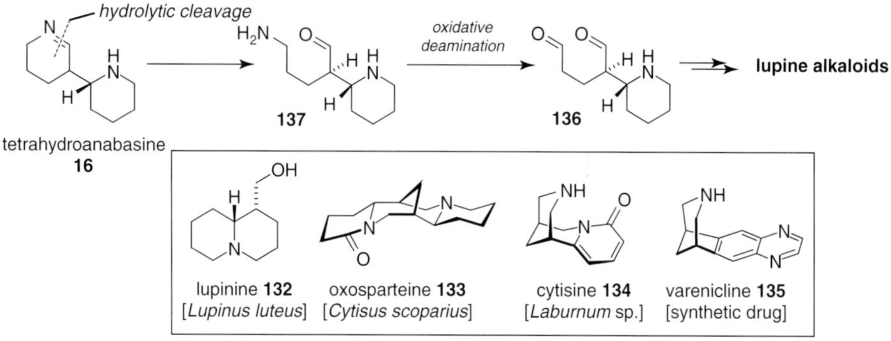

Scheme 1.35 Lupine alkaloids: origin and examples.

oxosparteine (**133**), and cytisine (**134**) are probably among the most well-known structures of lupine alkaloids (Scheme 1.35). The existence of multiple higher oxidized derivatives of lupine alkaloids and transformed skeletons thereof explains part of the diversity within this class of natural substances. They are of great interest for their biological properties such as nicotine-like properties. Varenicline (**135**) (Chantix®, Champix®), a synthetic drug that acts as a partial agonist of $\alpha_4\beta_2$-subtype of the nicotinic acetylcholine receptors, is directly inspired by the structure of natural cytisine. This drug is prescribed to treat smoking addiction in many countries in the USA and Europe. Many comprehensive reviews of interest have been published over the years covering many aspects of lupine alkaloids [74]. The suggestion that the quinolizidine is built up from L-lysine has now been widely confirmed by incorporation studies and chiral non-racemic intermediate **136** arising from selective hydrolysis of **16** (via intermediate **137**) appears to be a cornerstone in lupine alkaloid metabolism. The biosyntheses of alkaloids such as lupinine (**132**) or sparteine (**126**) are now textbook examples of L-lysine metabolism in plants.

1.4.2.2 **Biomimetic Synthesis of Lupine Alkaloids**

A unified biomimetic strategy starting from tetrahydroanabasine (**16**) was beautifully designed by Koomen and colleagues for the total synthesis of representative lupine alkaloids [75, 76]. Four natural substances, that is, lupinine (**132**), epilupinine (**138**), aminolupinane (**139**), and sparteine (**126**), as well as anabasine (**140**), were retrosynthetically traced back to the common precursor tetrahydroanabasine.

Mimic of the Fundamental Oxidative Deamination Step Topaquinone (2,4,5-trihydroxyphenylalanine) was identified as the covalently bound active site cofactor of copper-containing amine oxidases in the early 1990s [77]. These enzymes catalyze the fundamental two-electron oxidative deamination of primary amines into the corresponding aldehyde along with ammonia. The formation of a Schiff base that undergoes proton abstraction with aromatization as a driving force is implicated in the intimate mechanism of the enzyme. Several model studies of topaquinone or more generally quinoprotein enzymes have been

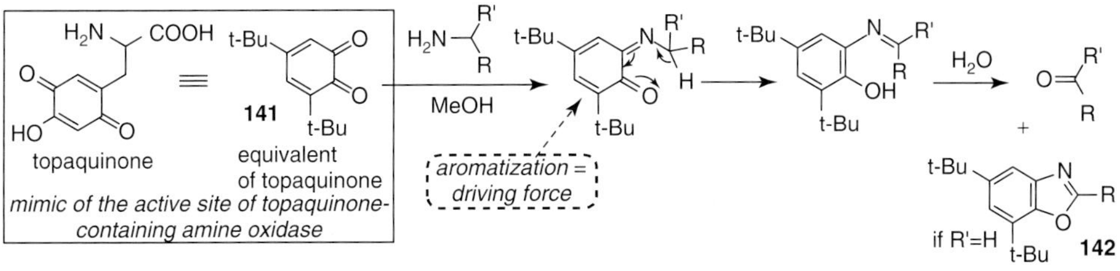

Scheme 1.36 Topaquinone and a topaquinone-mimic.

made but are beyond the scope of this chapter [77, 78]. Commercially available and stable di-*tert*-butyl-*o*-quinone **141** was successfully used as a topaquinone surrogate to achieve the desired oxidative deamination without any overoxidation to benzoxazoles of type **142** (Scheme 1.36). Quinone **141** was used to oxidize compound **143**, which resulted from the opening of tetrahydroanabasine (**16**) with methoxyamine (compound in which the stereochemistry was preserved), to give iminium or enamine **144** depending on the pH (Scheme 1.37).

Reduction of **144** gave a stable quinolizidine (Scheme 1.37), which was further reduced to aminolupinane (**139**, isolated as its acetate). Hydrolysis of the oxime of **145**, necessary to access lupinine, had to be performed under mild conditions to avoid epimerization of the intermediate aldehyde into the more stable epilupinine skeleton. Ozonolysis of **145** at low temperature followed by sodium borohydride reduction yielded exclusively lupinine (**132**). Under more vigorous conditions, a mixture of lupinine (**132**) and epilupinine (**138**) was obtained. The pathway to sparteine required participation of another tetrahydropyridine molecule (**15**). Reaction of **144** with **15** gave **146**, the oxime group of which was split under mild conditions to furnish aldehyde **147**, which is a direct precursor of the sparteine skeleton in which both the piperidine ring and the aldehyde occupy an axial position that can cyclize to **148**. Reduction of **148** furnished sparteine **126** but, as already observed with lupinine, oxime cleavage under more drastic conditions was accompanied with a loss of stereoselectivity, giving isomeric β-isosparteine along with sparteine. This synthesis clearly demonstrated that the whole pathway from tetrahydroanabasine (**16**) to polycyclic sparteine (**126**) can be efficiently mimicked in the laboratory. Finally, simple hydrolytic conditions from **144** surprisingly gave rise to anabasine (**140**) in a single step.[22)]

1.4.2.3 A Biomimetic Conversion of *N*-Methylcytisine into Kuraramine

Many examples of semi-synthetic conversions of lupine alkaloids have been disclosed over the years [74]; only one recent example will be given. Kuraramine (**149**) was isolated in 1981 from *Sophora flavescens* [79]. An obvious biosynthetic

22) Biosynthetically speaking, anabasine (**140**), for example, found in *Nicotiana sp.*, is derived from tetrahydropyridine and nicotinic acid and not from tetrahydroanabasine, nor does tetrahydroanabasine (**16**) derive from the reduction of anabasine (**140**).

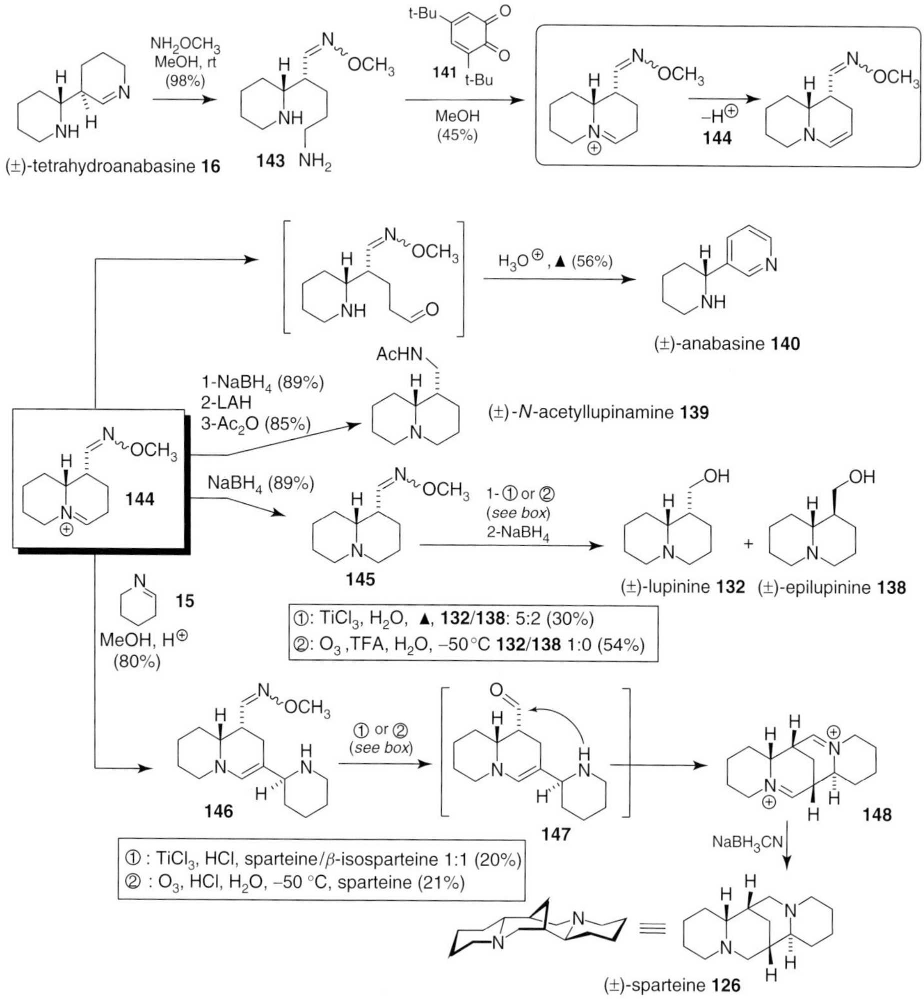

Scheme 1.37 Biomimetic unified access to lupine alkaloids.

relationship with *N*-methylcytisine (**150**) which could involve a N(1)–C(10) oxidative cleavage has been postulated and strengthened by a biomimetic conversion (Scheme 1.38). Indeed, in 2010, Gallagher *et al.* [80] first silylated **150** into **151**, which in turn was submitted to a Fleming–Tamao oxidation (**153**) followed by sodium borohydride reduction to give (–)-**149** in 18% overall yield.

1.4.3
Biomimetic Synthesis of *Nitraria* and *Myrioneuron* Alkaloids

Mainly from the steppes of Uzbekistan, intriguing alkaloids [such as sibirine (**154**), nitraramine (**155**), and nitrarine (**156**)] have been isolated from *Nitraria* species

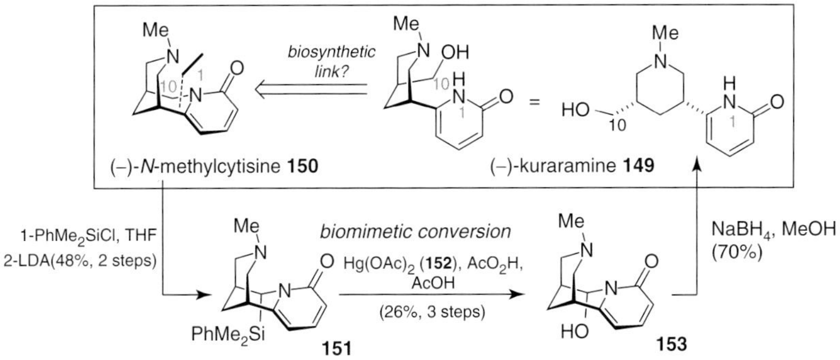

Scheme 1.38 Biomimetic synthesis of kuraramine from *N*-methylcytisine.

(Nitrariaceae). In the 1990s, an unusual lysine-derived metabolism was postulated by Koomen and colleagues to account for the biosynthesis of these often-complex molecules (Scheme 1.39) [81]. Often isolated as racemic mixtures, they may arise from an alternative opening/rearrangement of tetrahydroanabasine as compared to the "classical" *Lupinus* pathway (Scheme 1.39, pathway ① and Scheme 1.35 above). Central to this hypothesis is the formation of compound **157**, referred to as the "*key intermediate*" in *Nitraria* metabolism, by a retro-Michael reaction followed by an oxidative deamination (pathway ②).

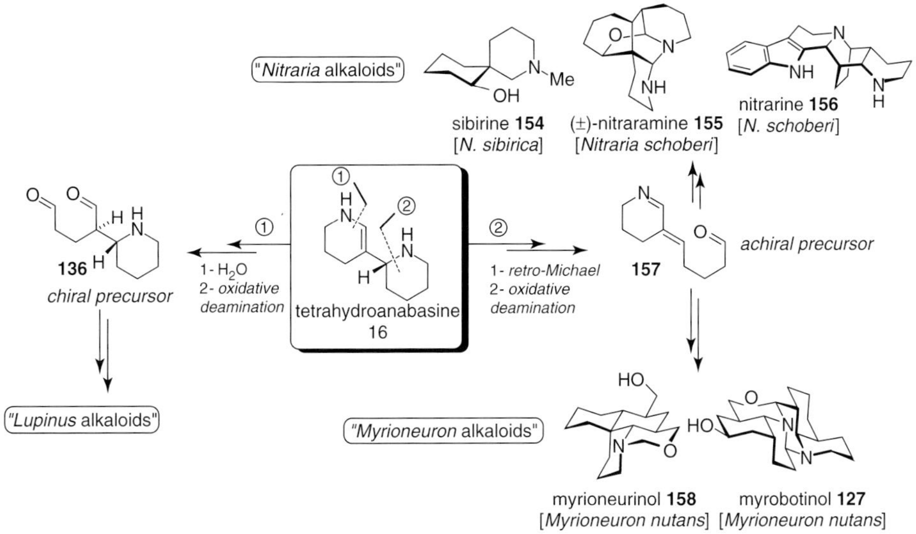

Scheme 1.39 *Nitraria* and *Myrioneuron* alkaloids: origin and selected examples.

1.4.3.1 Biomimetic Syntheses of Nitraramine

Nitraramine (**155**) is probably one of the most original and intricate structures among lysine-derived alkaloids (Scheme 1.40). It has been isolated from *Nitraria*

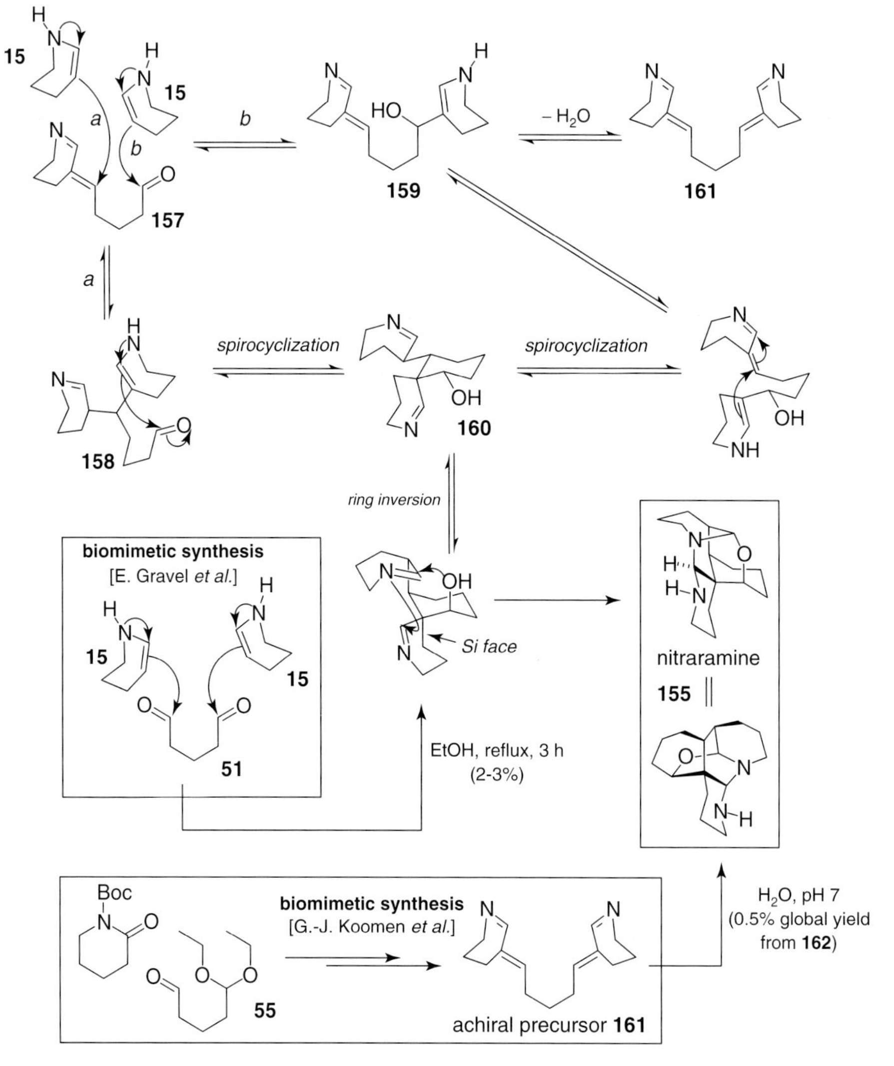

Scheme 1.40 Biosynthetic hypotheses of nitraramine and total biomimetic synthesis.

as a racemate. With several heterocycles (three peripheral cycles in a chair-like conformation and three central cycles in a boat-like conformation) and six contiguous stereogenic centers (one of which is a quaternary spiro center) this molecule, despite its rather modest molecular weight, can be seen as a real challenge for organic chemists.

The Koomen group proposed a biogenetic hypothesis according to which nitraramine might result from the assembly of lysine-derived simple precursors (Scheme 1.40), through a series of simple reactions. From key intermediate **157** as a pivotal achiral precursor, the addition of a piperideine molecule (**15**) could then afford compound **158** (or **159**). An intramolecular spirocyclization reaction could then give rise to the spiro quaternary center (intermediate **160**) found in the natural product. A ring inversion on the newly created cyclohexane of **160** is then needed to explain two ring-closures by imine trappings that end the cascade toward nitraramine **155**.

The Koomen group targeted compound **161** in their beautiful pioneering total synthesis of nitraramine in 1995 (Scheme 1.40). Achiral synthetic **161** was then treated in neutral conditions in water and afforded the natural compound in a global sequence of a dozen steps from **55** and a piperidone (0.5% yield) [82]. In 2005, a second straightforward synthesis strongly reinforced the Koomen hypotheses [83]. In view of the fact that **157** can be obtained from the reaction of enamine **15** with glutaraldehyde (**51**) followed by dehydration, we set up a biomimetic total synthesis of nitraramine by a one-pot sequence. In fact, from simple starting materials **15** and **51** the reaction cascade quickly takes place. By treating one equivalent of **15** with two equivalents of **51** in boiling ethanol, nitraramine (**155**) was obtained in a low yield that competed with the Koomen total synthesis and also with quantities available from natural sources.

Interestingly, this second total synthesis of nitraramine showed that the natural substance previously described as epinitraramine (**162**) was actually an artifact resulting from the protonation of nitraramine in the NMR tube (Scheme 1.41). In fact, ^{1}H NMR spectra of protonated nitraramine (with traces of hydrochloric acid liberated from $CDCl_3$) differ slightly from those of **155** as a free base. Thereby, the absence of **162** reflects the high stereoselectivity of the cascade reaction. Finally, the success of the total synthesis of **155**, despite low yields, raises interesting questions concerning the implication of enzymes in the biosynthesis of such alkaloids.

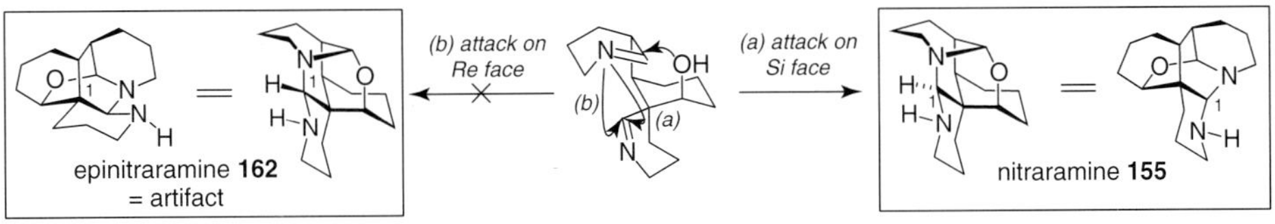

Scheme 1.41 Epinitraramine is an artifact.

1.4.3.2 Biomimetic Syntheses of Tangutorine

As seen above (Scheme 1.12), the self-condensation of glutaraldehyde (**51**) furnishes bicyclic compound **52**. A simple condensation of **52** with tryptamine under acidic conditions allowed the synthesis of **163**, which is a direct precursor of tangutorine (**164**), another alkaloid isolated from a *Nitraria* species (Scheme 1.42) [84]. Indeed, reduction of **163** with sodium borohydride gave tangutorine (**164**) after recrystallization to isolate the major diastereomer of the (75 : 25) mixture [85]. The reaction course with tryptamine could be rationalized by assuming the *in situ* formation of intermediate **165**. Interestingly, compound **166**, resulting from a direct Pictet–Spengler reaction of the aldehyde function of **52** with tryptamine, was also isolated from the reaction. It is reminiscent of the structure of many alkaloids from

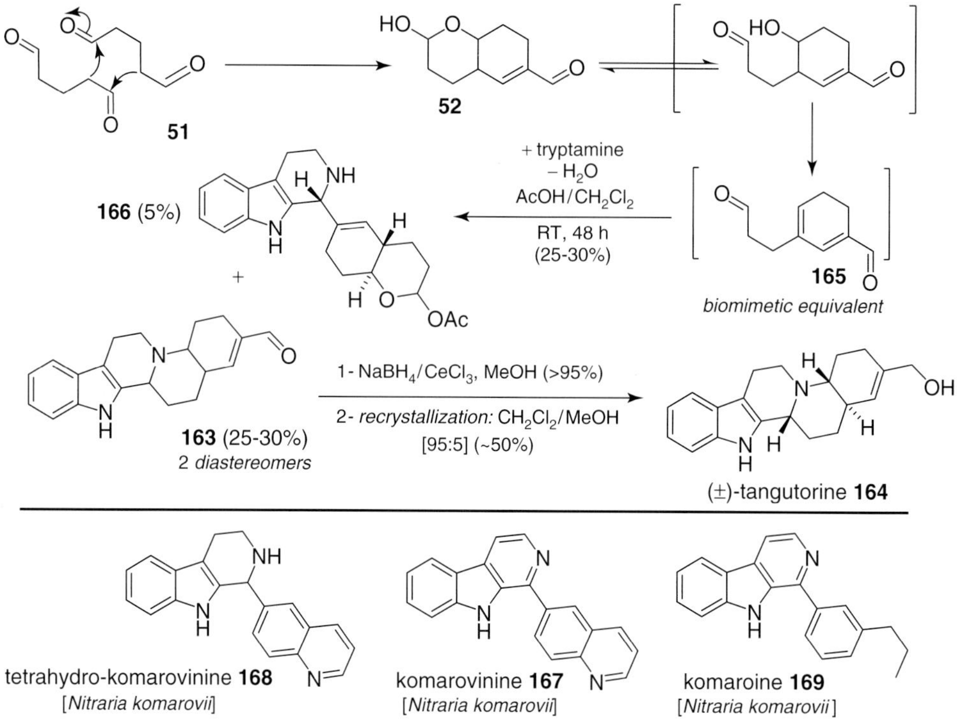

Scheme 1.42 Biomimetic synthesis of tangutorine.

the *Nitraria* genus in terms of carbon skeleton [see examples such as komarovinine (**167**), tetrahydro-komarovinine (**168**), and komaroine (**169**)], demonstrating thereby the impressive uniqueness of lysine metabolism in the *Nitraria* genus. It was therefore possible to postulate intermediate **165** as another achiral cornerstone metabolic intermediate along with already mentioned **157**. Both intermediates may, at least in part, contribute to explain the occurrence of many *Nitraria* alkaloids as racemates in Nature (this is especially the case for indolic *Nitraria* alkaloids).

This straightforward strategy also cast doubt on the previously proposed biosynthetic pathway to tangutorine (**164**), involving a complex rearrangement of a yohimbine-type indolomonoterpenic precursor, proposed by Jokela and colleagues (Scheme 1.43) [86]. These hypotheses appeared unlikely both in terms of chemical reactivity and chemotaxonomy.

In 2010, one of us published a comprehensive review covering the state of knowledge and detailing the isolation, structure determination (and revision), and the biomimetic syntheses of *Nitraria* alkaloids [87]; the interested reader is directed to this article as well as to older reviews [86]. Many achievements in biomimetic synthesis of *Nitraria* alkaloids have been disclosed by the Koomen group [88] and the Husson group [89]. Some of them beautifully addressed the issues of efficiency and selectivity in total synthesis.

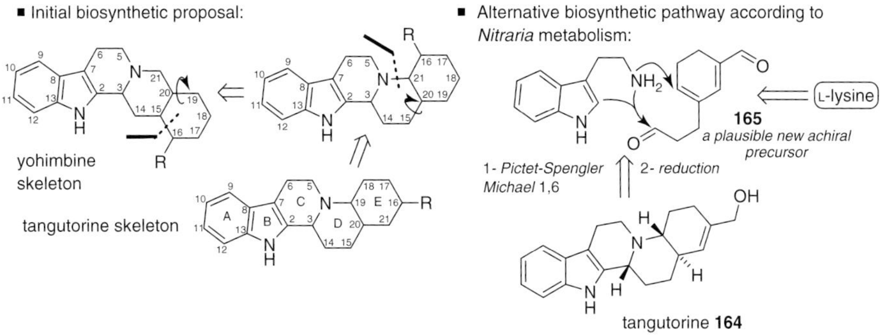

Scheme 1.43 Comparison of two biosynthetic proposals for tangutorine.

More recently, a phytochemical study of *Myrioneuron nutans* (a species collected in Vietnam from a small genus within the Rubiaceae family) by Bodo and colleagues at the Museum National d'Histoire Naturelle in Paris revealed the presence of a new class of alkaloids that was named "*Myrioneuron* alkaloids." Interestingly, these new alkaloids are closely related to the *Nitraria* alkaloids despite the taxonomic distance of the two families. Their biosynthesis is also discussed in the aforementioned review article. Representative structures [myrioneurinol (**158**) and myrobotinol (**127**)] are presented in Scheme 1.39 [90]; they constitute ideal targets for total synthesis inspired by biosynthetic pathways.

1.4.3.3 **Endocyclic Enamines Overview: Biomimetic Observations**

From five- to seven-membered rings, the propensity of enamines to dimerize is schematically represented in Figure 1.6. Only two cases with biosynthetic consequences seem to be favorable for dimerization (**16** and **38**). Especially at neutral pH (e.g., "physiological" conditions), dimerization is spontaneous and therefore likely. This intrinsic reactivity pattern has probably partly governed Darwinian selection and the subsequent development of two metabolic pathways from tetrahydroanabasine (Scheme 1.39) [91, 92].

1.4.4
Biomimetic Synthesis of Stenusine, the Spreading Agent of *Stenus comma*

Rove beetles *Stenus comma* (Coleoptera, Staphylinidae) synthesize the alkaloid stenusine (**170**) (Figure 1.7) in their pygidial glands as an escape mechanism on

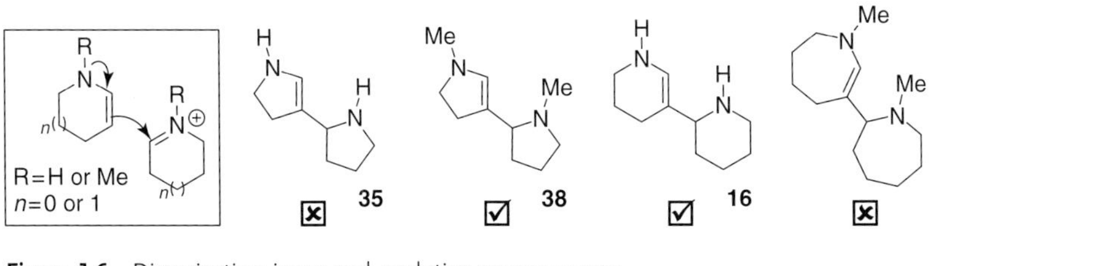

Figure 1.6 Dimerization issue and evolutive consequences.

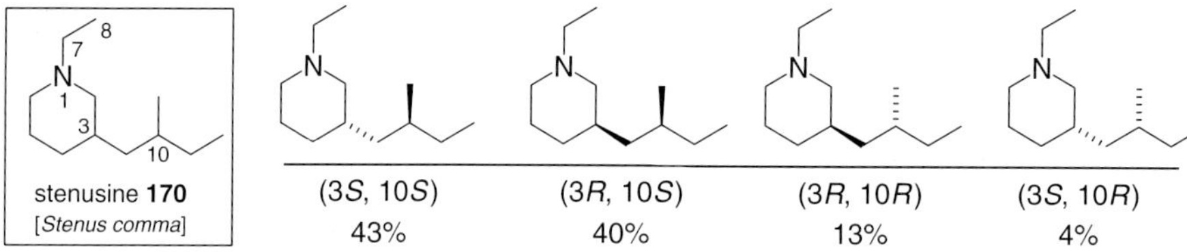

Figure 1.7 Stenusine: structure and stereochemical particularities.

water surfaces. The beetle propels itself over the water by expelling an oily substance with a high spreading capacity. Since the isolation and structure determination of stenusine (**170**) as the main component of the propulsion fluid, and a toxic chemical substance by Schildknecht *et al.* [93], numerous syntheses have been developed based on racemic and asymmetric strategies. Importantly, the enantioselective syntheses of isomeric (2*S*,3*S*)-stenusine and (2*S*,3*R*)-stenusine by Enders *et al.* [94] accompanied by chiral GC analysis for synthetic and natural samples revealed a strange fact. Indeed, stenusine is in fact a mixture of the four possible enantiomers in a ratio of $43:40:13:4 = (S,S):(S,R):(R,R):(R,S)$ with a predominance (83 : 17) of the two epimers with a (*S*)-configuration on the side chain. It is unusual that an organism produces a natural compound as a mixture of the two pairs of enantiomers in a particular ratio. Husson, Kunesch, and colleagues proposed a very appealing biogenetic scheme for stenusine (**170**) that explained both the origin and the stereochemical particularities (Scheme 1.44) [95]. The essence of the biogenetic scenario is that stenusine derives most likely from L-lysine and L-isoleucine (**171**) and could be considered as the biological condensation of two of their metabolites, namely piperideine (**15**) and methylbutyraldehyde (**172**). Intermediates resulting from the iminium/enamine equilibrium could explain the formation of enantiomers at the C3 center and the partial racemization on the side chain leading after reduction and *N*-ethylation to stenusine.

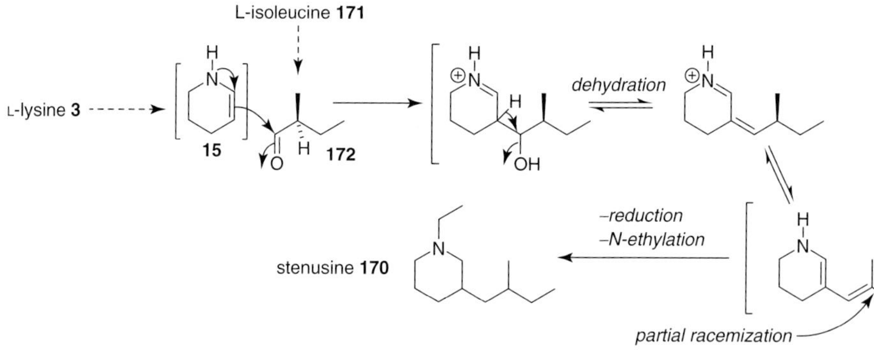

Scheme 1.44 Putative biosynthetic pathway to stenusine.

Phenyloxazolopiperidine **173**, a stable equivalent of piperideine (**15**), was used to test the chemical basis of the hypothesis (Scheme 1.45). Reaction of **173** with (*S*)-2-methylbutyraldehyde (**172**) in ethanol at room temperature was conducted.

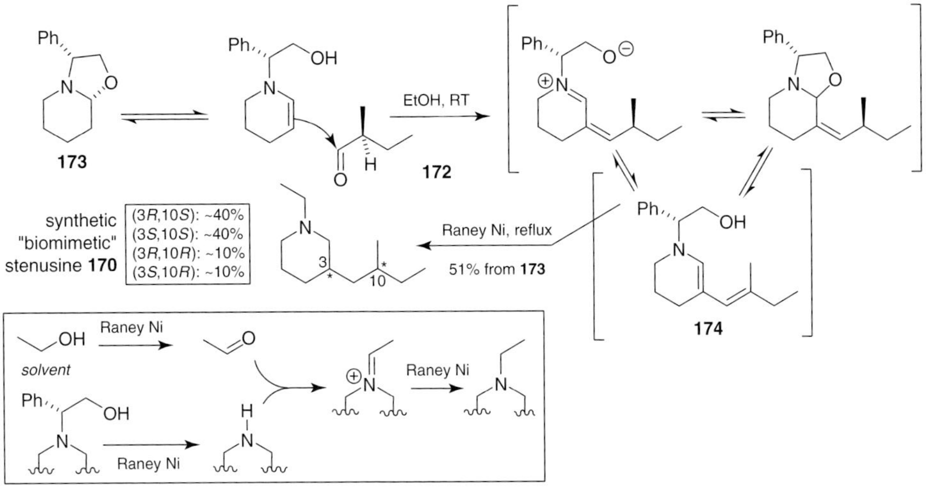

Scheme 1.45 Biomimetic synthesis of stenusine.

In the absence of air and after consumption of **173**, addition of Raney nickel to the mixture directly led to the formation of stenusine in an impressive succession of reactions (oxazolidine ring opening, condensation with aldehyde, dehydration, reduction, debenzylation, oxidation of ethanol to ethanal, Schiff base formation, and reduction to give the *N*-ethyl group). Perhaps the most significant feature of the reaction was the total lack of stereoselectivity when starting from chiral non-racemic **173** and *(S)*-**172**! The total synthesis was not shadowed for all that: in fact, comparison of the optical rotations of natural stenusine (**170**) and synthetic stenusine brought striking support to the biosynthetic hypothesis (Figure 1.8), which was confirmed when a ^{13}C NMR quantitative evaluation of the stereomeric composition of the Mosher acid salt of synthetic stenusine gave quite similar proportions.

The isolation of small amounts of aldehyde **175** or amounts up to 60% in boiling ethanol in the presence of air and molecular sieves provided strong support for the enticing sequence depicted in Scheme 1.45, and particularly for the occurrence enamine **174**. A mechanism involving formation of an oxetane (**176**) and double β-cleavage was proposed that could account for the formation of **175** (Scheme 1.46).

The story ended a few years later when Lusebrink and colleagues disclosed feeding experiments that clearly confirmed the biosynthetic scheme proposed earlier (Scheme 1.47) [96]. Deuterium-labeled amino acids and acetate combined

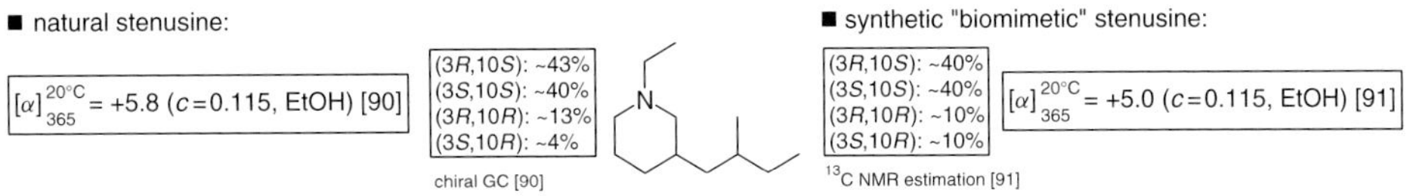

Figure 1.8 Natural and "biomimetic" stenusine: data comparisons.

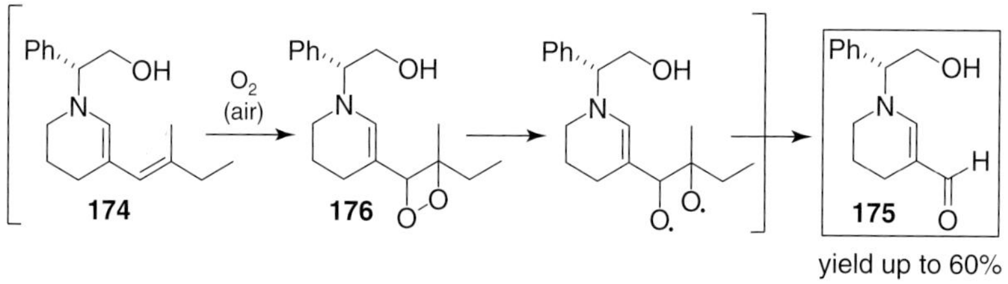

Scheme 1.46 Biomimetic synthesis of stenusine: side reaction.

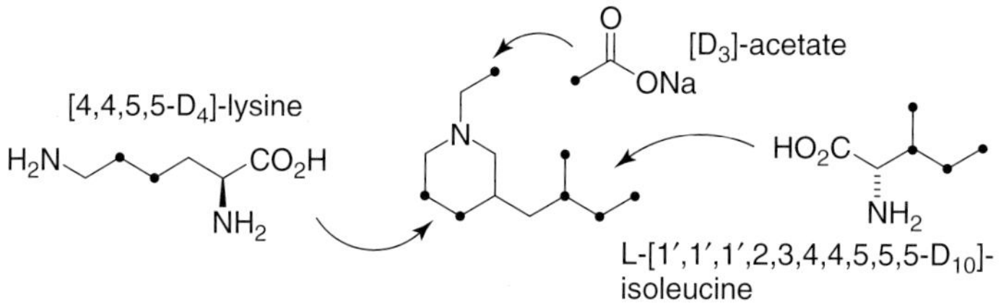

Scheme 1.47 Stenusine biosynthesis.

with GC/MS analysis showed that lysine forms the piperidine ring of stenusine, the side chain originates from isoleucine, and the *N*-ethyl group from acetate.[23)]

1.5 Pelletierine-Based Metabolism

1.5.1 Pelletierine: A Small Alkaloid with a Long History

Pelletierine (**129**)[24)] (Scheme 1.48) was isolated by Tanret in 1878 [97] from pomegranate (*Punica granatum* L.) (stem and root traditionally used as an anthelmintic against tapeworms) as a volatile, optically active compound along with three other alkaloids: methylpelletierine, pseudopelletierine (**177**), and isopelletierine (**178**). Pelletierine (**129**) was described as a colorless oil and its structure was regarded by Hess and Eichel in 1917 as an aldehyde: that is, 3-(2-piperidyl)propionaldehyde (**179**) [98]. However, its exact structure has long been debated. Among many studies worldwide, a Japanese team isolated pelletierine from pomegranate root bark following Hess' procedure, made a comparison with data provided by Hess and Eichel and concluded that the so-called pelletierine (**129**) was in fact isopelletierine (**178**) or 1-(2-piperidyl)-2-propanone [99]. NMR and IR studies by Gilman and Marion in 1961 were conducted on a

23) Piperidinic alkaloids isolated from insects usually have a polyketide and/or fatty acid origin; the case of stenusine is thereby remarkable. Alkaloids with a non-amino acid origin are discussed in Chapter 8.

24) Named in honor of Joseph Pelletier, French pharmacist and chemist (1788–1842).

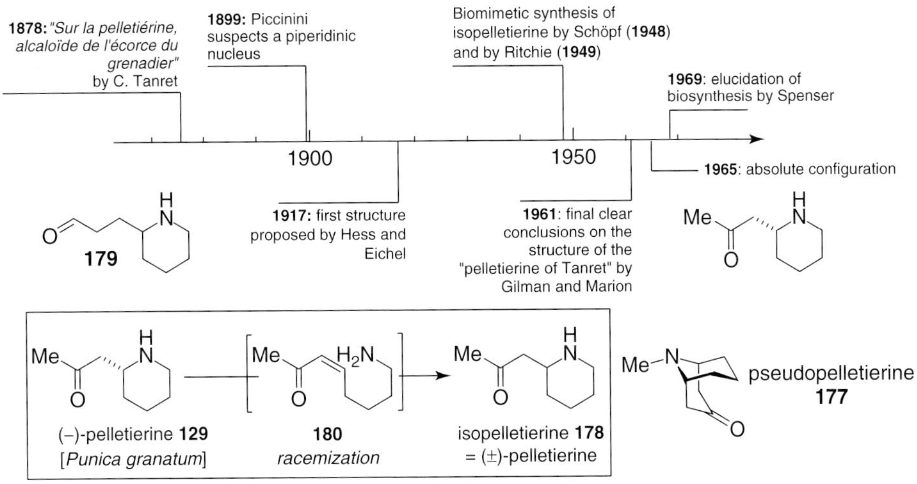

Scheme 1.48 Pelletierine story from its discovery.

sample of pelletierine sulfate isolated by Tanret himself and showed the presence of a ketone instead of an aldehyde, confirming that the pelletierine of Tanret was 1-(2-piperidyl)-2-propanone [100]. From this point, (−)-pelletierine (**129**) is referred to as (−)-1-(2-piperidyl)-2-propanone and isopelletierine (**177**) (a term that should be avoided) to racemic pelletierine. The curious history of this small molecule has been brilliantly analyzed with an historical perspective in the 1960s [101]. Later, both the absolute configuration (D) [102] and biosynthesis, via L-lysine (**3**), were disclosed [103].

Part of the questioning concerning the exact structure of pelletierine (**129**) found its roots in the epimerization that is now a well-known phenomenon with **129** and is believed to be a base-catalyzed reaction operating via retro-Michael intermediate **180**. Also found in pomegranate, pseudopelletierine (**177**), the existence of which unlike pelletierine has never been questioned, is the higher homolog of tropinone (**125**).

1.5.2 Biomimetic Synthesis of Pelletierine and Pseudopelletierine

1.5.2.1 Pelletierine (129)[25]

A Mannich reaction of acetoacetic acid followed by decarboxylation was early-on assumed to be at the origin of the propanone side chain of pelletierine (**129**) (Scheme 1.49). This was verified independently by Ritchie and colleagues and Schöpf in 1949 [104] and was proved to be biosynthetically totally correct 30 years

25) Compared to coniine, a similar structure in terms of complexity, pelletierine has been by far less studied in terms of asymmetric syntheses.

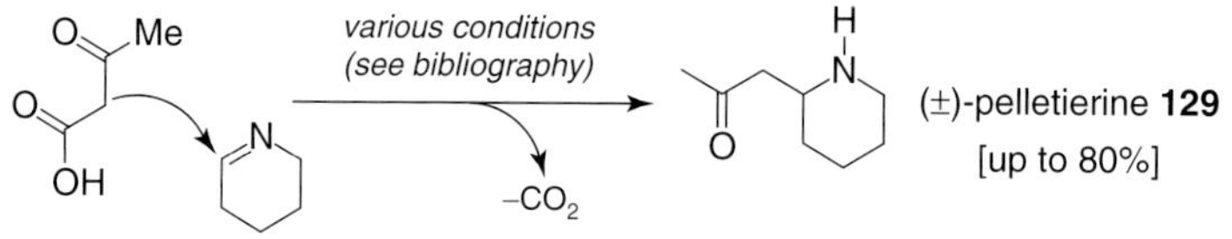

Scheme 1.49 Pelletierine biomimetic conditions.

later by incorporation experiments *in vivo* [102]. Many improvements of these pioneering syntheses were subsequently disclosed (including the study of competitive dimerization of piperideine (**15**) into tetrahydroanabasine (**16**) [105]) as well as other "non-biomimetic" syntheses, including asymmetric strategies [106]. Pelletierine was also used as a starting material for biomimetic Claisen–Schmidt reactions known as the "*pelletierine condensation*" by different groups [107].

1.5.2.2 Pseudopelletierine

Pseudopelletierine (**177**) was prepared by the same biomimetic Mannich-type reaction as for tropinone but starting from glutaraldehyde (**51**) by Robinson in 1924, Schöpf in 1937, and reinvestigated by Cope in 1951 [108].

1.5.3 *Lobelia* and *Sedum* Alkaloids

Lobelia and *Sedum* alkaloids are closely related to pelletierine in terms of structures and biosynthesis. Examples are given in Scheme 1.50 (**181–183**); the central piperidine ring of these alkaloids is known to derive from L-lysine. Interesting reviews have been published by Bates [109] and Felpin and Lebreton [110] to which the reader can refer advantageously. Numerous total syntheses have also been disclosed [111].

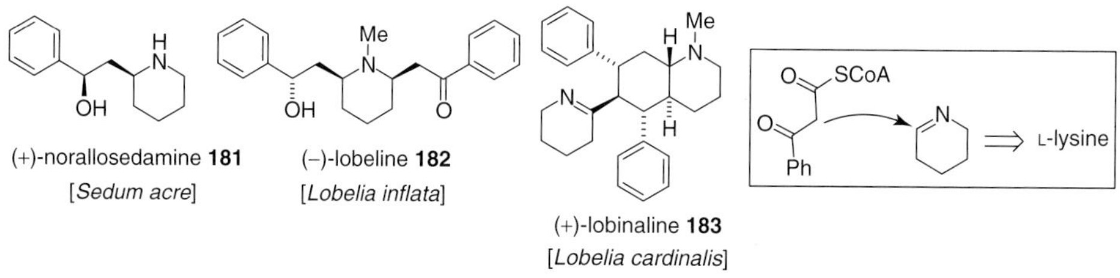

Scheme 1.50 *Sedum* and *Lobelia* alkaloids: selected structures and biosynthesis.

1.5.4 *Lycopodium* Alkaloids

1.5.4.1 Overview, Classification, and Biosynthesis

A significant number (>250 in 2010) of alkaloids with original and intricate structures are found in the *Lycopodium* genus. This genus has over 500 species

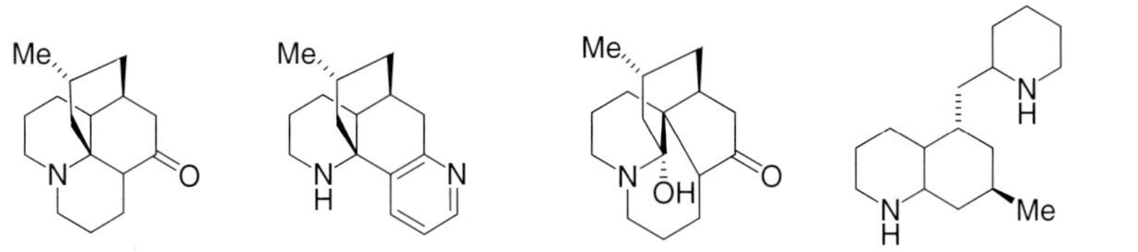

Figure 1.9 Examples of *Lycopodium* alkaloids.

and only around 10% of the species have been studied for their alkaloid content. Biological activities of such compounds can also be promising, as exemplified by well known huperzine A, which exhibits potent anticholinesterase activity. These alkaloids have attracted a great deal of interest from biogenetic and biological points of view as well as for providing challenging targets for total synthesis, especially because of their scarcity. W. A. Ayer has divided the *Lycopodium* alkaloids into four classes: lycopodine, lycodine, fawcettimine, and miscellaneous, with lycopodine (**184**), lycodine (**185**), fawcettimine (**186**), and phlegmarine (**187**) as representative compounds, respectively, for each class (Figure 1.9). Owing to the number of alkaloids now described, the classification has become tricky, especially with the isolation of highly complex structures (especially in the last ten years). Many comprehensive reviews have been published covering extraction, structure determination, biosynthetic hypotheses, total syntheses, and biological properties [112].

Based on biosynthetic considerations, another classification of *Lycopodium* alkaloids may be considered. In fact, even if the biosynthesis of the alkaloids is still not completely understood with very few feeding experiments conducted up to now, research groups have provided evidence for a lysine origin of these alkaloids. Furthermore, such as in the case of the biosynthesis of lycopodine (**184**) demonstrated by Spenser and colleagues (Scheme 1.51) [113], pelletierine units or analogs thereof are integrated in the structure. Therefore, *Lycopodium* alkaloids can be formally seen as pelletierine-type (propylpiperidine units[26] – $C_5N{-}C_3$) derived alkaloids and could then be classified according to the number of such units in the final natural substance, as proposed as early as 1960 by Conroy[27] [114]. As a consequence, some alkaloids may result from the condensation of two [e.g., nankakurine A (**188**), huperzine A (**189**), lyconadin A (**190**)], three [e.g., himeradine A (**191**), lycoperine A (**192**)], or even four [complanadine D (**193**)] pelletierine-type units as determined by the number of piperidine cycles in the molecules for the less rearranged skeletons (Scheme 1.52). At least chemically speaking, *Lycopodium* alkaloids can be

26) Whether this unit comes from pelletierine itself or from the *de novo* condensation of a piperideine of type **15** with C_3-acetone-type equivalent such as acetonedicarboxylic acid.

27) Conroy anticipated a $C_5{-}C_3$ common pattern in *Lycopodium* alkaloids but proposed a totally polyketide origin for it.

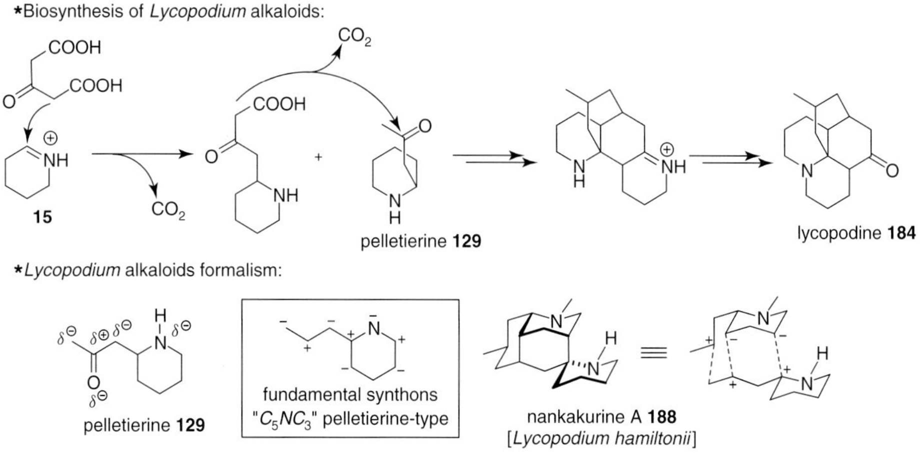

Scheme 1.51 Biosynthetic origin and analogy with pelletierine.

considered as being biosynthetically derived from simple more or less oxidized pelletierine-type units [see the example of nankakurine (**188**) in Scheme 1.51].

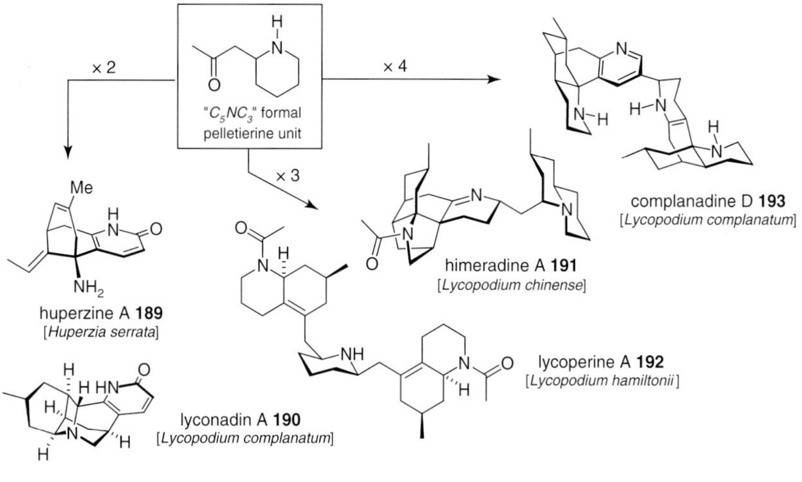

Scheme 1.52 A classification based on the formal analogy with pelletierine.

From the 1960s to the 1980s, the emergence of new strategies capable of addressing the *Lycopodium* alkaloid challenges permitted the total syntheses of complex structures by pioneering groups. This includes, among many others, work by Wenkert *et al.* [115], Evans [116], Heathcock *et al.* [117], and Schumann *et al.* [118], which often display interesting biomimetically related steps. Recent total syntheses of *Lycopodium* alkaloids were reviewed in 2009 (covering up to 2008) [112c, 119]. Most of them constitute the state of the art in total synthesis. Concerning biomimetic chemistry, we focus in this chapter on:

- Recent relevant "biomechanistic" conversions of alkaloids that permitted structure elucidation and the establishment of clear biosynthetic links. They constitute ideal examples of chemical predisposition in the world of natural substances. These studies exemplify magnificently the interconnection of numbers of *Lycopodium* alkaloids and highlight how simple reactions can explain part of both the diversity and complexity within this large family of alkaloids, which all derive from simple building blocks.
- Recent total synthesis that may feature at least one biomimetic step (usually a cascade reaction to form polycyclic skeletons).

1.5.4.2 Biomimetic Rearrangement of Serratinine into Serratezomine A

Serratezomine A (**194**) (Scheme 1.53) has been isolated from *Lycopodium serratum* var. *serratum* by the Kobayashi group [120], which proposed a biogenetic origin from serratinine (**195**), also found in this plant, through its *N*-oxide form **196** followed by a Polonovski-type fragmentation (Scheme 1.53).

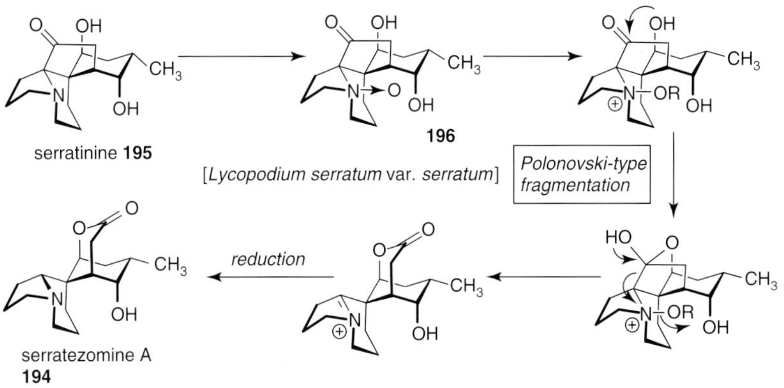

Scheme 1.53 Plausible biosynthetic correlation between serratinine and serratezomine A.

To verify this proposal (Scheme 1.54), the conditions of a modified Polonovski reaction (also known as the Polonovski–Potier reaction) were applied to serratinine (**195**): treatment with *m*-chloroperbenzoic (*m*-CPBA) acid followed by addition of trifluoroacetic anhydride and then sodium cyanoborohydride gave two compounds [121]. One of them was identified as serratezomine A (**194**) and was predominant at lower temperature (−50 and −20 °C) while the amount of the other (**197**) increased with temperature (0–20 °C). Formation of this latter indicates an acid-catalyzed-cleavage of the lactone ring.

1.5.4.3 Biomimetic Conversion of Serratinine into Lycoposerramine B

More recently, an alkaloid possessing an oxime function, lycoposerramine B (**198**, Scheme 1.55), was isolated from *Lycopodium serratum*. To confirm its structure, which was inferred by classical spectroscopic analysis, its semi-synthesis from serratinine (**195**) (for which the absolute configuration was known) was successfully attempted [122]. The retrosynthetic analysis (see box in Scheme 1.55) consisted of

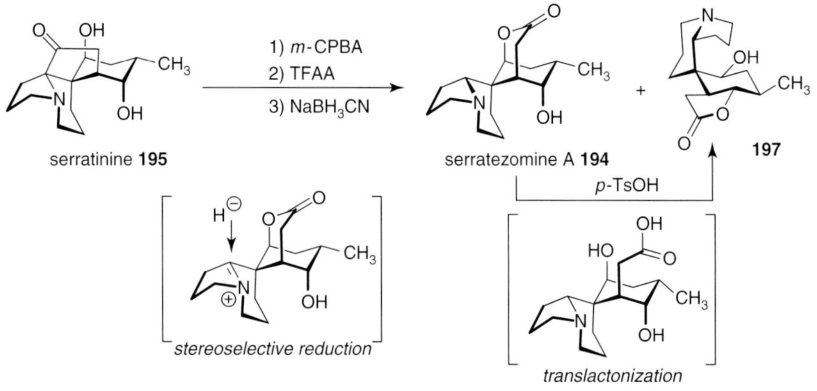

Scheme 1.54 Biomimetic conversion of serratinine into serratezomine A.

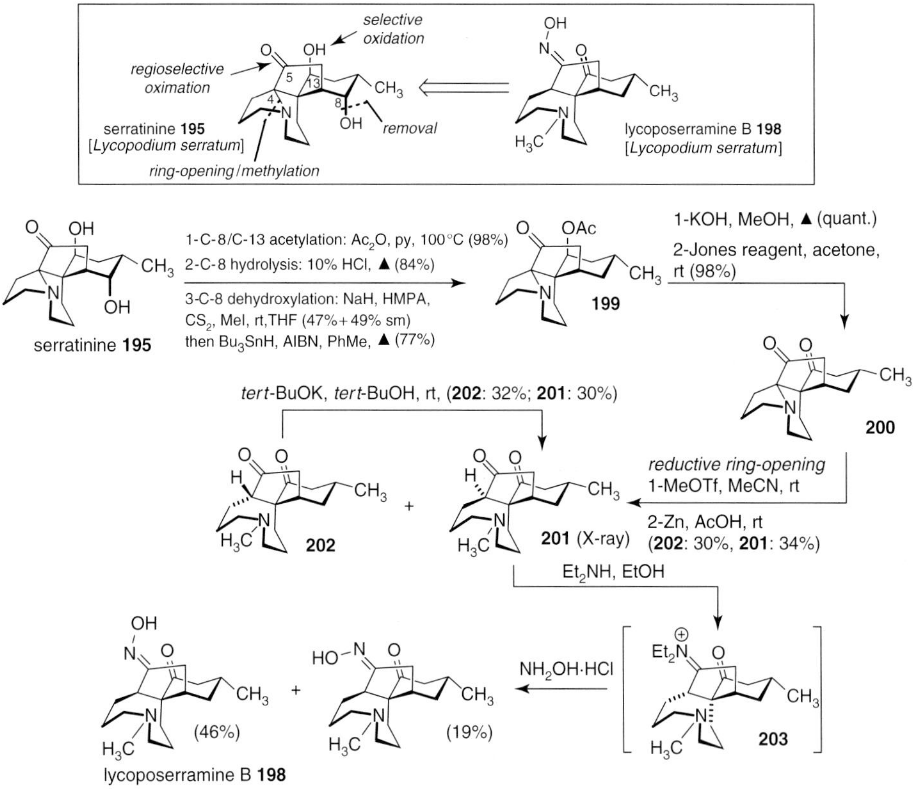

Scheme 1.55 Biomimetic conversion of serratinine into lycoposerramine B.

a removal of the hydroxyl group at C8, oxidation of the secondary hydroxyl at C13, ring-opening at the C4-N bond, and regioselective oximation at C5. A monoacetate was first prepared starting from serratinine (**195**) in a two-step sequence followed by a Barton–McCombie dehydroxylation. After acetate removal and oxidation to **200**, the reductive ring-opening reaction was then conducted and yielded two C4 epimeric compounds (**201** and **202**; the latter could be epimerized in basic conditions). To selectively convert the carbonyl at C5 into an oxime, **201** was first treated with diethylamine (steric hindrance at C13 probably explains the selectivity). On the putative intermediate **203**, hydroxylamine reacted preferentially with the more reactive iminium function at C5. The major isomer was identical with natural lycoposerramine B (**198**) (including CD spectra).

1.5.4.4 **Biomimetic Interrelations within the Lycoposerramine and Phlegmariurine Series**

Several alkaloids belonging to the fawcettimine group were isolated from *Lycopodium serratum* by the Takayama group, including several lycoposerramines [B (**198**) and C (**204**)] and related phlegmariurines A (**205**) and B (**206**) (Scheme 1.56) [123]. Lycoposerramine C could be converted into phlegmariurines A and B by ring opening in the presence of a base (whereas potassium *tert*-butylate furnished exclusively **205**, a mixture of **205** and **206** was obtained when treating with sodium methoxide). CD curves of semi-synthetic and natural phlegmariurine A were compared and shown to be identical, demonstrating thereby the same absolute configuration.

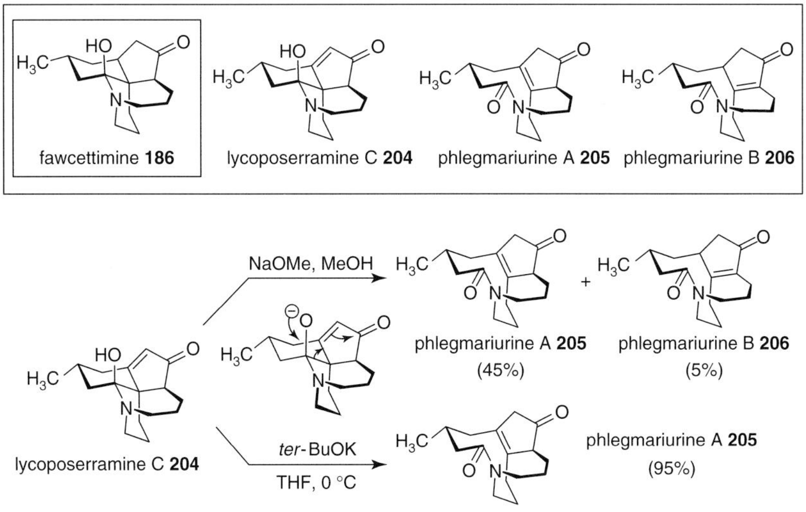

Scheme 1.56 Biomimetic conversions in the lycoposerramine/phlegmariurine series.

The structure of lycoposerramine D (**207**) (Scheme 1.57) revealed the presence of an isolated methylene carbon that corresponds to an extra carbon compared to common C_{16}-type *Lycopodium* alkaloids. A Mannich reaction involving iminium

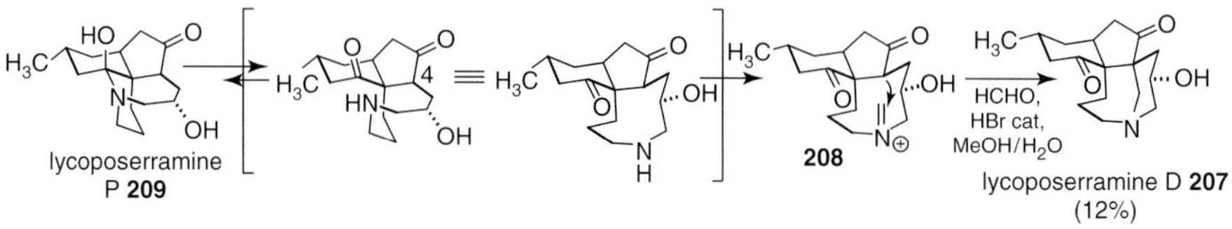

Scheme 1.57 Biomimetic conversion of lycoposerramine P into lycoposerramine D.

208 could easily explain the biosynthesis of lycoposerramine D (**207**) from lycoposerramine P (**209**), another alkaloid isolated and characterized concomitantly. This hypothesis was clearly ascertained as, when treated with formaldehyde, lycoposerramine P (**209**) yielded a semi-synthetic lycoposerramine D (**207**) that was totally identical to its natural counterpart [123a].

1.5.4.5 When Chemical Predisposition Does Not Follow Biosynthetic Hypotheses: Unnatural "*Lycopodium*-Like" Alkaloids

Yang and coworkers described an original work featuring a total synthetic dead end that unexpectedly gave interesting results (Scheme 1.58) [124]. The total synthesis of advanced intermediate **210** towards the total synthesis of lycojapodine A (**211**) was achieved but failed to give, through a possible biomimetic cascade, the desired carbinolamine lactone of the target molecule. Two competitive cyclizations can occur on such a diketone. But in fact, whatever the conditions tested, the initial 6/9 bicycle of **212** underwent the "wrong" pathway, leading after an impressive imine/enamine cascade to unnatural compound **213**. This unexpected

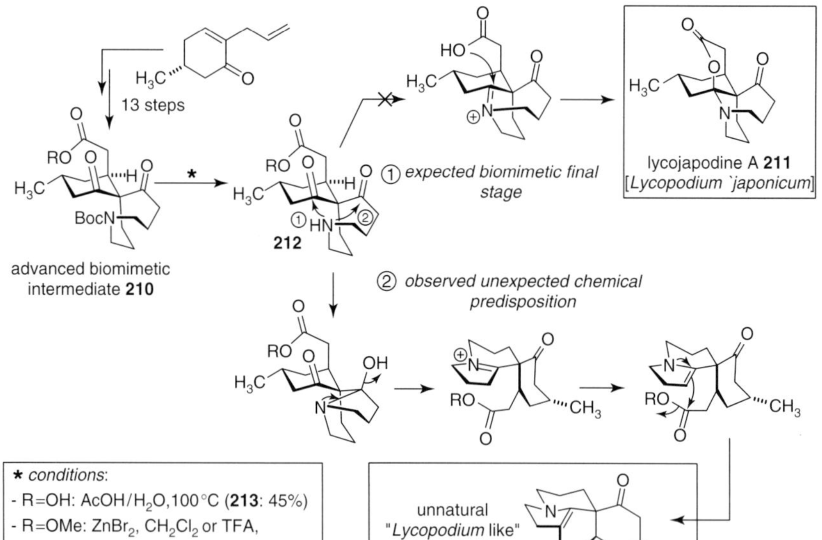

Scheme 1.58 Unexpected results: an interesting dead end in the course of the total synthesis of lycojapodine A.

chemoselectivity may facilitate a better understanding of the biosynthesis and rearrangements of complex *Lycopodium* alkaloids.

1.5.4.6 Total Synthesis of Cermizine C and Senepodine G

Among others, two small quinolizidine alkaloids, namely cermizine C (**214**) and senepodine G (**215**) (Scheme 1.59), were isolated from the club moss *Lycopodium cernuum* and *Lycopodium chinense*, respectively, as minor compounds (respectively 0.00008 and 0.00005%) by the Kobayashi group [125]. Simple biosynthetic hypotheses may be put forward to explain the formation of senepodine G, a possible direct precursor of cermizine C through iminium reduction.

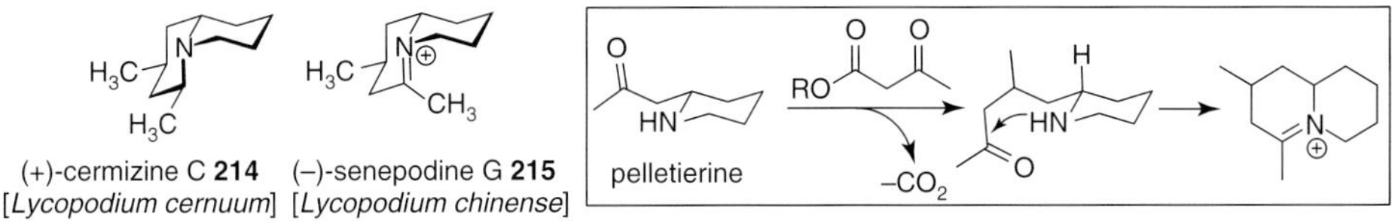

Scheme 1.59 Cermizine C and senepodine G: structure and plausible biosynthesis.

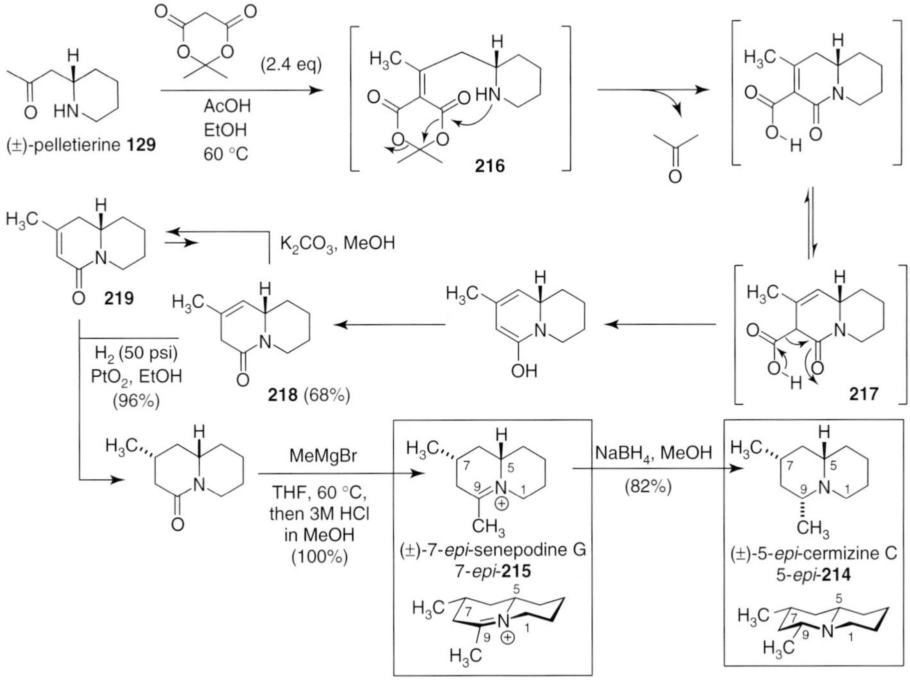

Scheme 1.60 Biomimetic synthesis of epi-senepodines from pelletierine.

Snider and colleagues set out to synthesize these two alkaloids [126]. Despite no mention of biosynthesis, their first approach started from pelletierine (**129**) and is worth noting as an interesting example of the conversion of one simple natural product into another (Scheme 1.60). A Knoevenagel reaction between the ketone of (±)-pelletierine and Meldrum's acid (a synthetic surrogate of biosynthetic acetate-derived building blocks) afforded intermediate **216**, which would cyclize to give directly the corresponding quinolizidine skeleton in a beautiful cascade of reactions. Indeed, **217** was not isolable but furnished β,γ-isomer **218**, which equilibrated in basic conditions to a mixture of α,β-isomer **219** and **218** (3 : 1). Hydrogenation of the double bonds of **218** and **219** followed by alkylation furnished (±)-7-epi-senepodine G (7-epi-**215**), which provided (±)-5-epi-cermizine C (5-epi-**214**) after stereospecific reduction using sodium borohydride (axial attack from the less hindered top face).

An alternative synthetic scheme starting from *(S)*-**220** via unsaturated lactam **221** enabled the synthesis of (−)-senepodine G (**215**) and (−)-cermizine C (**214**) (Scheme 1.61). Discrepancies in the comparison of optical rotation signs and magnitudes between natural and synthetic samples were also discussed by the authors. Very small amounts of natural compounds and the small value of the optical rotation clearly make the establishment of the absolute configuration of senepodine G and cermizine C difficult (although probably closely biosynthetically related, they were isolated from two different species).

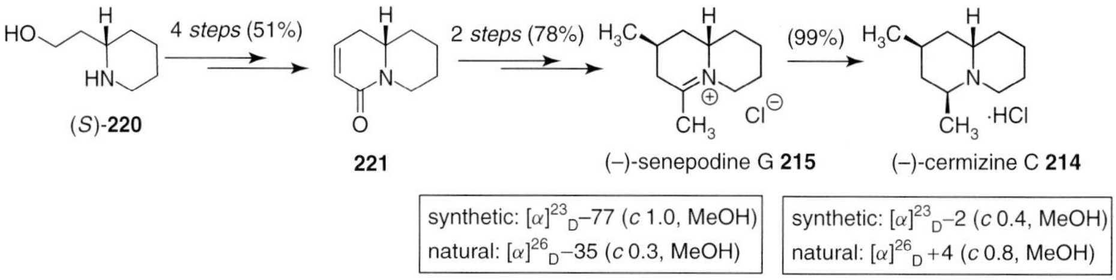

Scheme 1.61 Biomimetic synthesis of senepodine G and cermizine C.

1.5.4.7 **Biomimetic Steps in the Total Synthesis of Fastigiatine**

Fastigiatine (**222**, Scheme 1.62) was isolated as a minor alkaloid from *Lycopodium fastigiatum* in 1985 by the MacLean group [127]. The first total synthesis of this complex alkaloid was only accomplished in 2010 by the Shair group (Scheme 1.62) [128]. Despite being non-biomimetic in essence, the strategy beautifully demonstrated that such a total synthesis not only implied the development of highly efficient chemical transformations but also permitted interesting biosynthetic proposals, that is, ring formation implying an intermediate of type **223**. The approach relies on the readily available cyclopropane **224**, which was converted in ten steps into intermediate **225**. The latter underwent an impressive sequence of reactions including a transannular Mannich reaction in acidic conditions to give **226** featuring the core of the alkaloid. Two steps consisting of functional manipulations of the skeleton afforded (+)-fastigiatine (**222**) in 15 steps from **224** with an average 30% yield. The suitability of intermediate **227** to easily undergo an amazing cascade to build in one step the challenging skeleton of fastigiatine encouraged the author to propose an intermediate such as **223** as the plausible biosynthetic intermediate toward these molecules.

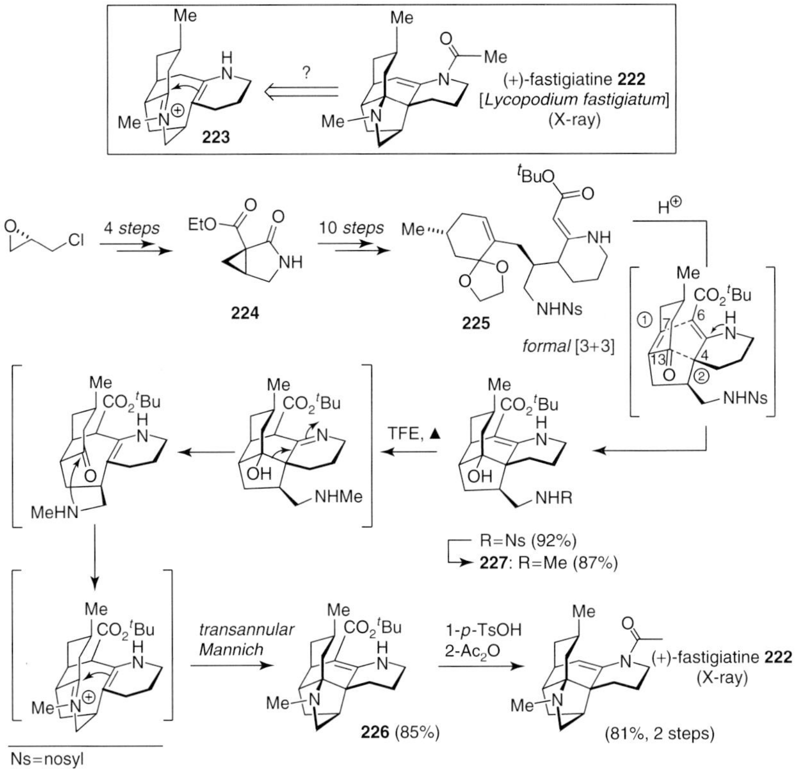

Scheme 1.62 A biomimetic cascade as a crucial step in the total synthesis of fastigiatine.

1.5.4.8 Biomimetic Steps in the Total Synthesis of Complanadine A

Earlier in 2010, a similar cascade reaction with evident biomimetic relevance was highlighted in the beautiful total synthesis of complanadine A (**228**) by Sarpong and colleagues (Scheme 1.63). [119c,d] who exploited a synthesis previously described by Schumann and Naumann [129]. *N*-Desmethyl α-obscurine (**229**) was prepared in a one-pot procedure from enamide **230** and enantiomerically pure **231**.

The next step consisted of an oxidation of lactam **232** to the corresponding pyridone **233** in a somewhat biomimetic way using lead tetraacetate. Such a pyridone nucleus is shared by many *Lycopodium* alkaloids such as huperzine A (**189**) and lyconadin A (**190**) (Scheme 1.52).

These few highlights of selected total syntheses conclude this section dedicated to *Lycopodium* alkaloids and also conclude this chapter devoted to the rich chemistry of alkaloids derived from ornithine and lysine. The exploration of the chemistry of these natural products is still ongoing. With the discovery of new complex structures – but in minute amounts – along with promising biological activities (e.g., neurotrophic properties for certain *Lycopodium* alkaloids), new synthetic strategies will undoubtedly be needed to secure enough material for biological purposes. Toward this end, no doubt biomimetic considerations will be able to

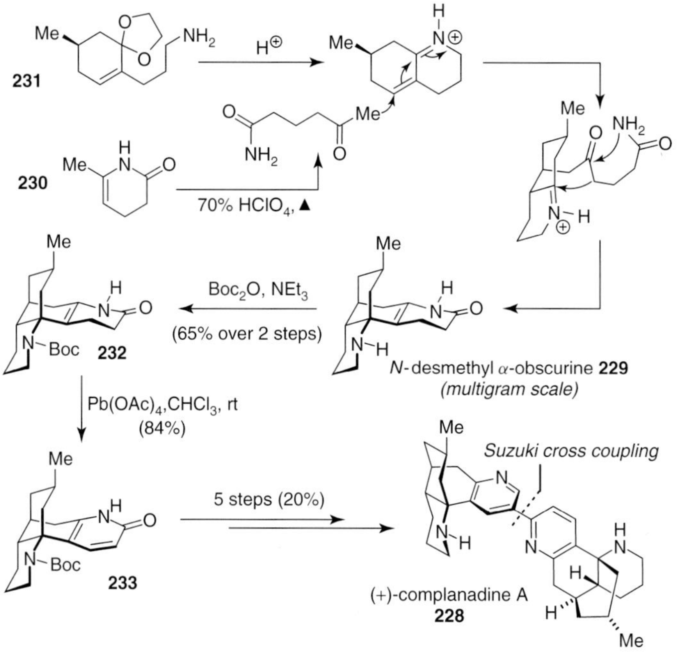

Scheme 1.63 Total synthesis of complanadine A by the Sarpong group.

answer synthetic problems. As mentioned above, organic synthesis is endowed with highly efficient biomimetic reactions such as the Mannich reaction and the Robinson–Schöpf condensation to tackle the access to more and more complex structures.

References

1. Wu, G. (2009) *Amino Acids*, **37**, 1–17.
2. Wu, G., Bazer, F.W., Davis, T.A., Kim, S.W., Li, P., Rhoads, J.M., Satterfield, M.C., Smith, S.B., Spencer, T.E., and Yin, Y. (2009) *Amino Acids*, **37**, 153–168.
3. Torruella, G., Suga, H., Riutort, M., Peretó, J., and Ruiz-Trillo, I. (2009) *J. Mol. Evol.*, **69**, 240–248 and references cited therein.
4. Azevedo, R.A. and Lea, P.J. (2001) *Amino Acids*, **20**, 261–279.
5. Dewick, P.M. (2009) *Medicinal Natural Product: A Biosynthetic Approach*, 3rd edn. John Wiley & Sons, Ltd, Chichester.
6. Hashimoto, T., Yukimune, Y., and Yamada, Y. (1989) *Planta*, **178**, 131–137.
7. Biastoff, S., Brandt, W., and Dräger, B. (2009) *Phytochemistry*, **70**, 1708–1718.
8. See among many others: Brown, A.M., Robins, D.J., Witte, L., and Wink, M. (1991) *Plant Physiol. (Life Sci. Adv.)*, **10**, 179–185.
9. (a) Tanner, J.J. (2008) *Amino Acids*, **35**, 719–730; (b) Verbruggen, N. and Hermans, C. (2008) *Amino Acids*, **35**, 753–759
10. (a) Vidal, J. (2009) in *Amino Acids, Peptides and Proteins in Organic Chemistry*, vol. 2 (ed. A.B. Hughes), Wiley-VCH

Verlag GmbH, Weinheim, pp. 35–92; (b) Ciufolini, M.A. and Xi, N. (1998) *Chem. Soc. Rev.*, **27**, 437–446.

11. See these four extensive review articles: (a) Guggisberg, A. and Hesse, M. (1983) in *The Alkaloids, Chemistry and Pharmacology*, vol. 22 (ed. A. Brossi), Academic Press Inc., New York, pp. 85–188; (b) Schäfer, A., Benz, H., Fiedler, W., Guggisberg, A., Bienz, S., and Hesse, M. (1994) in *The Alkaloids, Chemistry and Pharmacology*, vol. 45 (eds A. Brossi and G.A. Cordell), Academic Press Inc., San Diego, pp. 1–125; (c) Bienz, S., Detterbeck, R., Ensch, C., Guggisberg, A., Häusermann, U., Meisterhans, C., Wendt, B., Werner, C., and Hesse, M. (2002) in *The Alkaloids, Chemistry and Pharmacology*, vol. 58 (ed. G. Cordell), Academic Press Inc., San Diego, pp. 83–338; (d) Bienz, S., Bisseger, P., Guggisberg, A., and Hesse, M. (2005) *Nat. Prod. Rep.*, **22**, 647–648.
12. The following article is more focused on polyamine catabolism (especially in humans) but is worth reading: Seiler, N. (2004) *Amino Acids*, **26**, 217–233.
13. Dimitrov, V., Geneste, H., Guggisberg, A., and Hesse, M. (2001) *Helv. Chim. Acta*, **84**, 2108–2118.
14. Struve, C. and Christophersen, C. (2003) *Heterocycles*, **60**, 1907–1914.
15. Houen, G., Struve, C., Søndergaard, R., Friis, T., Anthoni, U., Nielsen, P.H., Christophersen, C., Petersen, B.O., and Duus, J.Ø. (2005) *Bioorg. Med. Chem.*, **13**, 3783–3796.
16. Crocker, S.J., Loeffler, R.S.T., Smith, T.A. and Sessions, R.B. (1983) *Tetrahedron Lett.*, **24**, 1559–1560 and references cited therein.
17. Feth, F., Wray, F., and Wagner, K.G. (1985) *Phytochemistry*, **24**, 1653–1655.
18. Wei, X., Sumithran, S.P., Deaciuc, A.G., Burton, H.R., Bush, L.P., Dwoskin, L.P., and Crooks, P.A. (2005) *Life Sci.*, **78**, 495–505.
19. (a) Leete, E., Kim, S.H., and Rana, J. (1988) *Phytochemistry*, **27**, 401–406; (b) see ref 17.
20. Leonard, N.J. and Cook, A.G. (1959) *J. Am. Chem. Soc.*, **81**, 5627–5631.
21. Swan, G.A. and Wilcock, J.D. (1974) *J. Chem. Soc. Perkin Trans. 1*, 885–891.
22. Franck, B. and Randau, D. (1966) *Angew. Chem. Int. Ed. Engl.*, **5**, 131.
23. Schöpf, C., Komzak, A., Braun, F., Jacobi, E., Bormuth, M.-L., Bullnheimer, M., and Hagel, I. (1948) *Justus Liebigs Ann. Chem.*, **559**, 1–42.
24. Kessler, H., Möhrle, H., and Zimmermann, G. (1977) *J. Org. Chem.*, **42**, 66–72.
25. Schöpf, C., Braun, F., Koop, H., Werner, G., Bressler, H., Neisius, K., and Schmadel, E. (1962) *Justus Liebigs Ann. Chem.*, **658**, 156–168.
26. Schöpf, C., Braun, F., and Komzak, A. (1956) *Chem. Ber.*, **89**, 1821–1833.
27. See among others: (a) Schöpf, C., Arm, H., and Krimm, H. (1951) *Chem. Ber.*, **84**, 690–699; (b) Schöpf, C. and Otte, K. (1956) *Chem. Ber.*, **89**, 335–340.
28. El-Shazly, A.M., Dora, G., and Wink, M. (2005) *Pharmazie*, **60**, 949–952.
29. Salame, R., Gravel, E., Retailleau, P., and Poupon, E. (2010) *Org. Biomol. Chem.*, **8**, 2522–2528.
30. Tashima, T., Imai, M., Kuroda, Y., Yagi, S., and Nakagawa, T. (1991) *J. Org. Chem.*, **56**, 694–697.
31. See for example: François, D., Lallemand, M.-C., Selkti, M., Tomas, A., Kunesch, N., and Husson, H.-P. (1998) *Angew. Chem. Int. Ed.*, **37**, 104–105.
32. (a) See for example: Roulland, E., Cecchin, F., and Husson, H.-P. (2005) *J. Org. Chem.*, **70**, 4474–4477 and references cited therein; (b) Review article: Husson, H.-P. and Royer, J. (1999) *Chem. Soc. Rev.*, **28**, 383–394.
33. Zacharius, R.M., Thompson, J.F., and Steward, F.C. (1952) *J. Am. Chem. Soc.*, **74**, 2949.
34. Review articles: (a) He, M. (2006) *J. Ind. Microbiol. Biotechnol.*, **33**, 401–407; (b) Broquist, H.P. (1991) *Annu. Rev. Nutr.*, **11**, 435–448.
35. Gatto, G.J., Boyne, M.T., Kelleher, N.L., and Walsh, C.T. (2006) *J. Am. Chem. Soc.*, **128**, 3838–3847 and references cited therein.
36. Zabriskie, T.M., Kelly, W.L., and Liang, X. (1997) *J. Am. Chem. Soc.*, **119**, 6446–6447.

37. See for example: Naranjo, L., Martin de Valmaseda, E., Bañuelos, O., Lopez, P., Riaño, J., Casqueiro, J., and Martin, J.F. (2001) *J. Bacteriol*, **183**, 7165–7172.
38. Matsumoto, S., Yamamoto, S., Sai, K., Maruo, K., Adachi, M., Saitoh, M., Nishizaki, T. (2003) *Brain Res.*, **980**, 179–184 and references cited therein.
39. Recent synthesis of L-pipecolic acid and analogs: Lemire, A. and Charette, A.B. (2010) *J. Org. Chem.*, **75**, 2077–2080 and references cited therein for previous work.
40. Kadouri-Puchot, C. and Comesse, S. (2005) *Amino Acids*, **29**, 101–130.
41. Aketa, K.-I., Terashima, S., and Yamada, S.-I. (1976) *Chem. Pharm. Bull.*, **24**, 621–631.
42. Rossen, K., Kolarovič, A., Baskakov, D., and Kiesel, M. (2004) *Tetrahedron Lett.*, **45**, 3023–3025.
43. (a) Pal, B., Ikeda, S., Kominami, H., Kera, Y., and Ohtani, B. (2003) *J. Catal.*, **217**, 152–159; (b) previous related work: Ohtani, B., Tsuru, S., Nishimoto, S., Kagiya, T., and Izawa, K. (1990) *J. Org. Chem.*, **55**, 5551–5553.
44. (a) Ober, D. and Kaltenegger, E. (2009) *Phytochemistry*, **70**, 1687–1695; (b) Robins, D.J. (1995) in *The Alkaloids, Chemistry and Pharmacology* vol. 46 (ed. G.A. Cordell), Academic Press Inc., San Diego, pp. 1–61.
45. Marson, C.M., Pink, J.H., Smith, C., Hursthouse, M.B., and Abdul Malik, K.M. (2000) *Tetrahedron Lett.*, **41**, 127–129.
46. (a) Gribble, G.W. and Soll, R.M. (1981) *J. Org. Chem.*, **46**, 2433–2434; (b) Gribble, G.W., Switzer, F.L., and Soll, R.M. (1988) *J. Org. Chem.* **53**, 3164–3170.
47. Lamberton, J.A. (1973) in *The Alkaloids*, vol. 14 (ed. R.H.F. Manske), Academic Press Inc., New York, pp. 325–346 and references cited in Reference [54].
48. Onaka, T. (1971) *Tetrahedron Lett.*, **12**, 4395–4398.
49. Bick, I.R.C., Gunawardana, Y.A.G.P., and Lamberton, J.A. (1985) *Tetrahedron*, **41**, 5627–5631.
50. Tufariello, J.J. and Ali, Sk.A. (1979) *Tetrahedron Lett.*, **20**, 4445–4448.
51. Chen, C.-K., Hortmann, A.G., and Marzabadi, M.R. (1988) *J. Am. Chem. Soc.*, **110**, 4829–4831.
52. Diker, K., El Biach, K., Döé de Maindreville, M., and Lévy, J. (1997) *J. Nat. Prod.*, **60**, 791–793.
53. (a) See for examples: grandisines A and B: Carroll, A.R., Arumugan, G., Quinn, R.J., Redburn, J., Guymer, G., and Grimshaw, P. (2005) *J. Org. Chem.*, **70**, 1889–1892; (b) grandisines C–G: Katavic, P.L., Venables, D.A., Forster, P.I., Guymer, G., and Carroll, A.R. (2006) *J. Nat. Prod.*, **69**, 1295–1299; (c) habbemines A and B: Katavic, P.L., Venables, D.A., Rali, T., and Carroll, A.R. (2007) *J. Nat. Prod.*, **70**, 866–868.
54. See for example, the total synthesis of grandisine D: Kurasaki, H., Okamoto, I., Morita, N., and Tamura, O. (2009) *Org. Lett.*, **11**, 1179–1181 and references cited therein.
55. Wu, J.-B., Cheng, Y.-D., Kuo, S.-C., Wu, T.-S., Iitaka, Y., Ebizuka, Y., and Sankawa, U. (1994) *Chem. Pharm. Bull.*, **42**, 2202–2204.
56. Twin, H., Wen, W.W.-H., Powell, D.A., Lough, A.J., and Batey, R.A. (2007) *Tetrahedron Lett.*, **48**, 1841–1844.
57. Snider, B.B. and Neubert, B.J. (2005) *Org. Lett.*, **7**, 2715–2718.
58. Baumgartner, B., Erdelmeier, C.A.J., Wright, A.D., Rali, T., and Sticher, O. (1990) *Phytochemistry*, **29**, 3327–3330.
59. Dätwyler, P., Ott-Longoni, R., Schöpp, E., and Hesse, M. (1981) *Helv. Chim. Acta*, **64**, 1959–1963.
60. Ott-Longoni, R., Viswanathan, N., and Hesse, M. (1980) *Helv. Chim. Acta*, **63**, 2119–2129.
61. (a) Tapiolas, D.M., Bowden, B.F., Abou-Mansour, E., Willis, R.H., Doyle, J.R., Muirhead, A.N., Liptrot, C., Llewellyn, L.E., Wolff, C.W.W., Wright, A.D., and Motti, C.A. (2009) *J. Nat. Prod.*, **72**, 1115–1120; (b) This latter publication includes the revision of the structure of previously described "eusynstyelamide" isolated from *Eusynstyela misakiensis*, see: Swersey, J.C., Ireland, C.M., Cornell, L.M., and Peterson, R.W. (1994) *J. Nat. Prod.*, **57**, 842–845.

62. (a) Casapullo, A., Finamore, E., Minale, L., and Zollo, F. (1993) *Tetrahedron Lett.*, **34**, 6297–6300; (b) Casapullo, A., Minale, L., Zollo, F., and Lavayre, J. (1994) *J. Nat. Prod.*, **57**, 1227–1233.
63. Snider, B.B., Song, F., and Foxman, B.M. (2000) *J. Org. Chem.*, **65**, 793–800.
64. Barykina, O.V. and Snider, B.B. (2010) *Org. Lett.*, **12**, 2664–2667.
65. Robinson, R.J. (1917) *J. Chem. Soc.*, **111**, 762–768.
66. Schöpf, C. (1937) *Angew. Chem.*, **50**, 779–787.
67. Humphrey, A.J. and O'Hagan, D. (2001) *Nat. Prod. Rep.*, **18**, 494–502.
68. Jönsson, D., Molin, H., and Undén, A. (1998) *Tetrahedron Lett.*, **39**, 1059–1062.
69. Mikami, K. and Ohmura, H. (2002) *Chem. Commun.*, 2626–2627.
70. See for example: Davis, F.A., Theddu, N., and Gaspari, P.M. (2009) *Org. Lett.*, **11**, 1647–1650.
71. Arend, M., Westermann, B., and Risch, N. (1998) *Angew. Chem. Int. Ed.*, **37**, 1044–1070.
72. Nicolaou, K.C. and Montagnon, T. (2008) *Molecules That Changed the World*, Wiley-VCH Verlag GmbH, Weinheim.
73. Nakano, H., Kosemura, S., Suzuki, T., Hirose, K., Kaji, R., and Sakai, M. (2009) *Tetrahedron Lett.*, **50**, 2003–2005.
74. (a) Saito, K. and Murakoshi, I. (1995) in *Studies in Natural Products Chemistry*, vol. 15 (ed. A.-U. Rahman), Elsevier Science, Amsterdam, pp. 519–549; (b) Ohmiya, S., Saito, K., and Murakoshi, I. (1995) in *The Alkaloids, Chemistry and Pharmacology*, vol. 47 (ed. G.A. Cordell), Academic Press Inc., San Diego, pp. 1–114; (c) Sekine, T., Saito, K., Minami, K., Arai, N., Suzuki, H., Koike, Y., and Murakoshi, I. (1993) *Yakugaku Zasshi*, **113**, 53–62.
75. Wanner, M.J. and Koomen, G.-J. (1996) *J. Org. Chem.*, **61**, 5581–5586.
76. Numerous total syntheses of lupine alkaloids have been published; refer to general review articles [74]. For the single example of sparteine, see recent "conventional" total syntheses: (a) Norcross, N.R., Melbardis, J.P., Ferris Solera, M., Sephton, M.A., Kilner, C., Zakharov, L.N., Astles, P.C., Warriner, S.L., and Blakemore, P.R. (2008) *J. Org. Chem.*, **73**, 7939–7951; (b) Hermet, J.-P.R., McGrath, M.J., O'Brien, P., Porter, D.W., and Gilday, J. (2004) *Chem. Commun.*, 1830–1831; (c) Smith, B.T., Wendt, J.A., and Aubé, J. (2002) *Org. Lett.*, **4**, 2577–2579; (d) See also, the beautiful biomimetic synthesis of sparteine by van Tamelen: van Tamelen, E.E. and Foltz, R.L. (1969) *J. Am. Chem. Soc.*, **91**, 7372–7377.
77. Selected publications: (a) Mure, M. and Klinman, J.P. (1995) *J. Am. Chem. Soc.*, **117**, 8698–9702; (b) Mure, M. and Klinman, J.P. (1995) *J. Am. Chem. Soc.*, **117**, 8707–8718; (c) Lee, Y. and Sayre, L.M. (1995) *J. Am. Chem. Soc.*, **117**, 11823–11828.
78. Largeron, M., Neudorffer, A., and Fleury, M.-B. (2003) *Angew. Chem. Int. Ed.*, **42**, 1026–1029.
79. Murakoshi, I., Kidoguchi, E., Haginiwa, J., Ohmiya, S., Higashiyama, K., and Otomasu, H. (1981) *Phytochemistry*, **20**, 1407–1409.
80. Frigerio, F., Haseler, C.A., and Gallagher, T. (2010) *Synlett*, 729–730.
81. (a) Wanner, M.J. and Koomen, G.-J. (1994) in *Studies in Natural Products Chemistry*, vol. 14 (ed. A.-U. Rahman), Elsevier, Amsterdam, pp. 731–768; (b) Wanner, M.J. and Koomen, G.J. (1994) *Pure Appl. Chem.*, **66**, 2239–2242; (c) Wanner, M.J. and Koomen, G.J. (1996) *Pure Appl. Chem.*, **68**, 2051–2056; (d) Wanner, M.J. and Koomen, G.J. (1997) *J. Indian Chem. Soc.*, **74**, 891–895; (e) Wanner, M.J. and Koomen, G.J. (1997) Invited lecture presented at the International Conference on Biodiversity and Bioresources: Conservation and Utilization, Phuket, Thailand, November 23–27, 1997 (see: *http://old.iupac.org/symposia/proceedings/phuket97/wanner.html*, accessed 26 August 2010).
82. Wanner, M.J. and Koomen, G.-J. (1995) *J. Org. Chem.*, **60**, 5634–5637.

83. Gravel, E., Poupon, E., and Hocquemiller, R. (2005) *Org. Lett.*, 7, 2497–2499.
84. Duan, J.-A., Williams, I.D., Che, C.-T., Zhou, R.-H., and Zhao, S.-X. (1999) *Tetrahedron Lett.*, **40**, 2593–2596.
85. Salame, R., Gravel, E., and Poupon, E. (2009) *Org. Lett.*, **11**, 1891–1894.
86. Putkonen, T., Tolvanen, A., Jokela, R., Caccamese, S., and Parrinello, N. (2003) *Tetrahedron*, **59**, 8589–8595.
87. Gravel, E. and Poupon, E. (2010) *Nat. Prod. Rep.*, **27**, 32–56.
88. See for examples: (a) nitraramine: Wanner, M.J. and Koomen, G.-J. (1995) *J. Org. Chem.*, **60**, 5634–5637; (b) nitrarine: Wanner, M.J. and Koomen, G.-J. (1994) *J. Org. Chem.*, **59**, 7479–7484.
89. See for examples: (a) biomimetic approach to nitrarine: François, D., Lallemand, M.-C., Selkti, M., Tomas, A., Kunesch, N., and Husson, H.-P. (1997) *J. Org. Chem.*, **62**, 8914–8916; (b) biomimetic synthesis of isonitramine and sibirine: François, D., Lallemand, M.C., Selkti, M., Tomas, A., Kunesch, N., and Husson, H.-P. (1998) *Angew. Chem. Int. Ed.*, **37**, 104–105.
90. See for examples: (a) myrioneurinol: Pham, V.C., Jossang, A., Sévenet, T., Nguyen, V.H., and Bodo, B. (2007) *Tetrahedron*, **63**, 11244–11249; (b) myrobotinol: Pham, V.C., Jossang, A., Sévenet, T., Nguyen, V.H., and Bodo, B. (2007) *J. Org. Chem.*, **72**, 9826–9829.
91. See, among others, pioneering work: (a) Leonard, N.J. and Cook, A.G. (1959) *J. Am. Chem. Soc.*, **81**, 5627–5631; (b) Leonard, N.J. and Hauck, F.P. Jr. (1957) *J. Am. Chem. Soc.*, **79**, 5279–5292.
92. Leonard, N.J. and Musker, K. (1959) *J. Am. Chem. Soc.*, **81**, 5631–5633.
93. (a) Schildknecht, H., Krauss, D., Connert, J., Essenbreis, H., and Orfanides, N. (1975) *Angew. Chem. Int. Ed. Engl.*, **14**, 427; (b) Schildknecht, H., Berger, D., Krauss, D., Connert, J., Gehlhaus, H., and Essenbreis, H. (1976) *J. Chem. Ecol.*, **2**, 1–11.
94. Enders, D., Tiebes, J., De Kimpe, N., Keppens, M., Stevens, C., Smagghe, G., and Betz, O. (1993) *J. Org. Chem.*, **58**, 4881–4884.
95. Poupon, E., Kunesch, N., and Husson, H.-P. (2000) *Angew. Chem. Int. Ed.*, **39**, 1493–1495.
96. Lusebrink, I., Dettner, K., and Seifert, K. (2008) *J. Nat. Prod.*, **71**, 743–745.
97. (a) Tanret, C. (1878) *C. R. Acad. Sci.*, **86**, 1270–1272; (b) Tanret, C. (1880) *C. R. Acad. Sci.*, **90**, 695–698.
98. Hess, K. and Eichel, A. (1917) *Ber. Dtsch. Chem. Ges.*, **50**, 1192–1199.
99. (a) Kuwata, S. (1960) *Bull. Chem. Soc. Jpn.*, **33**, 1668–1672; (b) Kuwata, S. (1960) *Bull. Chem. Soc. Jpn.*, **33**, 1672–1678.
100. Gilman, R.E. and Marion, L. (1961) *Bull. Soc. Chim. Fr.*, 1993–1995 and references cited therein.
101. Drillien, G. and Viel, C. (1963) *Bull. Soc. Chim. Fr.*, 2393–2340.
102. (a) Determination of absolute configuration: Beyerman, H.C., Maat, L., van Veen, A., Zweistra, A., and von Philipsborn, W. (1965) *Rec. Trav. Chim. Pays-Bas*, **84**, 1367–1379; (b) full chiroptical properties: Craig, J.C., Lee, S.Y.C., and Roy, S.K. (1978) *J. Org. Chem.*, **43**, 347–349.
103. (a) Gupta, R.N. and Spenser, I.D. (1969) *Can. J. Chem.*, **47**, 445–447; (b) Hemscheidt, T. and Spenser, I.D. (1990) *J. Am. Chem. Soc.*, **112**, 6360–6363 and references cited therein.
104. (a) Anet, E.F.L., Hughes, G.K., and Ritchie, E. (1949) *Nature*, **164**, 501; (b) Anet, E.F.L., Hughes, G.K., and Ritchie, E. (1950) *Aust. J. Sci. Res. A*, **3**, 336–341; (c) Schöpf, C. (1949) *Angew. Chem.*, **61**, 31
105. See for examples: (a) Wisse, J.H., De Klonia, H., and Visser, B.J. (1964) *Rec. Trav. Chim. Pays-Bas*, **83**, 1265–1272; (b) Quick, J. and Oterson, R. (1976) *Synthesis*, 745–756.
106. For recent asymmetric syntheses of pelletierine, see: (a) Takahata, H., Kubota, M., Takahashi, S., and Momose, T. (1996) *Tetrahedron: Asymmetry*, **7**, 3047–3054; (b) Turcaud, S., Martens, T., Sierecki, E., Pérard-Viret,

J., and Royer, J. (2005) *Tetrahedron Lett.*, **46**, 5131–5134; (c) Carlson, E.C., Rathbone, L.K., Yang, H., Collett, N.D., and Carter, R.G. (2008) *J. Org. Chem.*, **73**, 5155–5158.

107. Quick, J. and Meltz, C. (1979) *J. Org. Chem.*, **44**, 573–578 and references cited therein.
108. Cope, A.C., Dryden, H.L., Overberger, C.G., and D'Addieco, A.A. (1951) *J. Am. Chem. Soc.*, **73**, 3416–3418 and references cited therein.
109. Bates, R.W. and Sa-Ei, K. (2002) *Tetrahedron*, **58**, 5957–5978.
110. Felpin, F.-X. and Lebreton, J. (2004) *Tetrahedron*, **60**, 10127–10153.
111. See for example: Krishnan, S., Bagdanoff, J.T., Ebner, D.C., Ramtohul, Y.K., Tambar, U.K., and Stoltz, B.M. (2008) *J. Am. Chem. Soc*, **130**, 13745–13754 and references cited therein for recent syntheses of *Lobelia* and *Sedum* alkaloids.
112. Review articles: (a) Ma, X. and Gang, D.R. (2004) *Nat. Prod. Rep.*, **21**, 752–772; (b) Kobayashi, J. and Morita, H. (2005) in *The Alkaloids, Chemistry and Biology*, vol. 61 (ed. G.A. Cordell), Elsevier, San Diego, pp. 1–57; (c) Hirasawa, Y., Kobayashi, J., and Morita, H. (2009) *Heterocycles*, **77**, 679–729.
113. Hemscheidt, T. and Spenser, I.D. (1996) *J. Am. Chem. Soc.*, **118**, 1799–1800. See also the impressive number of discoveries from the Spenser group in the field of biosynthesis of alkaloids.
114. Conroy, H. (1960) *Tetrahedron Lett.*, **1**, 34–37.
115. Wenkert, E., Chauncy, B., Dave, K.G., Jeffcoat, A.R., Schell, F.M., and Schenk, H.P. (1973) *J. Am. Chem. Soc.*, **95**, 8427–8436.
116. Scott, W.L. and Evans, D.A. (1972) *J. Am. Chem. Soc.*, **94**, 4779–4780.
117. Heathcock, C.H., Kleinman, E.F., and Binkley, E.S. (1982) *J. Am. Chem. Soc.*, **104**, 1054–1068.
118. Schumann, D., Müller, H.-J., and Naumann, A. (1982) *Liebigs. Ann. Chem.*, 2057–2061.
119. For recent highly elegant first total synthesis of some complex *Lycopodium* alkaloids published from 2009, see: (a) (−)-deoxyserratinine: Yang, Y.-R., Lai, Z.-W., Shen, L., Huang, J.-Z., Wu, X.-D., Yin, J.-L., and Wei, K. (2010) *Org. Lett.*, **12**, 3430–3433; (b) (+)-sieboldine A: Canham, S.M., France, D.J., and Overman, L.E. (2010) *J. Am. Chem. Soc.*, **132**, 7876–7877; (c) (+)-complanadine A: Yuan, C., Chang, C.-T., Axelrod, A., and Siegel, D. (2010) *J. Am. Chem. Soc.*, **132**, 5924–5925; (d) (+)-complanadine A: Fischer, D.F. and Sarpong, R. (2010) *J. Am. Chem. Soc.*, **132**, 5926–5927; (e) (±)-lycoposerramine R: Bisai, V. and Sarpong, R. (2010) *Org. Lett.*, **12**, 2551–2553; (f) (+)-serratezomine: Chandra, A., Pigza, J.A., Han, J.-S., Mutnick, D., and Johnston, J.N. (2009) *J. Am. Chem. Soc.*, **131**, 3470–3471; (g) (+)-lyconadin A: Nishimura, T., Unni, A.K., Yokoshima, S. and Fukuyama, T. (2011) *J. Am. Chem. Soc.*, **133**, 418–419.
120. Morita, H., Arisaka, M., Yoshida, N., and Kobayashi, J. (2000) *J. Org. Chem.*, **65**, 6241–6245.
121. Morita, H. and Kobayashi, J. (2002) *J. Org. Chem.*, **67**, 5378–5381.
122. Katakawa, K., Kitajima, M., Aimi, N., Seki, H., Yamaguchi, K., Furihata, K., Harayama, T., and Takayama, H. (2005) *J. Org. Chem.*, **70**, 658–663.
123. (a) Takayama, H., Katakawa, K., Kitajima, M., Yamaguchi, K., and Aimi, N. (2002) *Tetrahedron Lett.*, **43**, 8307–8311; (b) Nakayama, A., Kogure, N., Kitajima, M., and Takayama, H. (2009) *Org. Lett.*, **11**, 5554–5557; (c) see also an interesting cascade of reaction in the first total synthesis of (+)-lycoflexine: Ramharter, J., Weinstabl, H., Mulzer, J. (2010) *J. Am. Chem. Soc.*, **132**, 14338–14339.
124. Yang, Y.-R., Shen, L., Wei, K., and Zhao, Q.-S. (2010) *J. Org. Chem.*, **75**, 1317–1320.
125. Morita, H., Hirasawa, Y., Shinzato, T., and Kobayashi, J. (2004) *Tetrahedron*, **60**, 7015–7023. From the same group, see also the chemical correlation between lyconadins B and F: Ishiuchi, K., Kubota, T., Ishiyama, H., Hayashi, S., Shibata, T., Kobayashi, J. (2011) *Tetrahedron Lett.*, **52**, 289–292.

126. Snider, B.B. and Grabowski, J.F. (2007) *J. Org. Chem.*, **72**, 1039–1042.
127. Gerard, R.V., MacLean, D.B., Fagianni, R., and Lock, C.J. (1986) *Can. J. Chem.*, **64**, 943–949.
128. Liau, B.B. and Shair, M.D. (2010) *J. Am. Chem. Soc.*, **132**, 9594–9595.
129. Schumann, D. and Naumann, A. (1983) *Liebigs Ann. Chem.*, 220–225.

2
Biomimetic Synthesis of Alkaloids Derived from Tyrosine: The Case of FR-901483 and TAN-1251 Compounds

Huan Liang and Marco A. Ciufolini

2.1
Introduction

Tyrosine is the biosynthetic precursor of such a large number of nitrogenous substances [1] that a general review of biomimetic syntheses of natural products derived from that amino acid would occupy considerable space. On the other hand, most of the compounds in question are plant alkaloids that have already been the subject of numerous reviews [2]. This chapter eschews any discussion of chemical syntheses of plant-derived agents, focusing instead on a small group of fungal metabolites discovered during the 1990s. Much of the impetus for the work described herein derived from a noteworthy paper that appeared in 1996 in the *Journal of Antibiotics*. In this publication, scientists at the Fujisawa (now Astellas) pharmaceutical company described a structurally novel natural product, FR-901483 (**1**, Figure 2.1) that displayed considerable immunosuppressive activity [3, 4]. The new substance was isolated from a culture of a Cladobotryum species and its structure and relative configuration were ascertained by X-ray diffractometry. Formula **1** depicts the molecule with its actual absolute configuration, which, however, was determined only later through chemical synthesis. The bioactivity of FR-901483 is intimately associated with the presence of the phosphate ester, which unfortunately is labile. Rapid dephosphorylation of **1** ensues even upon introduction into cellular cultures, resulting in formation of the corresponding alcohol, **2**, which is devoid of activity.

The new immunosuppressant appeared to be structurally related to the so-called TAN-1251 family of compounds (**3**–**6**, Figure 2.2), a group of substances with noteworthy antimuscarinic activity produced by a *Penicillium* species [5].

One can easily recognize that FR and TAN substances are formal dimers of tyrosine (Scheme 2.1). Moreover, the two types of compounds may share a common biosynthetic precursor in the form of aldehyde **7**. Indeed, TAN compounds could emerge upon cyclization of **7** to iminium ion **9**. By contrast, the framework of FR-901483 would result upon aldol cyclization to **8**.

An interesting question pertains to the biosynthetic origin of **7** (Scheme 2.2). This aldehyde could arise through oxidative cyclization of a tyrosinyl tyrosine, **11**,

Biomimetic Organic Synthesis, First Edition. Edited by Erwan Poupon and Bastien Nay.

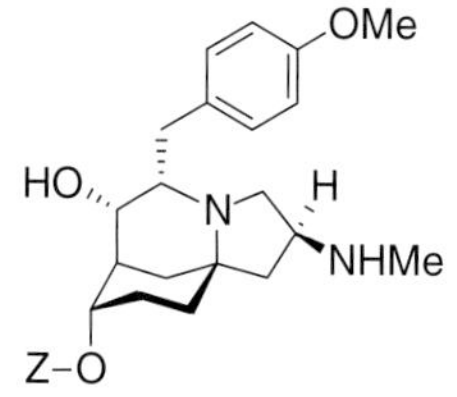

1 Z = P(O)(OH)$_2$ FR-901483
2 Z = H

Figure 2.1 Structure of FR-901483 (**1**) and of the corresponding diol (**2**, inactive).

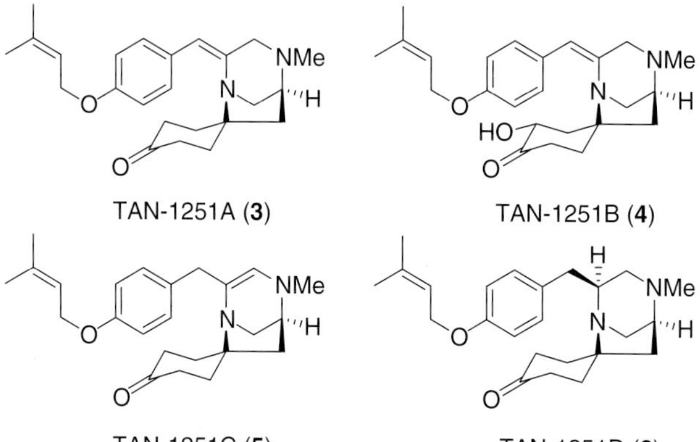

Figure 2.2 Structures of TAN-1251 compounds.

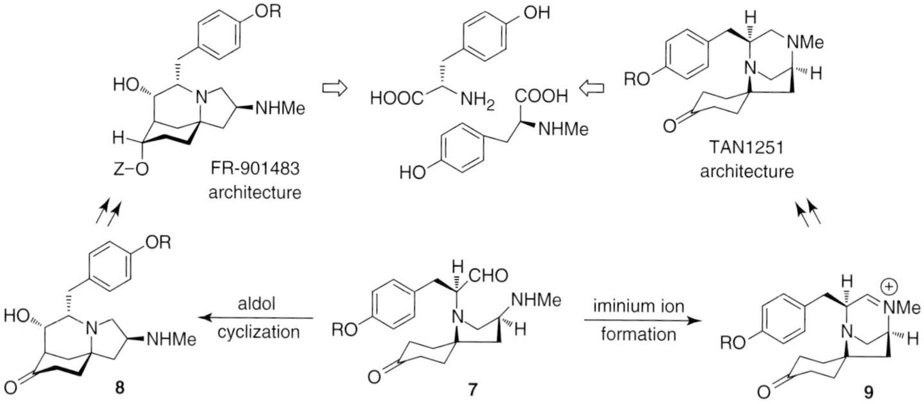

Scheme 2.1 FR-901483 and TAN-1251 as dimers of tyrosine.

to spirolactam **10**, followed by appropriate redox modifications. The conversion of **11** into **10** is envisioned to involve an initial oxidative activation of the phenol to furnish a reactive electrophilic intermediate, naively represented in Scheme 2.3 as **13**. The nitrogen atom of the amide subsequently intercepts **13**, thereby expressing *nucleophilic* reactivity. The overall process may be described as the

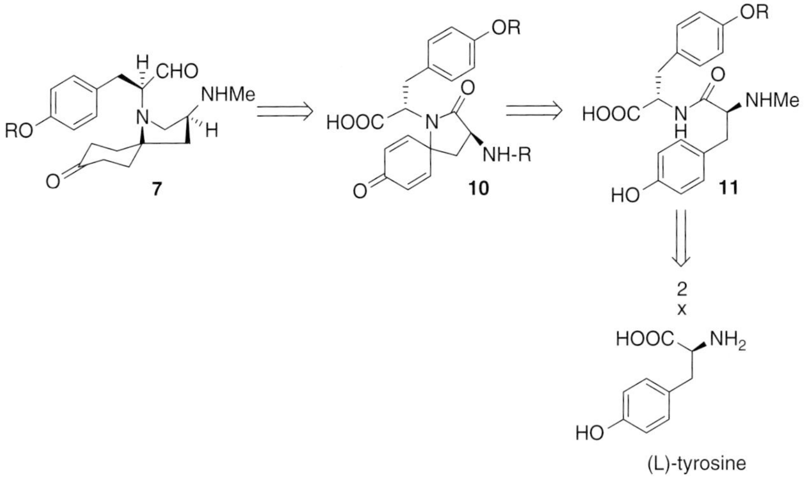

Scheme 2.2 Presumed biosynthetic origin of aldehyde **8**.

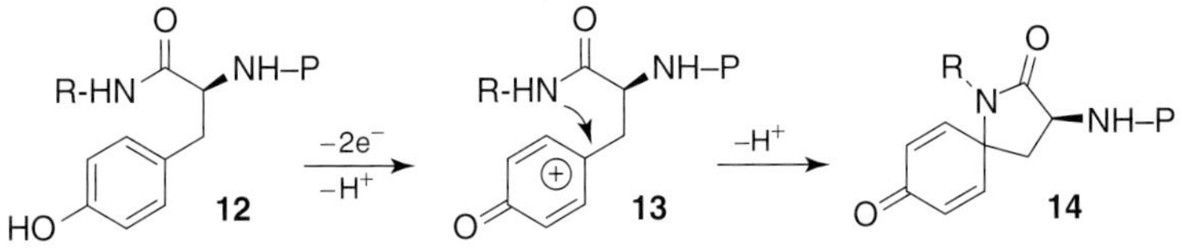

Scheme 2.3 Oxidative amidation of phenols.

"oxidative amidation" of a phenol. One may anticipate that if this transformation could be duplicated in the laboratory, the synthesis of **1–6** would be greatly simplified.

2.2 Biomimetic Total Syntheses of FR-901483 and TAN-1251 Compounds

The architectural novelty of **1–6** and their interesting biological activity have inspired a great deal of synthetic research, and various strategies, some biomimetic, some not, have been explored in order to construct the framework of these natural products. In the context of this chapter, a synthesis is regarded as biomimetic if it proceeds through an early intermediate, that is, a tyrosinyl tyrosine or an analog thereof (Scheme 2.4). Of course, in no way does this detract from the brilliancy and significance of alternative approaches explored in connection with total syntheses [6, 7] and synthetic studies [8, 9] of the foregoing natural products.

The criterion just stated qualifies as biomimetic the Snider, Sorensen, and Ciufolini syntheses of **1**, as well as the Snider, Ciufolini, and Honda syntheses of TAN-1251 substances. We shall now review these efforts in detail.

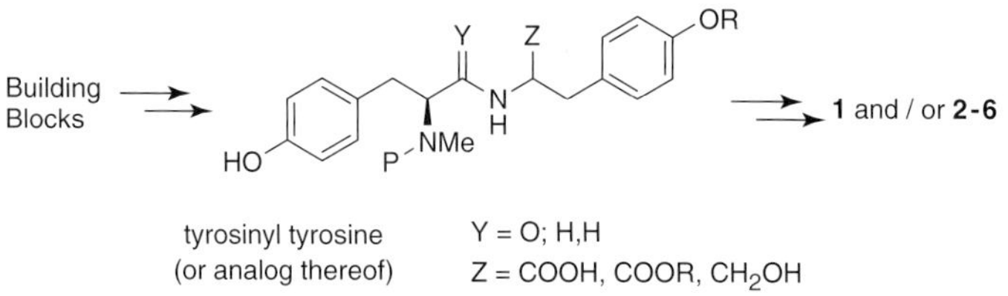

Scheme 2.4 Biomimetic synthesis: one that proceeds via a tyrosinyl tyrosine intermediate.

2.2.1
Snider Synthesis of FR-901483

The first total synthesis of FR-901483 was achieved in 1999 by Snider [10]. Key strategic aspects of this endeavor were the early construction of tyrosinyl tyrosine derivative **15** and regio- and stereoselective aldol cyclization thereof as an avenue to the tricyclic core of **1** (Scheme 2.5).

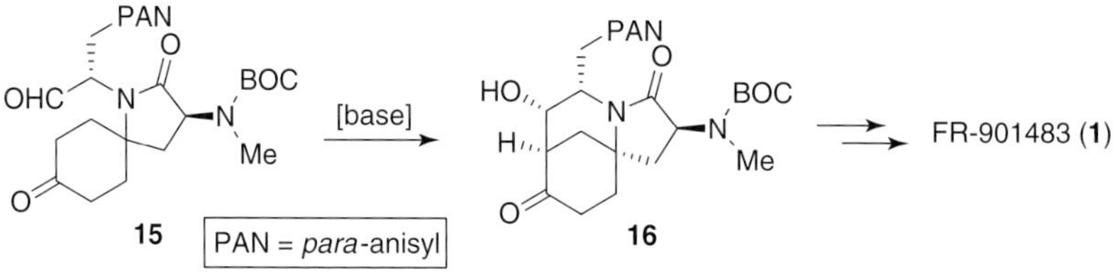

Scheme 2.5 Key steps in Snider's synthesis of FR-901483.

The assembly of **21** avoided the oxidative cyclization of a phenolic amide, because at that time no technology was available to effect such a transformation. Accordingly, its preparation relied instead on an ingenious 1,3-dipolar cycloaddition of tyrosine-derived nitrone **19** [11] to ethyl acrylate (Scheme 2.6). This step proceeded

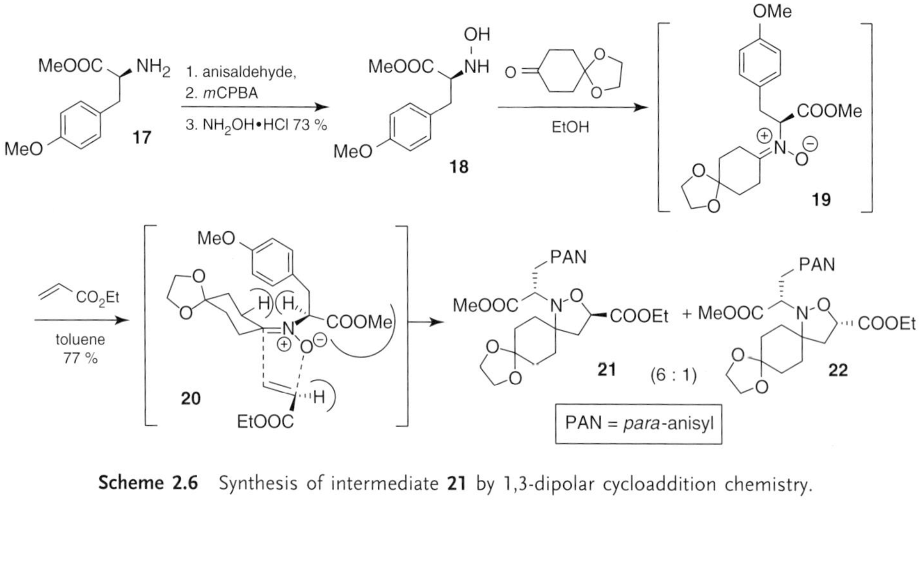

Scheme 2.6 Synthesis of intermediate **21** by 1,3-dipolar cycloaddition chemistry.

diastereoselectively to furnish a 6 : 1 mixture of adducts **21** (major) and **22** (minor). The stereochemical outcome of this step was rationalized by assuming that nitrone **19** reacts preferentially from conformation **20**, which minimizes allylic-type interactions and promotes the approach of the methyl acrylate molecule from the *Re* face of the imino linkage. Furthermore, the reaction proceeds with *exo* topology due to the absence of secondary orbital interactions.

Adducts **21** and **22** were not separated at this stage. Rather, hydrogenolysis of the N–O bond triggered formation of spirolactam **23**, which was advanced to **15** in a conventional fashion (Scheme 2.7). The crucial aldol cyclization of the latter presented several complex issues. The success of this step depends on: the sequential occurrence of a chemoselective enolization of the cyclohexanone without enolization of the aldehyde; regioselective formation of enolate **26** (Z = H, H, or O; P = protecting group); and diastereoselective addition thereof to the *Si*-face of the aldehyde to yield **28** (Scheme 2.8).

If enolization of the cyclohexanone segment **25** were indeed kinetically faster than that of the aldehyde, then the action of a gentle base could promote reversible,

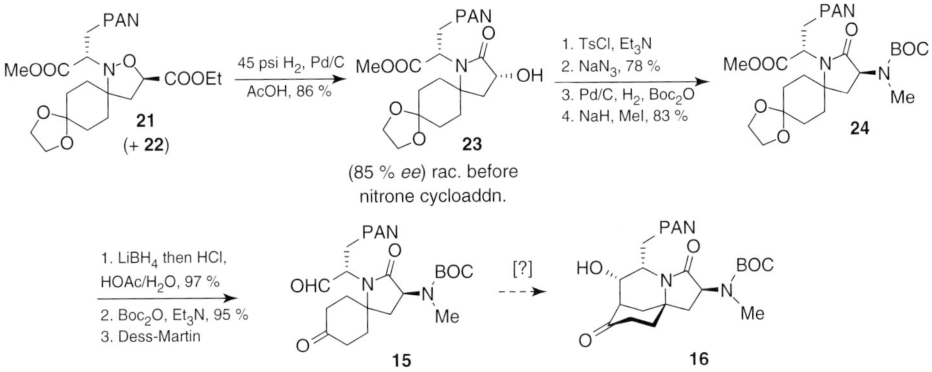

Scheme 2.7 Synthesis of aldol substrate **15**.

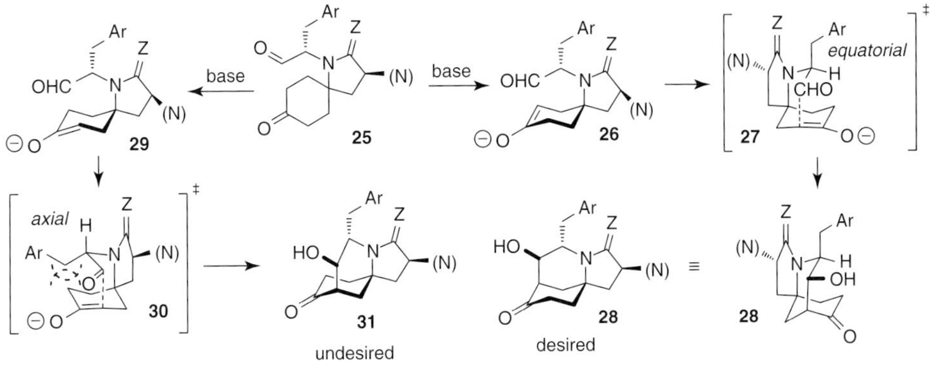

Scheme 2.8 Aspects of the aldol cyclization of **25**.

non-regioselective formation of the ketone enolate.[1] For a fixed (*S*)-configuration at the α-position of the aldehyde, aldol cyclization from the incorrect regioisomer of the cyclohexanone enolate, **29**, forces the 4-methoxybenzyl substituent into an axial orientation in transition state **30**. This generates severe compression against a methylene group of the enolate ring (dashed semicircles). No such problems exist in regioisomeric transition state **27**, where the substituent in question is pseudo-equatorial in the developing ring. This should favor selective formation of the desired **28**.

A potential complication is that for Z = O, the newly formed six-membered ring is an *N*-acylpiperidine. Substituents at the α-position of the nitrogen atom in such structures strongly prefer an axial orientation [12]. The incorrect aldol regioisomer might then become favored under thermodynamic conditions, and possibly even under kinetic conditions, if the transition state for the aldol cyclization were product-like. No such complication would affect aldol substrates in which Z is a pair of H atoms, and, indeed, the later Sorensen synthesis implemented precisely this second principle. Conversely, the question of diastereoselectivity in the aldol step is significantly more complex. Some aspects of this issue will be addressed later.

Snider found that the aldol cyclization of **15** and related substrates occurred regioselectively without epimerization of the aldehyde α-stereocenter and with useful levels of stereocontrol (Scheme 2.9). Model substrate **32** reacted with NaOMe in MeOH at room temperature to afford a mixture of three diastereomers in which compound **33**, which possesses the correct relative configuration of all ring stereocenters, was the major component. An interesting solvent effect on stereoselectivity was also unveiled, in that the action of *t*-BuOK in toluene afforded **34** as the major isomer and **36** as the minor one. The predominant formation

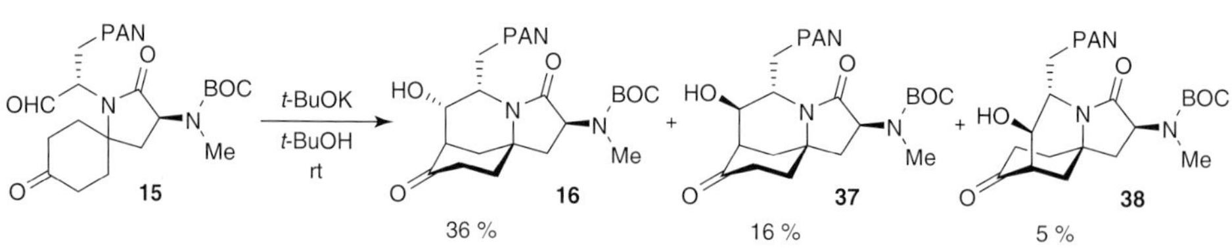

Scheme 2.9 Crucial aldol step in the Snider synthesis of FR-901483.

1) The lactam unit in 25 is much less C–H acidic than either the ketone or the aldehyde, and it was anticipated not to interfere with the operations leading to the tricyclic aldol intermediate.

of **34** and **36** in toluene was attributed to stabilization of a transient alkoxide by chelation with the ketone, while the selective formation of **33** in MeOH was ascribed to strong solvation of the alkoxide, presumably by hydrogen bonding to the protic solvent. In like manner, cyclization of FR-901483 precursor **15** with *t*-BuOK in *t*-BuOH afforded a mixture of **16** (36%), **37** (16%), and **38** (5%). Improved diastereoselectivity was observed in the aldol cyclization of **15** with *t*-BuOK/*t*-BuOH relative to MeONa/MeOH. Unfortunately, **16** and **38** were inseparable and were employed in subsequent steps as a mixture.

The resistance of the aldehyde to epimerization is consistent with earlier observations by Garner [13], and especially subsequent ones by Myers [14], who exploited this property in several brilliant total syntheses [15].

Release of the Boc group and LAH reduction afforded **39** (plus the corresponding product from **38**, Scheme 2.10). The virtually complete diastereoselectivity achieved during the reduction of the keto carbonyl is attributable to preferential approach of the reducing agent from the convex face of the molecule. Protection of the amino group as a Cbz derivative permitted the removal of the isomeric material derived from **39**. The reactivity of the OH groups in **39** differs significantly, permitting a highly selective conversion into **40**. Treatment of the latter with CsOAc/18-crown-6 resulted primarily in S_N2 displacement of the nosyl group by acetate ion (70% yield). An olefinic by-product (20% yield) arising through a competitive *E*2 reaction was also obtained. Release of the acetate and formation of dibenzyl phosphate derivative **42** set the stage for a final hydrogenolysis of all benzyl groups. The target **1** was best isolated and characterized as the mono-hydrochloride, which was most conveniently prepared by hydrogenolysis of the hydrochloride salt of O-deprotected **42**. The Snider synthesis of (−)-FR-901483 thus proceeded in 2% overall yield from *O*-methyltyrosine methyl ester (**17**) over 26 steps.

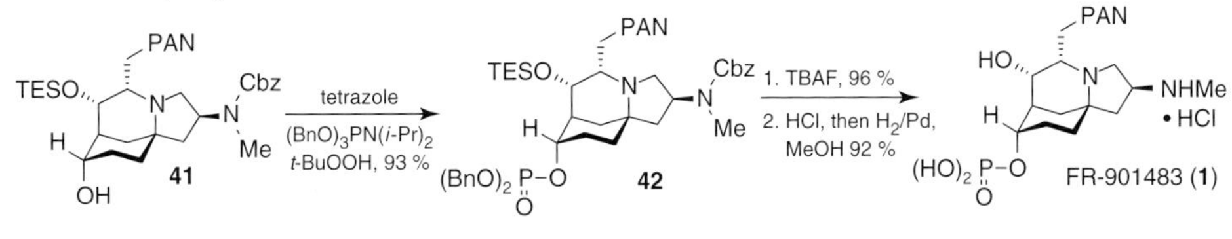

Scheme 2.10 Completion of the synthesis of FR-901483.

2.2.2
Snider Synthesis of TAN-1251 Substances

Snider also takes the credit for achieving the first total synthesis of TAN-1251 compounds, thereby ascertaining their absolute configuration [16]. The opening moves of this effort retrace substantially the route to the key intermediate for FR-901483 (Scheme 2.6). Thus, tyrosine benzyl ether **43** was converted into α-hydroxyl amino acid **44** (Scheme 2.11). Condensation with 1,4-cyclohexanedione mono 1,4-butyleneketal and 1,3-dipolar cycloaddition of the ensuing nitrone to ethyl acrylate afforded **45**. In accord with previous observations, this material was also formed as a 6 : 1 mixture of diastereomers. Without separation, these were advanced to **47**. Nucleophilic substitution of tosylate with azide ion in **47** delivered a 6 : 1 mixture of diastereomeric azides, which contrary to early intermediates were readily separated. The synthesis thus continued with the major azide diastereomer, which possesses the correct configuration. Straightforward manipulations ultimately produced amino-alcohol **49**.

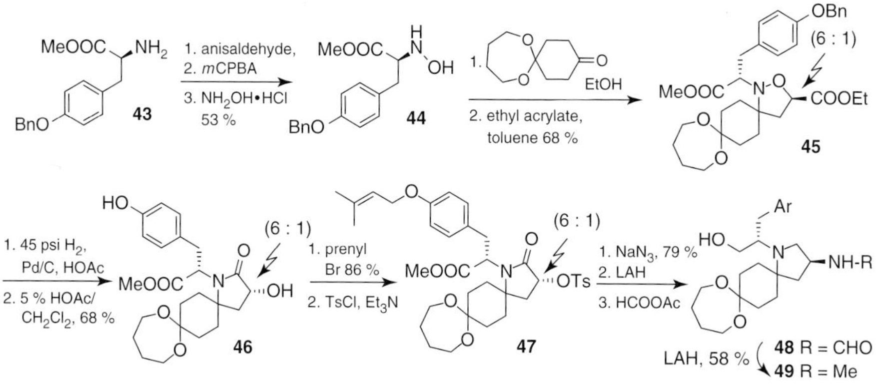

Scheme 2.11 Snider intermediate **49** for the synthesis of TAN-1251 compounds.

Exposure of **49** to DMSO/TFAA/Et_3N induced oxidation of the primary alcohol to the aldehyde and concomitant trifluoroacetylation of the amino group (Scheme 2.12). The free amino-aldehyde obtained upon release of the N-$COCF_3$ unit (methanolic K_2CO_3) instantly cyclized to **51**, which upon acidic hydrolysis of the ketal furnished TAN-1251C (**5**).

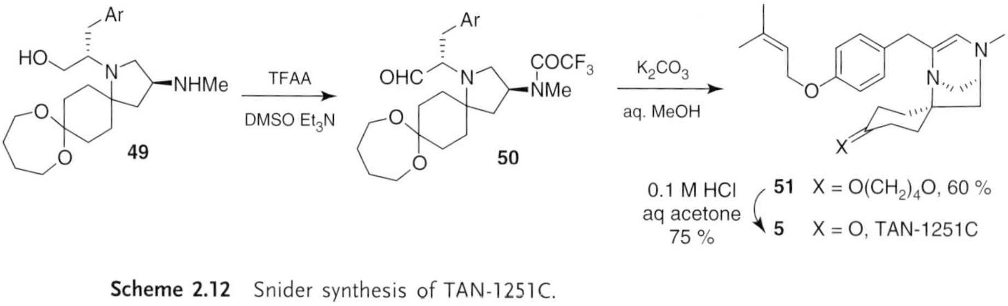

Scheme 2.12 Snider synthesis of TAN-1251C.

It is likely that TAN-1251C is the biosynthetic forerunner of the other members of the family. For instance, TAN-1251D (**6**) probably ensues upon reduction of an iminium ion formed by protonation of TAN-1251C (Scheme 2.13). Snider utilized

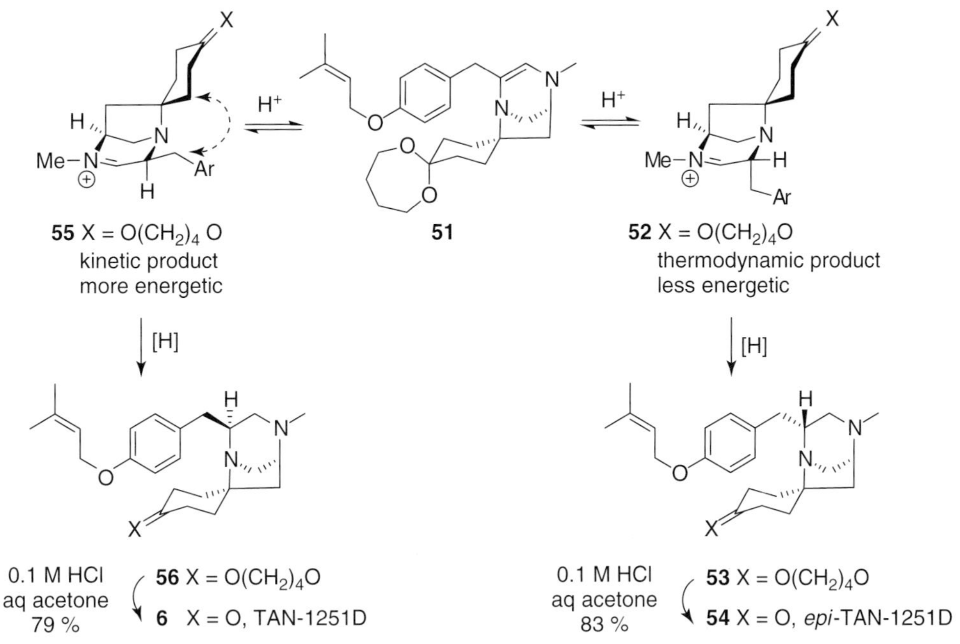

Scheme 2.13 Solvent effects in the reduction of TAN-1251C to TAN-1251D.

precisely such a transformation to convert TAN-1251C into TAN-1251D, and in the process he uncovered yet another remarkable solvent effect. Thus, reduction of **51** with $NaBH(OAc)_3$ in HOAc afforded 90% of a 1 : 9 mixture of ketals **56** and **53**. Acidic hydrolysis of the major product **53** surrendered epi-TAN-1251D (**54**) (83%). A reversal of selectivity occurred upon reduction of **51** with the more active reductant, $NaBH_3CN$, in the more polar solvent, MeOH (dropwise addition of HOAc), leading to a 2 : 1 mixture of **56** and **53**. An amelioration of selectivity in the desired sense was achieved by operating in even more polar solvents. Thus, the ratio of **56** : **53** increased to 6 : 1 when the reaction was carried out with $NaBH_3CN$ in CF_3CH_2OH, and it peaked at >25 : 1 by switching to the even more polar $(CF_3)_2CHOH$. Acidic hydrolysis of ketal **56** afforded TAN-1251D (**6**).

Snider offered the following rationale for the above phenomena. Stereochemical control in the reduction of **51** occurs during the protonation step, which determines the configuration of the benzyl-type substituent. Protonation from the convex face of the molecule is kinetically favored on grounds of reduced steric interaction with the approaching Brønsted acid. Protonation in this sense would form **55**, wherein the benzyl substituent occupies an equatorial position. However, MM2 calculations revealed that **55** contains 3 $kcal\,mol^{-1}$ more steric energy than the axial diastereomer **52**, probably because of a serious non-bonded interaction between the equatorial benzyl group and the spiro cyclohexane ring. Evidently, the energy penalty incurred in repositioning the benzyl substituent at an axial position is much smaller than the extent of the foregoing steric interaction.

Therefore, in media where the reduction of the iminium ion is slow and protonation is reversible, equilibration of iminium diastereomers can occur and the major iminium species present in the medium will be **52**. Reduction of the latter will then lead to an epimeric series of products. This is what happens with the weak reductant, $NaBH(OAc)_3$ in MeOH, a polar solvent of modest protonic acidity. In MeOH, the stronger reducing agent $NaBH_3CN$ evidently intercepts the transient iminium species at a rate comparable to that of equilibration; therefore, reduction of the kinetic iminium ion **55** becomes the major process. As the polarity and Brønsted acidity of the solvent increase, the ionic iminium species is stabilized relative to the electrostatically neutral enamine, and the equilibration of iminium diastereomers is retarded. Therefore, the ratio of diastereomers **56** : **53** steadily improves upon progressing from MeOH to CF_3CH_2OH and to $(CF_3)_2CHOH$.

TAN-1251A was reached starting with DDQ (2,3-dichloro-5,6-dicyano-1,4-benzoquinone) oxidation of **51** in dry CH_2Cl_2 (Scheme 2.14). This furnished eniminium ion **57**, which upon reduction with $NaBH_3CN$ in acidic MeOH afforded **58**. Hydrolysis of the ketal delivered fully synthetic **3**.

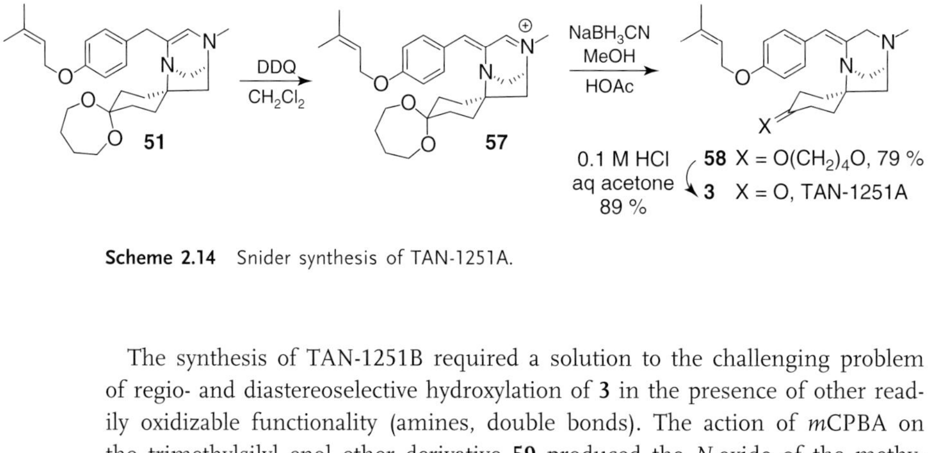

Scheme 2.14 Snider synthesis of TAN-1251A.

The synthesis of TAN-1251B required a solution to the challenging problem of regio- and diastereoselective hydroxylation of **3** in the presence of other readily oxidizable functionality (amines, double bonds). The action of *m*CPBA on the trimethylsilyl enol ether derivative **59** produced the *N*-oxide of the methylamino unit. The hydroxylation of the Na enolate of **3** (NaHMDS) with the Davis 10-(camphorsulfonyl)oxaziridine [17] produced inseparable mixtures of **60** and **61** in 40–50% yield. While no oxidation of the double bonds or the amines was detected, the reaction was taking place from the incorrect face of the enolate. Finally, it transpired that osmylation of enol ether derivatives **59** was moderately selective in the desired sense (Scheme 2.15). Thus, **3** was converted into a mixture of regioisomeric silyl enol ethers. Reaction of these with OsO_4, and NMO afforded a 2 : 1 : 8 : 4 mixture of **60**, **61**, TAN-1251B (**4**), and **62**, in 39% yield, plus a mixture of triols resulting from hydroxylation of both the enol ether and the prenyl group (23% yield), plus recovered **3** (9%). HPLC separation of this complex mixture provided pure **4**. Attempts to achieve a more regioselective enolization of **3** with chiral bases were fruitless.

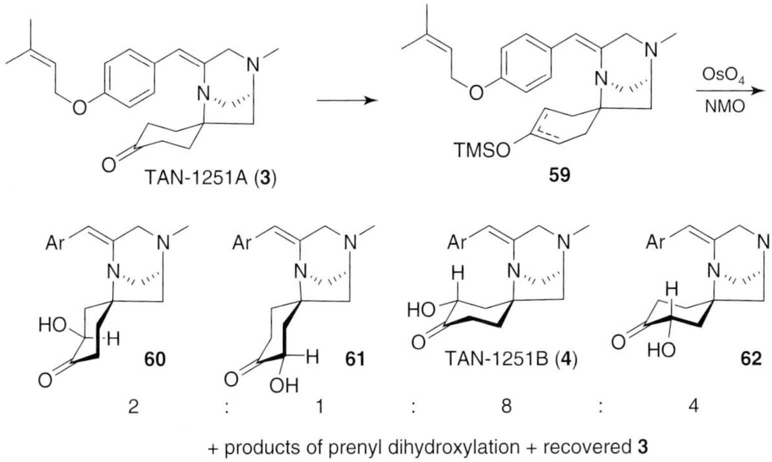

Scheme 2.15 Snider synthesis of TAN-1251B.

2.3 Oxidative Amidation of Phenols

The promise of a fully biomimetic synthesis of **1–6** awaited the development of methodology for the oxidative amidations of phenols according to the format of Scheme 2.3. This precise transformation was unknown until the late-1990s, and, indeed, it was believed not to be feasible. This lore was rooted in an important observation recorded in 1987 by Kita, who studied the oxidation of amides **63** with $PhI(OAc)_2$ (DIB), or $PhI(OCOCF_3)_2$ ("PIFA"), intending perhaps to create spirolactams **64** [18]. However, the reaction produced only lactone **67**, arguably through capture of electrophilic intermediate **65** by the carbonyl oxygen of the amide, followed by hydrolysis of the transient **66** upon aqueous workup (Scheme 2.16).

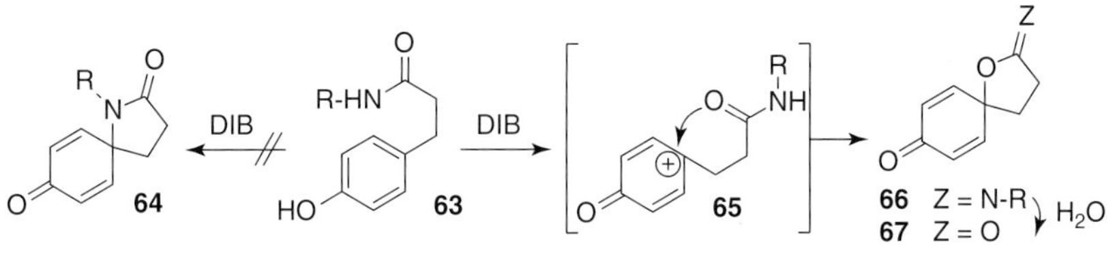

Scheme 2.16 Kita oxidation of phenolic amides leading to spirolactones.

Technology for oxidative amidation of phenols finally emerged in 1998 thanks to research carried out in our own laboratories [19]. First of all, we perceived a mechanistic analogy between the formation of lactone **67** upon oxidative activation of the phenol and the results obtained by Knapp during a study of the iodoamidation of olefins [20]. Knapp determined that the reaction of **68** with I_2 results in formation of **69** instead of the desired **71** (Scheme 2.17). This suggests that resonance

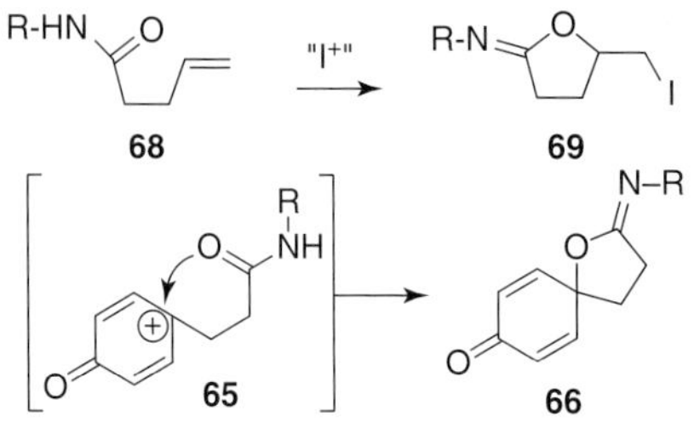

Scheme 2.17 Analogy between the Knapp iodocyclization of olefinic amides and the Kita lactone formation.

interactions between the N atom and the carbonyl system in an amide induce accumulation of electronic density on the oxygen atom, which therefore becomes nucleophilic at the expense of the N atom. The same effect accounts for the formation of **66** upon cyclization of presumed intermediate **65**. But when Knapp engaged imino ethers **70** in such iodocyclization reactions, the desired **71** was obtained in good yield (Scheme 2.18). Evidently, resonance interactions in **70** now promote accumulation of electronic density on the nitrogen atom, which therefore can express nucleophilic reactivity.

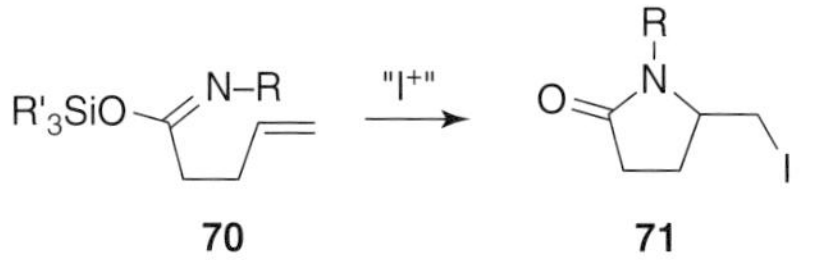

Scheme 2.18 Knapp iodolactamization reaction of olefinic imino ethers.

We surmised that an iminoether analog of the amide functionality should enable the desired spirolactamization reaction. In the end, oxazolines emerged as effective substrates for the target transformation [21]. The starting **74** were best prepared by Vorbrüggen condensation of a phenolic carboxylic acid with a suitable 1,2-aminoalcohol [22]. An advantage of this technique is that it required no protection of the phenol. Alternative procedures were less satisfactory in the present case [23]. Exposure of **74** to the action of DIB in fluoroalcohol solvents [CF_3CH_2OH (TFE) or $(CF_3)_2CHOH$ (HFIP)] according to Kita [24] afforded the long sought **77** [25] (Scheme 2.19).

Some oxazoline-derived products of oxidative amidation were observed to undergo spontaneous Michael cyclization to morpholine derivatives more or

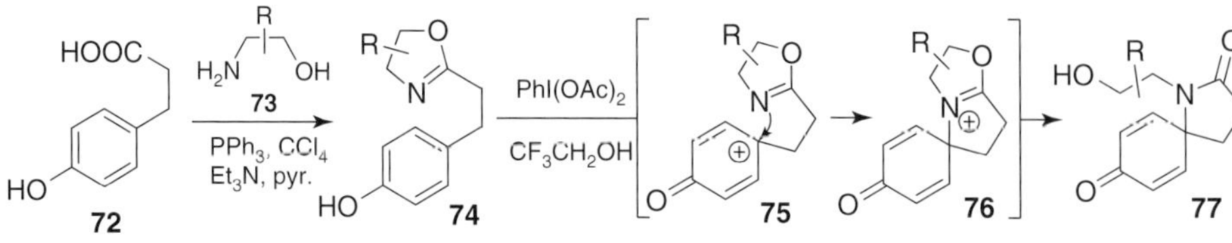

Scheme 2.19 Oxidative spirocyclization of phenolic oxazolines.

less rapidly, depending on structural details. In all cases, cyclization was highly diastereoselective. For instance, compound **78** gave **79** exclusively (Scheme 2.20; structure ascertained by X-ray crystallography). The stereoselective formation of **79** is attributable to the strong preference for the axial orientation on the part of alkyl substituents flanking the N atom in *N*-acyl piperidines and related six-membered heterocycles [12].

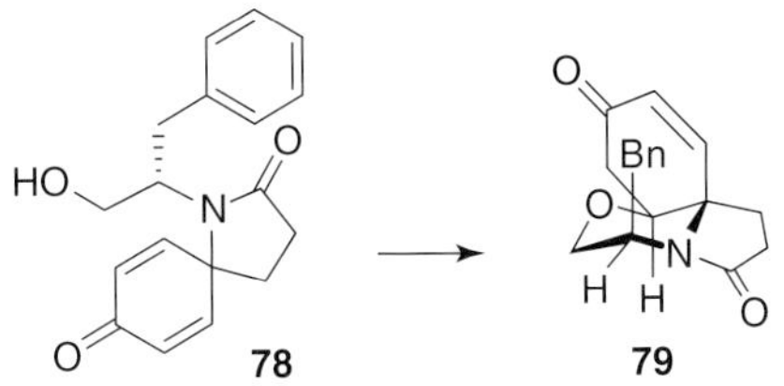

Scheme 2.20 Stereoselective cyclization of products of oxidative amidation.

Morpholine formation is useful in particular circumstances because it induces a stereocontrolled desymmetrization of the "locally symmetrical" dienone segment of the primary products, securing a specific configuration of the now stereogenic spiro center. In other instances, the proclivity to cyclize is problematic and must be suppressed. It is then convenient to acetylate the OH group prior to purification (cf. **83**→**84**, Scheme 2.21). Overall yields of acetates **84** from **82** are typically around 40–45%. Such moderate yields must be weighed against the fact that the reaction rapidly converts inexpensive amino acid-derived substances into valuable enantiopure intermediates.

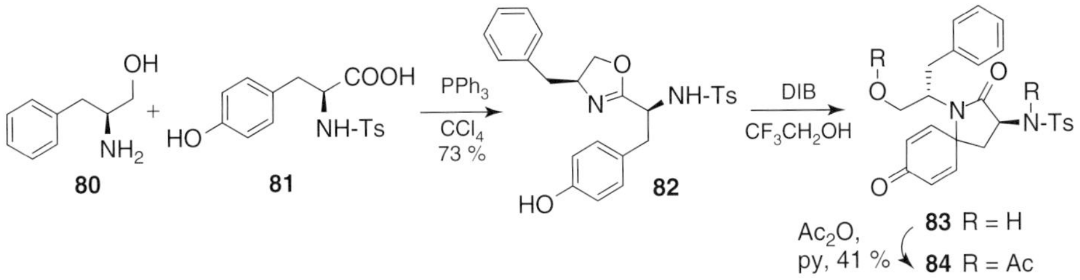

Scheme 2.21 Formation of spirolactam **84**.

A comment is in order regarding the choice of an *N*-sulfonyl protecting group for **82**. Experiments carried out with similar substrates that incorporated carbonyl-type N-protection furnished complex mixtures containing only some of the desired spirolactams. At least some of the by-products obtained from these reactions appeared to arise through interception of reactive intermediate **85** by the carbonyl unit (Scheme 2.22, pathway b). Iminocarbamate **86** thus formed could subsequently yield various secondary products. A sulfonyl protecting group is insufficiently nucleophilic to interact with the electrophilic moiety of **85**, maximizing formation of the desired spirolactam.

The chemistry of Scheme 2.19 constitutes what may be termed a "*first generation*" solution for the oxidative amidation of phenols. The methodology was later extended

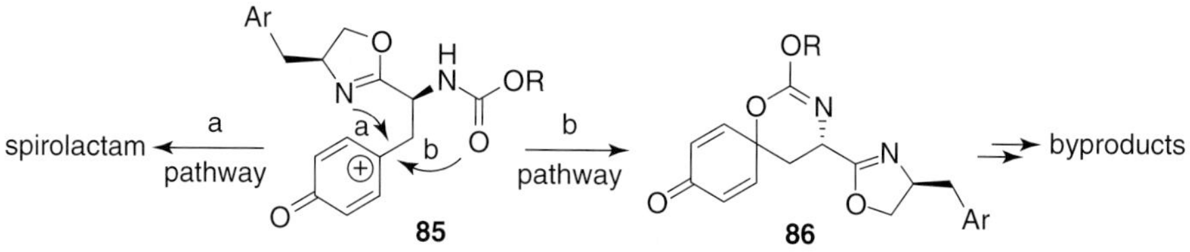

Scheme 2.22 By-products arising from substrates incorporating a carbamoyl group.

to the oxidative cyclization of phenolic sulfonamides [26] and to a bimolecular variant of the reaction, wherein a nitrile captures the presumed intermediate **90** in a Ritter mode [27]. These processes are outlined in Scheme 2.23; however, they are outside the scope of the present chapter.

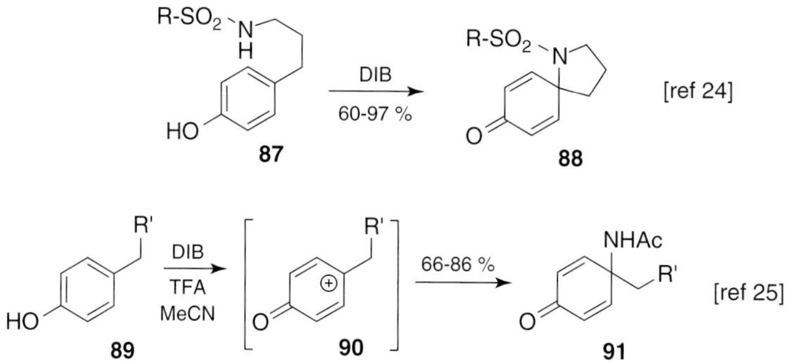

Scheme 2.23 Alternative modes of oxidative amidation of phenols.

A noteworthy variant of the above methodology achieves the oxidative cyclization of phenolic secondary amines. This chemistry was developed by Sorensen as the centerpiece of his biomimetic synthesis of FR-901483 [28]. A key step in this endeavor was the conversion of **92** into **93** upon exposure to DIB (Scheme 2.24). Notice that a sulfonamido protecting group was present on the spectator amino functionality in **92**. We presume that this choice was dictated by the difficulties adumbrated in Scheme 2.22.

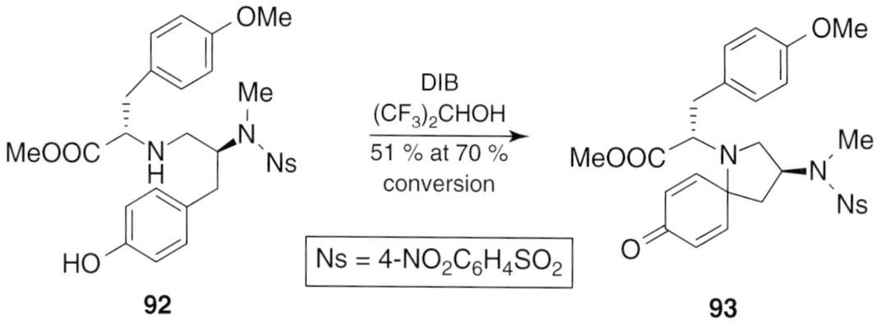

Scheme 2.24 Oxidative cyclization of a phenolic amine in the Sorensen synthesis of FR-901483.

A similar reaction described by Honda [29] accomplishes the DIB-mediated oxidative cyclization of **94**→**95** (Scheme 2.25). This transformation was central to a biomimetic synthesis of TAN-1251 substances.

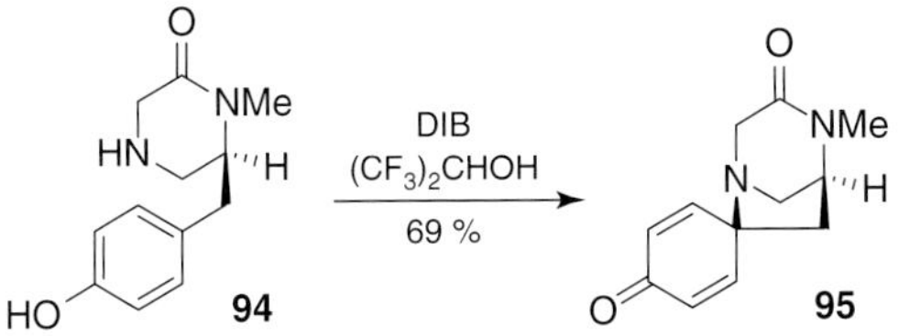

Scheme 2.25 Honda oxidative cyclization of phenolic amines.

We had an opportunity to examine the behavior of primary amines under Sorensen–Honda conditions, but we found them to be exceedingly poor substrates for the reaction [30]. Thus, cyclization of **96** furnished **97**, as a component of a complex mixture of products, in less than 10% yield (Scheme 2.26).

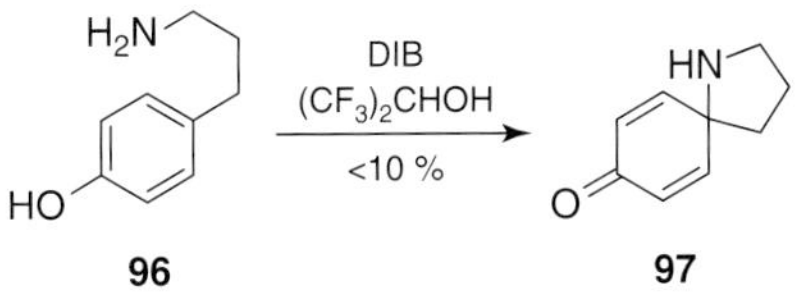

Scheme 2.26 Inefficient oxidative cyclization of phenolic primary amines.

We stress that the terminology "oxidative amidation of phenols" as used herein implies a mechanism that involves an initial oxidation of the phenol to an electrophilic species, which is then captured by the nitrogen atom of an appropriate amide equivalent. On the other hand, electronically complementary spirolactam syntheses involving the capture of an *electrophilic* N atom by an electron-rich aromatic ring have been known since the early 1980s, having been independently developed by Glover [31] and Kikugawa [32]. Notable examples of this chemistry evolve from *N*-alkoxy-acylnitrenium ions (Scheme 2.27).

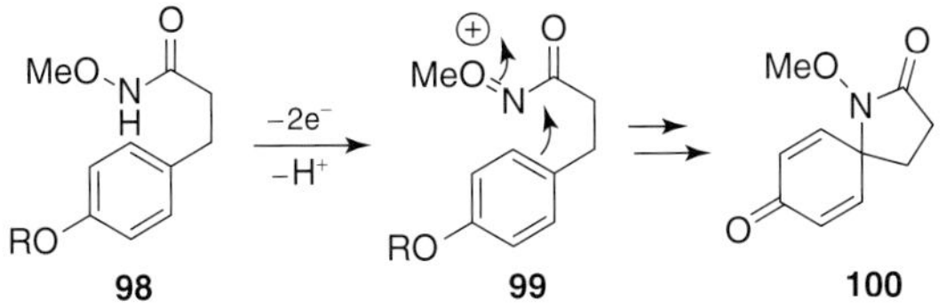

Scheme 2.27 Glover–Kikugawa reaction.

The synthetic potential of these processes has become apparent in recent times through the work of Wardrop and collaborators [33]. In particular, these workers disclosed a synthesis of TAN-1251A (**3**) based on an initial oxidative cyclization of tyrosine derivative **102**→**103** (Scheme 2.28). Whether this approach qualifies as "biomimetic" is a matter of debate. By the criterion enounced at the beginning of this chapter, the judgment would be in the negative (it does not involve a tyrosinyl

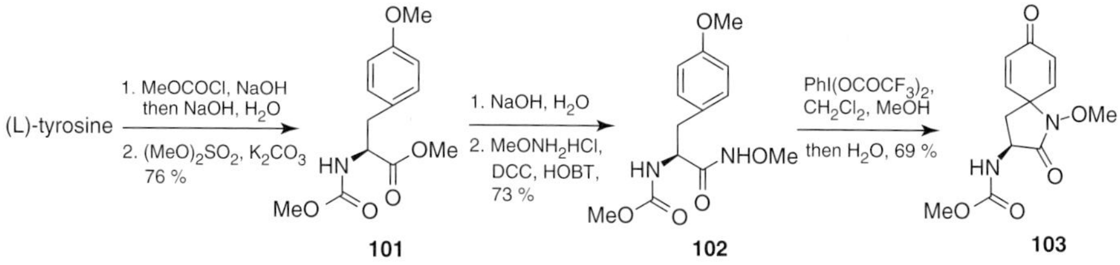

Scheme 2.28 Wardrop oxidative cyclization of **102**→**103**.

tyrosine equivalent). Regardless, the strategy is sufficiently interesting as to merit a brief parenthesis.

Thus, spirolactam **103** was elaborated to intermediate **107**, which underwent aldol condensation with **108** to provide **109** and subsequent alane reduction yielded (−)-TAN-1251A (**3**) (Scheme 2.29) [34]. Overall, **3** was prepared in only 13 steps from (L)-tyrosine. This is one of the more efficient syntheses of TAN compounds reported to date.

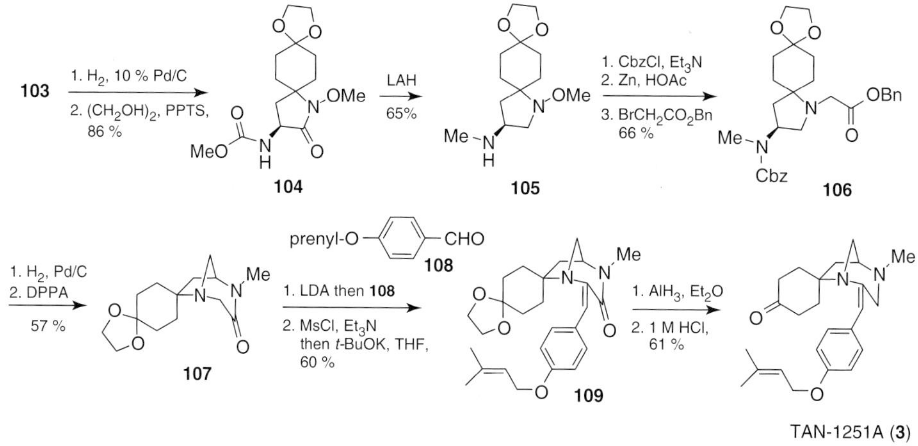

Scheme 2.29 Wardrop synthesis of TAN-1251A.

A similar strategy is apparent in a synthetic study toward FR-901483 [8c]. Oxidative cyclization of **99** (R = H) and reduction of the emerging **100** yielded **110**, which upon N-propargylation and enol silyl ether formation afforded compound **112** (Scheme 2.30). Reaction of the latter with tributyltin hydride under radical conditions, followed by acidic treatment, delivered a mixture of **115** (40%), **116** (2%), and **117** (45%). Tricyclic lactam **115** is structurally related to **1**. Its formation may be rationalized in terms of addition of tributyltin radical to the alkyne and radical cyclization of intermediate **113**, followed by protiodestannylation of the resultant vinyltin species upon acidic workup. By-product **116** is likely to form through reduction of **113** and destannylation, while pyrrolizidinone **117** presumably arises via intramolecular H-atom transfer of **113** leading to **114** and radical cyclization of the latter.

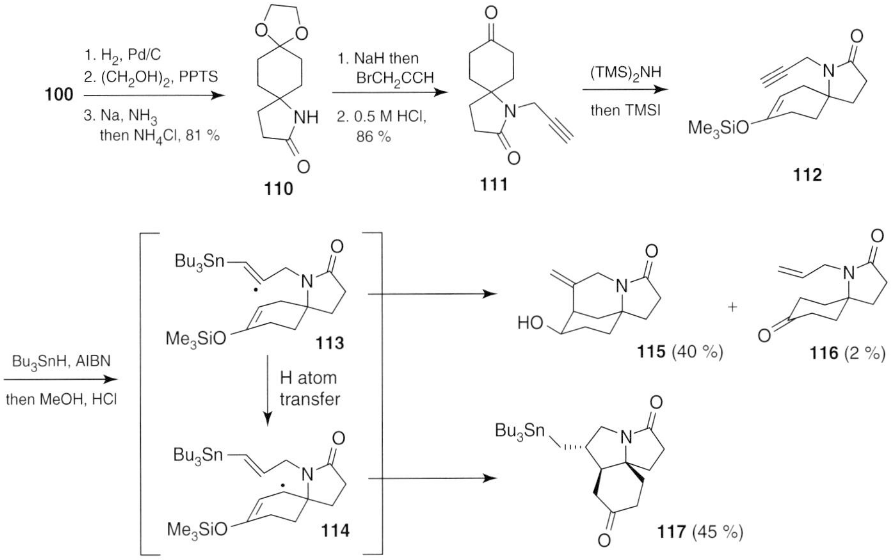

Scheme 2.30 Wardrop approach to FR-901483.

The straightforward sequence delineated in Scheme 2.31 then converted intermediate **115** into desmethylamino FR-901483 (**123**).

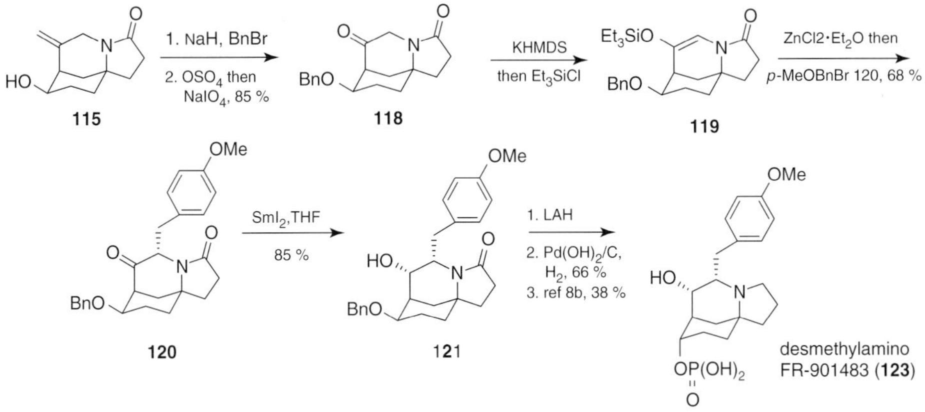

Scheme 2.31 Wardrop synthesis of desmethylamino FR-901483 (**123**).

2.4
Biomimetic Syntheses of FR-901483 and TAN-1251 Compounds via Oxidative Amidation Chemistry and Related Processes

The advent of methodology for the oxidative amidation of phenols enabled biomimetic syntheses of **1** and of TAN-1251 compounds through the actual union of two molecules of tyrosine. The Sorensen synthesis of FR-901483 was the first one to document this approach, and it was followed shortly thereafter by the Ciufolini synthesis of FR-901483 and of TAN-1251C. The 2005 Honda synthesis of TAN-1251C and TAN-1251D concludes the cycle of biomimetic syntheses relying on oxidative amidation chemistry.

2.4.1 Sorensen Synthesis of FR-901483

In 2000, Sorensen disclosed a synthesis of FR-901483 that relied on the oxidative cyclization of amine **92** (Scheme 2.24) as the key step [28]. The route to **92** started with the elaboration of tyrosine methyl ester hydrochloride into building blocks **125** and **129** (Scheme 2.32).

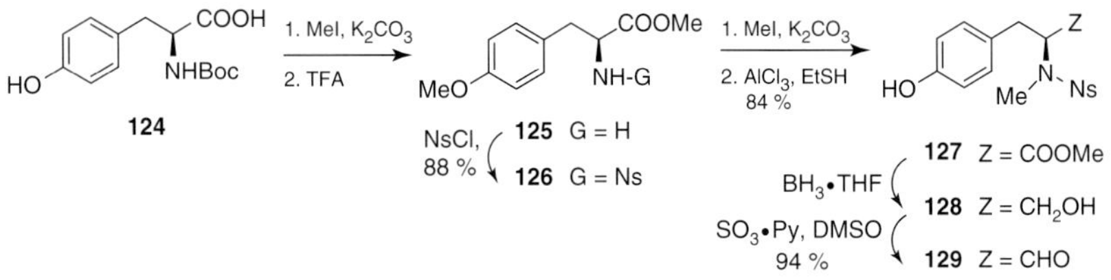

Scheme 2.32 Synthesis of building blocks **125** and **129**.

The sensitive aldehyde **129** was employed in crude form in a subsequent reductive amination with **125** (Scheme 2.33). This provided tyrosinyl tyrosine derivative **92**, which upon reaction with $PhI(OAc)_2$ in HFIP underwent oxidative cyclization to **93**. Exchange of the N-protection group, hydrogenation of the dienone **93** and redox manipulation of the resultant **130** delivered aldehyde **131**.

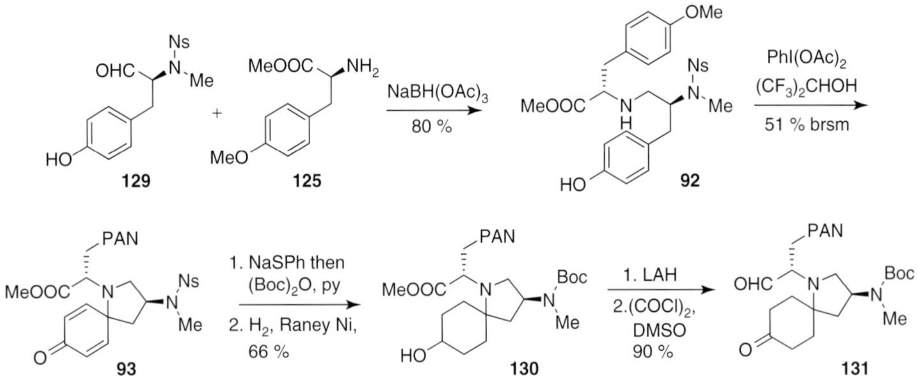

Scheme 2.33 Sorensen route to aldehyde **131**.

In contrast to the Snider substrate, **15**, the stereoselective aldol cyclization of **131** occurred best in methanol containing MeONa (Scheme 2.34). Notice that **131** differs from **15** in that it lacks a carbonyl group on the pyrrolidine segment.

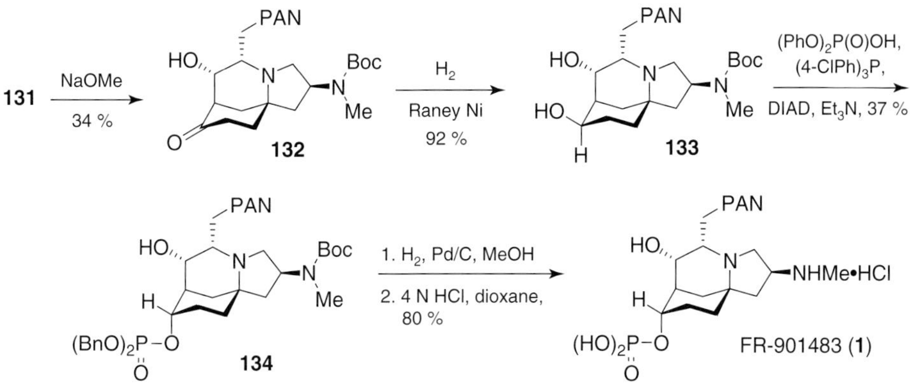

Scheme 2.34 Sorensen synthesis of FR-901483.

Tricyclic intermediate **132** was thus isolated in 34% yield from the ensuing mixture of products. Stereoselective reduction of the ketone from the convex face was accomplished by hydrogenation, and the resultant **133** underwent a Mitsunobu reaction with dibenzyl hydrogen phosphate to afford **134**. In accord with Snider, the considerably less reactive OH group adjacent to the benzyl-type substituent did not interfere. Deblocking of **134** produced the target **1**, which was thus reached in 18 steps from *N*-Boc tyrosine **124**.

2.4.2
Honda Synthesis of TAN-1251 Substances

A key aspect of the Honda synthesis of TAN-1251 compounds is the reduction of dienones of the type **95** to the corresponding cyclohexanones (Scheme 2.35). This step was problematic and it necessitated the implementation of a variant of a procedure developed earlier by Buchwald [35]. Thus, treatment of **95**

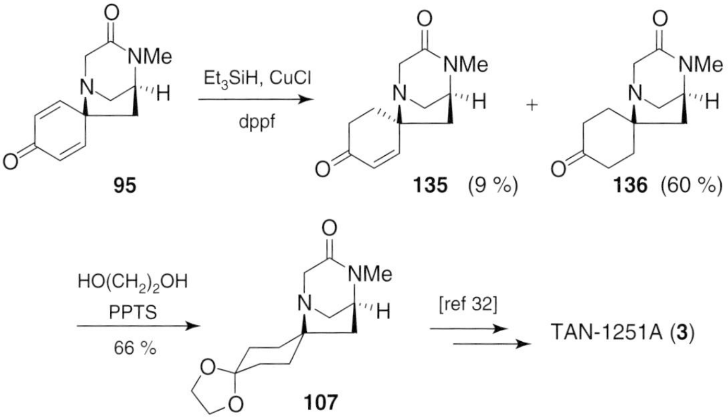

Scheme 2.35 Honda formal synthesis of TAN-1251A (**3**).

with 3 equiv. of Et_3SiH in the presence of 20 mol% of CuCl and 20 mol% of diphenyl(phosphino)ferrocene (dppf) accomplished reduction to **136** in 60% yield. The reaction also afforded some enone **135** (9% yield). Ketalization of **136** produced a substance identical to the Wardrop TAN-1251A precursor, **107**, thereby realizing a formal synthesis of that natural product [29, 34].

In 2005, Honda reported actual total syntheses of TAN-1251C and TAN-1251D based on his oxidative cyclization reaction [36]. The route to **5**→**6** commenced with the assembly of intermediate **143** by a method that retraced the Sorensen avenue to **92** (Scheme 2.32). Thus, (L)-tyrosine methyl ester was transformed into aldehyde **140**, which upon reductive amination with **141** afforded **142** (Scheme 2.36). Release of the Boc group and cyclization of the resultant compound surrendered piperazinone **144**. The crucial oxidative cyclization of **145** with DIB in HFIP delivered **146** in 49% yield. This material was reduced to the corresponding cyclohexanone as seen earlier (Scheme 2.35). A sequence involving ketalization, hydrogenolysis of the benzyl group and *O*-prenylation of the free phenol led to intermediate **148**, which was the precursor of both TAN-1251C and TAN-1251D (Scheme 2.36).

The action of DIBAL upon **148** produced enamine **149**, presumably through elimination from a carbinolamine-type intermediate formed upon delivery of a hydride to the carbonyl system (Scheme 2.37). Aqueous hydrolysis of the ketal liberated totally synthetic TAN-1251C. By contrast, LAH in the presence of $AlCl_3$ converted **148** into **150**, which upon ketal hydrolysis furnished TAN-1251D.

2.4.3
Ciufolini Synthesis of FR-901483 and TAN-1251C

The 2001 Ciufolini syntheses of FR-901483 and TAN-1251C [37] relied on the oxidative cyclization of phenolic oxazolines obtained through the union of a molecule of *N*-tosyltyrosine, **152**, with one of tyrosinol methyl ether, **151**. Quantities of **152** of excellent chemical and optical purity were prepared from tyrosine ethyl ester according to a 1915 procedure by Emil Fischer (Scheme 2.38) [38]. Notably, while tosylation of tyrosine ester produced the anticipated sulfonamide very selectively, that of free tyrosine gave only the *N,O*-ditosyl derivative [39]. The choice of a tosylamide, instead of a more readily cleavable Fukuyama nitrosulfonamide [40], was motivated by a desire to retain the original N-protecting group during later transformation requiring reductants or nucleophiles. The *N*-tosyl group would be cleaved during hydride reduction of the pyrrolidinone to the corresponding pyrrolidine.

By contrast, the elaboration of tyrosinol **151** by literature methods [41] incurred significant erosion of optical integrity. Fortunately, the route shown in Scheme 2.38 afforded a product of excellent optical and chemical quality.

Merger of **151** and **152** in the manner of Vorbrüggen afforded oxazoline **153**, which underwent oxidative cyclization and acetylation to provide **155** (Scheme 2.39). Notice that the latter step results also in *N*-acylation of the tosylamide. However, this is inconsequential, because the acetyl groups are released simultaneously at a

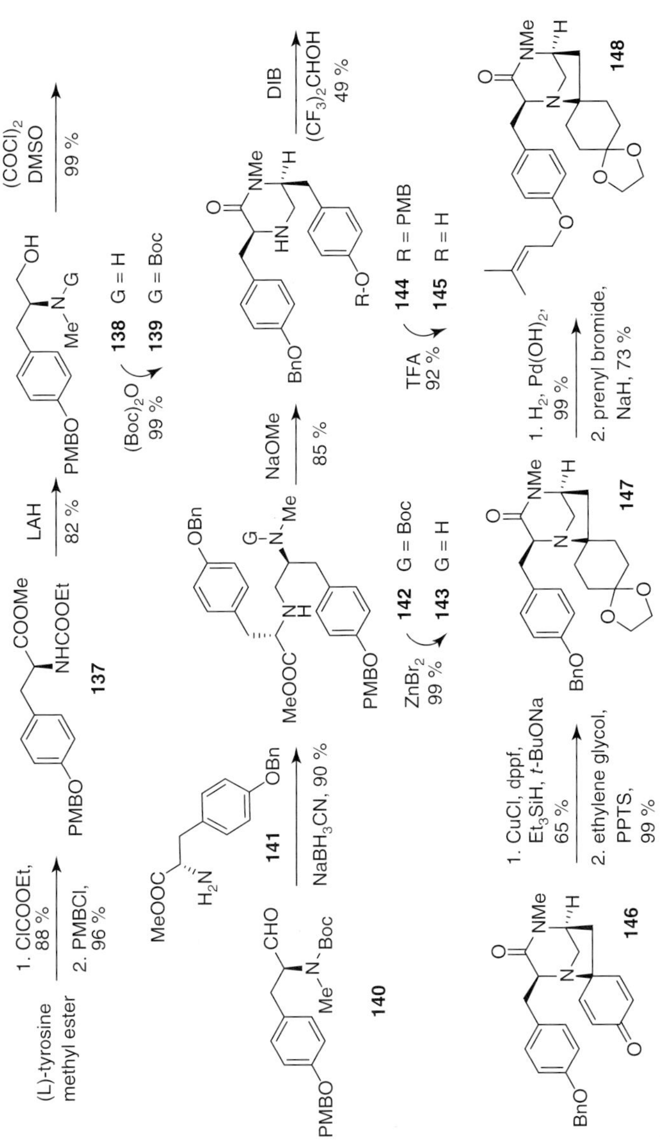

Scheme 2.36 Honda route to **148**, precursor of TAN-1251 substances.

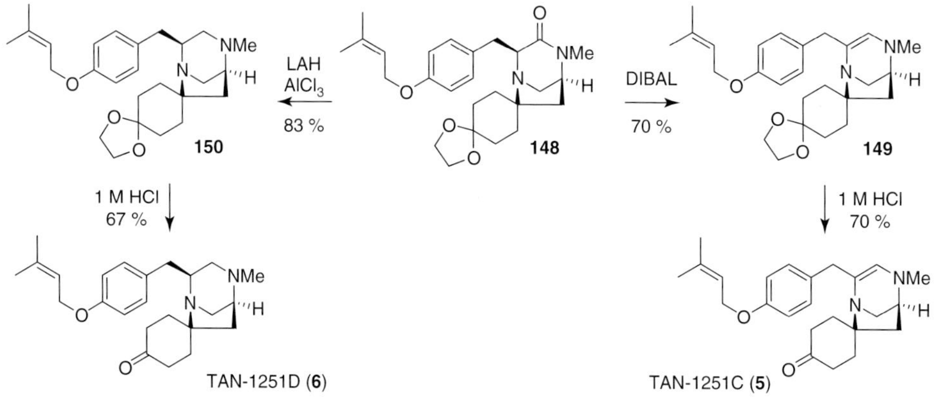

Scheme 2.37 Honda synthesis of TAN-1251C (**5**) and D (**6**).

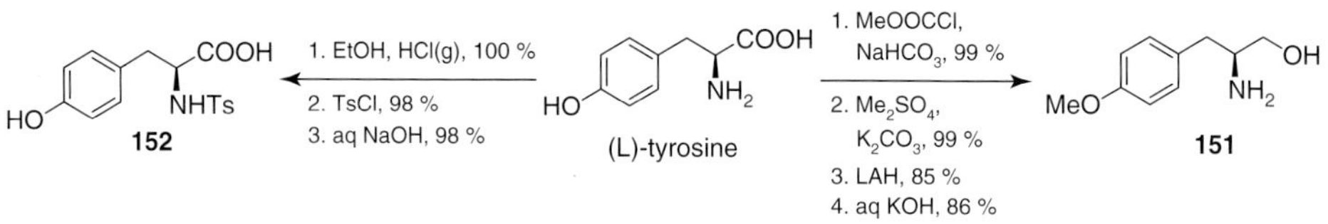

Scheme 2.38 Preparation of building blocks **151** and **152**.

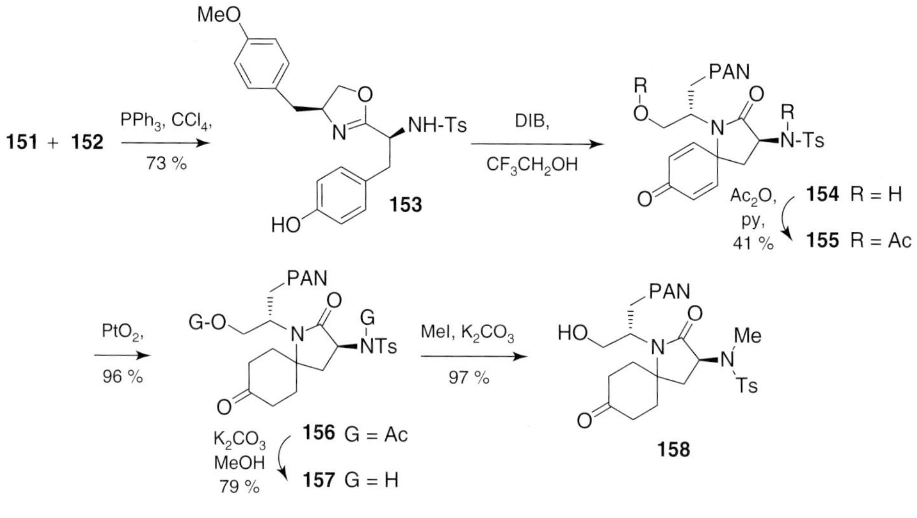

Scheme 2.39 Ciufolini synthesis of alcohol **158**.

later stage. Saturation of the dienone was carried out by hydrogenation over PtO_2 (Adams catalyst). Palladium or rhodium catalysts were unsatisfactory, because they promoted various degrees of reductive aromatization of **155** through C–N bond cleavage. Hydrogenation in the presence of supported platinum, for example, Pt(C), resulted in a variable extent of ketone reduction to the corresponding cyclohexanol. No such problem was observed with the Adams catalyst. Simultaneous *N*- and *O*-deacetylation of **156** prepared the molecule for selective methylation of the tosylamide, which was readily accomplished with MeI in the presence of K_2CO_3. The emerging **158** is the forerunner of both TAN-1251C and of FR-901483.

The assembly of **5** (Scheme 2.40) commenced with *O*-demethylation of **158** and prenylation of the intermediate phenol. The resultant **159** underwent stereoselective LAH reduction to aminodiol **160**. The latter is recognized as the Snider TAN-1251 intermediate. Whereas elaboration of **160** as prescribed by Snider (oxidation to keto aldehyde **162** with DMSO/TFAA followed by treatment with methanolic K_2CO_3) did effect conversion into TAN-1251C, we found the overall yield of the final sequence disappointing. We suspected that the basic treatment required to release the *N*-trifluoroacetyl group may have diverted a portion of **162** into aldol manifolds. Considerable improvement in overall efficiency was observed upon protection of the amino group as a 2,2,2-trichloroethyl (Troc) carbamate, followed by Ley oxidation [42] of the resulting **160**–**162** and final Troc release with Cd/Pb couple [43]. Because Snider had shown that all other TAN-1251 compounds are accessible from TAN-1251C (**5**), a synthesis of **5** is tantamount to a formal synthesis of **3**, **4**, and **6**.

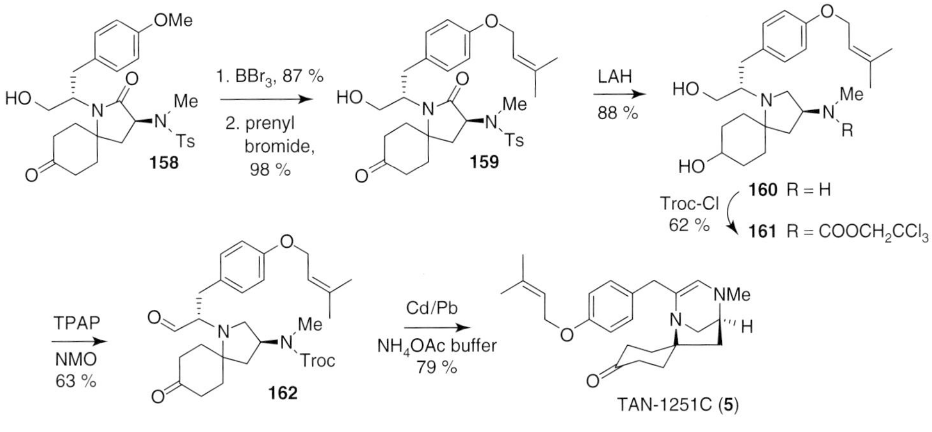

Scheme 2.40 Ciufolini synthesis of TAN-1251C (**5**).

The synthesis of FR-901483 commenced with the oxidation of **158** to keto aldehyde **163** (Scheme 2.41), which is destined to undergo a crucial aldol cyclization. The Snider synthesis of **1** had already appeared in print at this juncture of our own work: the important observations recorded by the Brandeis team greatly assisted us in our endeavor. An aspect of our work that merits some space here pertains to the results of earlier computational (MM+) and experimental results. First, an MM+

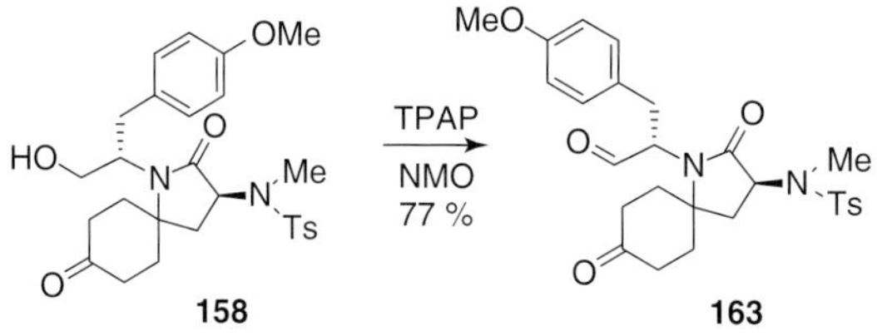

Scheme 2.41 Oxidation of **158** to the aldol substrate **163**.

simulation carried out with simplified model substrate **164** failed to detect a distinct preference for any one regio- or stereoisomer of the aldol product. An estimate of the relative energy demand of transition states **165** (leading to the desired *regioisomer*) and **168** (producing the incorrect regioisomer) indicated that **165a,b** are less energetic than **168** by about 0.8 kcal mol^{-1}. However, such a small energy difference was below the confidence level of the calculation ($\pm$1–2 kcal mol^{-1}). Furthermore, **165a** and **165b** were calculated to be isoenergetic, as were the two diastereomers of either regioisomer of the aldol product. This implied that no substrate-directed diastereocontrol could be anticipated, and that the aldol step was likely to become non-stereoselective if carried out under conditions of thermodynamic reversibility (Scheme 2.42).

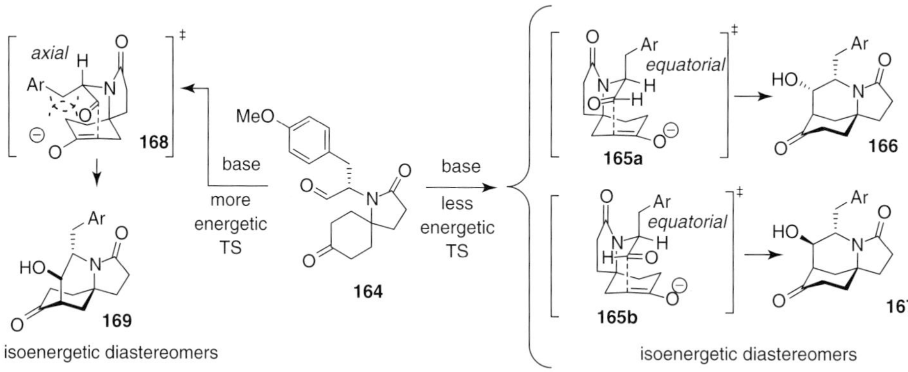

Scheme 2.42 Pathways of aldol cyclization.

Turning to experiment, we found that cyclization of **163** using DBU in CH_2Cl_2 formed **170** stereoselectively (Scheme 2.43, stereoisomeric products were also formed). This result reaffirmed the fact that the enolization of the cyclohexanone is kinetically faster than that of the aldehyde. More importantly, it indicated that the formation of the major product was taking place in accord with the so-called Seebach rule [44]. This induced us to suspect that Seebach forces, rather than the chelation metal alkoxide intermediate invoked by Snider, may determine stereoselectivity in nonprotic solvents. Such a reactivity model is valid only in nonprotic solvents: polar ones may induce erosion, or even reversal, of Seebach diastereoselectivity, presumably due to solvation of reactive species through H-bonding. Faced with the choice of advancing **170** into the synthesis through an inversion of configuration of

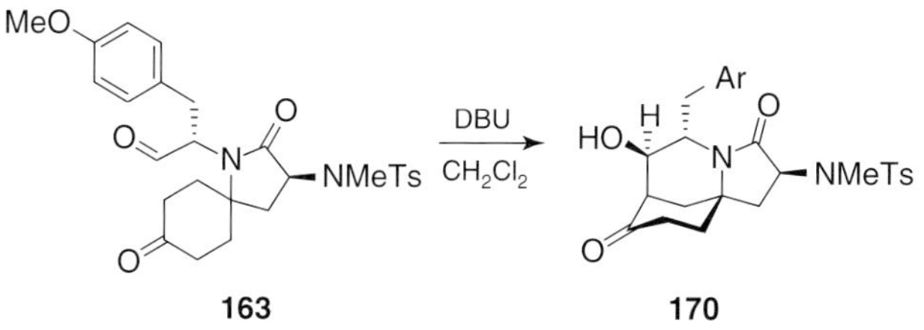

Scheme 2.43 Stereoselective aldol cyclization of **163** promoted by DBU.

the newly formed carbinol, or relying on the then-newly published work of Snider and endeavor to create the correct aldol diastereomer by operating in protic solvents, we opted for the second solution. One of the reasons that determined our choice was the anticipated (and in retrospect unjustified) difficulty of inducing an S_N2 reaction at the level of a secondary center flanked on either side by branched carbon atoms. In fact, Funk subsequently demonstrated that this reaction is quite feasible [6a].

The Snider aldol conditions (*t*-BuOH, *t*-BuOK), so successful with substrate **15**, proved to be unsatisfactory with ketoaldehyde **163**. First, the compound suffered from modest solubility in *t*-BuOH, necessitating the use of a 3 : 1 mixture of *t*-BuOH/THF to effect aldol cyclization. The resulting decrease in polarity and proticity of the medium was deleterious to yields and selectivity: **171** emerged as component of a mixture of products in a poor 21% yield after chromatography (Scheme 2.44). A switch to EtOH/EtONa and then to MeOH/MeONa progressively improved yields and stereoselectivities. This again contrasted with the behavior of the Snider substrate, but it was consonant with the Sorensen findings. Evidently, optimal conditions for this step are intimately dependent on the structure of the substrate.

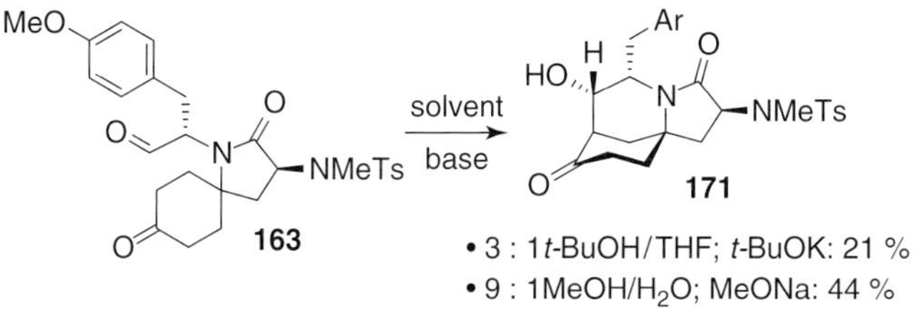

Scheme 2.44 Improved stereoselectivity of the aldol cyclization of **163** in protic media.

The fact that media of greater polarity[2] and hydrogen bonding ability[3] had a favorable influence on the outcome of the reaction induced us to examine the effect of added water. A methanolic solution of **163** was found to remain homogeneous upon dilution with up to 10 vol.% of water. Addition of solid NaOMe to such an aqueous/methanolic solution triggered rapid and diastereoselective aldol

2) Dielectric constants at 25 °C: t-BuOH = 10.9; EtOH = 24.3; MeOH = 32.6 [45].

3) The OH group is more accessible in the less sterically hindered MeOH and EtOH relative to the more sterically encumbered *t*-BuOH. This should result in stronger H-bonds.

cyclization leading to **171** in 44% chromatographed yield. We speculate that if the biosynthesis of FR-901483 (perhaps via intermediate of type **171**) indeed involves aldol cyclization of a species related to **163**, then the occurrence of such an event within an enzyme possessing a hydrophilic active site hosting numerous water molecules may well assist in the formation of the correct aldol diastereomer.

The synthesis was completed rapidly and uneventfully from **171** (Scheme 2.45). Vigorous LAH reduction produced **172** (82%) plus a small amount of the epimeric alcohol. Mitsunobu reaction of **172** with dibenzyl phosphate according to Sorensen yielded extremely polar **173**, which was best converted into the *N*-Cbz derivative **174** to effect purification. Hydrogenolysis in the presence of aqueous HCl provided the bis-hydrochloride salt of **1**, which was identical in all respects to material prepared from an authentic sample of FR-901483, kindly provided by the Fujisawa Pharmaceutical Company. In summary, the synthesis of FR-901483 required 17 steps from (L)-tyrosine and resulted in a 1.3% overall yield of **1**, while that of TAN-1251C proceeded in 16 linear steps from (L)-tyrosine and in 4% overall yield.

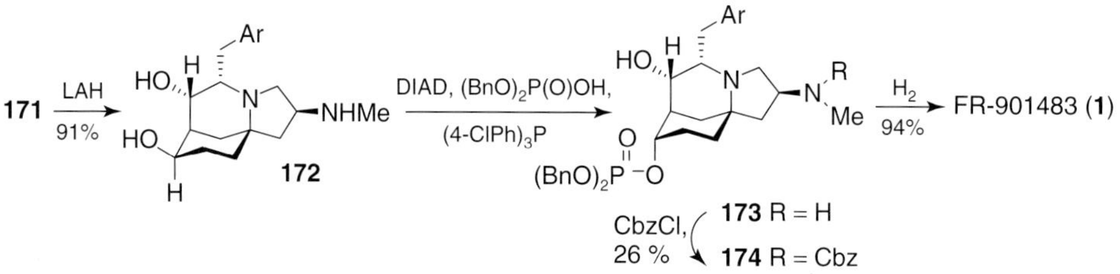

Scheme 2.45 Ciufolini synthesis of FR-901483 (**1**).

The oxidative amidation of phenols remains the centerpiece of several ongoing synthetic endeavors in our laboratory. Present efforts retain very little of the original "biomimetic" flavor that stimulated the development of the technology. Still, our initial investigations spawned a great deal of new chemistry that has proved, and continues to prove, quite valuable in the synthesis of diverse nitrogenous substances. Undeniably, biosynthetic hypotheses remain a major source of inspiration for the development of new reactions and for the progress of organic chemical technology.

References

1. Stanforth, S.P. (2006) *Natural Product Chemistry at a Glance*, Blackwell Publishers, Malden.
2. Leading reviews: (a) Misra, N., Luthra, R., Singh, K.L., and Kumar, S. (1999) *Comprehensive Natural Product Chemistry*, vol. 4, Elsevier, Amsterdam, pp. 25–59; (b) Tillequin, F. (2007) *Phytochem. Rev.*, **6**, 65–79; (c) Sato, F., Inui, T., and Takemura, T. (2007) *Curr. Pharm. Biotechnol.*, **8**, 211–218; (d) Liscombe, D.K. and Facchini, P.J. (2008) *Curr. Opin. Biotechnol.*, **19**, 173–180.
3. Sakamoto, K., Tsujii, E., Abe, F., Nakanishi, T., Yamashita, M., Shigematsu, N., Izumi, S., and Okuhara, M. (1996) *J. Antibiot.*, **49**, 37–44.
4. Tsujii, E., Nakanishi, T., Takase, S., Yamashita, M., Izumi, S., and

Okuhara, M. (1993) PCT Int. Appl. WO 9, 312,125.

5. Shirafuji, H., Tsubotani, S., Ishimaru, T., and Harada, S. (1991) PCT Int. Appl. WO 9, 113,887.
6. Total syntheses of FR-901483: (a) Maeng, J.-H. and Funk, R.L. (2001) *Org. Lett.*, **3**, 1125–1128; (b) Kan, T., Fujimoto, T., Ieda, S., Asoh, Y., Kitaoka, H., and Fukuyama, T. (2004) *Org. Lett.*, **6**, 2729–2731; (c) Ieda, S., Asoh, Y., Fujimoto, T., Kitaoka, H., Kan, T., and Fukuyama, T. (2009) *Heterocycles*, **79**, 721–738
7. Total syntheses of TAN-1251 compounds: (a) Carson, C.A. and Kerr, M.A. (2009) *Org. Lett.*, **11**, 777–779; (b) Nagumo, S., Nishida, A., Yamazaki, C., Murashige, K., and Nawahara, N. (1998) *Tetrahedron Lett.*, **39**, 4493–4496; (c) Nagumo, S., Nishida, A., Yamazaki, C., Matoba, A., Murashige, K., and Kawahara, N. (2002) *Tetrahedron*, **58**, 4917–4924; (d) Nagumo, S., Matoba, A., Ishii, Y., Yamaguchi, S., Akutsu, N., Nishijima, H., Nishida, A., and Kawahara, N. (2002) *Tetrahedron*, **58**, 9871–9877
8. Synthetic studies toward FR-901483: (a) Bonjoch, J., Diaba, F., Puigbo, G., Sole, D., Segarra, V., Santamaria, L., Beleta, J., Ryder, H., and Palacios, J.-M. (1999) *Bioorg. Med. Chem.*, **7**, 2891–2897; (b) Brummond, K.M. and Lu, J. (2001) *Org. Lett.*, **3**, 1347–1349; (c) Wardrop, D.J. and Zhang, W. (2001) *Org. Lett.*, **3**, 2353–2356; (d) Suzuki, H., Yamazaki, N., and Kibayashi, C. (2001) *Tetrahedron Lett.*, **42**, 3013–3015; (e) Bonjoch, J., Diaba, F., Puigbo, G., Peidro, E., and Sole, D. (2003) *Tetrahedron Lett.*, **44**, 8387–8390; (f) Panchaud, P., Ollivier, C., Renaud, P., and Zigmantas, S. (2004) *J. Org. Chem.*, **69**, 2755–2759; (g) Brummond, K.M. and Hong, S.-P. (2005) *J. Org. Chem.*, **70**, 907–916; (h) Gotchev, D.B. and Comins, D.L. (2006) *J. Org. Chem.*, **71**, 9393–9402; (i) Kaden, S. and Reissig, H.-U. (2006) *Org. Lett.*, **8**, 4763–4766; (j) Diaba, F., Ricou, E., and Bonjoch, J. (2006) *Tetrahedron: Asymmetry*, **17**, 1437–1443; (k) Kropf, J.E., Meigh, I.C., Bebbington, M.W.P., and Weinreb, S.M. (2006) *J. Org. Chem.*, **71**, 2046–2055; (l) Simila, S.T.M. and Martin, S.F. (2007) *J. Org. Chem.*, **72**, 5342–5349; (m) Asari, A., Angelov, P., Auty, J.M., and Hayes, C.J. (2007) *Tetrahedron Lett.*, **48**, 2631–2634; (n) Diaba, F., Ricou, E., Sole, D., Teixido, E., Valls, N., and Bonjoch, J. (2007) *Arkivoc*, **iv**, 320–330
9. Synthetic studies toward TAN-1251 compounds: (a) Seike, H. and Sorensen, E.J. (2008) *Synlett*, 695–701; (b) Nagumo, S., Akutsu, N., Yamazaki, C., Kawahara, N., and Nishida, A. (1998) *Tennen Yuki Kagobutsu Toronkai Koen Yoshishu*, **40**, 595–600; (c) Auty, J.M.A., Churcher, I., and Hayes, C.J. (2004) *Synlett*, 1443–1445
10. (a) Snider, B.B. and Lin, H. (1999) *J. Am. Chem. Soc.*, **121**, 7778–7786; (b) Snider, B.B., Lin, H., and Foxman, B.M. (1998) *J. Org. Chem.*, **63**, 6442–6443.
11. (a) Available in three steps from *O*-methyl tyrosine methyl ester (**17**) according to the procedure of: Grundke, G., Keese, W., and Rimpler, M. (1987) *Synthesis*, 1115–1116.
12. (a) Chow, Y.L., Colón, C.J., and Tan, J.N.S. (1968) *Can. J. Chem.*, **46**, 2821–2825; (b) Lunazzi, L., Cerioni, G., Foresti, E., and Macciantelli, D. (1980) *J. Chem. Soc. Perkin Trans. 2*, 717–723; (c) Johnson, F. (1968) *Chem. Rev.*, **68**, 375–413
13. (a) Garner, P. (1984) *Tetrahedron Lett.*, **25**, 5855–5858; (b) review: Jurczak, J. and Golebiowski, A. (1989) *Chem. Rev.*, **89**, 149–164
14. (a) Myers, A.G. and Kung, D.W. (1999) *J. Am. Chem. Soc.*, **121**, 8401–8402; (b) Myers, A.G., Zhong, B., Movassaghi, M., Kung, D.W., Lanman, B.A., and Kwon, S. (2000) *Tetrahedron Lett.*, **41**, 1359–1362; (c) Myers, A.G., Kung, D.W., and Zhong, B. (2000) *J. Am. Chem. Soc.*, **122**, 3236–3237; (d) Myers, A.G., Zhong, B., Kung, D.W., Movassaghi, M., Lanman, B.A., and Kwon, S. (2000) *Org. Lett.*, **2**, 3337–3340
15. (a) Myers, A.G. and Kung, D.W. (1999) *J. Am. Chem. Soc.*, **121**, 10828–10829; (b) Myers, A.G. and Kung, D.W. (2000) *Org. Lett.*, **2**, 3019–3022; (c) Myers, A.G.

and Lanman, B.A. (2002) *J. Am. Chem. Soc.*, **44**, 12969–12971; (d) Charest, M.G., Lerner, C.D., Brubaker, J.D., Siegel, D.R., and Myers, A.G. (2005) *Science*, **308**, 395–398; (e) Charest, M.G., Siegel, D.R., and Myers, A.G. (2005) *J. Am. Chem. Soc.*, **127**, 8292–8293; (f) Kwon, S. and Myers, A.G. (2005) *J. Am. Chem. Soc.*, **127**, 16796–16797.

16. Snider, B.B. and Lin, H. (2000) *Org. Lett.*, **2**, 643–646.
17. (a) Davis, F.A. and Haque, M.S. (1986) *J. Org. Chem.*, **51**, 4083–4085; (b) Davis, F.A. and Chen, B.-C. (1993) *Chem. Rev.*, **92**, 919–934.
18. (a) Tamura, Y., Yakura, T., Haruta, J.-I., and Kita, Y. (1987) *J. Org. Chem.*, **52**, 3927–3930; See also: (b) Kita, Y., Tohma, H., Kikuchi, K., Inagaki, M., and Yakura, T. (1991) *J. Org. Chem.*, **56**, 435–438.
19. (a) Review: Ciufolini, M.A., Braun, N.A., Canesi, S., Ousmer, M., Chang, J., and Chai, D. (2007) *Synthesis*, 3759–3772.
20. Knapp, S. (1996) in *Advances in Heterocyclic Natural Product Synthesis*, vol. 3 (ed. W.H. Pearson), JAI Press, Greenwich, pp. 57–98.
21. Oxazolines also participate effectively in iodolactamization reactions: Kurth, M.J. and Bloom, S.H. (1989) *J. Org. Chem.*, **54**, 411–414.
22. (a) Vorbrüggen, H. and Krolikiewicz, K. (1981) *Tetrahedron Lett.*, **22**, 4471–4474; (b) Vorbrüggen, H. and Krolikiewicz, K. (1993) *Tetrahedron*, **49**, 9353–9372.
23. For example, the Wipf method: (a) Wipf, P. and Miller, C.P. (1992) *Tetrahedron Lett.*, **33**, 907–910; (b) Wipf, P., Kim, Y., and Goldstein, D.M. (1995) *J. Am. Chem. Soc.*, **117**, 11106–11112; (c) Wipf, P. and Li, W. (1999) *J. Org. Chem.*, **64**, 4576–4577.
24. Kita, Y., Takada, T., Gyoten, M., Tohma, H., Zenk, M.H., and Eichhorn, J. (1996) *J. Org. Chem.*, **61**, 5857–5864, see also Reference [17].
25. (a) Braun, N.A., Ciufolini, M.A., Peters, K., and Peters, E.-M. (1998) *Tetrahedron Lett.*, **39**, 4667–4670; (b) Braun, N.A., Ousmer, M., Bray, J.D., Bouchu, D., Peters, K., Peters, E.-M., and Ciufolini, M.A. (2000) *J. Org. Chem.*, **65**, 4397–4408.
26. (a) Canesi, S., Belmont, P., Bouchu, D., Rousset, L., and Ciufolini, M.A. (2002) *Tetrahedron Lett.*, **43**, 5193–5195; (b) Canesi, S., Bouchu, D., and Ciufolini, M.A. (2004) *Angew. Chem. Int. Ed.*, **43**, 4336–4338.
27. (a) Canesi, S., Bouchu, D., and Ciufolini, M.A. (2005) *Org. Lett.*, **7**, 175–177; (b) Liang, H. and Ciufolini, M.A. (2008) *J. Org. Chem.*, **73**, 4299–4301; (c) Mendelsohn, B., Lee, S., Kim, S., Teyssier, F., Aulakh, V.S., and Ciufolini, M.A. (2009) *Org. Lett.*, **11**, 1539–1542; (d) Mendelsohn, B.A. and Ciufolini, M.A. (2009) *Org. Lett.*, **11**, 4736–4739.
28. (a) Scheffler, G., Seike, H., and Sorensen, E.J. (2000) *Angew. Chem. Int. Ed.*, **39**, 4593–4596.
29. Mizutani, H., Takayama, J., Soeda, Y., and Honda, T. (2002) *Tetrahedron Lett.*, **43**, 2411–2414.
30. Canesi, S. (2004) La Cylindricine C: Synthèse et Méthodologie, Dissertation, University Claude Bernard Lyon 1.
31. (a) Glover, S.A., Goosen, A., McCleland, C.W., and Schoonraad, J.L. (1984) *J. Chem. Soc. Perkin Trans. 1*, 2255–2260; (b) Glover, S.A. and Scott, A.P. (1989) *Tetrahedron*, **45**, 1763–1776.
32. (a) Kikugawa, Y. and Kawase, M. (1984) *J. Am. Chem. Soc.*, **106**, 5728–5729; (b) Kawase, M., Kitamura, T., and Kikugawa, Y. (1989) *J. Org. Chem.*, **54**, 3394–3403; (c) Kikugawa, Y. and Kawase, M. (1990) *Chem. Lett.*, 581–584; (d) Kikugawa, Y., Shimada, M., and Matsumoto, K. (1994) *Heterocycles*, **37**, 293–301.
33. (a) Wardrop, D.J., Burge, M.S., Zhang, W., and Ortiz, J.A. (2003) *Tetrahedron Lett.*, **44**, 2587–2591; (b) Wardrop, D.J., Landrie, C.L., and Ortiz, J.A. (2003) *Synlett*, 1352–1354; (c) Wardrop, D.J. and Burge, M.S. (2004) *Chem. Commun.*, 1230–1231; (d) Wardrop, D.J., Zhang, W., and Landrie, C.L. (2004) *Tetrahedron Lett.*, **45**, 4229–4231; (e) Wardrop, D.J. and Burge, M.S. (2005) *J. Org. Chem.*, **70**, 10271–10284; (f) Wardrop, D.J., Bowen, E.G., Forslund, R.E., Sussman, A.D., and Weerasekera, S.L. (2010) *J. Am. Chem. Soc.*, **132**, 1188–1189.

34. (a) Wardrop, D.J. and Basak, A. (2001) *Org. Lett.*, **3**, 1053–1056.
35. Moritani, Y., Appella, D.H., Jurkauska, V., and Buchwald, S.L. (2000) *J. Am. Chem. Soc.*, **122**, 6797–6798.
36. (a) Mizutani, H., Takayama, J., and Honda, T. (2005) *Synlett*, 328–330; (b) Mizutani, H., Takayama, J., Soeda, Y., and Honda, T. (2004) *Heterocycles*, **62**, 343–355.
37. (a) Ousmer, M., Braun, N.A., and Ciufolini, M.A. (2001) *Org. Lett.*, **3**, 765–767; (b) Ousmer, M., Braun, N.A., Bavoux, C., Perrin, M., and Ciufolini, M.A. (2001) *J. Am. Chem. Soc.*, **123**, 7534–7538; (c) Ciufolini, M.A. (2005) *Il Farmaco*, **60**, 627–641; (d) Ciufolini, M.A., Canesi, S., Ousmer, M., and Braun, N.A. (2006) *Tetrahedron*, **62**, 5318–5337.
38. Fischer, E. and Lipschitz, W. (1915) *Ber. Dtsch. Chem. Ges.*, **48**, 360–378.
39. McChesney, E.V. and Swann, W.K. Jr. (1937) *J. Am. Chem. Soc.*, **59**, 1116–1118.
40. Kan, T. and Fukuyama, T. (2004) *Chem. Commun.*, 353–359.
41. (a) Abarbri, M., Guignard, A., and Lamant, M. (1995) *Helv. Chim. Acta*, **78**, 109–121; (b) Jung, M.E., Jachiet, D., and Rohloff, J.R. (1989) *Tetrahedron Lett.*, **30**, 4211–4214, the material thus produced was essentially racemic.
42. Ley, S.V., Normand, J., Griffith, W.P., and Marsden, S.P. (1994) *Synthesis*, 639–666.
43. Dong, Q., Anderson, C.E., and Ciufolini, M.A. (1995) *Tetrahedron Lett.*, **36**, 5681–5682.
44. Seebach, D. and Golinski, J. (1981) *Helv. Chim. Acta*, **64**, 1413–1423.
45. *CRC Handbook of Chemistry and Physics* 59th edn (ed. R. C. Weast) (1978) CRC Press, Boca Raton, Florida.

3
Biomimetic Synthesis of Alkaloids Derived from Tryptophan: Indolemonoterpene Alkaloids

Sylvie Michel and François Tillequin

3.1
Introduction

3.1.1
Indolemonoterpene Alkaloids

Tryptophan-derived indolemonoterpene and related quinoline alkaloids represent a group of circa 3000 different compounds isolated to date from higher plants [1]. Many are active principles used in therapeutics that were introduced in clinics during the nineteenth century, following their isolation from plants that were used in medicine at that time, such as quinine (**1**) from *Cinchona* bark or strychnine (**2**) from *Strychnos nux-vomica* seeds (Figure 3.1). The isolation, in the 1950s of the anti-psychotic and anti-hypertensive reserpine (**3**) from *Rauwolfia serpentina*, followed by that of the anticancer vinblastine (**4**) from the aerial parts of *Catharanthus roseus* stimulated a new expansion in research on indolemonoterpene alkaloids, including exploration of the biosynthetic pathways giving rise to these natural products [2] and, since the early 1970s, the biomimetic syntheses of these compounds [3], that is, synthetic attempts toward their assembly inspired by the biosynthetic lines, without recourse to the enzymatic machinery present in Nature. Following a brief overview of the structures and botanical distribution of indolemonoterpene alkaloids, this chapter will describe a few selected highlights concerning their biomimetic synthesis, chosen to illustrate useful methods and significant achievements toward the obtainment of the main skeletons encountered in this series of natural products.

3.1.2
Classification and Botanical Distribution

Only a few indolemonoterpene alkaloids, which have been almost exclusively obtained from species belonging to the genus *Aristotelia*, arise from the condensation of tryptamine with a non-rearranged (geranyl) monoterpene unit [4]. Most of them derive, from a biosynthetic point of view, from the initial condensation of a

Biomimetic Organic Synthesis, First Edition. Edited by Erwan Poupon and Bastien Nay.

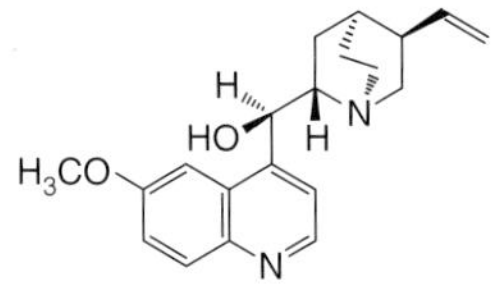

1 Quinine

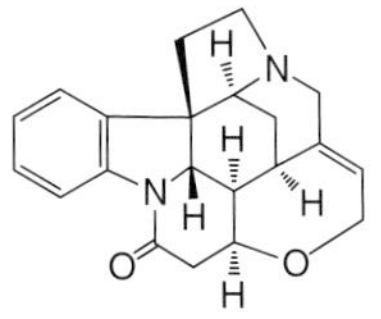

2 Strychnine

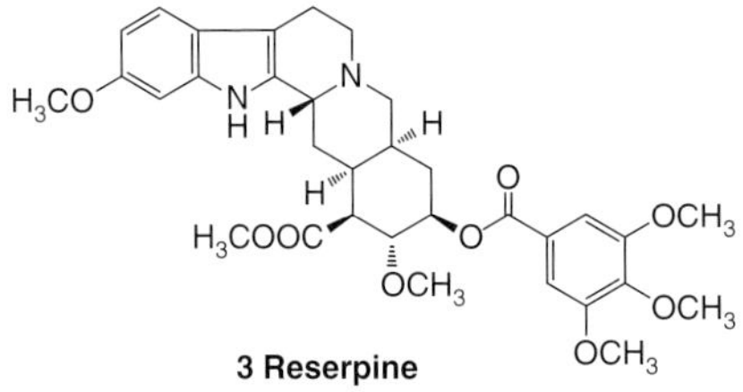

3 Reserpine

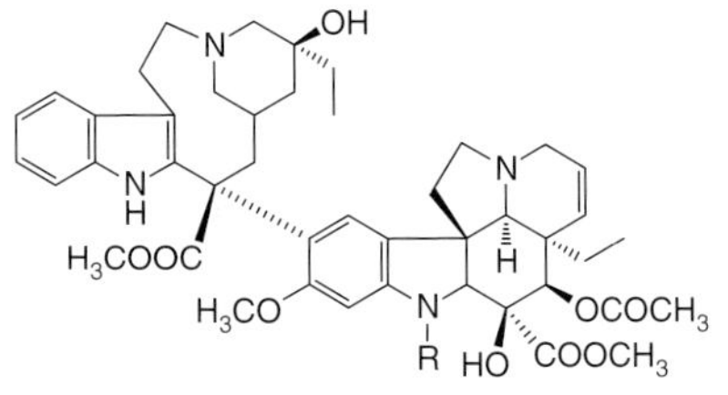

4 R = CH_3 **Vinblastine**
11 R = CHO **Vincristine**

Figure 3.1 Selected examples of indolemonoterpene alkaloids.

tryptamine (**5**) unit with the iridoid glycoside secologanin (**6**), under the influence of the enzyme strictosidine synthase, to give stereospecifically strictosidine (**7**) (Scheme 3.1) [5].

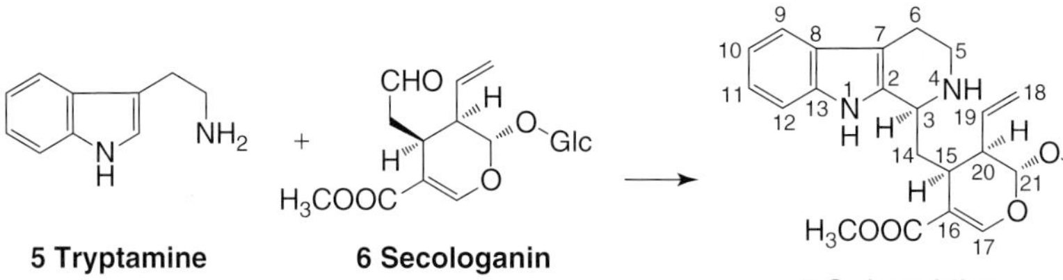

5 Tryptamine **6 Secologanin** **7 Strictosidine**

Scheme 3.1 Biosynthesis of strictosidine (**7**).

This glycoside is the common precursor to most indolemonoterpene alkaloids, despite the impressive array of their final structures. These alkaloids are essentially encountered in plant families belonging to the order Gentianales (Apocynaceae, Loganiaceae, and Rubiaceae) and in some species belonging to the related order Cornales (Cornaceae, Nyssaceae, Alangiaceae...).

Compounds of this group with a terpene moiety remaining in an unrearranged form constitute the *Corynanthe* and *Strychnos* alkaloids (corynane-strychnane or type I series), whereas rearrangement of the terpenoid part gives rise to the *Iboga* alkaloids (ibogane or type II series, with fragmentation of the C15–C16 bond and creation of a C14–C17 bond) and the *Aspidosperma* alkaloids (aspidospermane or type III series, with fragmentation of the C15–C16 bond and creation of a C17–C20 bond) (Scheme 3.2) [6].

Rearrangements permitting ring expansion lead to quinoline alkaloids, exemplified by camptothecin (**8**) and quinine (**1**), whereas ring cleavage leads to ring-opened alkaloids such as rhazinilam (**9**) (Figure 3.2).

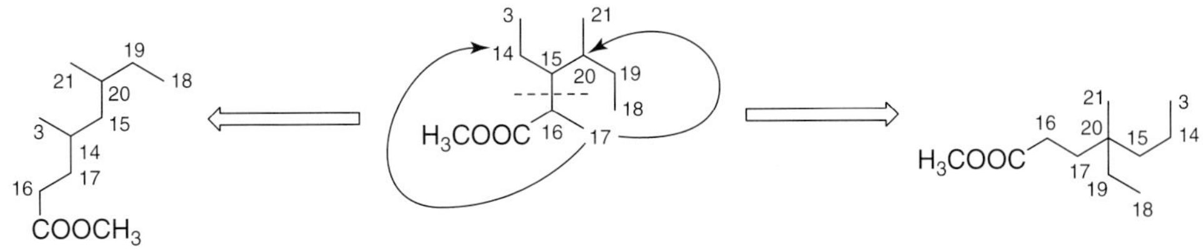

Iboga-type alkaloids | *Corynanthe* and *Strychnos*-type alkaloids | *Aspidosperma*-type alkaloids

Scheme 3.2 Main skeletons of secologanin-derived indolemonoterpene alkaloids.

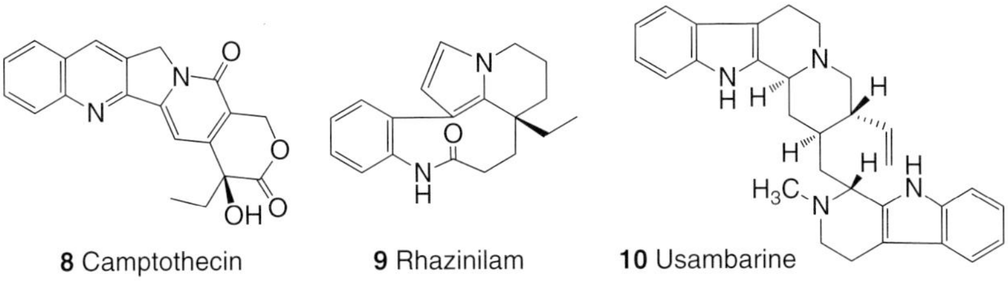

8 Camptothecin **9** Rhazinilam **10** Usambarine

Figure 3.2 Structures of camptothecin (**8**), rhazinilam (**9**), and usambarine (**10**).

Subsequent condensation of an indolemonoterpene with a second tryptamine unit gives rise to quasi-dimeric alkaloids, exemplified by usambarine (**10**). Dimerization involving two indolomonoterpene units leads to dimeric alkaloids, illustrated by vinblastine (**4**) and vincristine (**11**), both isolated from the aerial parts of *Catharanthus roseus* and currently widely used in cancer chemotherapy.

3.2
Biomimetic Synthesis of Indolomonoterpene Alkaloids with a Non-rearranged Monoterpene Unit: *Aristotelia* Alkaloids

The *Aristotelia* alkaloids constitute a small group of some 50 alkaloids. All of them have been isolated in minute amounts from plants belonging to the eleocarpaceous genus *Aristotelia* indigenous in the Southern hemisphere: Australia, New Zealand, and Chile.

A biogenetic scheme leading to *Aristotelia* alkaloids, through a linkage between tryptamine and an unarranged monoterpene unit such as geraniol, was first hypothesized by Bick [7] and extended later on by Bick and Hesse [8]. Accordingly, nucleophilic attack of tryptamine (**5**) on the α-terpinyl cation arising from the cyclization of linalyl or neryl diphosphate should give rise to an α-terpinyltryptamine (**12**) unit possessing the same skeleton as the natural alkaloid (+)-fruticosimine (**13**), which only differs in its oxidation level. Dehydrogenation of the putative α-terpinyltryptamine unit **12** should generate an aldimine intermediate (**14**), which should cyclize upon protonation to makomakine (**15**), aristoteline (**16**), or hobartine (**17**), which are the major alkaloids isolated from *Aristotelia* species (Scheme 3.3).

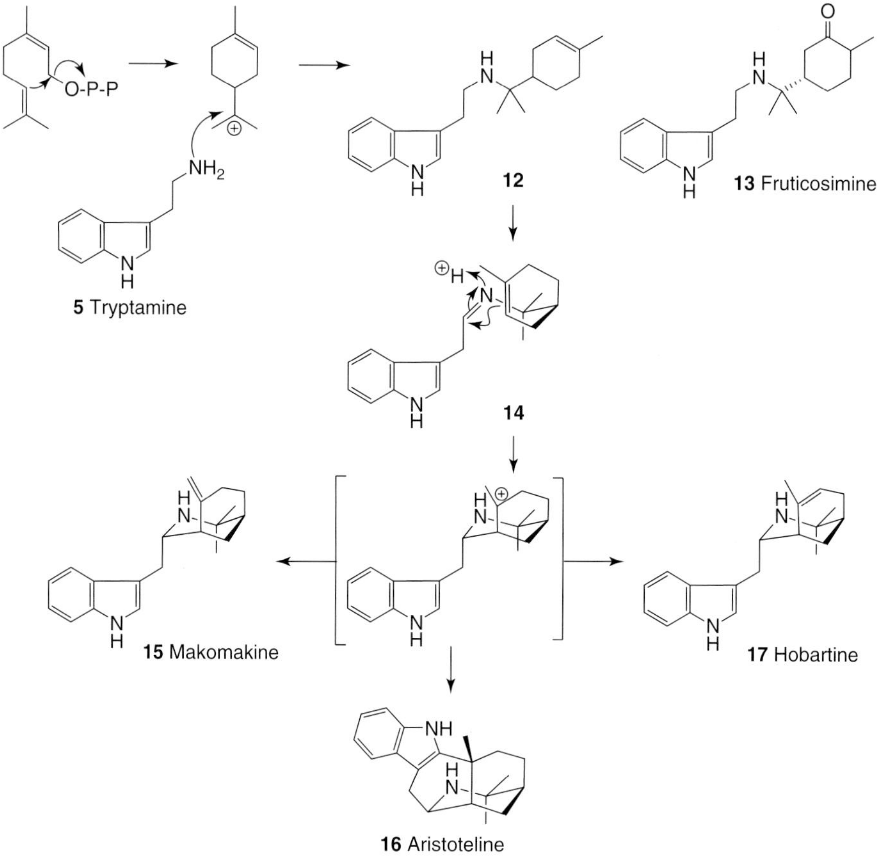

Scheme 3.3 Biosynthesis of *Aristotelia* alkaloids.

Following a pioneering approach by Lévy and coworkers involving isatin as source of indole nucleus [9], the first fully biomimetic synthesis of the *Aristotelia* alkaloids makomakine (**15**), aristoteline (**16**), and hobartine (**17**) was performed by Stevens and Kenney [10]. Thus, mercury(II) nitrate-initiated Ritter reaction [11] of indol-3-ylacetonitrile (**18**), bringing the carbon and nitrogen atoms of tryptamine, with (−)-β-pinene (**19**) gave the corresponding imine, which was reduced with sodium borohydride in alkaline medium to afford (+)-makomakine (**15**). Further treatment of **15** with concentrated hydrochloric acid gave (+)-aristoteline (**16**) (Scheme 3.4). Similarly, Ritter condensation of (+)-α-pinene (**20**) with **18** gave (±)-hobartine (**17**) (Scheme 3.5). Notably, the enantiospecificity observed in the synthesis of **15** and **16** was not observed in that of **17**, since the allylic mercurial intermediate can cyclize at either of the two enantiomeric sites in the latter case (Scheme 3.6).

Following the work of Stevens, several syntheses of *Aristotelia* alkaloids were performed, through an initial condensation of *(S)*-(−)-α-terpinylamine (**21**)

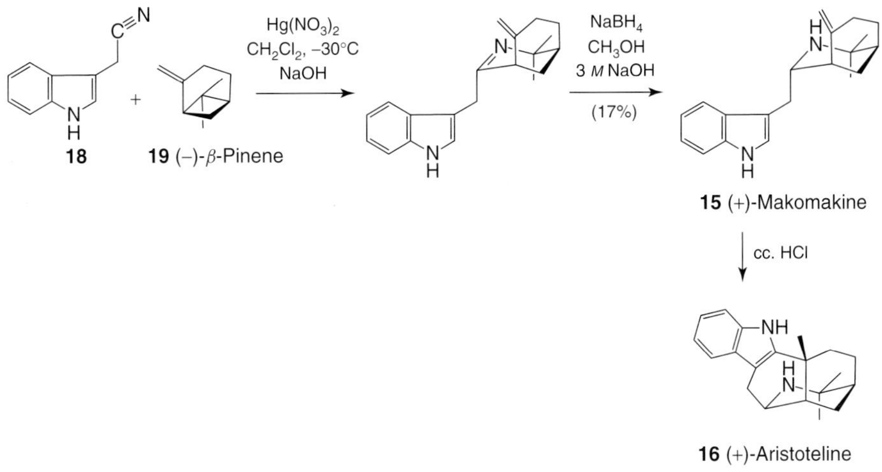

Scheme 3.4 Biomimetic synthesis of (+)-makomakine (**15**) and (+)-aristoteline (**16**).

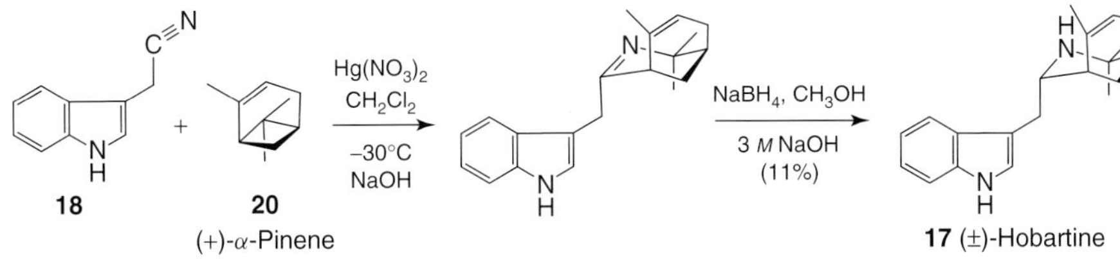

Scheme 3.5 Biomimetic synthesis of (±)-hobartine (**17**).

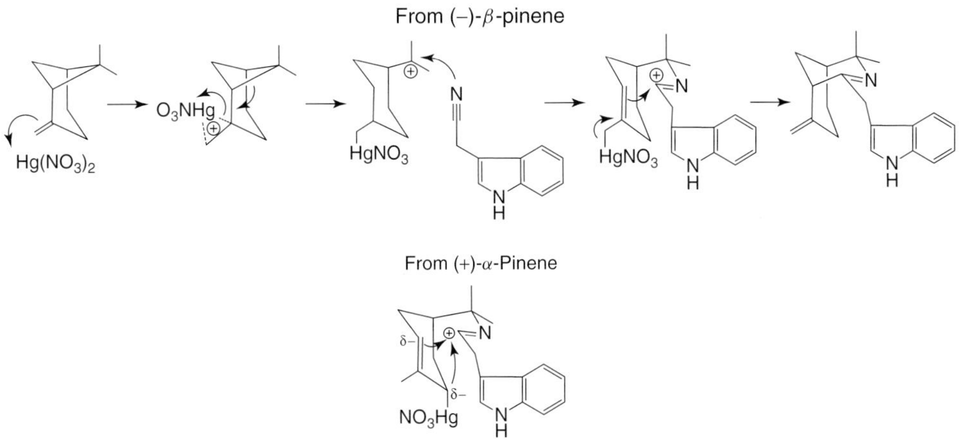

Scheme 3.6 Comparison of the mechanisms involved in the syntheses of (+)-makomakine (**15**) and (+)-aristoteline (**16**) from (−)-β-pinene, and that of (±)-hobartine (**17**) from (+)-α-pinene.

with various indole substrates, including tryptophyl bromide (**22**) [12] and indol-3-ylacetaldehyde (**23**) (Figure 3.3) [13].

This latter approach proved particularly successful, since most of the *Aristotelia* alkaloids isolated to date could be prepared in the group of Borschberg [14], by condensation of unprotected or protected indol-3-ylacetaldehyde (**23**) with α-terpinylamine (**21**), terpinyl-derived amino-alcohols, or synthetic equivalents such as phenylthioterpinylamines. The use of these latter reagents permitted the obtainment of *Aristotelia* alkaloids at various oxidation levels, exemplified by aristoserratine (**24**) [15], aristofruticosine (**25**) [16], and serratenone (**26**) (Figure 3.4) [17, 18].

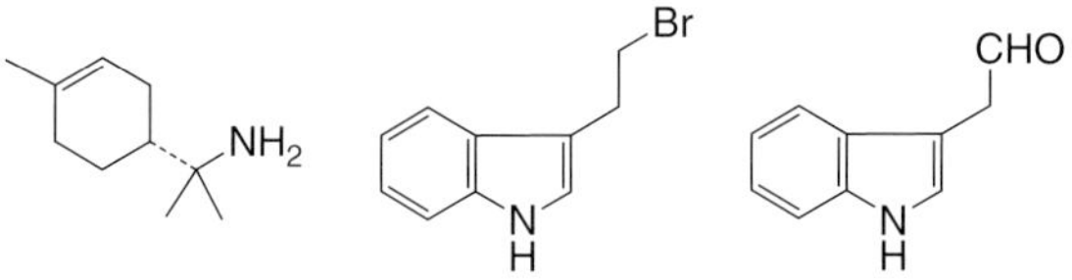

21 **22** **23**

Figure 3.3 Structures of *(S)*-(−)-α-terpinlamine (**21**), trptophl bromide (**22**), and indol-3-ylacetaldehyde (**23**).

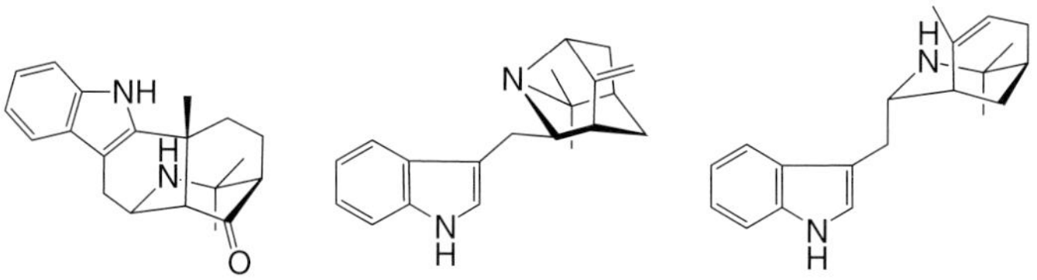

24 Aristoserratine **25** Aristofruticosine **26** Serratenone

Figure 3.4 Structures of aristoserratine (**24**), aristofruticosine (**25**), and serratenone (**26**).

3.3
Biomimetic Synthesis of Secologanin-Derived Indolomonoterpene Alkaloids

3.3.1
Strictosidine, Vincoside, and Simple *Corynanthe* Alkaloids: Heteroyohimbines and Yohimbines

The isolation of the iridoid secologanin (**6**) in large amounts from *Lonicera* species permitted the development of the pioneering work performed in the group of R.T. Brown towards the biomimetic synthesis of indolomonoterpene alkaloids [19]. The first achievements, taking place in the early 1970s, implied the condensation of tryptamine (**5**) with secologanin (**6**), to give strictosidine (**7**) and/or its C3 epimer vincoside (**27**), followed by the hydrolysis of the sugar moiety with β-glucosidase and subsequent rearrangement and/or reduction to give monoterpenoid indole alkaloids (Figure 3.5). However, this initial approach suffered several drawbacks, such as the facile reaction of N4 of **7** and **27** with the carbomethoxy group at C16 to give the corresponding lactams.

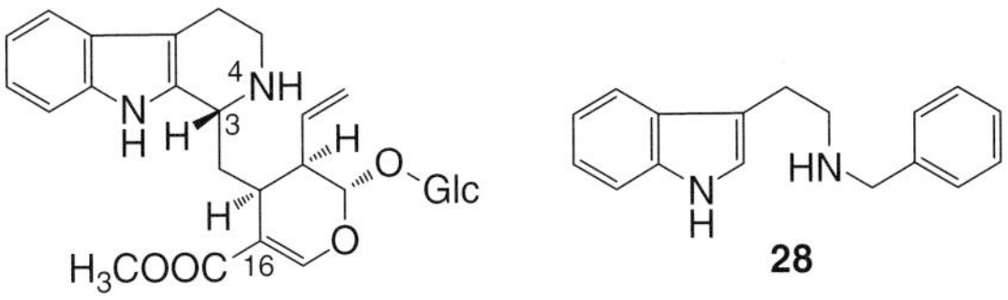

27 Vincoside

28

Figure 3.5 Structures of vincoside (**27**) and N_b-benzyltrptamine (**28**).

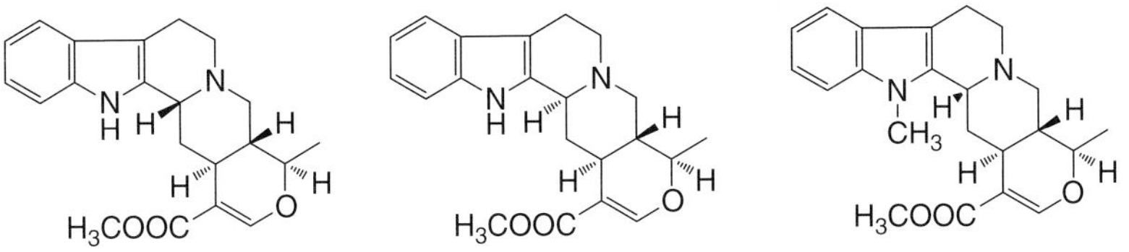

29 3-Iso-19-epiajmalicine **30** 19-Epiajmalicine **31** 1-Methyltetrahydroalstonine

Figure 3.6 Structures of 3-iso-19-epiajmalicine (**29**), 19-epiajmalicine (**30**), and 1-methyltetrahydroalstonine (**31**).

A benzyl protecting group at N4, introduced by using N_b-benzyltryptamine (**28**) as starting material, circumvented this problem and afforded the first biomimetic syntheses of several heteroyohimbine alkaloids, exemplified by 3-iso-19-epiajmalicine (**29**), 19-epiajmalicine (**30**) [20], and 1-methyltetrahydroalstonine (**31**) (Figure 3.6) [21].

Finally, a more efficient and elegant approach, developed more recently in the same group, implied an alteration of the initial sequence, which permitted the selective generation of a single C3 epimer and also avoided the use of the *N*-benzyl protecting group. In those biogenetically inspired syntheses, secologanin was first selectively converted into variously protected aglucone aldehydes. Pictet–Spengler condensation with tryptamine was carried out afterwards, as one of the last steps of the envisioned synthesis. This strategy proved particularly efficient, since it permitted the group of R.T. Brown to successfully prepare several series of non-rearranged indolemonoterpene alkaloids, including heteroyohimbines and yohimbines, but also anthirine derivatives, rearranged *Aspidosperma* alkaloids, and ring-expanded compounds in the camptothecin series [22].

The synthesis of ajmalicine (**32**) illustrates the approach towards heteroyohimbine alkaloids by this modified strategy. Secologanin (**6**) was first acetylated (Scheme 3.7). The free aldehyde group on the aglucone part was then protected as an ethylene acetal, by treatment with ethylene glycol in acidic medium. Zemplèn deacetylation of the sugar moiety gave the dihydropyran aldehyde **33**. When performed at pH 5, hydrolysis of the sugar part of **33** with β-glucosidase afforded **34**. Condensation of **34** with tryptamine (**5**) followed by reduction of the imine intermediate gave **35**, which was deprotected and stereospecifically cyclized to **36** in acidic medium. Finally, epimerization at C3 afforded the desired ajmalicine (**32**) (Scheme 3.7) [23].

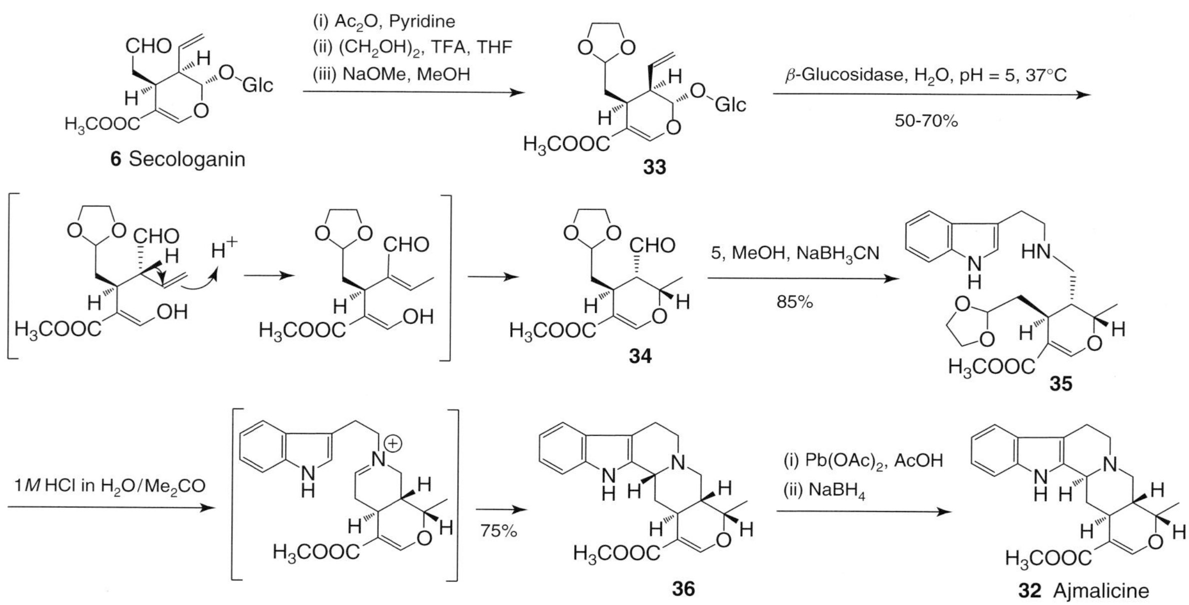

Scheme 3.7 Biomimetic synthesis of the heteroyohimbine alkaloid ajmalicine (**32**) from secologanin (**6**).

When hydrolysis of the sugar part of the dihydropyran aldehyde **33** was carried out at pH 7 in the presence of β-glucosidase, the cyclohexene carbaldehyde **37** was obtained as the major reaction product, providing an entry toward yohimbine alkaloids. Thus, reaction of **37** with tryptamine (**5**), followed by reduction, afforded **38**, which was cyclized into (−)-3-iso-19,20-dehydro-β-yohimbine (**39**) by treatment under acidic conditions [24]. This latter compound could be further converted into deserpidine analog **40** and reserpine analog precursor **41** (Scheme 3.8) [25].

3.3.2
Antirhine Derivatives

A straightforward biogenetically inspired entry to the antirhine group of alkaloids followed the observation that the treatment of secologanin ethylene acetal (**33**) with baker's yeast (*Saccharomyces cerevisiae*) at pH 6.4 gave in high yield the lactol **42**, in equilibrium with the open form **43**, through hydrolysis of the glucose unit followed by reduction of the aldehyde group at C1. Pictet–Spengler condensation with tryptamine (**5**), carried out by heating in pH 3.5 aqueous buffer and acetone, afforded stereospecifically 16-methoxycarbonyl-16,17-dehydroantirhine (**44**) as major reaction product (Scheme 3.9) [26].

For the synthesis of antirhine itself, the lactol **42** was first heated in alkaline medium to afford the decarbomethoxy lactone **45**. Condensation of **45** with tryptamine (**5**) gave the amide **46**, which was reduced to the corresponding amine with lithium aluminum hydride. Acid-catalyzed hydrolysis of the acetal protecting group occurred with simultaneous Pictet–Spengler cyclization to afford the desired antirhine (**47**) (Scheme 3.10).

3.3.3
Conversion of the *Corynanthe* Skeleton into the *Strychnos* Skeleton

The group of Stephen Martin described an elegant conversion of the *Corynanthe* into the *Strychnos* skeleton, in the continuation of an enantioselective total synthesis of geissoschizine (**48**) (Figure 3.7) [27].

In this approach, the treatment of the racemic synthetic *Corynanthe* alkaloid **49** with *tert*-butyl hypochlorite afforded a mixture of the corresponding epimeric α-and β-chloroindolenines **50** (Scheme 3.11). At this step, the addition of a Lewis acid, such as $SnCl_4$, prior to the conversion into chloroindolenines played a crucial role in obtaining correct control of the stereochemistry. Indeed, without addition of Lewis acid, the attack of the chloronium ion proceeded essentially by the less hindered convex α-face of the molecule, giving the α-chloroindolenine as major reaction product, which proved inert in the following step. Initial addition of a Lewis acid permitted complexation of the electron lone pair on the basic nitrogen atom, which is projected on the αface, ensuring an increased β attack of the oxidative reagent. More specifically, the use of tin tetrachloride permitted the formation of the β-chloroindolenine as predominant reaction product, accompanied by only trace amounts of the α-isomer. Further deprotonation in strong alkaline medium

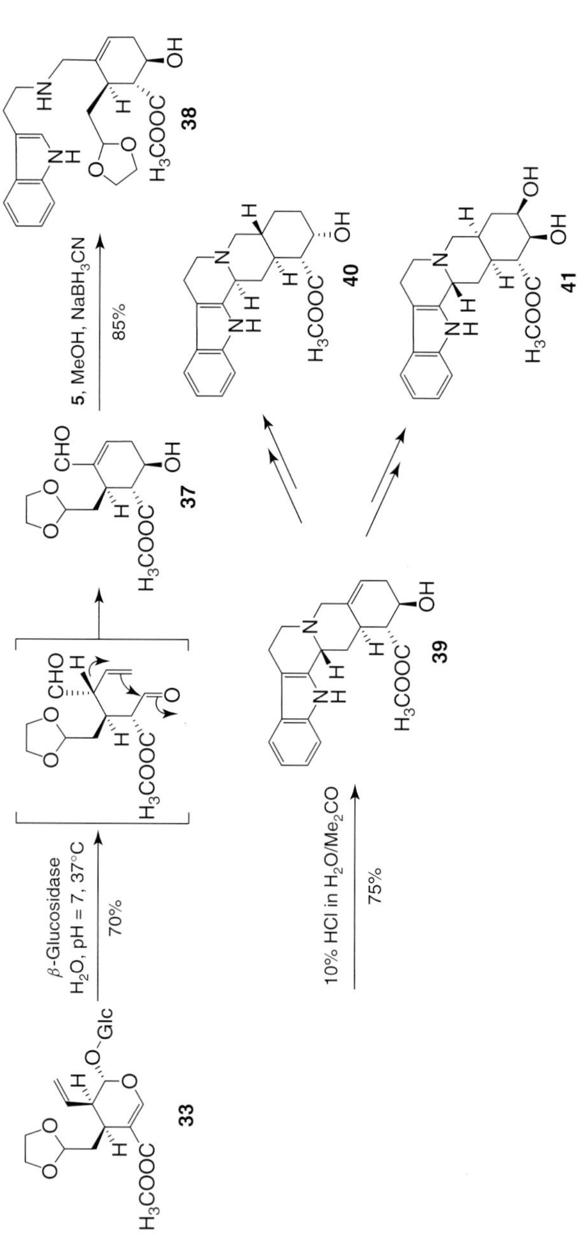

Scheme 3.8 Biomimetic synthesis of yohimbine alkaloids.

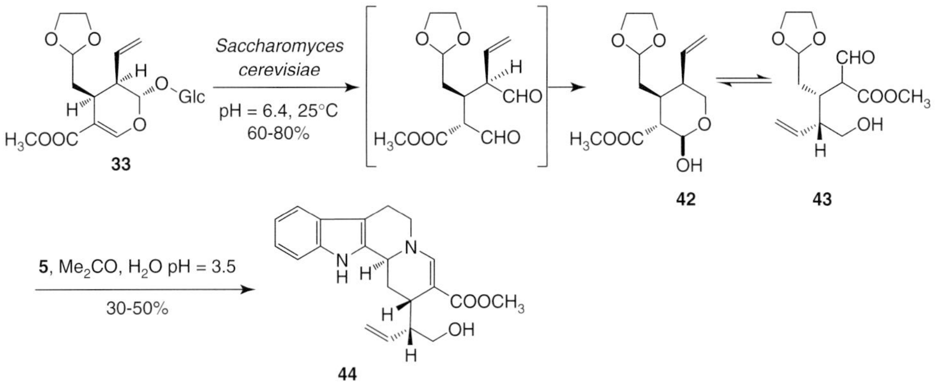

Scheme 3.9 Biomimetic synthesis of antirhine derivatives.

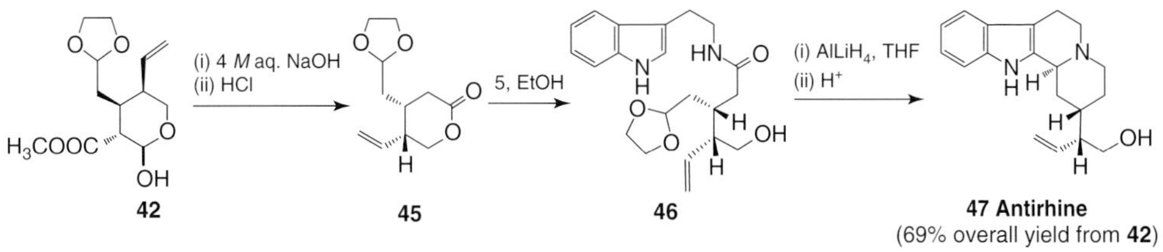

Scheme 3.10 Biomimetic synthesis of antirhine (**47**).

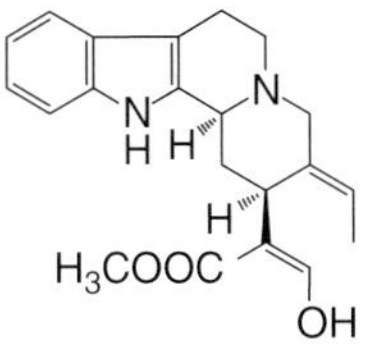

48 Geissoschizine

Figure 3.7 Structure of geissoschizine (**48**).

of the crude β-chloroindolenine at C16, to give an enolate that could cyclize onto C2, ensured the skeletal reorganization, affording the *Strychnos* alkaloid (±)-akuammicine (**51**) in a good 52% overall yield from **49** (Scheme 3.11) [28].

The same reaction sequence, starting from **52**, a protected hydroxylated analog of **49**, gave **53**, which was further deprotected to 18-hydroxyakuammicine (**54**), providing an entry to the total biomimetic synthesis of strychnine (**2**). Indeed, the conversion of **54** into strychnine (**2**) can be achieved through the intermediacies of the dihydro-derivative **55** and Wieland–Gümlich aldehyde (**56**), according to standard procedures (Scheme 3.12) [29]. More recently, the same type of intermediates also permitted a biomimetic entry to the sarpagan alkaloids [30].

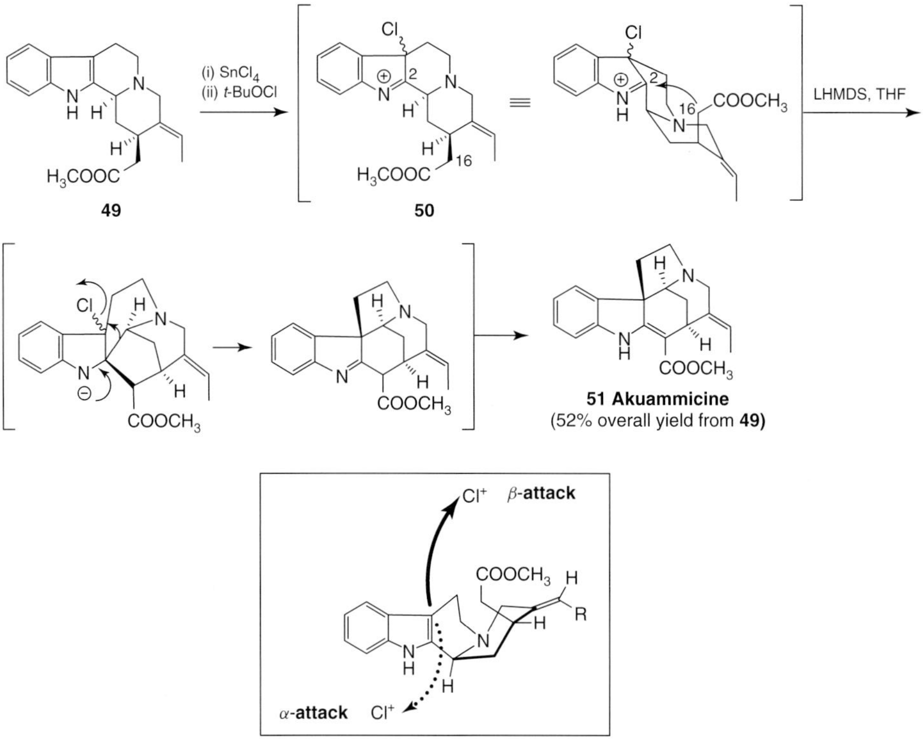

Scheme 3.11 Biomimetic synthesis of (±)-akuammicine (**51**).

3.3.4
Fragmentation and Rearrangements of *Corynanthe* Alkaloids: Ervitsine-, Ervatamine-, Olivacine-, and Ellipticine-Type Alkaloids

The use of simple or conjugated iminium ions derived from a tetrahydro- or dihydro-pyridine units as intermediates proved an efficient method to construct the backbone of several alkaloids that arise from a classical skeleton through a fragmentation reaction followed by a rearrangement, such as ervitsine, ervatamine, olivacine, ellipticine, and related alkaloids. The groups of Husson [31] and Bosch [32] have in particular studied this methodology, which reproduces the key step invoked in the biosynthesis of a large number of indolemonoterpene alkaloids [33].

For instance, the key step of the pioneering biomimetic synthesis of ellipticine (**57**) by Langlois *et al.* [34] (Scheme 3.13) involves cyclization of the imminium **58**, followed by oxidation, according to the biogenetic hypothesis previously published by Potier and Janot, which involves fragmentation of the C5–C6 bond of the *Corynanthe* alkaloid stemmadenine (**59**) followed by rearrangement [35] (Scheme 3.14).

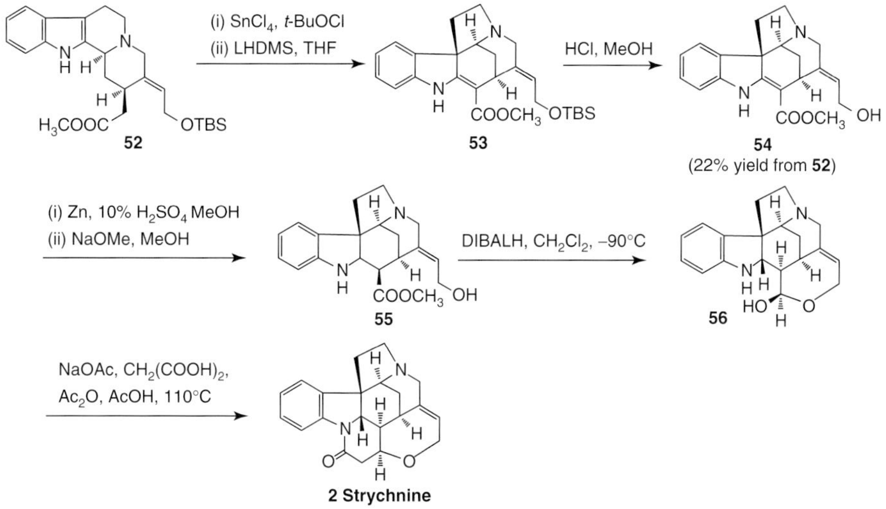

Scheme 3.12 Biomimetic synthesis of strychnine (**2**).

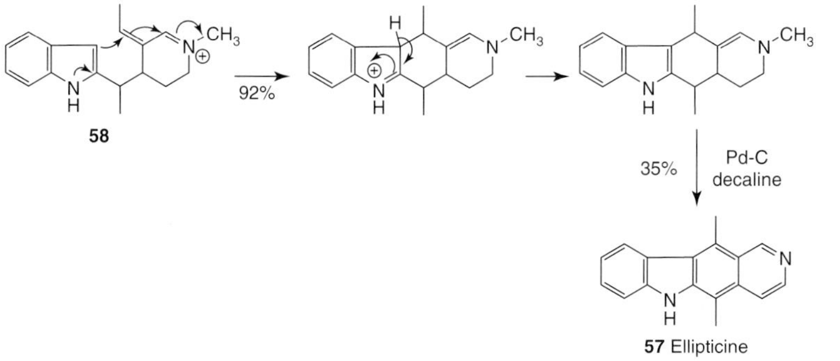

Scheme 3.13 Key step of the biomimetic synthesis of ellipticine (**57**).

The enantioselective total synthesis of $N_{(a)}$-methylervitsine (**60**) more recently developed by Bennasar *et al.* [36] is a typical example of an asymmetric version of this approach. Addition of the enolate of 2-acetyl-1-methylindole (**61**) onto the *N*-methylpyridinium salt **62**, bearing a chiral auxiliary derived from (*S*)-*O*-methylprolinol, afforded the intermediate 1,4-dihydropyridine **63**. Conversion into the imminium cation **64**, upon treatment with Eschenmoser's salt, permitted a direct cyclization into the tetracycle **65**, obtained as a mixture of diastereoisomers, in which the 15β isomer predominated. Cope elimination performed on the corresponding *N*-oxides gave methylene derivatives at C16, from which the major diastereoisomer **66** could be separated by crystallization. Removal of the chiral auxiliary required prior reduction to the secondary alcohol **67**, which afforded **68** upon reaction with methyllithium. Re-oxidation to the corresponding 2-acylindole followed by final stereoselective elaboration of the (20*E*)-ethylidene side chain gave the desired (−)-$N_{(a)}$-methylervitsine (**60**) (Scheme 3.15).

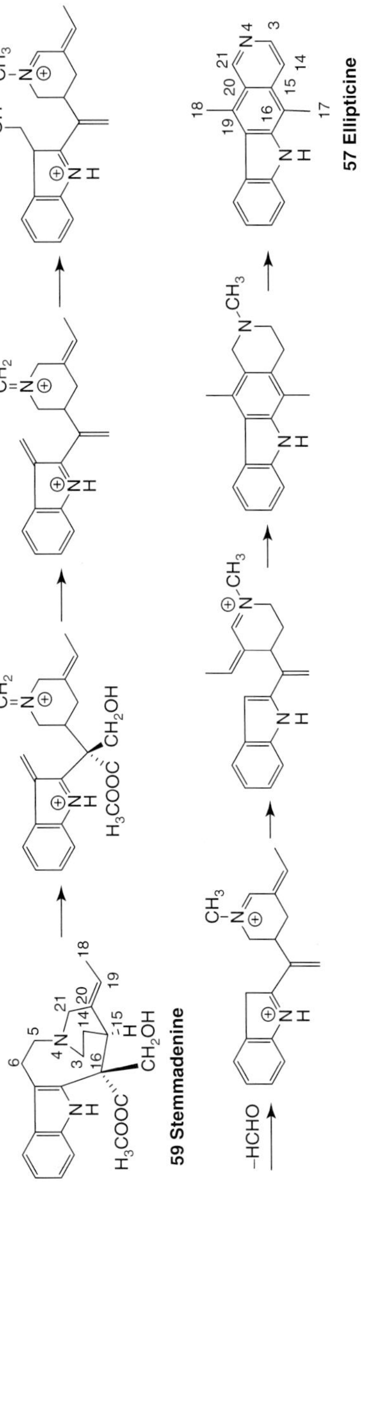

Scheme 3.14 Hypothesis of ellipticine Biosynthesis according to Potier and Janot.

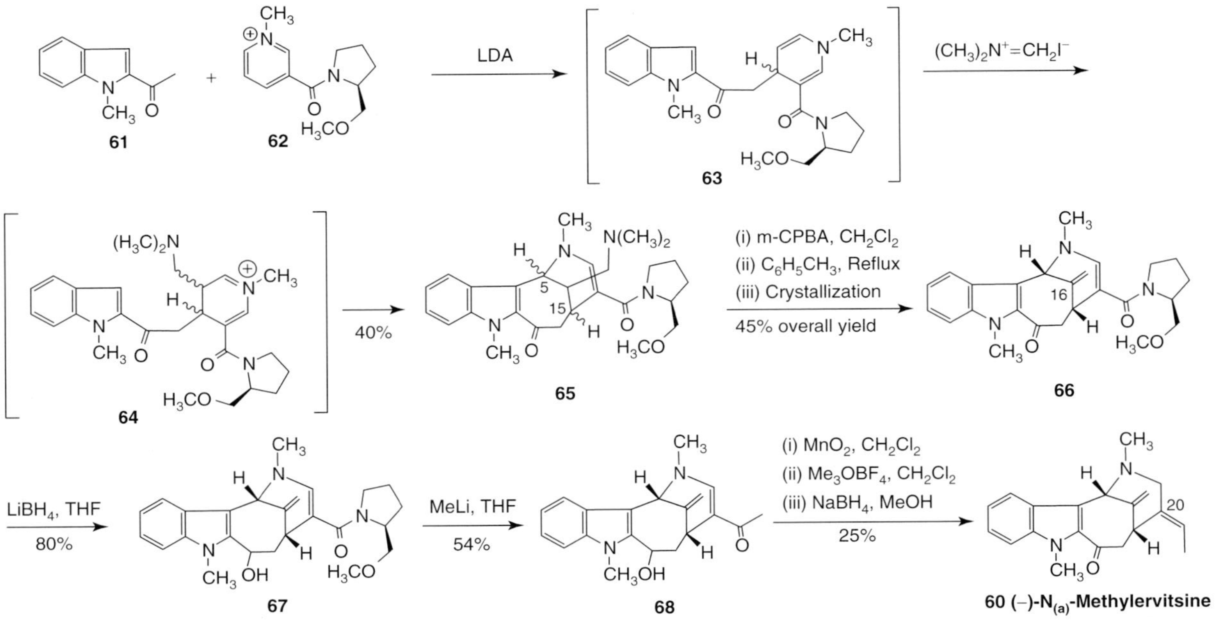

Scheme 3.15 Biomimetic synthesis of (–)-$N_{(a)}$-methylervitsine (**60**).

3.3.5
Iboga and *Aspidosperma* Alkaloids

From a biogenetic viewpoint, there is general agreement to consider that both *Iboga* and *Aspidosperma* alkaloids derive from the *Corynanthe* precursor stemmadenine (**59**), through a retro-aldol fragmentation sequence, leading to the intermediate diene dehydrosecodine (**69**), followed by a Diels–Alder type cycloaddition, which should give rise to either the ibogane skeleton, for example, catharanthine (**70**), or the aspidospermane system, exemplified by tabersonine (**71**) (Scheme 3.16) [37].

Based on this biogenetic sequence, Martin Kuehne and his group developed a highly elegant and flexible approach to *Iboga* and *Aspidosperma* alkaloids, involving dehydrosecodine synthetic equivalents as key intermediates [38]. The syntheses of catharanthine (**70**) and tabersonine (**71**), from the common 15-oxosecodine intermediate **72**, can be considered as typical examples of this synthetic strategy [39]. Indeed, keto compound **72** appears as a direct precursor of *Aspidosperma* alkaloids, whereas a stabilized derivative of its tautomeric enol form can be considered as a precursor of the *Iboga* skeleton. Thus, the oxosecodine **72** was prepared by condensation of indoloazepine **73** with 1-chloro-2-ethylpenta-1,4-dien-3-one (**74**), which was obtained by addition of vinylmagnesium bromide followed by Swern oxidation of the Vilsmeier reaction product of butyraldehyde and *N,N*-dimethylformamide (Scheme 3.17).

To construct the *Iboga* skeleton, the oxosecodine precursor **72** was treated with *t*-butyldimethylsilyl chloride or triflate in alkaline medium to give the corresponding *t*-butyldimethylsilyl enol ether, which could not be isolated but underwent spontaneous Diels–Alder cyclization to afford the catharanthine derivative **75**. Further conversion of **75** into catharanthine (**70**) was ensured by transformation into the corresponding oxocoronaridines (**76a**, **76b**) with fluoride ion, conversion into thiones **77a** and **77b** by reaction with phosphorus pentasulfide, *S*-methylation to *S*-methylcatharanthine (**78**), and final desulfurization with Raney nickel (Scheme 3.18).

To access the *Aspidosperma* series, the oxosecodine precursor **72** was heated in refluxing toluene to give 15-oxovincadifformine (**79**) as unique Diels–Alder reaction product, obtained in almost quantitative yield. Bromination of ketone **79**, followed by sodium borohydride reduction, afforded the bromohydrin **80**, which was converted into tabersonine (**71**) through a radical elimination process upon treatment with $TiCl_4$-$AlLiH_4$ (Scheme 3.19).

3.3.6
Fragmentation and Rearrangements of *Aspidosperma* Alkaloids: *Vinca* Alkaloids and Rhazinilam

Based on a biogenetic hypothesis of formation of the vincamine skeleton through oxidation of an *Aspidosperma* precursor such as vincadifformine (**81**) [40], Jean Lévy described an industrial process for the partial synthesis of the clinically interesting alkaloid vincamine from tabersonine, which can be extracted in high yield from

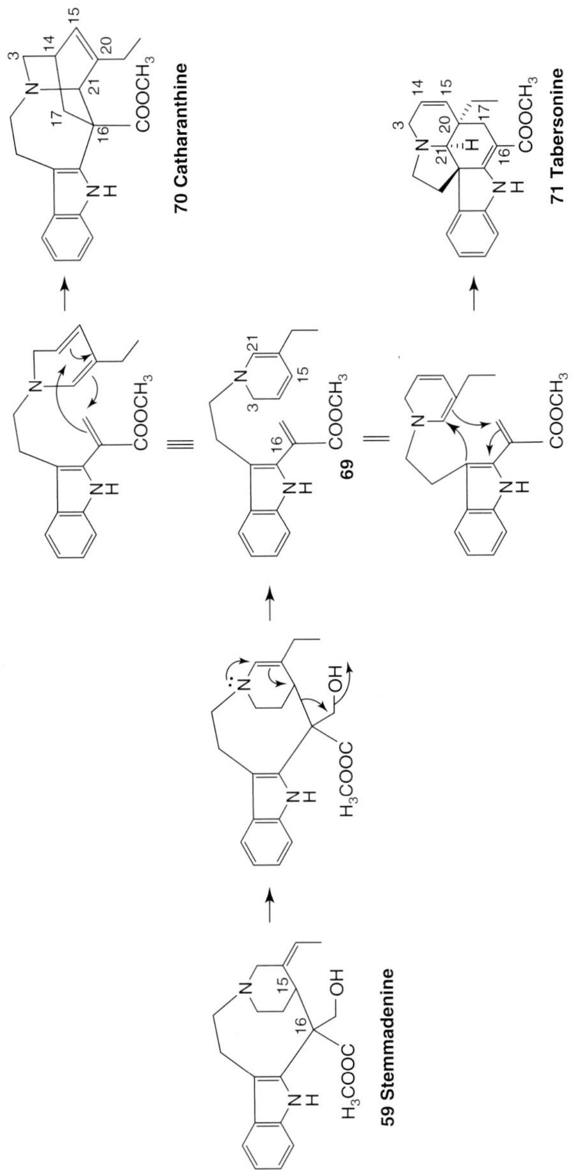

Scheme 3.16 Biosynthesis of *Iboga* and *Aspidosperma* alkaloids.

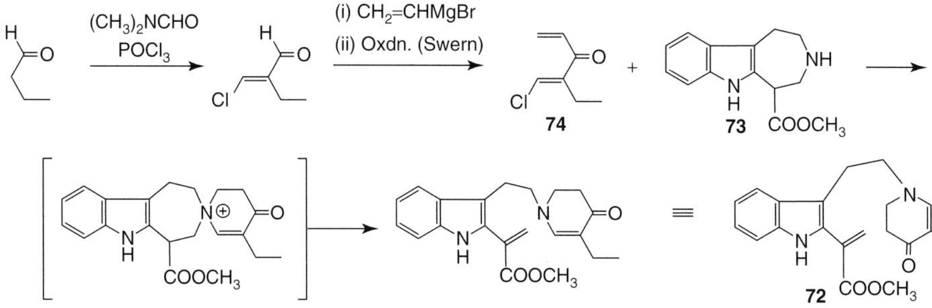

Scheme 3.17 Synthesis of the of *Iboga* and *Aspidosperma* precursor **72**.

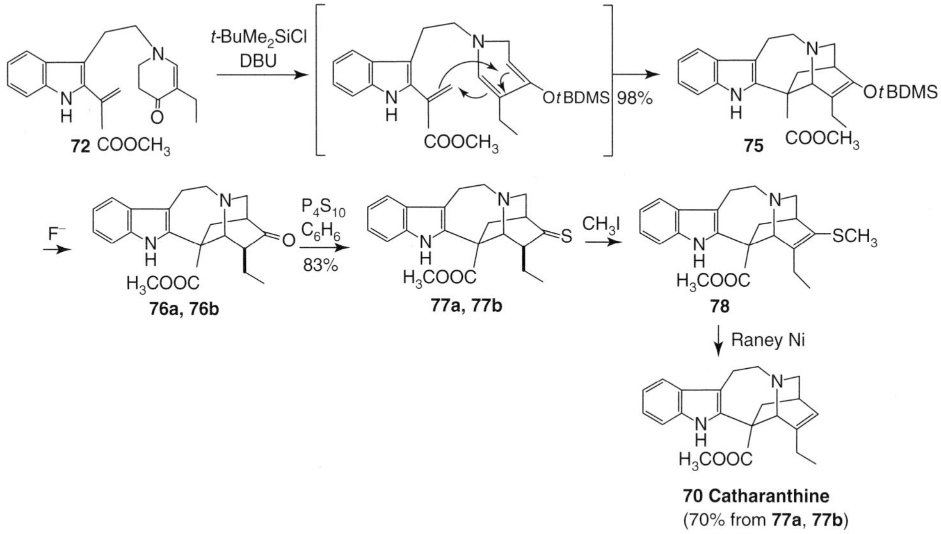

Scheme 3.18 Biomimetic synthesis of the *Iboga* alkaloid catharanthine (**70**).

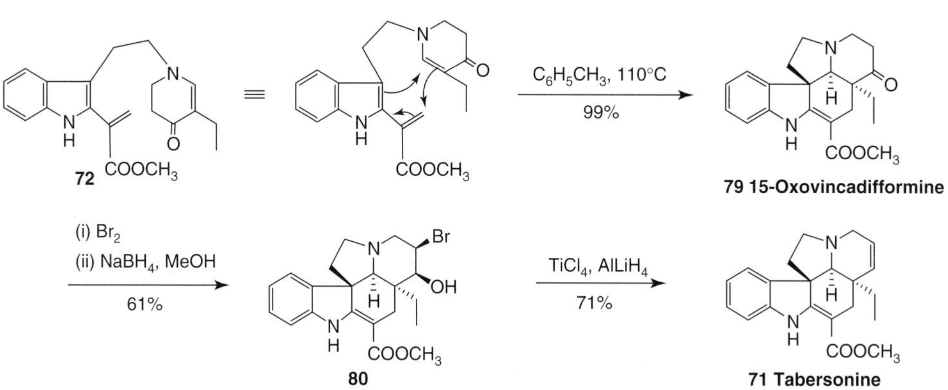

Scheme 3.19 Biomimetic synthesis of the *Aspidosperma* alkaloid tabersonine (**71**).

the seeds of the African trees *Voacanga africana* and *Voacanga thouarsii* [41]. Vincadifformine (**81**), obtained by catalytic hydrogenation of tabersonine C14 = C15 double bond, was first converted into the corresponding hydroxyindolenine *N*-oxide **82**, upon prolonged treatment with a suitable peracid. Reduction of the *N*-oxide with triphenylphosphine and rearrangement in acidic medium, performed in a one-pot procedure, smoothly afforded the desired vincamine (**83**), accompanied by its epimer at C16 (Scheme 3.20).

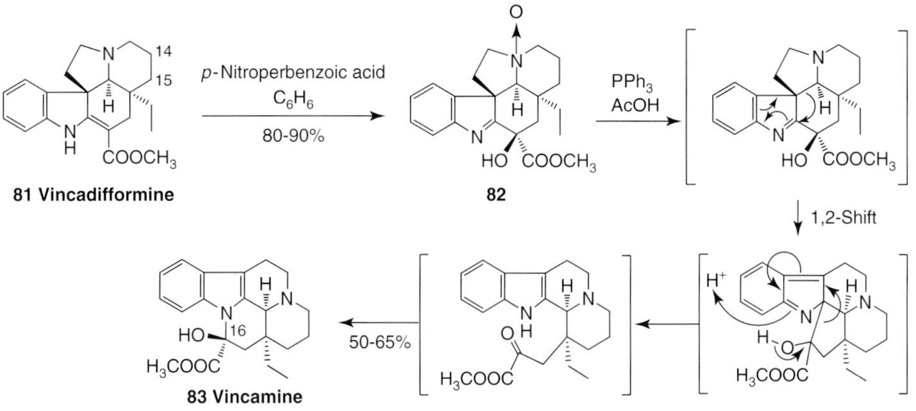

Scheme 3.20 Biomimetic conversion of vincadifformine (**81**) into vincamine (**83**).

A similar reaction sequence, performed later on starting from tabersonine (**71**), gave a mixture of 14,15-dehydrovincamine (**84**) and 16-epi-14,15-dehydrovincamine (**85**) in good yields, accompanied by smaller amounts (circa 5%) of the ring-opened *N*-oxide **86**, resulting from the oxidative cleavage of the C2–C3 bond of the indole moiety, that possessed the same chromophore as rhazinilam (**9**). Catalytic reduction of **86** to **87**, followed by saponification, decarboxylation, and lithium aluminum hydride reduction of the intermediate alcohol **88** afforded a biomimetic entry to rhazinilam (**9**) (Scheme 3.21) [42].

3.4 Biomimetic Synthesis of Secologanin-Derived Quinoline Alkaloids

The intriguing structure of the anticancer quinoline alkaloid camptothecin (**8**), isolated from the bark of *Camptotheca acuminata* (Nyssaceae) and the seeds of *Nothapodytes foetida* (Icacinaceae), was early-on shown to derive biogenetically from strictosidine (**7**), on the basis of both chemical considerations [43] and biosynthetic studies [44]. The major breakthrough towards the synthesis of the camptothecin chromophore was the observation of the facile oxidative rearrangement of indoloquinolizidine-derived lactams (e.g., **89**) to the corresponding quinoloindolizidinones (e.g., **90**), through a ring-opened intermediate (e.g., **91**) [45]. A biomimetic total synthesis of camptothecin, using tryptamine (**5**) and secologanin

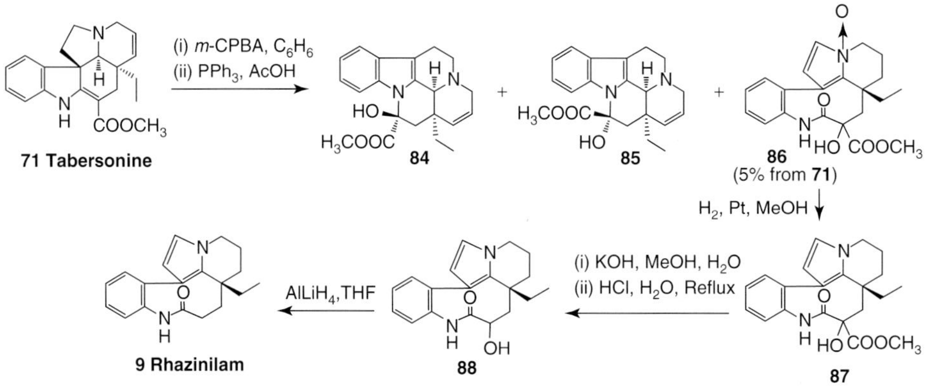

Scheme 3.21 Biomimetic conversion of tabersonine (**71**) into rhazinilam (**9**).

(**6**) as starting materials, was recently developed on the basis on this rearrangement [46]. When performed at pH 4, the condensation of **5** and **6** gave a 3 : 2 mixture of vincoside (**27**) and strictosidine (**7**). Upon heating in alkaline medium, vincoside (**27**) was converted into the corresponding lactam (**92**), which was also obtained, together with its epimer at C3, starting from the crude vincoside–strictosidine mixture. Sodium periodide oxidation, performed after protection of the sugar moiety by acetylation, gave the expected ring-opened intermediate **91**. Rearrangement of **91** in alkaline medium afforded the protected glycoside **90**, possessing the same carbon skeleton as camptothecin. Conversion of the quinolone into the corresponding quinoline was ensured by successive chlorination with thionyl chloride, catalytic hydrogenolysis and hydrogenation, and re-aromatization by use of DDQ (2,3-dichloro-5,6-dicyano-1,4-benzoquinone) to afford **93**. Deprotection of the sugar moiety, followed by treatment with β-glucosidase, gave the lactol **94**, which was oxidized to the corresponding lactone **95**. Finally, oxidation at C20 by oxygen in the presence of cupric chloride afforded the desired alkaloid camptothecin (**8**) (Scheme 3.22).

3.5
Biomimetic Synthesis of Dimeric Indolomonoterpene Alkaloids

3.5.1
Anhydrovinblastine and the Anticancer Vinblastine Series

The presence in only minute amounts (circa 5–50 g per ton of dried *Catharanthus roseus* aerial parts) and the clinical importance of the dimeric alkaloids vinblastine (**4**) and vincristine (**11**), currently used in cancer chemotherapy, strongly stimulated synthetic research toward their biomimetic synthesis, starting from the two monomeric units catharanthine (**70**) and vindoline (**96**), which can be isolated in relatively high yield (circa 1%) from the plant material. This dimerization sequence

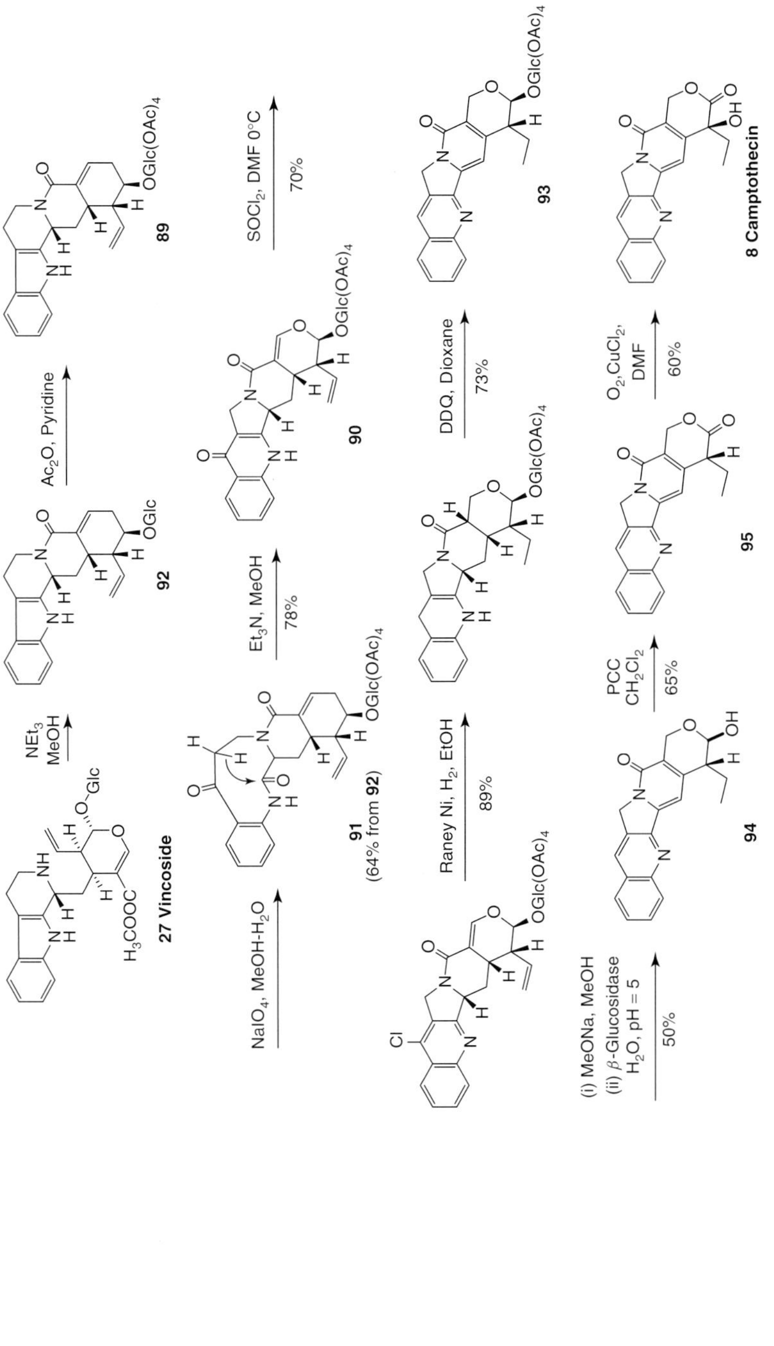

Scheme 3.22 Biomimetic synthesis of camptothecin (**8**).

was successfully conducted by the group of Pierre Potier, using a modification of the classical Polonovski fragmentation reaction. Indeed, treating a *N*-oxide with trifluoroacetic anhydride, instead of acetic anhydride in the initial sequence, permitted blocking of the reaction at the iminium stage [47, 48].

Thus, catharanthine *N*-oxide, prepared by treatment of catharanthine (**70**) with *p*-nitroperbenzoic acid, was acylated by trifluoroacetic anhydride to generate the strongly electrophilic iminium **97**. Subsequent nucleophilic attack by the electron-rich aromatic ring of vindoline (**96**), followed by reduction of the conjugated iminium, led to the desired anhydrovinblastine (**98**), having the same (16′*S*) configuration as the natural biologically active compounds, when the whole reaction sequence was performed at −50 °C (Scheme 3.23) [49].

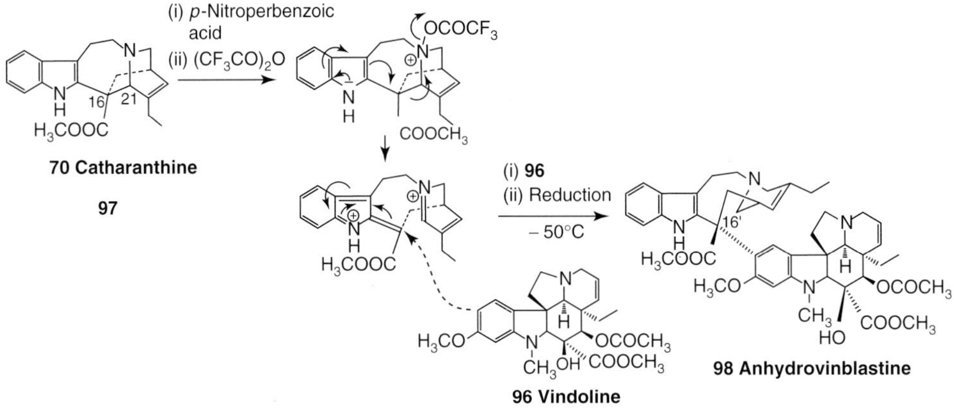

Scheme 3.23 Biomimetic synthesis of anhydrovinblastine (**98**).

In contrast, when the reaction temperature was allowed to rise to 0 °C, only the biologically inactive (16′*R*)-epimer isoanhydrovinblastine (**99**) was obtained, due to conversion of the kinetic iminium **97** into the thermodynamic conformer **100** (Scheme 3.24).

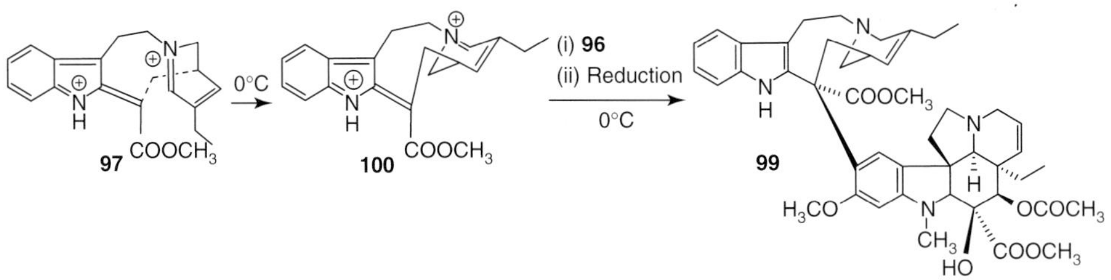

Scheme 3.24 Formation of isoanhydrovinblastine (**99**).

More recently, an efficient alternative method for coupling the two monomeric units present in anhydrovinblastine (**98**) and vinblastine (**4**) with an excellent stereocontrol of the C16′ chiral center has been developed in the group of Boger, based on the use of a $FeCl_3$ oxidation of the catharanthine unit in the presence of a non-nucleophilic cosolvent, such as 2,2,2-trifluoroethanol [50].

3.5.2
Strellidimine

Another type of dimerization mechanism is involved in the formation of strellidimine (**101**), a dimeric ellipticine alkaloid, isolated from the bark of the African Loganiaceae *Strychnos dinklagei*, together with several monomeric ellipticine derivatives, including 3,14-dihydroellipticine (**102**) and 10-hydroxyellipticine (**103**) [51]. The biomimetic synthesis of strellidimine, performed to confirm the structure of the natural alkaloid, involved first the oxidation of 10-hydroxyellipticine (**103**) to the corresponding highly electrophilic quinone-imine **104**, upon treatment with hydrogen peroxide in the presence of horseradish peroxidase (HRP). Nucleophilic attack by the basic nitrogen atom of 3,14-dihydroellipticine (**102**) followed by rearrangement afforded the desired bisindole alkaloid strellidimine (**101**) in almost quantitative yield (Scheme 3.25) [52].

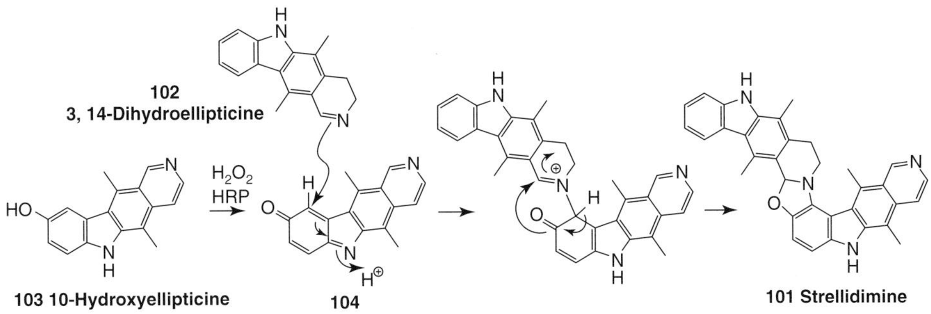

Scheme 3.25 Biomimetic synthesis of strellidimine (**101**).

3.6
Conclusion

In conclusion, the complex structures of natural compounds, particularly well illustrated by the monoterpene indole alkaloids, in connection with the versatile reactivity of both the indole nucleus and the iridoid moiety, which includes two carbon–carbon double bonds and two masked aldehyde groups, have stimulated the imagination of synthetic chemists and biochemists. The biomimetic syntheses of these compounds permit us to better understand and rationalize the mechanisms involved in their formation and remain a complex challenge for some of their representatives. In that way, Nature is a major source of inspiration for organic chemists, both in terms of reaction mechanisms and structural diversity.

References

1. (a) Saxton, J.E. (ed.) (1983) *Indoles: The Monoterpenoid Indole Alkaloids*, The Chemistry of Heterocyclic Compounds, vol. 25, Part 4, John Wiley & Sons, Ltd, Chichester; (b) Tillequin, F., Michel, S., and Seguin, E. (1993) in *Alkaloids and Sulphur Compounds* (eds P.M. Dey, J.B. Harborne, and P.G. Waterman), Methods in Plant Biochemistry, vol. 8, Academic Press, London, pp. 309–371.
2. Atta-ur-Rahman and Basha, A. (1983) *Biosynthesis of Indole Alkaloids*, Clarendon Press, Oxford.
3. Takayama, H. and Sakai, S.-I. (1998) in *The Alkaloids, Chemistry and Pharmacology*, vol. 50 (ed. G.A. Cordell), Academic Press, New York, pp. 415–452; (b) Scholz, U. and Winterfeldt, E. (2000) *Nat. Prod. Rep.*, **17**, 349–366.
4. Bick, I.R.C. and Hai, M.A. (1985) in *The Alkaloids, Chemistry and Pharmacolog*, vol. 24 (ed. A. Brossi), Academic Press, New York, pp. 113–151.
5. (a) Nagakura, N., Rüffer, M., and Zenk, M.H. (1979) *J. Chem. Soc. Perkin Trans. 1*, 2308–2312; (b) Loris, E.A., Pahjikar, S., Ruppert, M., Barleben, L., Unger, M., Schübel, H., and Stöckigt, J. (2007) *Chem. Biol.*, **14**, 979–985.
6. Le Men, J. and Taylor, W.I. (1965) *Experientia*, **21**, 508–510.
7. Bick, R.I.C., Hai, M.A., and Preston, N.W. (1979) *Heterocycles*, **12**, 1563–1565.
8. Kburz, R., Schöpp, E., Bick, R.I.C., and Hesse, M. (1981) *Helv. Chim. Acta*, **64**, 2555–2561.
9. Mirand, C., Massiot, G., and Lévy, J. (1982) *J. Org. Chem.*, **47**, 4710–4711.
10. Stevens, R.V. and Kenney, P.M. (1983) *J. Chem. Soc., Chem. Commun.*, 384–386.
11. Delpech, B. and Khuong-Huu, Q. (1978) *J. Org. Chem.*, **43**, 4898–4900.
12. Gribble, G.W. and Barden, T.C. (1985) *J. Org. Chem.*, **50**, 5900–5902.
13. Darbre, T., Nussbaumer, H.-J., and Borschberg, H.-J. (1984) *Helv. Chim. Acta*, **67**, 1040–1052.
14. (a) Borschberg, H.-J. (1991) *Chimia*, **45**, 329–341; (b) Borschberg, H.-J. (1992) in *Studies in Natural Products Chemistry*, vol. 11 (ed. Atta-ur-Rahman Elsevier, Amsterdam, pp. 277–334; (c) Borschberg, H.-J. (1994) in *Indoles: The Monoterpenoid Indole Alkaloids*, The Chemistry of Heterocyclic Compounds, vol. 25, Part 4, Supplementary Volume (ed. J.E. Saxton John Wiley & Sons, Ltd, Chichester, pp. 15–56; (d) Borschberg, H.-J. (1996) in *The Alkaloids, Chemistry and Pharmacology*, vol. 48 (ed. G.A. Cordell Academic Press, New York, pp. 192–248; (e) Borschberg, H.-J. (2005) *Curr. Org. Chem.*, **9**, 1465–1491.
15. Burkard, S. and Borschberg, H.-J. (1989) *Helv. Chim. Acta*, **72**, 254–263.
16. Beerli, R. and Borschberg, H.-J. (1991) *Helv. Chim. Acta*, **72**, 110–116.
17. Burkard, S. and Borschberg, H.-J. (1991) *Helv. Chim. Acta*, **74**, 275–289.
18. Galli, R., Dobler, M., Güller, R., Stahl, R., and Borschberg, H.-J. (2002) *Helv. Chim. Acta*, **85**, 3400–3413.
19. Brown, R.T. (1980) in *Indole and Biogenetically Related Alkaloids* (eds J.D. Phillipson and M.H. Zenk), Academic Press, London, pp. 171–184.
20. Brown, R.T. and Chaple, C.L. (1974) *J. Chem. Soc., Chem. Commun.*, 740–742.
21. Brown, R.T., Chaple, C.L., Platt, R., and Spencer, H. (1974) *J. Chem. Soc., Chem. Commun.*, 929–930.
22. Brown, R.T. (2000) *J. Indian Chem. Soc.*, **77**, 609–616.
23. Brown, R.T., Dauda, B.E.N., Pratt, S.B., and Richards, P. (2002) *Heterocycles*, **56**, 51–58.
24. Brown, R.T., Pratt, S.B., and Richards, P. (2000) *Tetrahedron Lett.*, **41**, 5627–5630.
25. Binns, F., Brown, R.T., and Dauda, B.E.N. (2000) *Tetrahedron Lett.*, **41**, 5631–5635.
26. Brown, R.T., Dauda, B.E.N., Jameson, S.B., and Santos, C.A.M. (2000) *Tetrahedron Lett.*, **41**, 8861–8865.
27. Martin, S.F., Chen, K.X., and Eary, C.T. (1999) *Org. Lett.*, **1**, 79–81.
28. Ito, M., Clark, C.W., Mortimore, M., Goh, J.B., and Martin, S.F. (2001) *J. Am. Chem. Soc.*, **123**, 8003–8010.

29. (a) Knight, S.D., Overman, L.E., and Pairaudeau, G. (1993) *J. Am. Chem. Soc.*, **115**, 9293–9294; (b) Knight, S.D., Overman, L.E., and Pairaudeau, G. (1995) *J. Am. Chem. Soc.*, **117**, 5776–5788; (c) Anet, F.A.L. and Robinson, R. (1953) *Chem. Ind. (London)*, 245.
30. Deiters, A., Chen, K., Eary, C.T., and Martin, S.F. (2003) *J. Am. Chem. Soc.*, **125**, 4541–4550.
31. (a) Husson, A., Langlois, Y., Riche, C., Husson, H.-P., and Potier, P. (1973) *Tetrahedron*, **29**, 3095–3098; (b) Besseliévre, R. and Husson, H.-P. (1976) *Tetrahedron Lett.*, **17**, 1873–1876; (c) Husson, H.-P. (1980) in *Indole and Biogeneticallty Related Alkaloids* (eds J.D. Phillipson and M.H. Zenk), Academic Press, London, pp. 185–200.
32. (a) Bennasar, M.-L., Vidal, B. and Bosch, J. (1993) *J. Am. Chem. Soc.*, **115**, 5340–5341; (b) Bennasar, M.-L., Vidal, B., and Bosch, J. (1996) *J. Org. Chem.*, **61**, 1916–1917; (c) Bennasar, M.-L., Vidal, B., and Bosch, J. (1997) *J. Org. Chem.*, **62**, 3597–3609.
33. Andriantsiferana, M., Besseliévre, R., and Husson, H.-P. (1977) *Tetrahedron Lett.*, **18**, 2587–2590.
34. Langlois, Y., Langloisy, N., and Potier, P. (1975) *Tetrahedron Lett.*, **16**, 955–958.
35. Potier, P. and Janot, M.M. (1973) *C. R. Acad. Sci. (Paris)*, **276**, 1727–1728.
36. Bennasar, M.-L., Zulaica, E., Alonso, Y., Mata, I., Molins, E., and Bosch, J. (2001) *Chem. Commun.*, 1166–1167.
37. Scott, A.I., Lee, S.L., Wan, W., Hirata, T., Guéritte, F., Baxter, R.L., Nordlov, H., Dorschel, C.A., Mizukawi, H., and Mackenzie, N.E. (1981) *Heterocycles*, **20**, 1257–1274.
38. (a) Kuehne, M.E., Kirkemo, C.L., Matsko, T.H., and Bohnert, J.C. (1980) *J. Org. Chem.*, **45**, 3259–3265; (b) Kuehne, M.E., Okuniewicz, F.J., Kirkemo, C.L., and Bohnert, J.C. (1982) *J. Org. Chem.*, **47**, 1335–1343; (c) Kuehne, M.E., Bohnert, J.C., Bornmann, W.G., Kirkemo, C.L., Kuehne, S.E., Seaton, P.J., and Zebovitz, T.C. (1985) *J. Org. Chem.*, **50**, 919–923; (d) Kuehne, M.E. and Podhorez, D.E. (1985) *J. Org. Chem.*, **50**, 924–929; (e) Kuehne, M.E. and Zebovitz, T.C. (1987) *J. Org. Chem.*, **52**, 4331–4339; (f) Kuehne, M.E. and Pitner, J.B. (1989) *J. Org. Chem.*, **54**, 4553–4569.
39. Kuehne, M.E., Bornmann, W.G., Earle, W.G., and Marko, I. (1986) *J. Org. Chem.*, **51**, 2913–2927.
40. Wenkert, E. and Wickberg, B. (1965) *J. Am. Chem. Soc.*, **87**, 1580–1589.
41. Lévy, J. (1971) Procédé de préparation de la vincamine naturelle à partir de la tabersonine et derivés indoliques nouveaux, Patent BE 7, 616,28 [to Omnium Chimique, SA].
42. Hugel, G., Gourdier, B., Lévy, J., and Le Men, J. (1980) *Tetrahedron*, **36**, 511–513.
43. (a) Hutchinson, C.R., O'Loughlin, G.J., Fraser, S.B., and Brown, R.T. (1974) *Chem. Commun.*, 928; (b) Winterfeldt, E. and Radunz, H. (1971) *J. Chem. Soc., Chem. Commun.*, 374–375.
44. (a) Hutchinson, C.R., Heckendorf, A.H., Sraughn, J.L., Daddona, P.E., and Cane, D.E. (1979) *J. Am. Chem. Soc.*, **101**, 3358–3369; (b) Hutchinson, C.R. (1981) *Tetrahedron*, **37**, 1047–1065.
45. (a) Warneke, J. and Winterfeldt, E. (1972) *Chem. Ber.*, **105**, 2120–2125; (b) Boch, M., Korth, T., Nelke, J.M., Pike, D., Radunz, H., and Winterfeldt, E. (1972) *Chem. Ber.*, **105**, 2126–2142; (c) Krohn, K. and Winterfeldt, E. (1975) *Chem. Ber.*, **108**, 3030–3042; (d) Hutchinson, C.R., Hsia, M.-T.S., Heckendorf, A.H., and O'Loughlin, G.J. (1976) *J. Org. Chem.*, **41**, 3493–3494.
46. Brown, R.T., Jianli, L., and Santos, C.A.M. (2000) *Tetrahedron Lett.*, **41**, 859–862.
47. Ahond, A., Cavé, Ad., Kan-Fan, C., and Potier, P. (1970) *J. Chem. Soc., Chem. Commun.*, 517.
48. Potier, P. (1980) in *Indole and Biogenetically Related Alkaloids* (eds J.D. Phillipson and M.H. Zenk Academic Press, London, pp. 159–169.
49. Langlois, N., Langlois, Y., Guéritte, F., and Potier, P. (1976) *J. Am. Chem. Soc.*, **98**, 7017–7024.
50. (a) Ishikawa, H., Colb, D.A., Seto, S., Va, P., Tam, A., Kakei, H.,

Rayl, T.J., Hwang, I., and Boger, D.L. (2009) *J. Am. Chem. Soc.*, **131**, 4904–4916; (b) Va, P., Campbell, E.L., Robertson, W.M., and Boger, D.L. (2010) *J. Am. Chem. Soc.*, **132**, 8489–8495.

51. (a) Michel, S., Tillequin, F., and Koch, M. (1980) *Tetrahedron Lett.*, **21**, 4027–4030; (b) Michel, S., Tillequin, F., Koch, M., and Aké-Assi, L. (1982) *J. Nat. Prod.*, **45**, 489–494.

52. Michel, S., Tillequin, F., and Koch, M. (1987) *J. Chem. Soc., Chem. Commun.*, 229–230.

4
Biomimetic Synthesis of Alkaloids Derived from Tryptophan: Dioxopiperazine Alkaloids

Timothy R. Welch and Robert M. Williams

4.1
Introduction

Countless secondary metabolic indole alkaloids produced in both marine and terrestrial fungi are derived from tryptophan. Our crude attempts to synthesize some of the vast array of structurally diverse natural alkaloids only serves to showcase the efficiency and elegance with which Nature is able assemble the same molecules. Still, we strive to mimic and in turn better understand the mechanisms inside the cell that are able to produce molecular architecture of such synthetic complexity.

Moreover, it has often been found advantageous to exploit Nature's evolutionary creative design of alkaloids in search of new compounds of therapeutic potential. The rich subclass of biologically active, tryptophan-derived dioxopiperazines found in nature has inspired medicinal chemists to use the dioxopiperazine core in drug design efforts to mimic the interactions of natural peptides while reducing susceptibility to metabolic amide bond cleavage. Furthermore, a dioxopiperazine should pay a small entropic penalty upon binding to a target in comparison to an analogous peptide, as a direct consequence of the reduced conformational mobility inherent in the dioxopiperazine ring. In this chapter, we present a brief review of a select group of (partly) biomimetic syntheses of tryptophan-derived dioxopiperazine alkaloids (Figure 4.1). In most of these syntheses, a single step or key transformation has been deemed to constitute the "biomimetic" aspect of that particular work. As the actual biosynthetic pathways to most, if not all, of the alkaloid natural products covered here are either unknown or known only in part we have attempted, where appropriate, to point out the particular biomimetic step or transformation.

4.2
Prenylated Indole Alkaloids

Birch, Wright, and Russell first isolated brevianamide A from *Penicillium brevicompactum* in 1969 [1–3]. Several years later, Birch and coworkers determined that brevianamide A was biosynthetically derived from tryptophan, proline, and

Biomimetic Organic Synthesis, First Edition. Edited by Erwan Poupon and Bastien Nay.

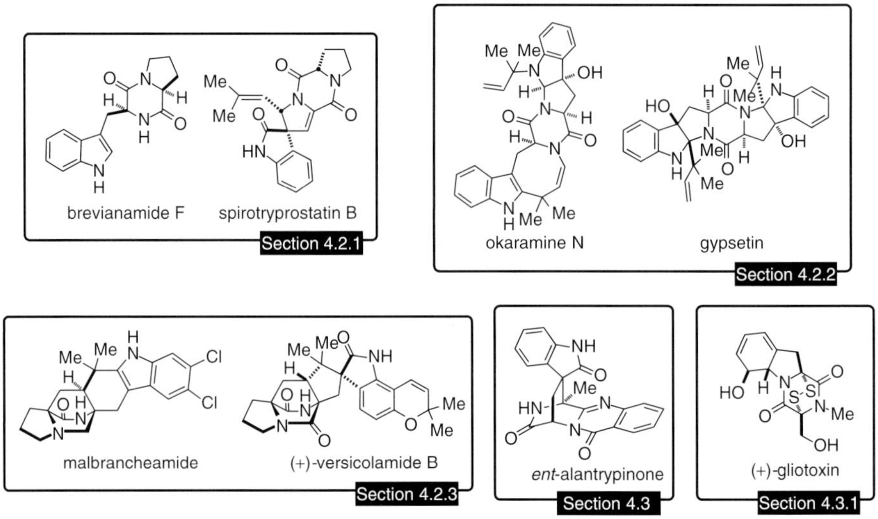

Figure 4.1 Representative molecules discussed in this chapter.

mevalonic acid through feeding experiments (Scheme 4.1) [4]. Furthermore, Birch showed that radiolabeled brevianamide F was incorporated into **5**. It was postulated at the time and later supported with experimental evidence by Williams and coworkers that deoxybrevianamide E was also a biosynthetic precursor [5].

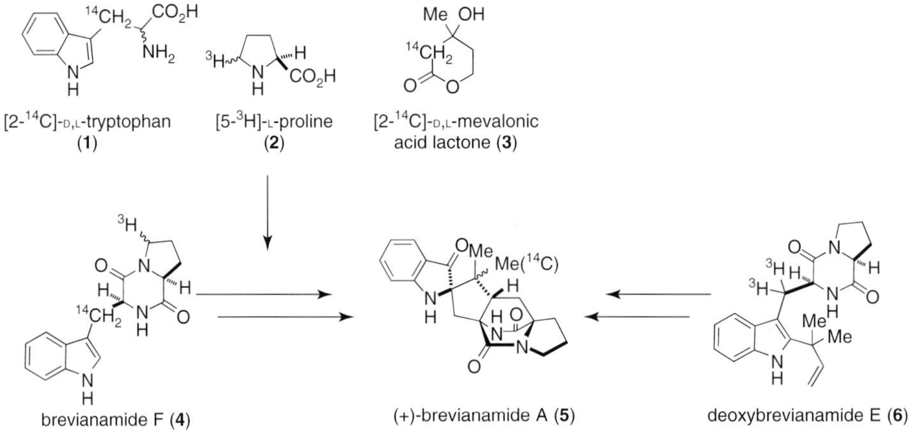

Scheme 4.1 Proposed biosynthesis of the brevianamides.

Since these early studies, Williams has developed a proposal for the biosynthesis of the brevianamides, with that of brevianamide E shown in Scheme 4.2. Derived from tryptophan and proline, **6** was proposed to undergo oxidation to hydroxyindolenine **7**. Irreversible nucleophilic ring closure was supposed to lead to brevianamide E (**8**), supported by incorporation of [8-$^3H_2$]**6** into **8** in significant radiochemical yield [5].

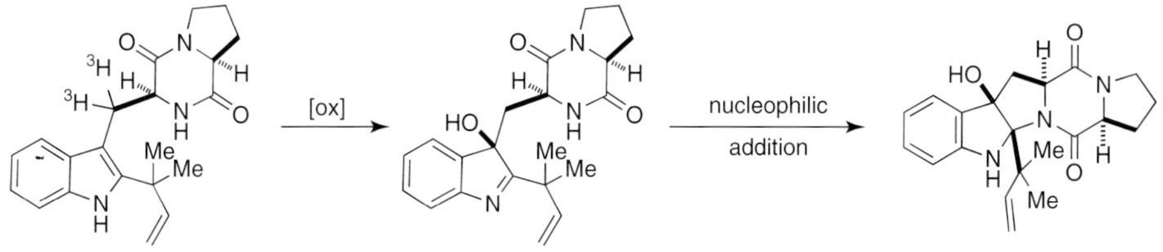

deoxybrevianamide E (**6**) **7** brevianamide E (**8**)

Scheme 4.2 Proposed biosynthesis of brevianamide E.

4.2.1 Dioxopiperazines Derived from Tryptophan and Proline

Brevianamide F (**12**) was first isolated in 1972 and is one of the simplest tryptophan-derived dioxopiperazine natural products. It is readily synthesized through amino acid coupling of *N*-Boc-tryptophan (**9**) to proline ethyl ester (**10**), Boc-deprotection, and ring closure, in modest overall yield (Scheme 4.3) (R.M. Williams, unpublished results).

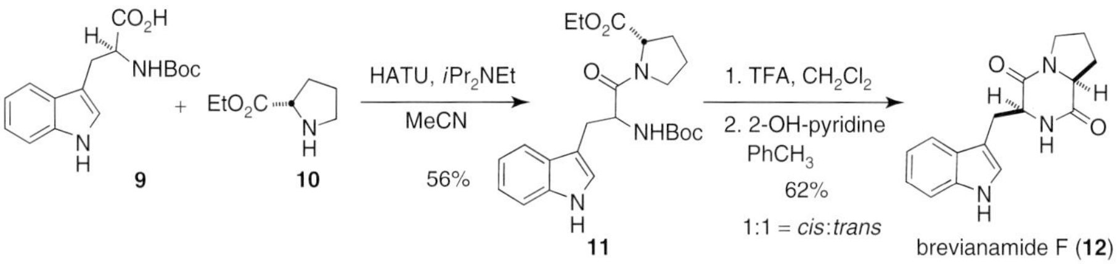

9 **10** **11** brevianamide F (**12**)

Scheme 4.3 Biomimetic synthesis of brevianamide F.

Brevianamide F lacks only the reverse prenyl group found in deoxybrevianamide E, which has been synthesized by Kametani and coworkers *en route* to brevianamide E (Scheme 4.4) [6]. *N*-Benzyloxycarbonyl-L-proline (**13**) was subjected to Schotten–Baumann conditions with dimethyl aminomalonate to give amide **14**. Debenzyloxycarbonylation of **14** followed by heating with catalytic 2-hydroxypyridine effected cyclization to dioxopiperazine **15** in 93% yield. Condensation with indole **16** gave a separable mixture of diastereomers, individually hydrolyzed to the corresponding free acids (**17**). Heating of the desired diastereomer in dioxane gave deoxybrevianamide E (**18**) and its epimer (**19**) in 29 and 55% yield, respectively. Irradiation of methanolic **18**, containing Rose Bengal in the presence of oxygen, followed by addition of dimethyl sulfide, resulted in the biomimetic hydroxylation of deoxybrevianamide E, furnishing brevianamide E (**8**) and **20** as a separable mixture of diastereomers.

Nineteen years later, a more efficient synthesis of brevianamide E was completed by Danishefsky and coworkers [7]. The synthesis commenced with C3 chlorination of the known phthaloylated tryptophan derivative **21**, followed by addition of fresh prenyl-9-borabicyclo[3.3.1]nonane (prenyl-9-BBN) to the resultant 3-chloroindolenine (Scheme 4.5). Hydrazinolysis in ethanol provided amino ester

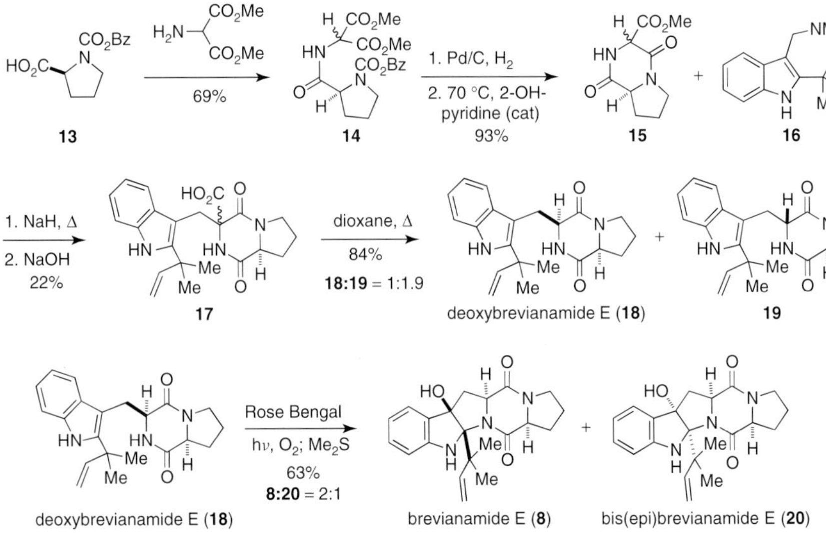

Scheme 4.4 Kametani's total synthesis of deoxybrevianamide E and brevianamide E.

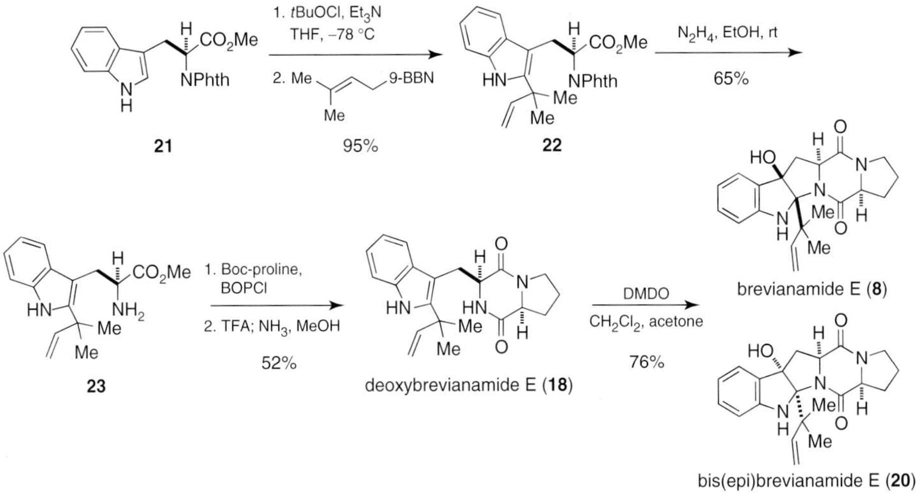

Scheme 4.5 Danishefsky's total synthesis of brevianamide E.

23 in 65% yield, which was coupled to *N*-Boc-L-proline, deprotected, and cyclized to afford deoxybrevianamide E (**18**) in 52% yield. Compound **18** was elaborated to brevianamide E (**8**) and bis(epi)brevianamide E (**20**) in a ratio of ~1 : 5 upon treatment with dimethyldioxirane (DMDO) in a biomimetic oxidative cyclization sequence reminiscent of the original Kametani work discussed above.

The structural similarities between deoxybrevianamide E and the then newly isolated natural products tryprostatin A and B did not escape notice of Danishefsky and coworkers. While prenylation failed in attempts to use a reverse prenylborane nucleophile directly as in the method used to synthesize **22** above, a solution was found in treating chloroindolenine **24** with tri(*n*-butyl)prenylstannane and BCl_3 to afford the desired prenyl functionality at C_2 in excellent yield (Scheme 4.6). Phthalimide deprotection, peptide coupling, Boc-deprotection, and cyclization were achieved to afford tryprostatin B (**30**) in 43% overall yield [7].

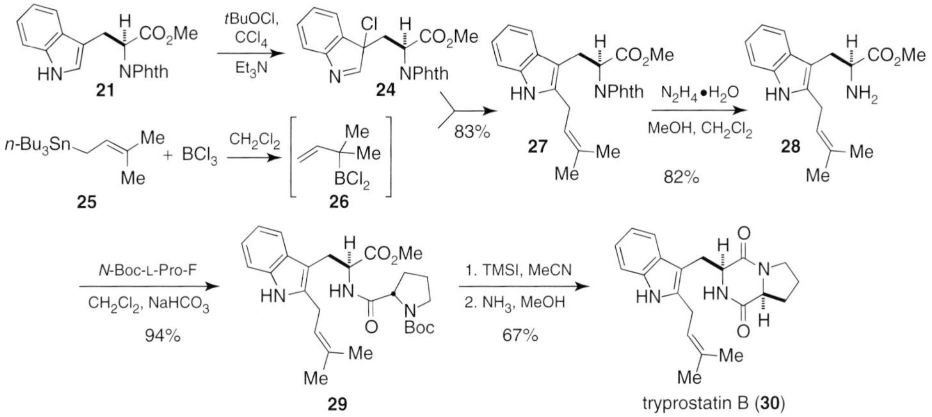

Scheme 4.6 Biomimetic total synthesis of tryprostatin B.

Danishefsky and coworkers also completed the total synthesis of the spirooxindole spirotryprostatin B [8]. L-Tryptophan methyl ester was converted into the oxindole derivative **32**, followed by addition of prenyl aldehyde under basic conditions to afford an inseparable four-component mixture of spirooxindoles (**33–36**, Scheme 4.7). Peptide coupling and subsequent treatment of the mixture with lithium bis(trimethylsilyl)amide (LHMDS) followed by selenylation presumably gave phenyl selenide mixture **38**. Oxidative elimination produced a mixture from which **39** was separated and elaborated to spirotryprostatin B (**40**) via Boc-deprotection and base-induced cyclization.

Danishefsky took a markedly different approach in the synthesis of spirotryprostatin A [9]. A potentially biomimetic Pictet–Spengler reaction of tryptophan derivative **42** with thioaldehyde **41** as a masked isoprene equivalent gave the desired *cis*-tetrahydrocarboline (**43**) with marginal selectivity (Scheme 4.8). *N*-Bromosuccinimide (NBS)-mediated oxidative rearrangement proceeded via intermediate **44** to the oxindole was followed by deprotection of the carbamate to give amine **45**. The modest yield of the sequence (57%) reflects the susceptibility of the oxindole to electrophilic aromatic bromination under the *spiro*-rearrangement conditions. Peptide coupling and Troc-deprotection resulted in cyclization to the dioxopiperazine, after which oxidation and sulfoxide elimination revealed the prenyl group to afford selectively spirotryprostatin A (**48**).

The synthesis of notoamide J was completed by Williams and coworkers, starting with the Boc-protection of 7-hydroxyindole (Scheme 4.9) [10]. Chlorination at C3 was followed by reverse prenylation of the resultant 3-chloroindolenine **51** to afford

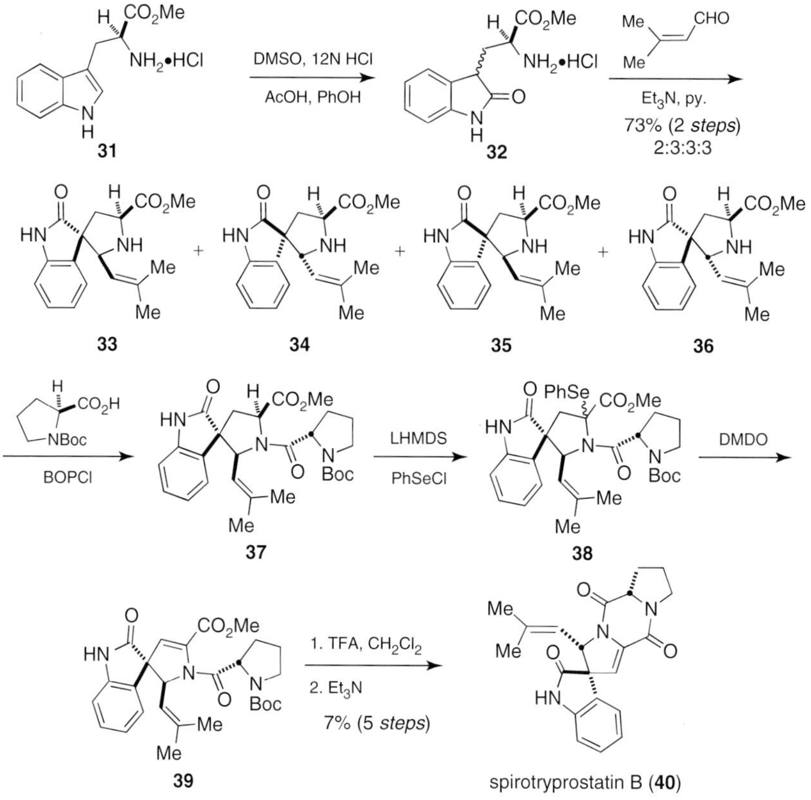

Scheme 4.7 Danishefsky's synthesis of spirotryprostatin B.

52. The corresponding gramine was prepared by treating **52** with formaldehyde and dimethylamine, and subsequent Somei–Kametani coupling and imine hydrolysis gave tryptophan derivative **54** in good yield. Protection of the free amine as the Boc-carbamate and ester hydrolysis gave **55**, which was coupled to proline ethyl ester in the presence of *O*-(7-azabenzotriazol-1-yl)-*N,N,N′,N′*-tetramethyluronium hexafluorophosphate (HATU) to afford amide **56**. Cyclization to dioxopiperazine **57** was followed by a biomimetic oxidation sequence accompanied by pinacol-type rearrangement to the oxindoles notoamide J (**58**) and 3-epi-notoamide J (**59**) in a 2 : 1 separable mixture.

4.2.2
Dioxopiperazine Derived from Tryptophan and Amino Acids other than Proline

Corey and coworkers designed a succinct synthesis of okaramine N, featuring a Pd-promoted dihydroindoloazocine formation [11]. Readily available tryptophan derivative **60** was reduced to indoline **61** and subsequently prenylated via copper(I)-catalyzed alkylation with butyne **62** (Scheme 4.10). Treatment with

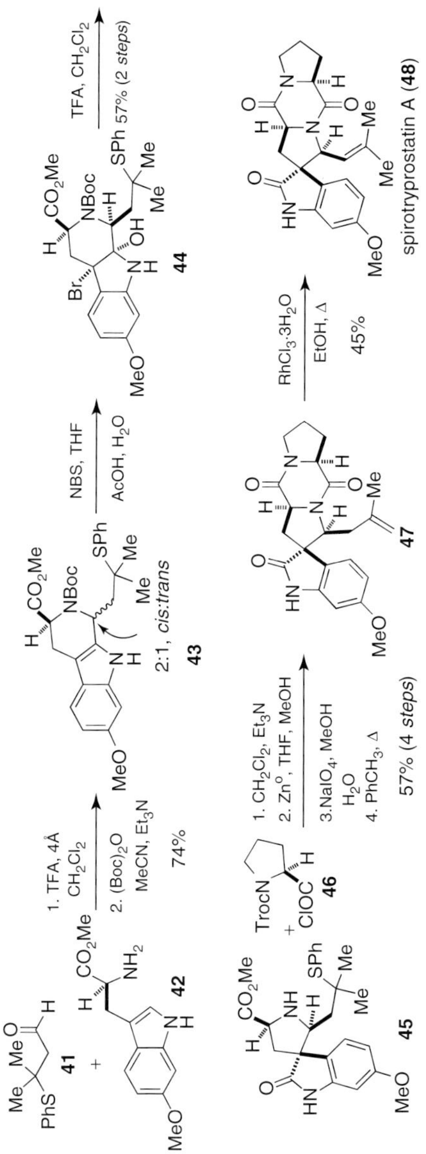

Scheme 4.8 Danishefsky's synthesis of spirotryprostatin A.

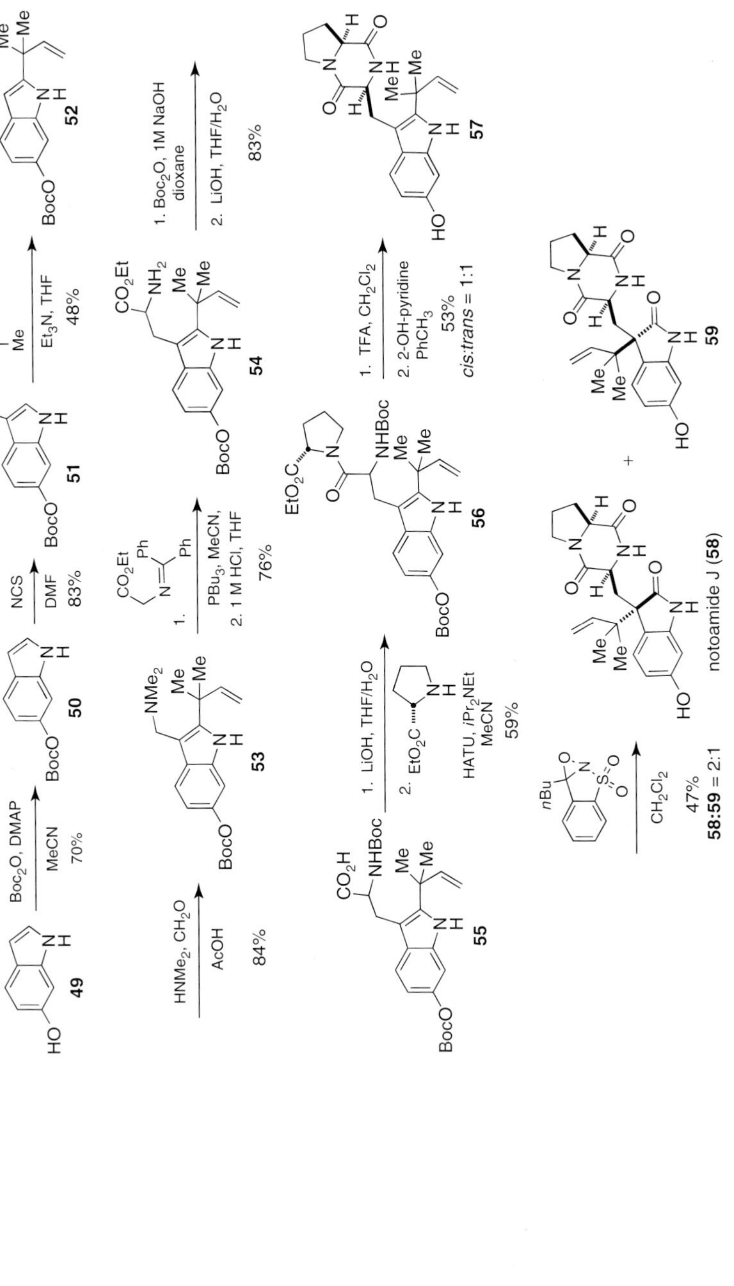

Scheme 4.9 Williams' biomimetic synthesis of notoamide J.

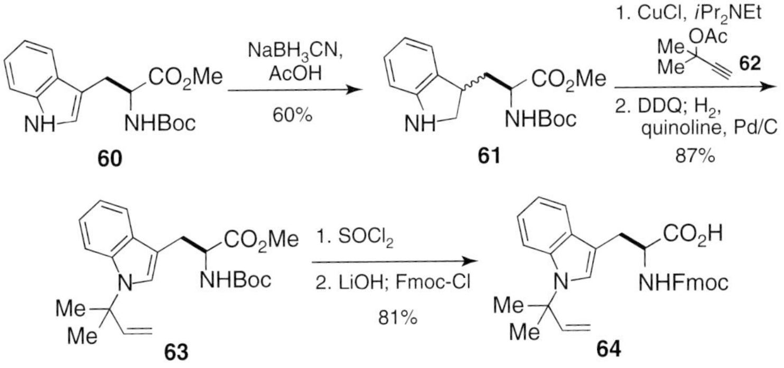

Scheme 4.10 Synthesis of the *N*-reverse-prenylated tryptophan derivative.

2,3-dichloro-5,6-dicyano-1,4-benzoquinone (DDQ) effected the dehydrogenation of the indoline, and the resulting tryptophan derivative **63** was deprotected, saponified, and reprotected as the N-Fmoc derivative (**64**).

The synthesis of okaramine N was completed through reductive amination of 3-methyl-buten-2-al onto L-tryptophan methyl ester, followed by coupling of the resultant product (**65**) with acid **64** to form the desired tetracycle (Scheme 4.11). Treatment of **66** with $Pd(OAc)_2$ provided the eight-membered ring (**67**) in modest yield (38%). The free amine obtained upon Fmoc cleavage underwent cyclization to form dioxopiperazine **68** in 95% yield. A potentially biomimetic oxidative cyclization was effected by treating **68** with *N*-methyltriazolinedione (MTAD), which reacted selectively with the N-unsubstituted indole subunit, and after photooxidation and

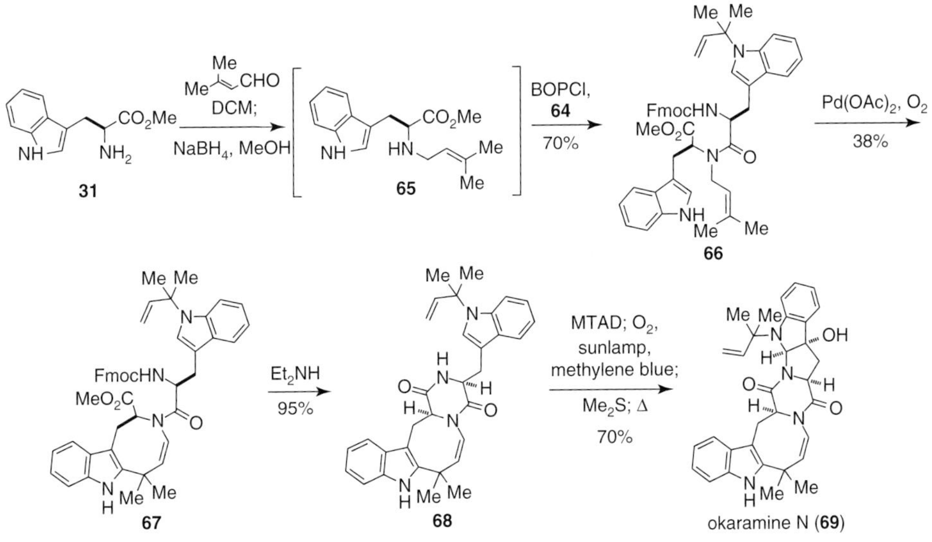

Scheme 4.11 Completion of the total synthesis of okaramine N.

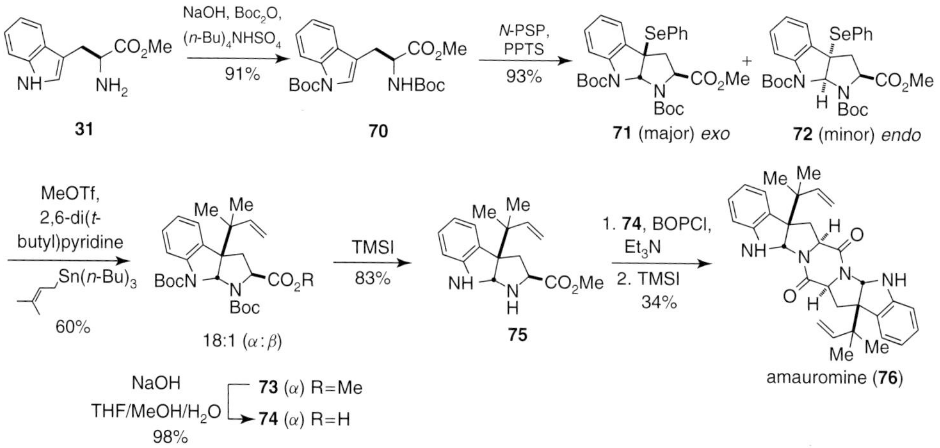

Scheme 4.12 Danishefsky's total synthesis of amauromine.

then reduction formed a hydroxylated octacycle that was directly converted into okaramine N (**69**) via thermolysis in 70% yield.

Amauromine was synthesized as shown in Scheme 4.12, starting with the bis(Boc) protection of L-tryptophan [12]. Conversion into the selenide by reaction with *N*-phenylselenophthalimide (*N*-PSP) in the presence of pyridinium *p*-toluenesulfonate (PPTS) afforded a mixture of **71** and **72** (9 : 1). Photolysis of the mixture in the presence of prenyltri(*n*-butyl)tin produced a mixture of reverse prenylated pyrrolodinoindolines, the desired major diastereomer of which was separated by crystallization from hexanes. The Boc groups were globally cleaved using iodotrimethylsilane (TMSI) to afford the ester (**75**). Coupling of **75** with acid **74** gave the amide, which readily cyclized to the desired dioxopiperazine (**76**, amauromine) upon treatment with TMSI.

Danishefsky and coworkers completed a total synthesis of gypsetin in the same fashion as their effort on brevianamide E [7]. The reverse prenylated amine, **23**, was synthesized as shown above (Scheme 4.5). Boc protection and cleavage of the methyl ester afforded acid **77**, which was coupled to amine **23**, deprotected, and cyclized to dioxopiperazine **78** (Scheme 4.13). Treatment with DMDO effected the biomimetic oxidative conversion into the natural product gypsetin (**79**).

4.2.3
Bicyclo[2.2.2]diazaoctanes

In 1970, Sammes proposed a hetero-Diels–Alder cycloaddition to be the biosynthetic origin of the bicyclo[2.2.2]diazaoctane core found in brevianamides A and B (Scheme 4.14) [13].

Support for this proposal was observed upon treating dihydroxypyrazine **82** with dimethyl acetylenedicarboxylate (**83**) or with norbornadiene (**84**) to give cycloadducts **85** or **86**, respectively (Scheme 4.15) [13].

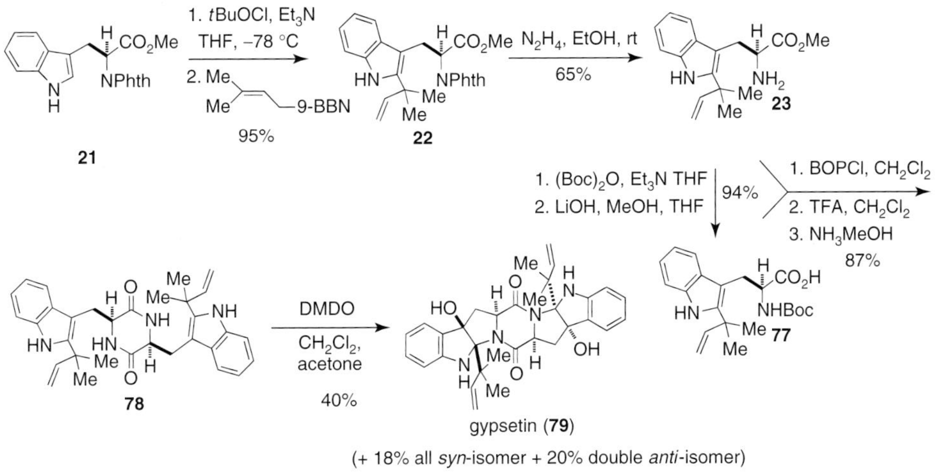

Scheme 4.13 Synthesis of gypsetin.

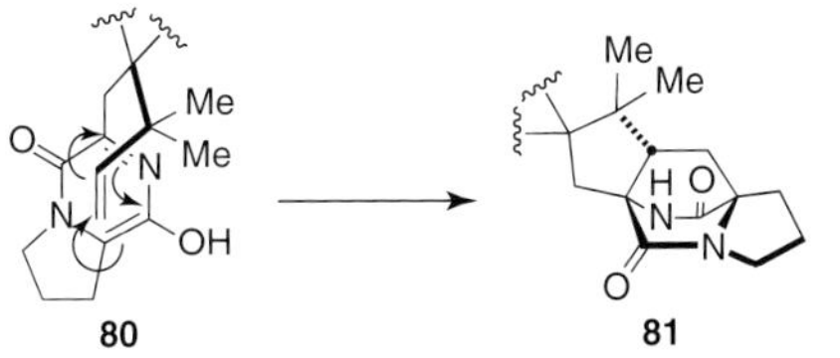

Scheme 4.14 Proposed hetero-Diels–Alder formation of bicyclo[2.2.2]diazaoctanes.

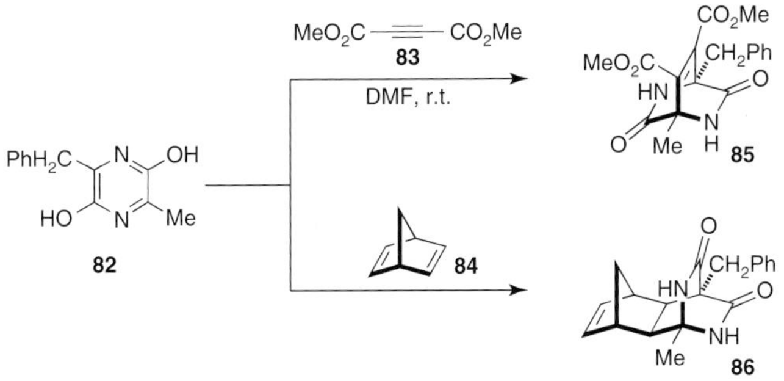

Scheme 4.15 Sammes' model study of proposed cycloaddition.

Williams and coworkers expanded on the pioneering work of Sammes with the following biosynthetic proposal for the brevianamides [5]. Deoxybrevianamide E (**18**) was thought to undergo oxidation to hydroxyindolenine **7**, which could undergo a pinacol-type rearrangement to indoxyl **87** (Scheme 4.16). Subsequent two-electron oxidation and enolization to azadiene **88**, followed by an intramolecular hetero-Diels–Alder reaction was, following from the original proposal of Sammes, envisioned to give the natural products (+)-brevianamide A and (+)-brevianamide

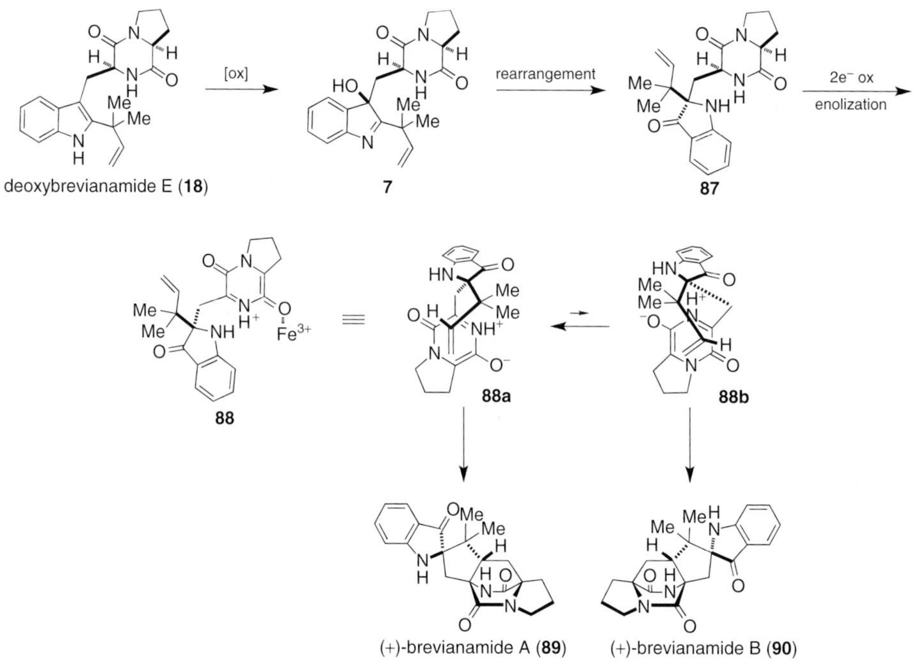

Scheme 4.16 Biosynthetic proposal for the brevianamides.

B. The pseudo-enantiomorphic relationship between the two natural products was envisaged to arise from the equilibrium between conformers **88a** and **88b**, which undergo cycloaddition to give **89** and **90**, respectively. *Ab initio* studies of the two transition states demonstrated that **88a** is the more stable of the two, which is consistent with the observed product ratios of **89** and **90**. The theoretical insights published by Domingo *et al.* lend support to the proposal of a biosynthetic intramolecular Diels–Alder reaction of intermediate **88** [14]. Unfortunately, experimental support for this pathway has been elusive to secure and thus it remains a speculative biogenetic construction.

The total synthesis of D,L-brevianamide B demonstrated the first congruent application of a biomimetic Diels–Alder reaction used to form the bicyclo[2.2.2]diazaoctane core common to the series of prenylated dioxopiperazines to be discussed in this section [15]. 9-Epi-deoxybrevianamide E (**19**) was converted into the lactim ether (**91**) and oxidized to give the Diels–Alder precursor **92** (Scheme 4.17). Treatment with aqueous methanolic KOH induced tautomerization to azadiene **93**, which underwent a potentially biomimetic intramolecular Diels–Alder cycloaddition to give a mixture of diastereomers (**94** and **95**, 2 : 1). Oxidation, pinacol-type rearrangement, and lactim ether deprotection of the minor diastereomer (**95**) afforded brevianamide B (**90**) in 65% overall yield from **96**. This study was one of the first to experimentally support the biogenetic origin of the core bicyclo[2.2.2] ring system as arising via a dioxopiperazine that undergoes a net two-electron oxidation to an azadiene moiety.

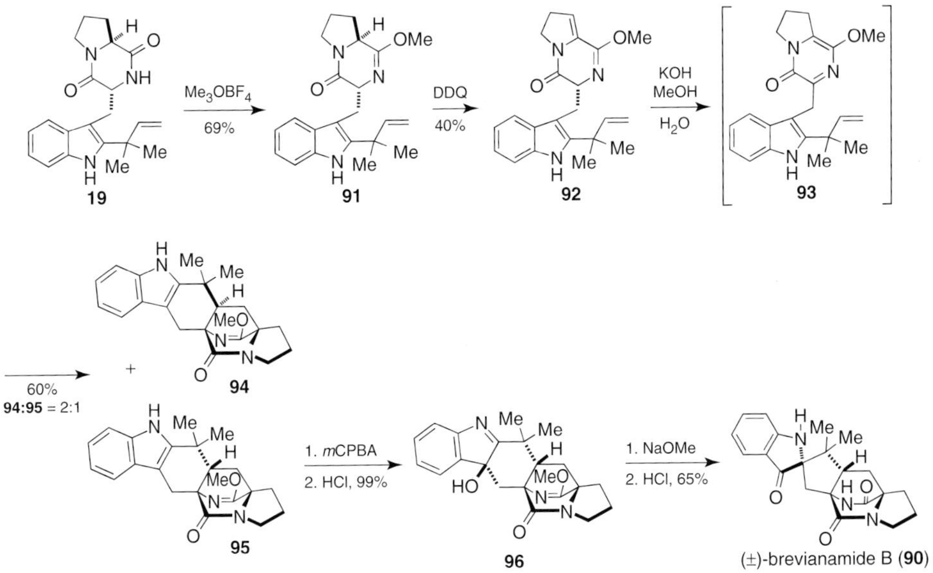

Scheme 4.17 Application of the proposed biomimetic Diels–Alder reaction.

Williams and coworkers also applied the intramolecular Diels–Alder reaction to the racemic synthesis of VM55599 [16]. The reverse-prenylated tryptophan derivative **97** was coupled to β-methyl-β-hydroxyproline ethyl ester **98** to afford dipeptide **99** (Scheme 4.18). *N*-Boc deprotection and cyclization afforded dioxopiperazine **100**, and elimination with thionyl chloride gave enamide **101**. Formation of the lactim ether (**102**) and treatment with aqueous KOH gave azadiene **103**, which spontaneously suffered intramolecular Diels–Alder reaction to give a separable mixture of all four possible racemic diastereomers (**104–107**, 2.6 : 3.7 : 1.0 : 1.6, respectively). Cleavage of the lactim ether and diisobutylaluminum hydride (DIBAL-H) reduction gave VM55599 in 73% yield from **105**.

The synthesis of VM55599 allowed for assignment of the absolute stereochemistry of the molecule, which places the methyl group at the β-position of the proline residue *syn*- to the bridging isoprene moiety [16]. In stark contrast, the analogous methyl group of paraherquamide A is *anti* to the bridging isoprene unit. Scheme 4.19 outlines a possible unified biosynthesis for VM55599 and paraherquamide A, arising from dimethylallyl pyrophosphate (DMAPP), L-isoleucine, and L-tryptophan. If a Diels–Alder cycloaddition is to be invoked, approach of the isoprene moiety must occur from the same face as the methyl group on the proline ring for synthesis of VM55599, and from the opposite face to the methyl group to give paraherquamide A. The diastereofacial selectivity of the Diels–Alder reaction gave a preponderance of the *syn*-relative stereochemistry *alpha* to the *gem*-dimethyl group in both molecules. As VM55599 is a very minor metabolite of *Penicillium* sp. IMI332995, it is plausible that cyclization of **112**→**114** is preferred and further metabolization gives paraherquamide A, whereas the minor cycloaddition via **113** produces VM55599 as a dead-end shunt metabolite. As shown in Scheme 4.18

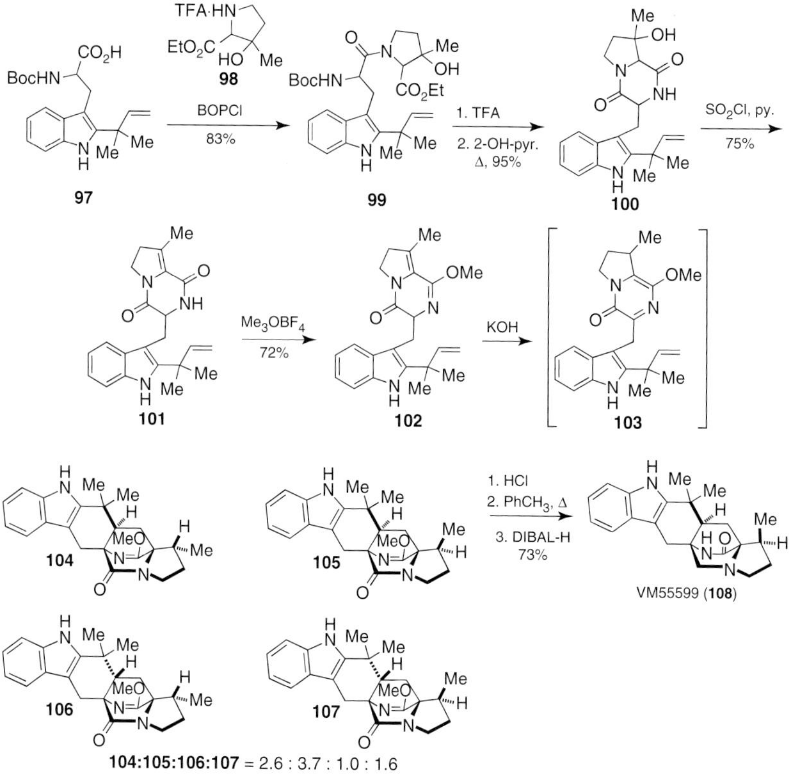

Scheme 4.18 Williams' biomimetic total synthesis of VM55599.

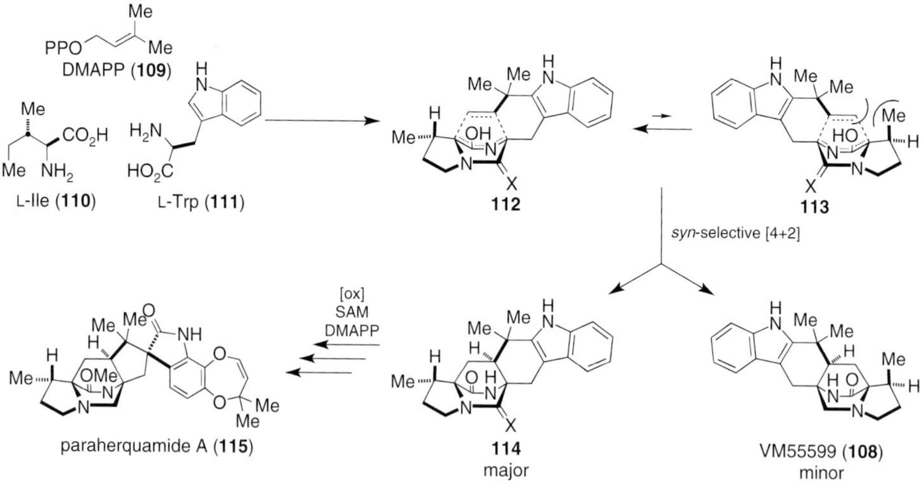

Scheme 4.19 Proposed biosynthesis of paraherquamide A and VM55599.

above, the intrinsic diastereofacial bias of the Diels–Alder reaction is modest at best, giving a slight excess (1.47 : 1) of cycloaddition from the same face as the methyl group, and favoring the *syn*-relative stereochemistry to the extent of 2.4 : 1. Such observations suggest that the biosynthesis may rely on protein organization of the precyclization conformers to stereoselectively produce the *syn*-isomers.

A report from Liebscher and coworkers showed promise in terms of improving the diastereoselectivity of the Diels–Alder cycloaddition, using neutral conditions to prepare the azadiene in contrast to the basic conditions employed by Williams [17]. Compound **118** was prepared by a Horner–Wadsworth–Emmons reaction of aldehyde **116** with phosphonate **117** (Scheme 4.20). Treatment of **118** with neat acetyl chloride for 20 days gave the Diels–Alder product **119** as a single diastereomer in 48% yield.

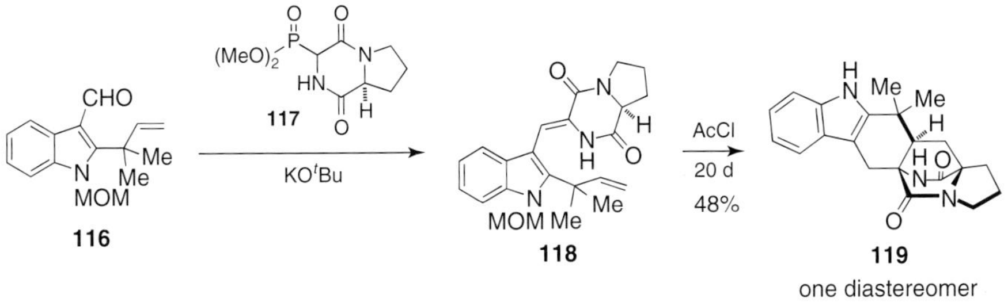

Scheme 4.20 Liebscher's Diels–Alder work.

The precedent set by Liebscher's work was applied to an asymmetric total synthesis of VM55599 [18]. The loss of stereochemistry observed at the proline methyl group in **101** above led Williams to employ a dehydrotryptophan derivative as the Diels–Alder azadiene precursor, allowing for the preparation of VM55599 in an enantioselective fashion. Williams has demonstrated that the β-methylproline residue of paraherquamide A and VM55599 are biosynthetically derived from L-Ile. In a biomimetic construct, L-Ile was converted into optically pure β-methylproline **120** using Hoffman–Löffler–Freytag conditions in 45% yield (Scheme 4.21).

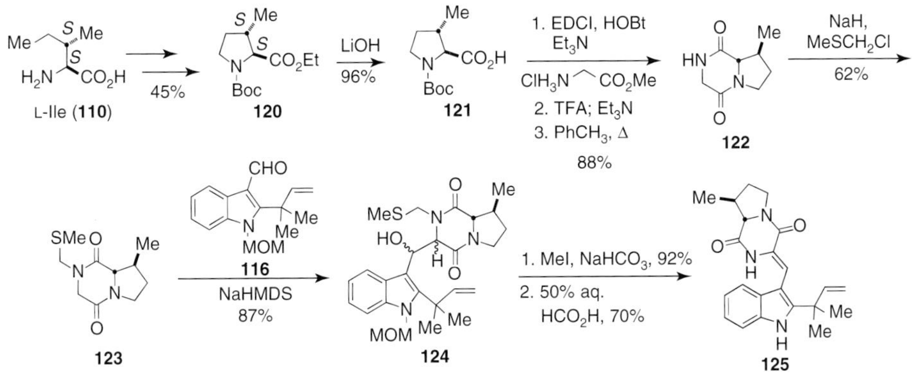

Scheme 4.21 Key dioxopiperazine synthesis.

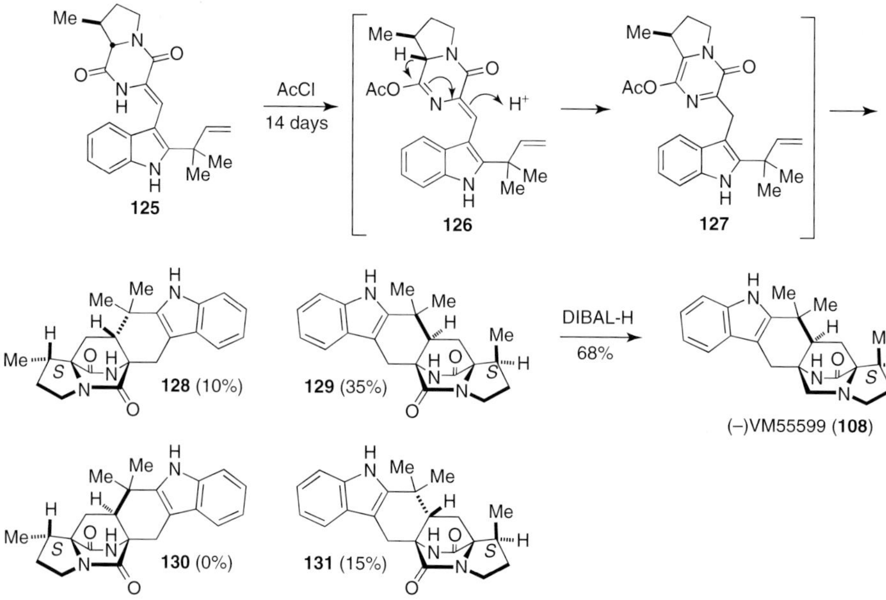

Scheme 4.22 Completion of the asymmetric total synthesis of (−)-VM55599.

Cleavage of the ethyl ester, peptide coupling to glycine methyl ester hydrochloride, Boc deprotection, and cyclization gave dioxopiperazine **122** in good overall yield. Protection of the secondary amide and subsequent condensation with aldehyde **116** gave an epimeric mixture of dioxopiperazines, which gave selectively *(Z)*-isomer **125** following deprotection and dehydration.

Following Liebscher's protocol, **125** was treated with acetyl chloride for 14 days, yielding a mixture of three diastereomers (Scheme 4.22). The reaction is thought to proceed by initial acylation to the *O*-acyl lactim **126**, tautomerization to azadiene **127**, which suffers intramolecular Diels–Alder reaction from three of the four possible diastereomeric transition states, followed by loss of acetate to give compounds **128–131**. The major diastereomer **129** was treated with excess DIBAL-H to effect reduction to (−)-VM55599 (**108**). Interestingly, cycloadduct **130** was not observed from the cycloaddition reaction. *Penicillium* sp. produces paraherquamide A in large excess over VM55599 (>600 : 1), so it is surprising that this provocative biogenetic precursor to paraherquamide A is not observed in laboratory cycloaddition reactions. Despite the structural differences between the laboratory and biological Diels–Alder precursors, the intrinsic facial selectivity of the cyclization does not appear to mimic the bias toward paraherquamide stereochemistry one would expect given the observed product ratios of the fungal metabolites.

As many of the bicyclo[2.2.2]diazaoctane natural products differ only at the substitution of the indole ring, a convergent approach utilizing a Fischer indole synthesis was undertaken in an alternative racemic synthesis of brevianamide B [19]. This work had the objective of validating other possible biosynthetic pathways

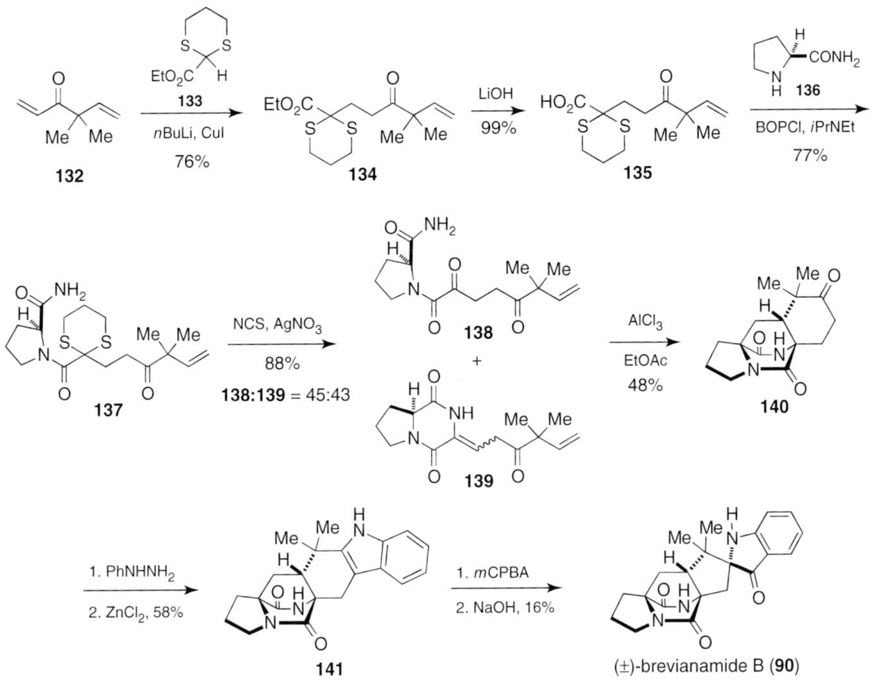

Scheme 4.23 Concise synthesis of brevianamide B.

to reach the oxidation state of the azadiene. In the present instance, the α-ketoamide species (**138**) was posed to serve as a surrogate for the possible biosynthetic oxidative deamination of the tryptophan moiety and coupling to a proline amide species (Scheme 4.23). Conjugate addition of carboxylate **133** to ketone **132** gave ester **134** in 76% yield. Saponification, peptide coupling with L-proline amide, and dithiane deprotection gave a mixture of the uncyclized amide **138** and dioxopiperazine **139**. Aluminum trichloride was added to the mixture to give the Diels–Alder cycloadduct (**140**) in exclusively the *anti*-configuration, whereas mixtures of both the *syn*- and *anti*-cycloadducts were observed in previous syntheses of VM55599 and brevianamide B discussed above. A Fischer indole synthesis was completed by treatment of **140** with phenyl hydrazine followed by $ZnCl_2$, affording **141** in good yield. Oxidation and pinacol-type rearrangement of this known intermediate gave brevianamide B. While an appealing convergent approach towards the synthesis of other related bicyclo[2.2.2]diazaoctanes, the utility of this strategy is limited to those natural products containing the *anti*-stereochemistry observed in formation of compound **140**, the remainder of which must be able to withstand the harsh conditions of the Fischer indole synthesis.

The same biomimetic Diels–Alder disconnection was exploited in the total synthesis of stephacidin A [20]. Starting with reverse prenylated tryptophan derivative **142** (prepared in eleven steps from 6-hydroxyproline), dipeptide **144** was prepared through bis(2-oxo-3-oxazolindinyl)phosphinic chloride (BOPCl) mediated coupling

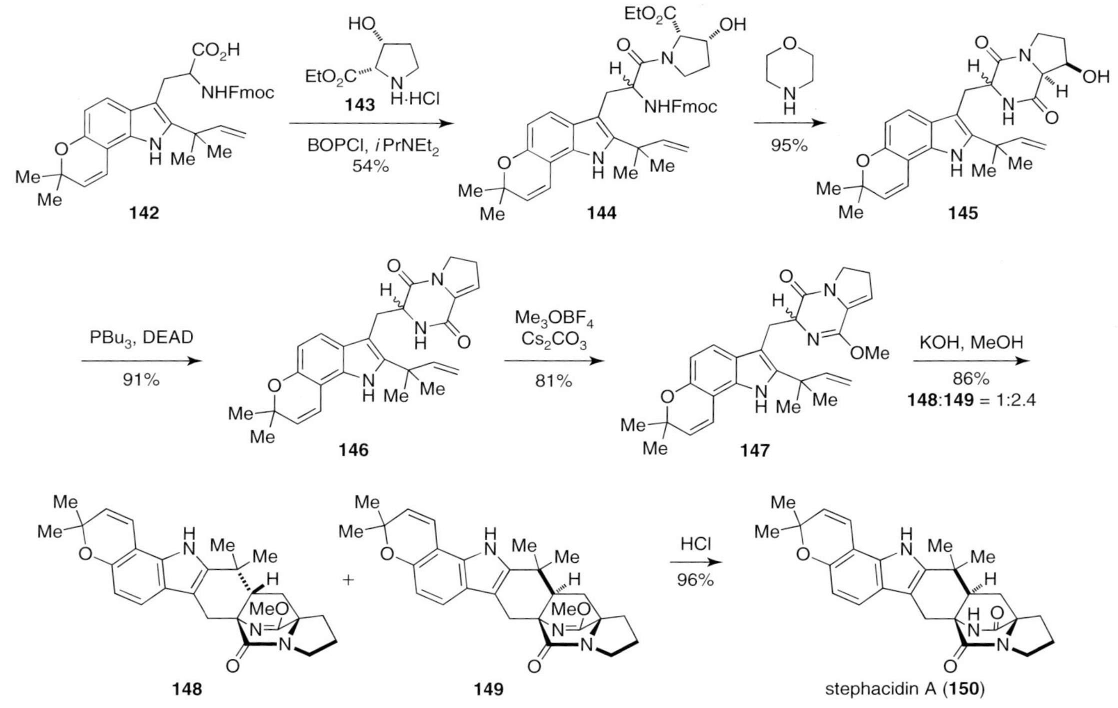

Scheme 4.24 Synthesis of stephacidin A.

with *cis*-3-hydroxyproline ethyl ester (**143**, Scheme 4.24). Fmoc deprotection resulted in the cyclization to dioxopiperazine **145**, which upon treatment with tributyl phosphine and diethyl azodicarboxylate (DEAD) underwent Mitsunobu dehydration to give enamide **146**. Formation of the lactim ether (**147**) was followed by intramolecular Diels–Alder reaction to give a mixture of epimers enriched with the *syn*-isomer (**149**, 2.4 : 1). Deprotection of **149** gave stephacidin A (**150**) in excellent yield.

Williams found that the bicyclo[2.2.2]diazaoctane core could be accessed directly from compound **145** by treating it with excess PBu_3 and DEAD, effecting the dehydration, tautomerization, and Diels–Alder reaction in one pot to afford stephacidin A and its epimer (Scheme 4.25) [21].

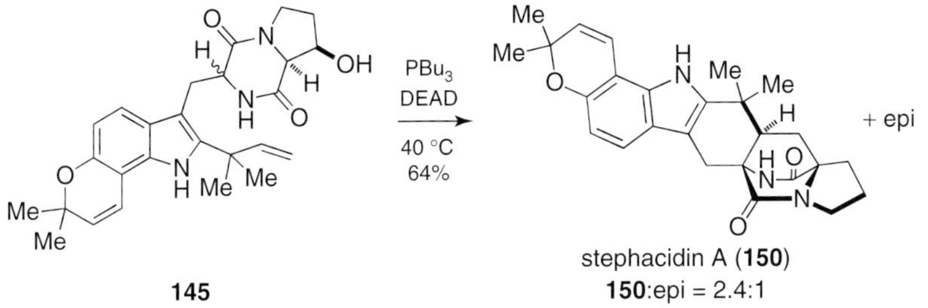

Scheme 4.25 Improved biomimetic synthesis of stephacidin A.

Myers and coworkers, in the course of their total synthesis of avrainvillamide, discovered that synthetic (−)-**152** spontaneously dimerized to stephacidin B under several mild conditions, including addition of triethylamine or exposure to silica gel [22, 23]. Baran and coworkers later determined that stephacidin A could be readily converted into avrainvillamide through reduction to indoline **151** followed by Somei oxidation (Scheme 4.26). In accord with Myers' observations, dimerization to stephacidin B occurred readily upon exposure to silica gel, triethylamine, or on evaporation from dimethyl sulfoxide (DMSO) [24–26]. Myers has further demonstrated that the observed biological activity of stephacidin B may be due to the formation of **152** from **153** *in vivo* [22].

Stephacidin A is of particular biogenetic interest, as both enantiomers have been isolated in nature: (+)-stephacidin A from *Aspergillus ochraceus* [27] and from a marine-derived *Aspergillus* sp. [28] and (−)-stephacidin A from terrestrial Aspergillus versicolor [29–31]. Operating under the assumption that the biosynthesis of stephacidin A proceeds through a common, *achiral* intermediate, Williams and coworkers have proposed notoamide S (**154**) as the point of divergence in the two biosyntheses [32]. Oxidation of **154** could give achiral azadiene **155**, which is postulated to undergo intramolecular Diels–Alder reaction to give either (+)- or (−)-stephacidin A depending on the enantiofacial selectivity of the reaction (Scheme 4.27).

Tryptophan derivative **142** used in the above synthesis of stephacidin A was also employed in the biomimetic total synthesis of marcfortine C (**164**) [33]. Pipecolic acid derivative **156** was coupled to acid **142** to form amide **157**, which underwent cyclization to the dioxopiperazine following Fmoc-deprotection to give **158** as

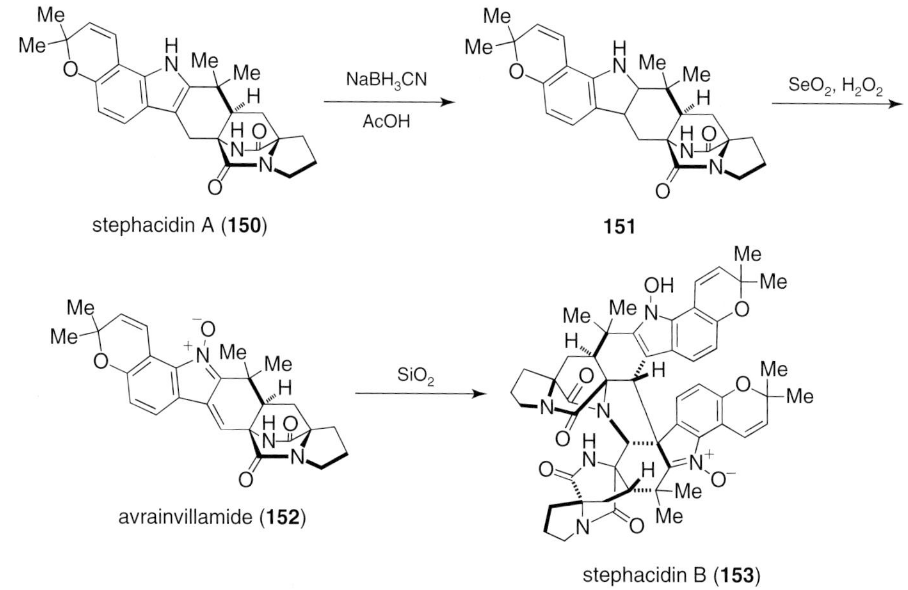

Scheme 4.26 Biomimetic conversion of stephacidin A into stephacidin B.

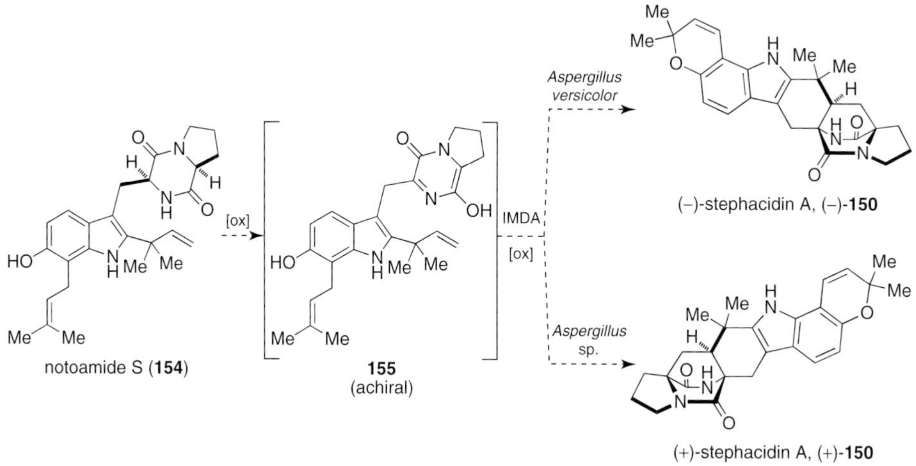

Scheme 4.27 Proposed biosynthesis of (+)- and (−)-stephacidin A through notoamide S.

an inconsequential mixture of diastereomers (Scheme 4.28). The biomimetic Diels–Alder reaction preferred the *syn*-diastereomer **160** in 2.4-fold excess, as expected. Excess DIBAL-H selectively reduced the tertiary amide of **160**, and amine salt formation followed by a biomimetic oxidative rearrangement gave marcfortine C (**164**).

Malbrancheamide and malbrancheamide B were synthesized via the biomimetic Diels–Alder reaction of enamides **165** and **166** to afford both *syn*-cycloadducts **167**

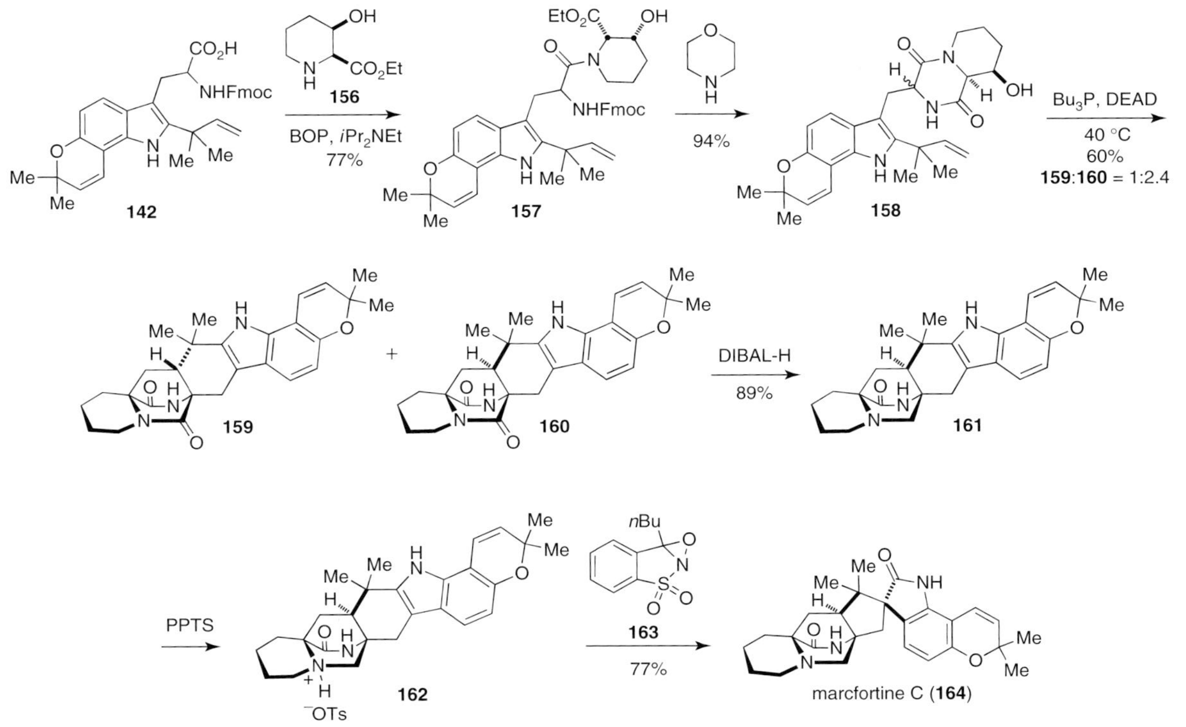

Scheme 4.28 Total synthesis of marcfortine C.

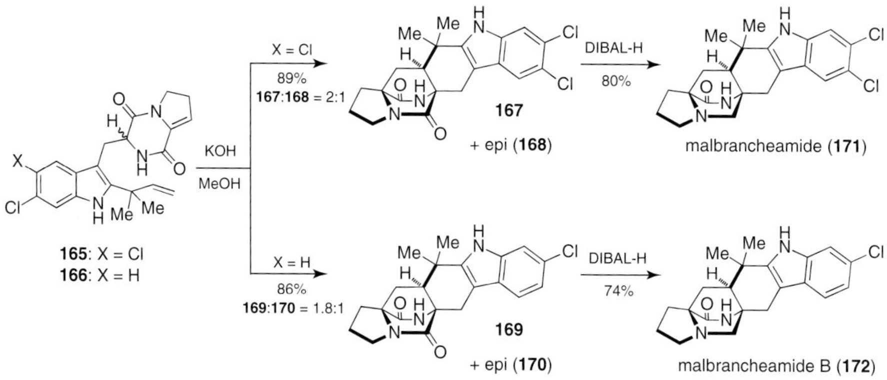

Scheme 4.29 Completion of the total syntheses of the malbrancheamides.

and **169**, as well as the *anti*-epimers **168** and **170** (Scheme 4.29) [34]. Treatment of the *syn*-cycloadducts with excess DIBAL-H gave either malbrancheamide or malbrancheamide B.

As the malbrancheamides were the first of this family of prenylated indole alkaloids to possess a halogenated indole ring, Williams and coworkers probed the biosynthesis to establish the timing of the chlorination event [35]. Malbrancheamide is proposed to arise from tryptophan, proline, and dimethylallyl diphosphate, leading to deoxybrevianamide E (**18**, Scheme 4.30). Oxidation of **18** could give intermediate **174**, which is expected to suffer intramolecular Diels–Alder reaction

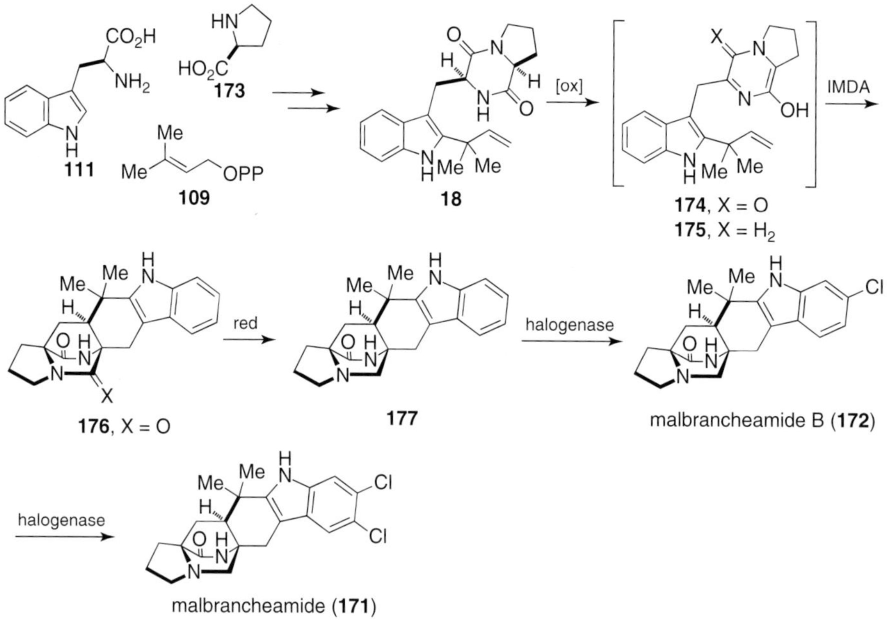

Scheme 4.30 Proposed biosynthesis of the malbrancheamides.

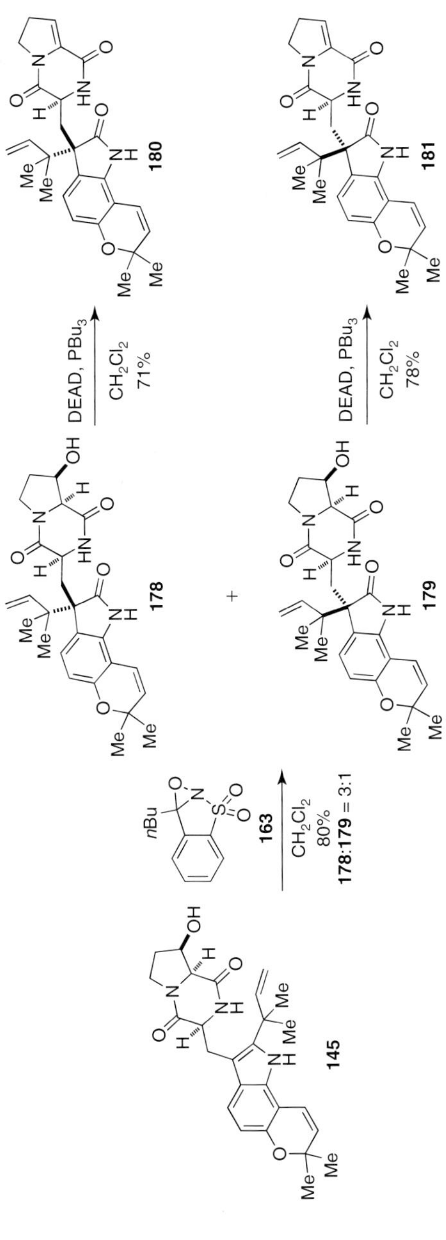

Scheme 4.31 Synthesis of key oxindoles.

(IMDA) to cycloadduct **176**. Reduction of the tertiary amide would provide premalbrancheamide (**177**). Alternatively, **18** could suffer reduction of the tertiary amide to **175**, providing **177** directly upon cycloaddition. Premalbrancheamide is proposed to undergo subsequent halogenation events to give both malbrancheamide B and malbrancheamide. In feeding studies, labeled dioxopiperazine **176** and premalbrancheamide (**177**) were added to separate cultures of *Malbranchea aurantiaca*, but interestingly only premalbrancheamide **177** was incorporated into malbrancheamide B. This suggests that reduction of the tertiary amide must precede Diels–Alder construction of the bicyclo[2.2.2]diazaoctane core through an intermediate analogous to monooxopiperazine **175**.

Williams has reported that–like stephacidin A and notoamide B, which are produced in Nature as distinct enantiomers–the minor metabolite versicolamide B is likewise produced as distinct enantiomers in different strains of *Aspergillus* sp. The asymmetric syntheses of both (+)- and (−)-versicolamide B (**182**) have recently been accomplished by deploying compound **145**, previously used in the synthesis of stephacidin A (Scheme 4.31). Oxaziridine oxidation of **145** and pinacol-type rearrangement consequent to oxidation of the indole gave a 3 : 1 separable mixture of **178** and **179** [36]. As previously reported, treatment with tributyl phosphine and DEAD induced Mitsunobu dehydration to give enamides **180** and **181**.

Treatment of **180** and **181** individually with potassium hydroxide in methanol induced the intramolecular Diels–Alder reaction to afford either (+)- or (−)-versicolamide B (**182**), along with the minor *anti*-diastereomers (+)-**183** or (−)-**183** (Scheme 4.32). As previous biomimetic Diels–Alder reactions containing indole-based azadienes display *syn*-selectivity (typically ~2.5 : 1 *syn* : *anti*), the exclusive selectivity for the *anti*-products in the versicolamide B syntheses is of particular interest. The authors suggest that the *anti*-preference stems from a stable transition state leading to the *anti*-cycloadduct when an oxindolic azadiene is employed, whereas the *syn*- and *anti*-transition states arising from an indolic azadiene are of roughly equal stability [14, 37].

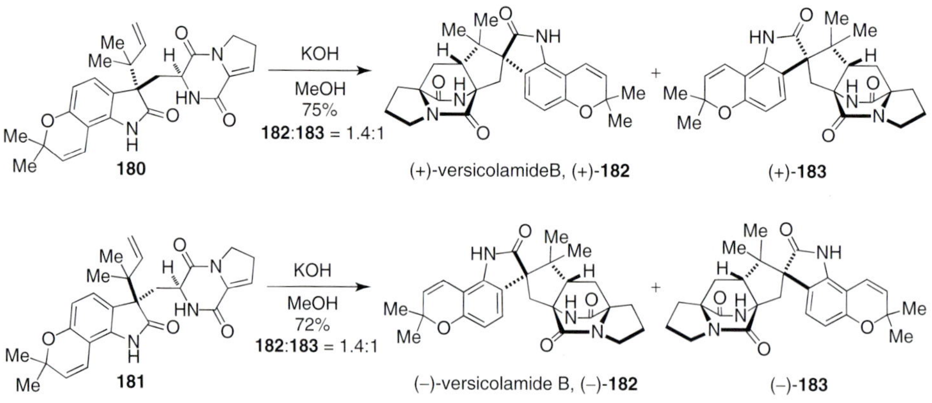

Scheme 4.32 Completion of the asymmetric syntheses of the versicolamides.

While the above syntheses have differed in the approach to the azadiene, all share a common biomimetic intramolecular Diels–Alder reaction to give the bicyclo[2.2.2]diazaoctane core featured in the molecules presented in this section.

4.3
Non-prenylated Indole Alkaloids

Hart and coworkers have expressed interest in the biosynthesis of the fumiquinazoline family of alkaloid natural products, and reported a biomimetic synthesis of *ent*-alantrypinone as the initial effort at a synthetic program to access more complex, related substances [38]. L-Tryptophan methyl ester was coupled to isatoic anhydride to afford amide **184** in good yield (Scheme 4.33). Acylation of **184** with acyl chloride **186** (prepared in two steps from *S*-methyl-L-cysteine, **185**) under Schotten–Baumann conditions furnished diamide **187**, which underwent cyclodehydration to iminobenzoxazine **188**. Treatment with excess Li[Me_3AlSPh] effected the rearrangement to quinazolinone **189**, and Fmoc deprotection was accompanied by amide formation to give **190**. Oxidation of **190** provided the sulfoxide, which was converted into enamide **191** upon heating in benzene with triphenylphosphine. Trifluoroacetic acid induced the conversion into bicycle **192**, presumably through intramolecular electrophilic attack of an intermediate *N*-acyliminium ion onto indole. Conversion into the oxindole proceeded via oxidative rearrangement of **192** with NBS to give the polybrominated indolinone, which was hydrogenolyzed over platinum on carbon to give *ent*-alantrypinone (**193**), along with *ent*-17-epi-alantrypinone (**194**).

Three dimeric tryptophan-derived dioxopiperazines have succumbed to biomimetic total syntheses, all completed by Movassaghi and coworkers, namely, (+)-WIN 64821, (−)-ditryptophenaline, and (+)-11,11′-dideoxyverticillin A [39, 40]. Syntheses of the former two began with cleavage of the Boc carbamate to effect the cyclization to dioxopiperazine **196** (Scheme 4.34) [40]. Treatment with bromine then gave a separable mixture of two diastereomers, *endo*-(+)-**197** and *exo*-(−)-**198**. *endo*-Bromide (+)-**197** was carried on to (+)-WIN 64821 (**201**) by, first, treatment with tris(triphenylphosphine)cobalt chloride to afford the dimerized product **199**, which was globally deprotected upon exposure to samarium diiodide. (+)-WIN 64821 was thus obtained in 75% yield. Similarly, (−)-ditryptophenaline (**202**) was synthesized following methylation of *exo*-(−)-**198**, dimerization, and deprotection to give the product in 79% yield.

4.3.1
Epidithiodioxopiperazines

(+)-11,11′-Dideoxyverticillin A (**211**) differs from dimers **201** and **202** in that it is likely derived from L-tryptophan and L-alanine, rather than from L-phenylalanine. Additionally, the dioxopiperazines are bridged by a disulfide, adding a difficult challenge to the synthetic construction of the molecule. Similar to their previous

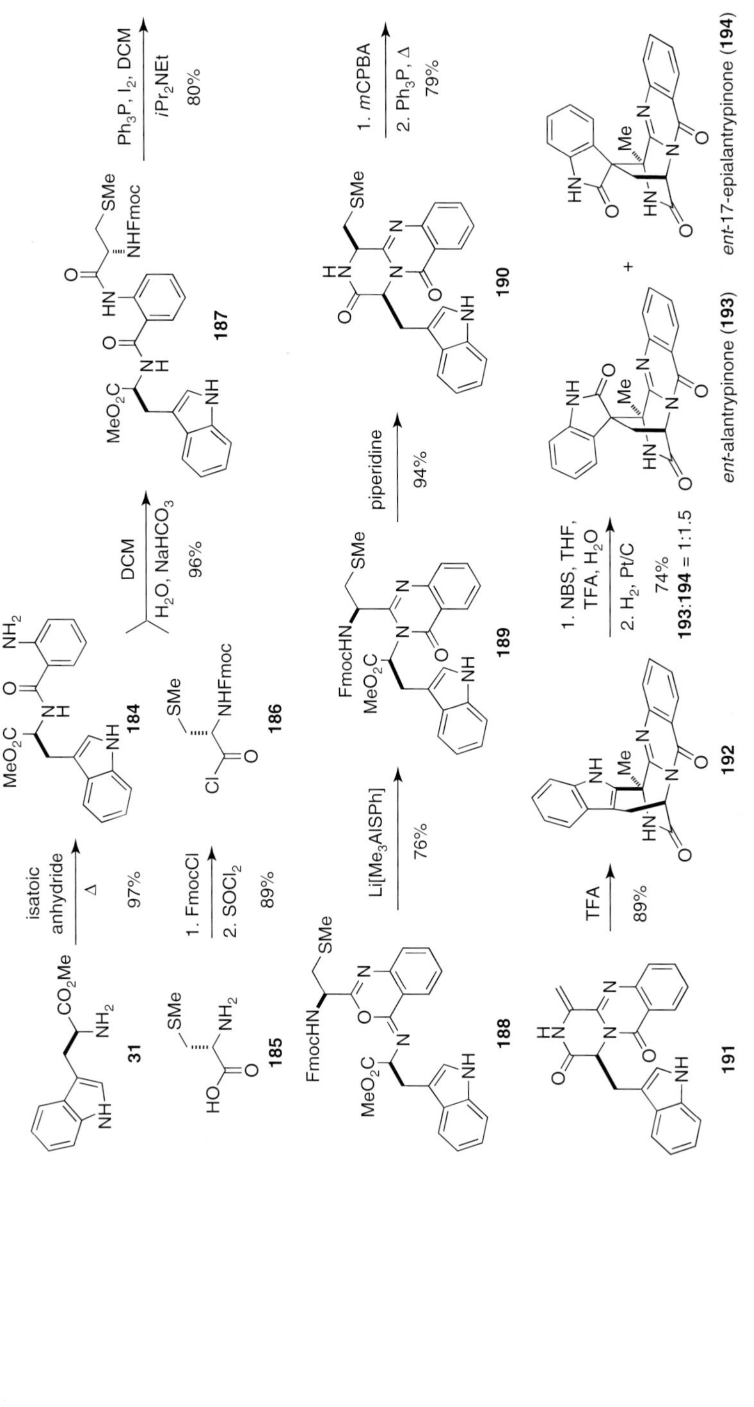

Scheme 4.33 Synthesis of *ent*-alantrypinone.

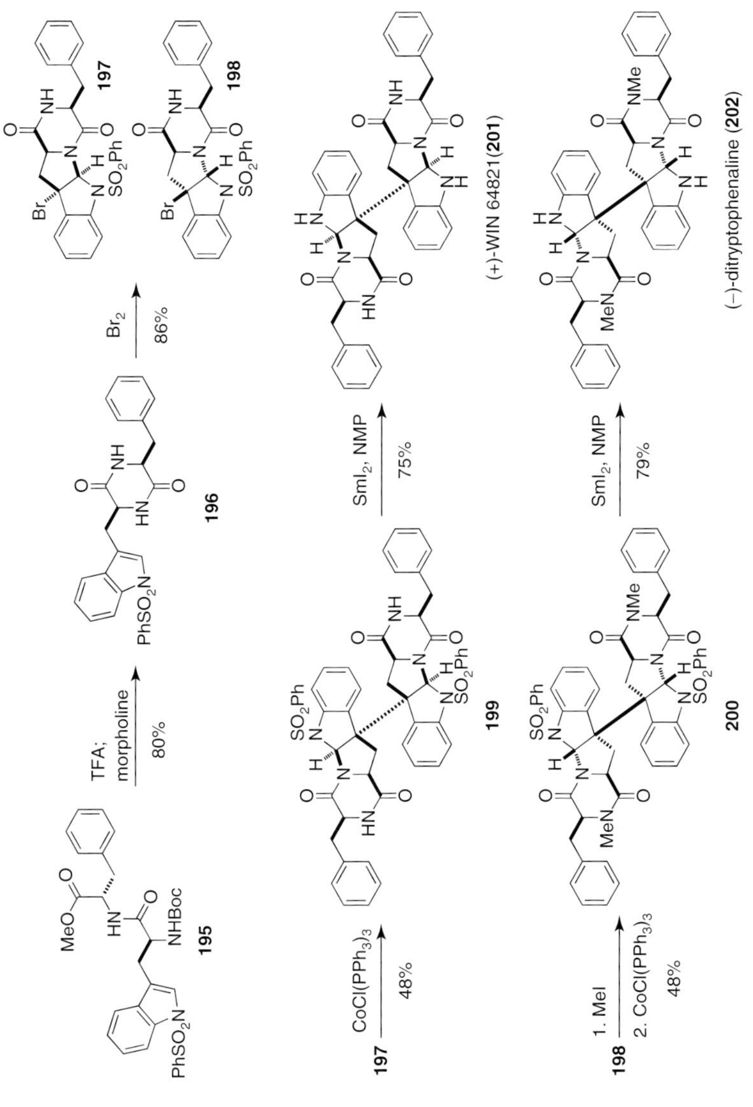

Scheme 4.34 Concise total syntheses of (+)-WIN 64821 and (−)-dityptophenaline.

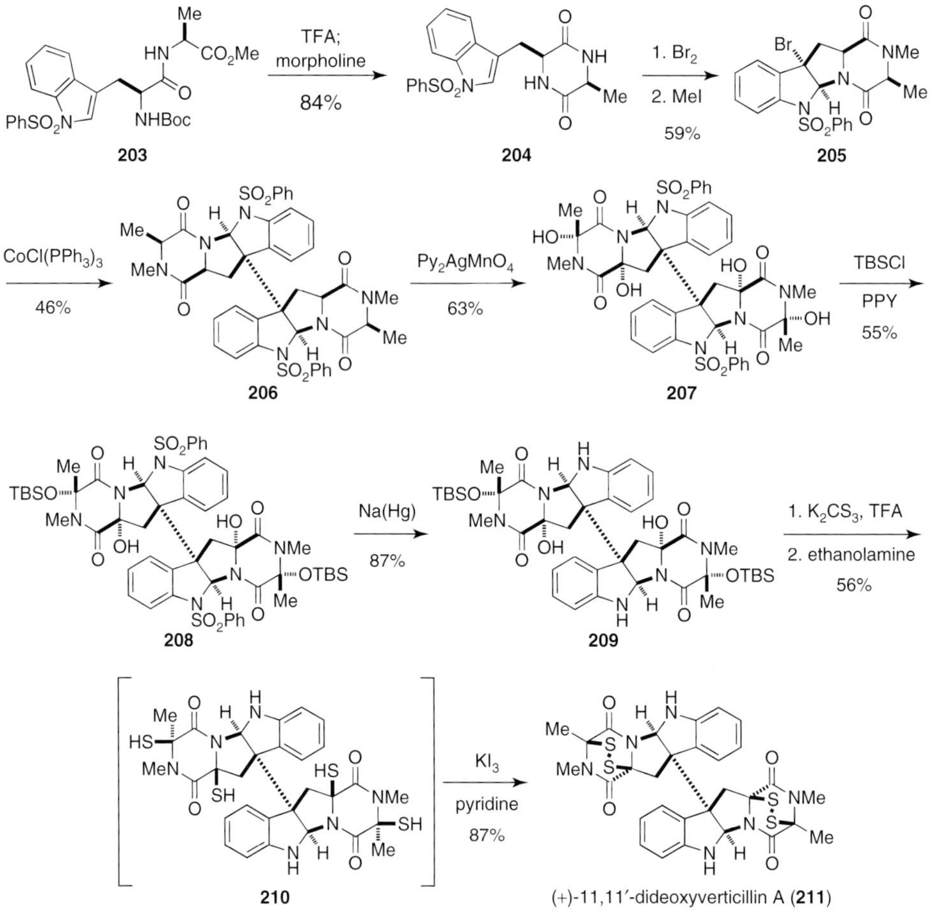

Scheme 4.35 Biomimetic total synthesis of (+)-11,11′-dideoxyverticillin A.

work, Movassaghi and coworkers showed that cleavage of the *N*-Boc carbamate was accompanied by cyclization to dioxopiperazine **204** (Scheme 4.35) [39]. Exposure of **204** to bromine produced the 3-bromopyrroloindoline, and the amides were subsequently methylated upon treatment with iodomethane. Reductive dimerization with the cobalt(I) complex as before gave the desired dimeric intermediate **206**. The dimer was oxidized with bis(pyridine)-silver(I) permanganate to octacycle **207**, and exposure to Fu's *(R)*-(+)-4-pyrrolidinopyridinyl(pentamethylcyclopentadienyl)iron (PPY) catalyst with t-butyldimethylsilyl chloride (TBSCl) gave selectively the alanine-derived protected hemiaminals of **208**. Removal of the benzenesulfonyl groups with sodium amalgam revealed diaminodiol **209**. Treatment of **209** with K_2CS_3 followed by ethanolamine gave diaminotetrathiol **210**, which readily oxidized to (+)-11,11′-dideoxyverticillin A (**211**) when partitioned between aqueous hydrochloric acid and dichloromethane and treated with potassium triiodide.

The total synthesis of sporidesmin A was completed by Kishi and coworkers in 1973. In a series of communications, Kishi described a novel strategy for the synthesis of epidithiodioxopiperazines using a dithioacetal moiety as a protecting group for the disulfide bridge [41–43]. Thus protected, the dithioacetal is stable to acidic, basic, and reducing conditions, allowing for the introduction of thiol groups at an early stage in a total synthesis. Synthesis of the sporidesmins began with the treatment of dioxopiperazine **212** with the dithiane derivative of *p*-anisaldehyde in the presence of acid to afford dithioacetal-protected dioxopiperazine **213** (Scheme 4.36) [43]. Condensation with acid chloride **214** and subsequent methoxymethyl deprotection gave compound **215**. Treatment of ketone **215** with DIBAL-H at −78 °C resulted in stereoselective reduction to the alcohol, which was then converted into acetate **216** in 80% yield. Cyclization to the diacetate (**217**) proceeded upon addition of iodosobenzene diacetate, and hydrolysis of the acetates gave the corresponding diol. Treatment of the diol with *m*-chloroperbenzoic acid (*m*CPBA) afforded an intermediate sulfoxide, which decomposed to the disulfide upon exposure to strong Lewis acid, revealing (±)-sporidesmin A (**218**).

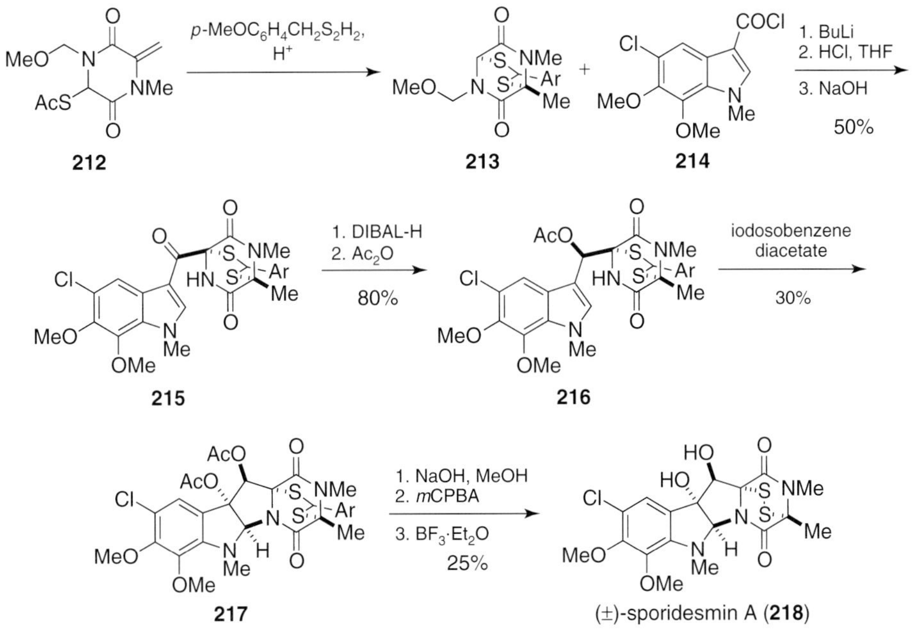

Scheme 4.36 Total synthesis of (±)-sporidesmin A.

The biosynthesis of gliotoxin is believed to proceed through the intramolecular nucleophilic ring-opening of a phenylalanine-derived arene oxide and has been the subject of considerable speculation and interest. Kishi and coworkers drew inspiration from these biogenetic hypotheses in devising a brilliant total synthesis of gliotoxin. The total synthesis of (±)-gliotoxin was completed in 1976 utilizing the same disulfide protecting strategy as deployed above for the sporidesmins, and was re-engineered in 1981 by the same route starting from optically pure

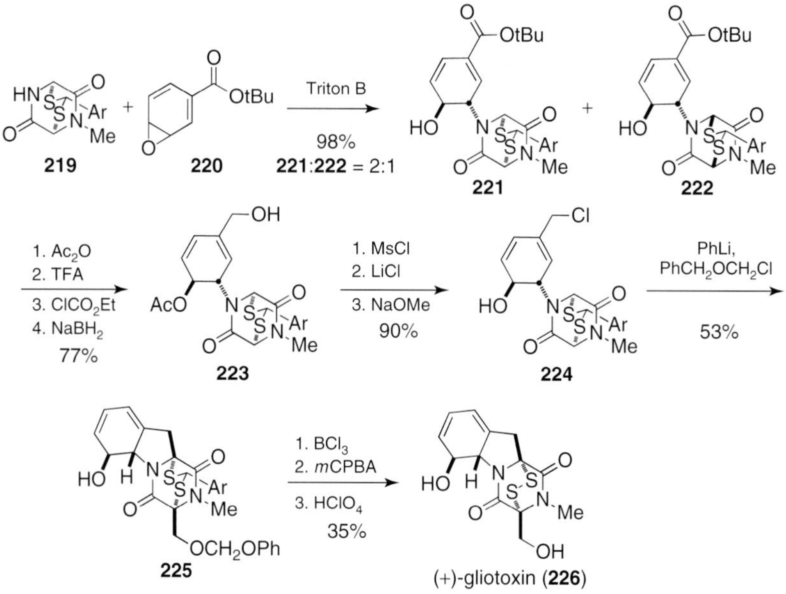

Scheme 4.37 Kishi's total synthesis of (±)- and (+)-gliotoxin.

dithioacetal **219** obtained from resolution (Scheme 4.37) [44, 45]. Coupling of **219** with *t*-butoxy arene oxide **220** in the presence of triton B afforded **221** and **222** in a 2 : 1 ratio. Acylation, deprotection, mixed anhydride formation, and reduction gave alcohol **223** in 77% yield. Alcohol **223** was converted into the chloride following mesylation, and then deprotected to reveal alcohol **224**. The key stereoselective cyclization–alkylation reaction was achieved upon addition of phenyllithium to **224** and phenoxymethyl chloride, affording cycloadduct **225** in modest yield (53%). The primary alcohol was revealed upon removal of the benzyl ether, and the thioacetal oxidatively removed to afford either (±)- or (+)-gliotoxin (**226**).

4.4 Conclusion

Dioxopiperazine alkaloids cover an astonishing array of molecular architecture and, with that, corresponding synthetic challenges to construct such substances. The biosynthesis of many of the natural substances touched on in this chapter has, in some instances, been studied going back several decades, and many workers have sought to exploit insights from Nature's strategic bond constructions in a synthetic laboratory context. Advances in whole genome sequencing have brought new and invigorated interest in elucidating the biosynthesis of structurally intriguing and biomedically relevant secondary metabolites. The insights to be gained from educated guess work on what specific compounds might lie along a possible biosynthetic pathway, traditionally accomplished by isolation and structural elucidation to map metabolite co-occurrence in conjunction with isotopically labeled precursor incorporation experiments, is in the process of giving way to a much higher resolution picture of secondary metabolism in microorganisms and plants. With the advent of powerful new genomics and proteomics tools to study and manipulate secondary metabolite production, advances in our understanding of Nature's creative synthetic palette will surely explode in the coming years. The fruits of these insights will undoubtedly be extensively exploited by synthetic chemists working at the forefront of complex molecule synthesis. In addition, many new natural products–the biosynthetic intermediates themselves, often isolated in trace amounts if at all–will provide and constitute worthy new synthetic targets and substrates for various important applications.

Acknowledgment

R.M.W. is grateful to the National Institutes of Health (GM068011; CA70375; CA85419) for financial support.

References

1. Birch, A.J. and Russell, R.A. (1972) *Tetrahedron*, **28**, 2999–3008.
2. Birch, A.J. and Wright, J.J. (1969) *J. Chem. Soc., Chem. Commun.*, 644–645.
3. Birch, A.J. and Wright, J.J. (1970) *Tetrahedron*, **26**, 2329–2344.
4. Baldas, J., Birch, A.J., and Russell, R.A. (1974) *J. Chem. Soc., Perkin Trans. 1*, 50–52.
5. Sanz-Cervera, J.F., Glinka, T., and Williams, R.M. (1993) *Tetrahedron*, **49**, 8471–8482.
6. Kametani, T., Kanaya, N., and Ihara, M. (1980) *J. Am. Chem. Soc.*, **102**, 3972–3975.
7. Schkeryantz, J.M., Woo, J.C.G., Siliphaivanh, P., Depew, K.M., and Danishefsky, S.J. (1999) *J. Am. Chem. Soc.*, **121**, 11964–11975.
8. von Nussbaum, F. and Danishefsky, S.J. (2000) *Angew. Chem., Int. Ed.*, **39**, 2175–2178.
9. Edmondson, S., Danishefsky, S.J., Sepp-Lorenzino, L., and Rosen, N. (1999) *J. Am. Chem. Soc.*, **121**, 2147–2155.
10. Finefield, J.M. and Williams, R.M. (2010) *J. Org. Chem.*, **75**, 2785–2789.
11. Baran, P.S., Guerrero, C.A., and Corey, E.J. (2003) *J. Am. Chem. Soc.*, **125**, 5628–5629.
12. Depew, K.M., Marsden, S.P., Zatorska, D., Zatorski, A., Bornmann, W.G., and Danishefsky, S.J. (1999) *J. Am. Chem. Soc.*, **121**, 11953–11963.
13. Porter, A.E.A. and Sammes, P.G. (1970) *J. Chem. Soc., Chem. Commun.*, 1103.
14. Domingo, L.R., Sanz-Cervera, J.F., Williams, R.M., Picher, M.T., and Marco, J.A. (1997) *J. Org. Chem.*, **62**, 1662–1667.
15. Williams, R.M., Sanz-Cervera, J.F., Sancenon, F., Marco, J.A., and Halligan, K. (1998) *J. Am. Chem. Soc.*, **120**, 1090–1091.
16. Stocking, E.M., Sanz-Cervera, J.F., and Williams, R.M. (2000) *J. Am. Chem. Soc.*, **122**, 1675–1683.
17. Jin, S.D., Wessig, P., and Liebscher, J. (2001) *J. Org. Chem.*, **66**, 3984–3997.
18. Sanz-Cervera, J.F. and Williams, R.M. (2002) *J. Am. Chem. Soc.*, **124**, 2556–2559.
19. Adams, L.A., Valente, M.W.N., and Williams, R.M. (2006) *Tetrahedron*, **62**, 5195–5200.

20. Greshock, T.J., Grubbs, A.W., Tsukamoto, S., and Williams, R.M. (2007) *Angew. Chem. Int. Ed.*, **46**, 2262–2265.
21. Greshock, T.J. and Williams, R.M. (2007) *Org. Lett.*, **9**, 4255–4258.
22. Herzon, S.B. and Myers, A.G. (2005) *J. Am. Chem. Soc.*, **127**, 5342–5344.
23. Myers, A.G. and Herzon, S.B. (2003) *J. Am. Chem. Soc.*, **125**, 12080–12081.
24. Baran, P.S., Guerrero, C.A., Ambhaikar, N.B., and Hafensteiner, B.D. (2005) *Angew. Chem. Int. Ed.*, **44**, 606–609.
25. Baran, P.S., Guerrero, C.A., Hafensteiner, B.D., and Ambhaikar, N.B. (2005) *Angew. Chem. Int. Ed.*, **44**, 3892–3895.
26. Baran, P.S., Hafensteiner, B.D., Ambhaikar, N.B., Guerrero, C.A., and Gallagher, J.D. (2006) *J. Am. Chem. Soc.*, **128**, 8678–8693.
27. Qian-Cutrone, J., Huang, S., Shu, Y.Z., Vyas, D., Fairchild, C., Menendez, A., Krampitz, K., Dalterio, R., Klohr, S.E., and Gao, Q. (2002) *J. Am. Chem. Soc.*, **124**, 14556–14557.
28. Kato, H., Yoshida, T., Tokue, T., Nojiri, Y., Hirota, H., Ohta, T., Williams, R.M., and Tsukamoto, S. (2007) *Angew. Chem. Int. Ed.*, **46**, 2254–2256.
29. Deyrup, S.T., Swenson, D.C., Gloer, J.B., and Wicklow, D.T. (2006) *J. Nat. Prod.*, **69**, 608–611.
30. Mudur, S.V., Gloer, J.B., and Wicklow, D.T. (2006) *J. Antibiot.*, **59**, 500–506.
31. Shim, S.H., Swenson, D.C., Gloer, J.B., Dowd, P.F., and Wicklow, D.T. (2006) *Org. Lett.*, **8**, 1225–1228.
32. McAfoos, T.J., Li, S., Tsukamoto, S., Sherman, D.H., and Williams, R.M. (2010) *Heterocycles*, **82**, 461–472.
33. Greshock, T.J., Grubbs, A.W., and Williams, R.M. (2007) *Tetrahedron*, **63**, 6124–6130.
34. Miller, K.A., Welch, T.R., Greshock, T.J., Ding, Y.S., Sherman, D.H., and Williams, R.M. (2008) *J. Org. Chem.*, **73**, 3116–3119.
35. Ding, Y.S., Greshock, T.J., Miller, K.A., Sherman, D.H., and Williams, R.M. (2008) *Org. Lett.*, **10**, 4863–4866.
36. Miller, K.A., Tsukamoto, S., and Williams, R.M. (2009) *Nat. Chem.*, **1**, 63–68.
37. Domingo, L.R., Zaragoza, R.J., and Williams, R.M. (2003) *J. Org. Chem.*, **68**, 2895–2902.
38. Hart, D.J. and Magomedov, N.A. (2001) *J. Am. Chem. Soc.*, **123**, 5892–5899.
39. Kim, J., Ashenhurst, J.A., and Movassaghi, M. (2009) *Science*, **324**, 238–241.
40. Movassaghi, M., Schmidt, M.A., and Ashenhurst, J.A. (2008) *Angew. Chem. Int. Ed.*, **47**, 1485–1487.
41. Kishi, Y., Fukuyama, T., and Nakatsuk, S. (1973) *J. Am. Chem. Soc.*, **95**, 6492–6493.
42. Kishi, Y., Fukuyama, T., and Nakatsuk, S. (1973) *J. Am. Chem. Soc.*, **95**, 6490–6492.
43. Kishi, Y., Nakatsuk, S., Fukuyama, T., and Havel, M. (1973) *J. Am. Chem. Soc.*, **95**, 6493–6495.
44. Fukuyama, T. and Kishi, Y. (1976) *J. Am. Chem. Soc.*, **98**, 6723–6724.
45. Fukuyama, T., Nakatsuka, S., and Kishi, Y. (1981) *Tetrahedron*, **37**, 2045–2078.

5
Biomimetic Synthesis of Alkaloids with a Modified Indole Nucleus

Tanja Gaich and Johann Mulzer

5.1
Introduction

Indole alkaloids make up the largest group of alkaloids, with more than 2000 members. They exhibit a huge structural diversity and an often mind dazzling molecular complexity. Owing to their architectural attractiveness and their pronounced pharmacological activities they have been under extensive investigations over recent decades, including the elucidation of their biosynthetic pathways, which have been pioneered by Battersby, Arigoni, Scott, and others. The landmark works were the unraveling of the biosynthetic pathway of monoterpenoid indole alkaloids (e.g., strictosidine **9** as the first biosynthetic node of the series), especially the monoterpenic part **7** derived from geraniol diphosphate **5** (highlighted in Scheme 5.1) [1].

These early biosynthetic studies relied on the administration of isotopically labeled common starter units, followed by isolation and structural characterization. These findings inspired synthetic chemists not only to "mimic" but also predict these biogenetic transformations and probe them *in vitro*." The synthetic elegance, ease, and conciseness that can be achieved by applying a biomimetic route or even just one "biomimetic key step" has led to a myriad of awe-inspiring total syntheses. Some of these biomimetic syntheses lead to natural products in which the indole nucleus is lost and an indoline, indolenine, indoxyl, or indolinone ring system is obtained instead (Figure 5.1).

This chapter is focused on such structures with a "broken indole nucleus." We are only able to cover a small fraction of these syntheses, and therefore the selection was conducted to provide the reader with an overview of no more than 13 different structure types (Figure 5.2).

Biomimetic Organic Synthesis, First Edition. Edited by Erwan Poupon and Bastien Nay.

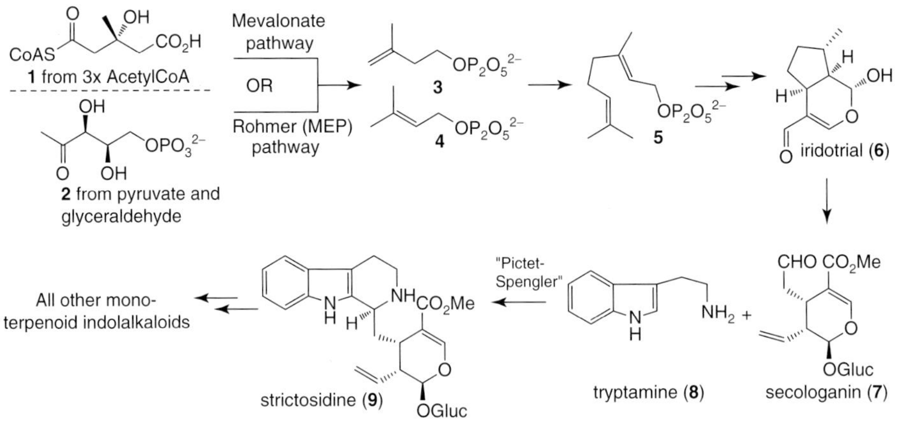

Scheme 5.1 Biosynthesis of monoterpenoid indole alkaloids.

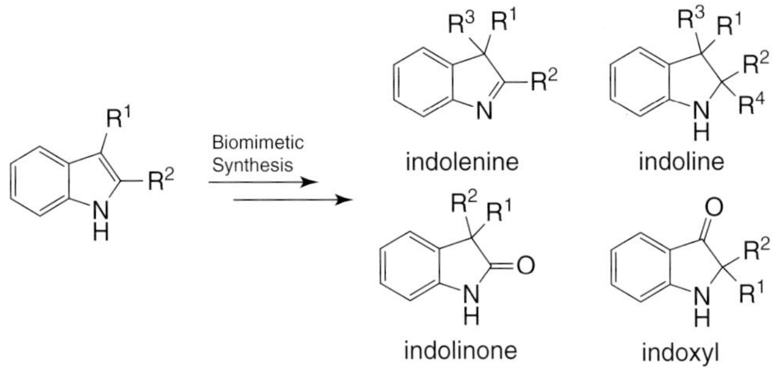

Figure 5.1 Conversion of the indole nucleus into first derivatives.

5.2 Individual Examples

5.2.1 (±)-Camptothecin

Camptothecin was discovered in 1966 by M. E. Wall and M. C. Wani in a systematic screening of natural products for anticancer drugs [2, 3]. It was isolated from the bark and stem of *Camptotheca acuminata* (Camptotheca, Happy tree), a tree native in China. Camptothecin showed remarkable anticancer activity in preliminary clinical trials, by inhibiting topoisomerase of type I enzyme, and thereby causing apoptosis. Its disadvantageous low solubility and (high) adverse drug reaction has initiated numerous syntheses of camptothecin derivatives, two of which have been approved, and are used as anticancer drugs (Figure 5.3) [4].

Camptothecin has a planar pentacyclic ring structure that includes a pyrrolo[3,4-β]-quinoline moiety (rings A, B, and C), a conjugated pyridone moiety (ring D), and one chiral center at position 20 within the α-hydroxy lactone ring

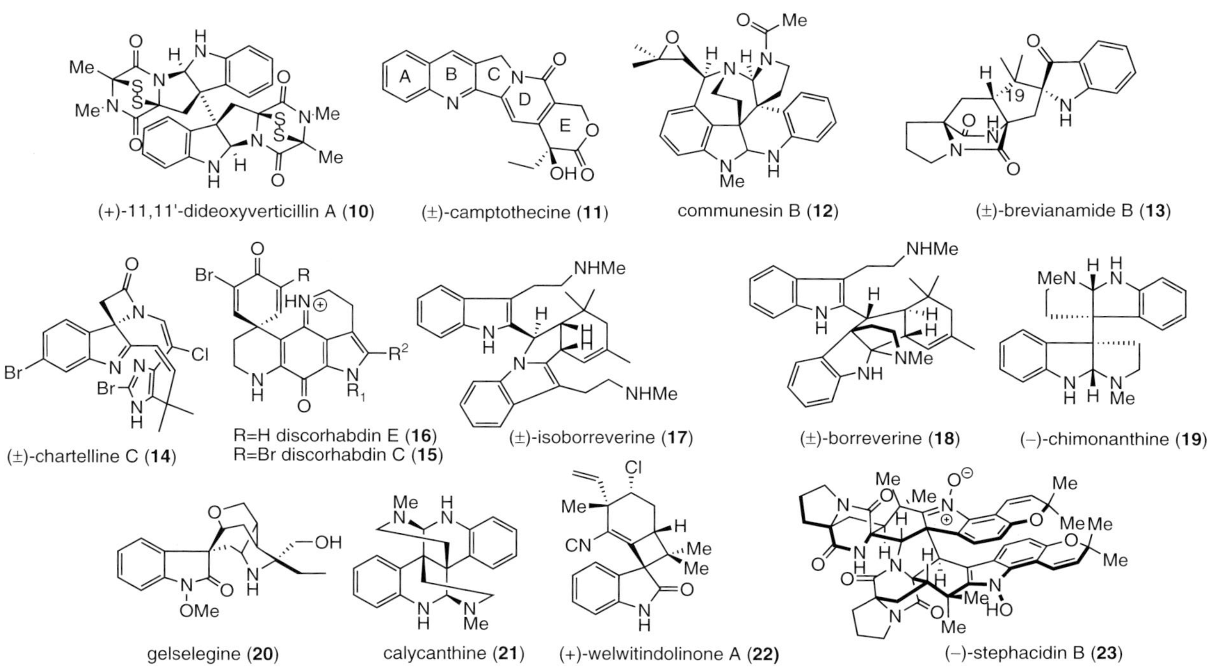

Figure 5.2 Overview of structures discussed in this chapter.

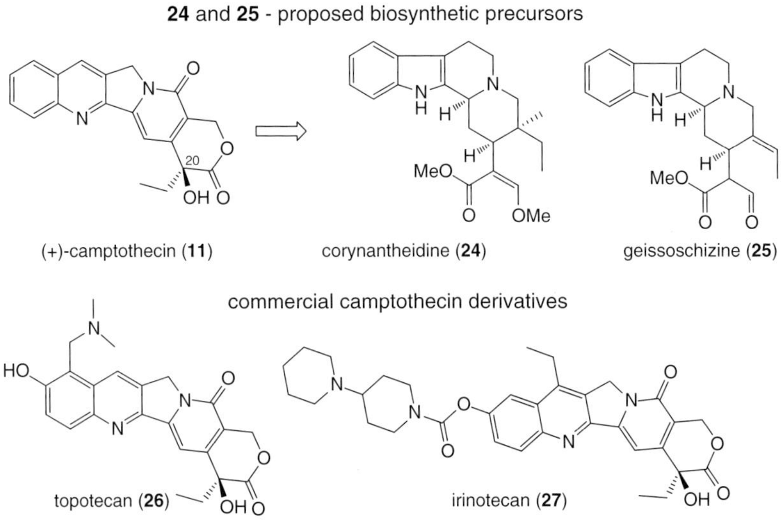

Figure 5.3 Camptothecin (**11**) with its proposed biosynthetic precursors and derivatives used as anticancer drugs.

with *(S)*-configuration (ring E). Its planar structure is thought to be one of the most important factors in topoisomerase I inhibition.

Biosynthetically, **11** is unique, because it is the first reported example of an alkaloid containing the pyrrolo[3,4-β]quinoline unit. In 1967 Wenkert [5], suggested that **11** might be formed *in vivo* from an indole alkaloid of the corynantheidine type (**24**), whereas Winterfeldt suggested a biosynthetic relationship between **11** and geissoschizine (**25**) [6]. It was Hutchinson *et al.*, in brilliant work in 1974, who finally solved the biosynthetic maze by isotope labeling experiments [7]. They were able to show that strictosidine (**9**) is unequivocally the biosynthetic precursor via which Nature accesses structure **11** (Scheme 5.2). As camptothecin (**11**) appears to be more of a quinoline than of an indole alkaloid, the question arose how **9** is converted into **11**. The first event after the formation of strictosidine (**9**) is the lactam formation of strictosamide (**28**). After this an oxidative cleavage of the indole nucleus with subsequent recyclization takes place to give the ABC-ring moiety of **11**. The last events in the biosynthesis are the placements of the correct oxidation states at rings D and E.

Soon after Hutchinson's biosynthesis was published, Winterfeldt and coworkers completed the first biomimetic total synthesis of this molecule [8], following Hutchinson's findings (Scheme 5.3). Their synthesis started from known compound **31**, which was converted into **32** via diisopropyl-carbodiimide (DIC) mediated amide bond formation. Claisen ester condensation with potassium *tert*-butoxide furnished compound **33** (86% yield). This 1,3-dicarbonyl compound was then methylated with diazomethane, and a subsequent Michael addition/elimination reaction with di-*tert*-butyl malonate gave compound **34**. Biomimetic oxidation and recyclization under mild basic conditions afforded quinolone **35** in impressive 75%

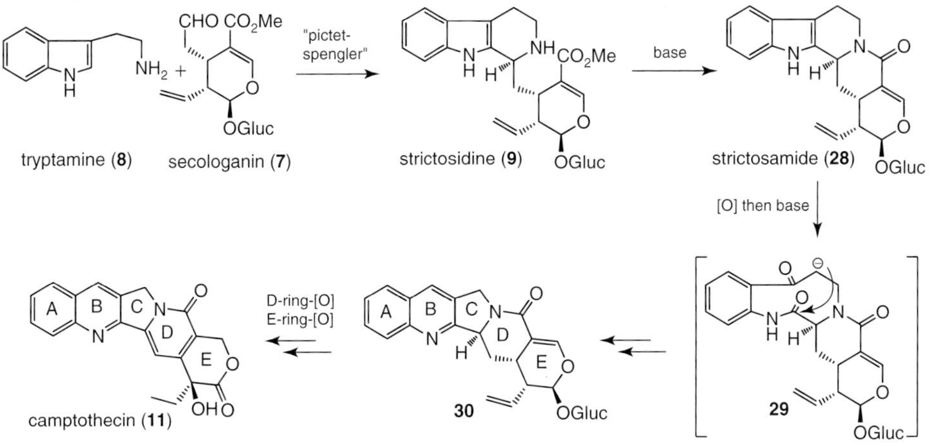

Scheme 5.2 Biosynthesis of camptothecin (11) by Hutchinson *et al.*

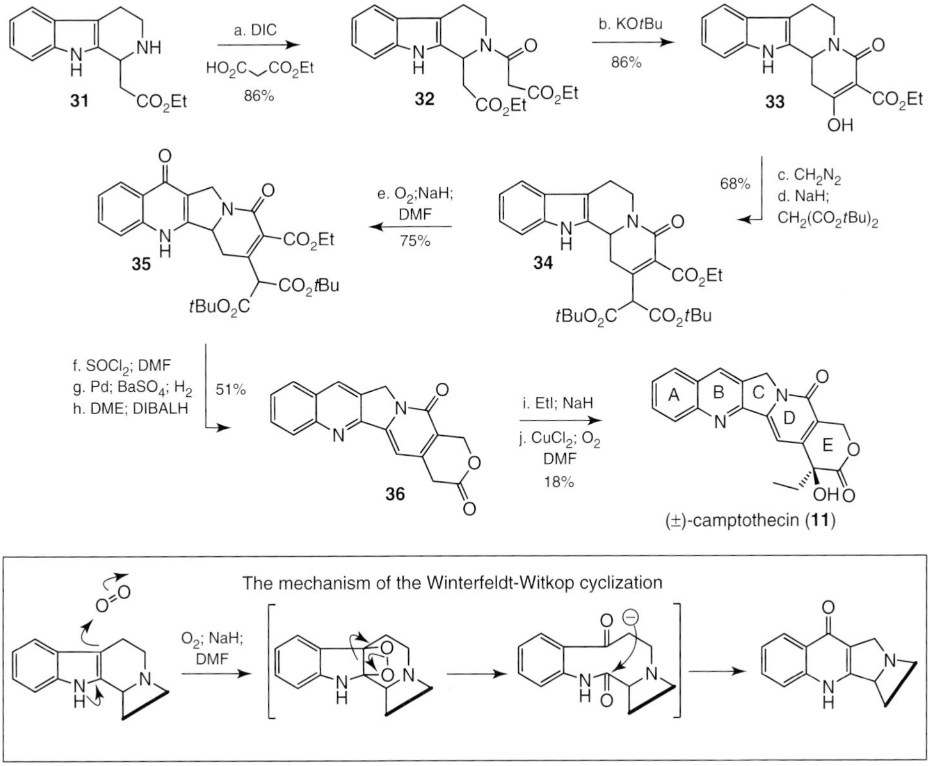

Scheme 5.3 Biomimetic synthesis of camptothecin (11) by Winterfeldt *et al.*

yield. Conveniently, spontaneous oxidation of ring D to the final oxidation state occurred. Chlorination of the quinolone moiety, followed by dechlorination under palladium catalysis and reduction of the ester functionality, led to compound **36**. This was alkylated and oxidized via copper(II) catalysis with oxygen to give the tertiary alcohol of (±)-camptothecin. This very elegant synthesis applies all the biosynthetic key reactions, namely, the formation of the quinoline from the indole moiety, the spontaneous oxidation of ring D, and the introduction of the tertiary alcohol in the final step of the total synthesis. The synthesis is straightforward and very short, even enabling facile access to camptothecin derivatives. However, owing to its biological importance, and although many synthetic approaches and syntheses have been achieved, camptothecin is still the subject of numerous synthetic investigations. Thereby, progress has been made with respect to oxidation conditions, which were developed for base-sensitive substrates [9].

5.2.2
(±)-Discorhabdins C and E

Discorhabdins, like makaluvamines and damirones, belong to the group of pyrroloiminoquinone alkaloids. Over the last 20 years they have been regularly isolated from sponges generally confined to two families, Latrunculidae and Acarnidae [10]. The sponges producing these natural products have been collected from shallow and deep-water habitats, ranging from tropical regions of the Caribbean Sea to the cold water regions of New Zealand and the Antarctica. Discorhabdins, also known as prianosins, are the major pigments in these beautiful sponges. They exhibit extremely high activity against P-388 lymphotic leukemia cell lines *in vitro* (IC_{50} values 30–1500 ng ml^{-1}). The mode of action of these natural products is the strong inhibition of the topoisomerase II enzyme, which plays a vital role in cell division and other fundamental processes in eukaryotic cells. Discorhabdins induce apoptosis and thereby serve as potent chemotherapeutic agents. The most active congener appears to be discorhabdin C (**15**).

Munro *et al.* proposed a biosynthesis that includes all pyrroloiminoquinone alkaloids (makaluvamines, damirone, and discorhabdins), which shows how these alkaloids are biogenetically connected (Scheme 5.4) [11]. Tryptamine **7** is the starting point of the proposed biosynthesis. Oxidation of **7** gives pyrroloiminoquinone **37**. 1,4-Addition of tyramine, and subsequent oxidation leads to makaluvamines (**39**). Intramolecular oxidative phenol coupling leads to the carboskeleton of discorhabdins (**40**). Sulfur-containing discorhabdins can, in principle, be accessed in two ways, either via makaluvamine F (**42**) or via discorhabdins (**40**) and a late sulfur introduction.

Numerous approaches to discorhabdins have been reported [12]. Owing to lack of space we focus here on the biomimetic synthesis of racemic discorhabdin C by Heathcock *et al.* in 1999 (Scheme 5.5) [13].

For the pyrroloiminoquinone synthesis, Heathcock *et al.* started from nitrophenol **45**. Benzylation of the phenolic alcohol and reduction of the nitro group to the aromatic amine was followed by iodination of the aromatic ring and alkylation of

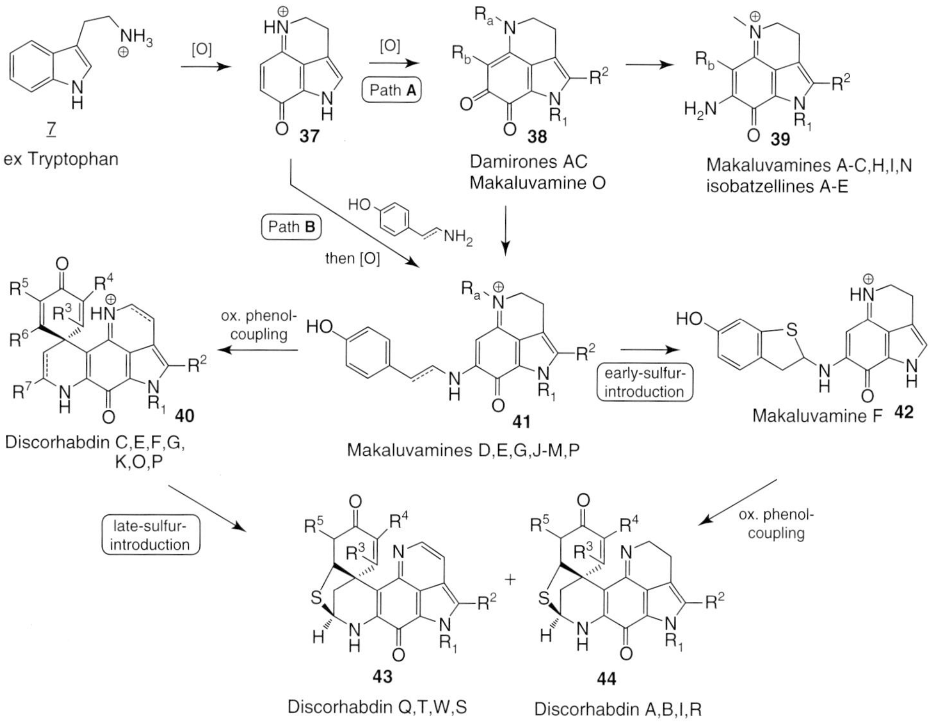

Scheme 5.4 Biosynthesis of discorhabdins proposed by Munro *et al.*

the amine, giving **46**. Palladium-catalyzed indole formation via a Heck coupling reaction and subsequent standard functional group manipulation gave Boc-protected indole **47**. This was oxidized with Frémy's salt [$Na_2NO(SO_3)_2$] to yield quinone **48**. Compound **48** was treated with trifluoroacetic acid (TFA) to deprotect the primary amine and subsequently cyclized to yield desired pyrroloiminoquinone **49** in eleven steps. Pyrroloiminoquinone **49** underwent a 1,4-addition elimination reaction with amines **A** and **B**, affording makaluvamines **50A** and **50B** in 54% yield. Application of biomimetic oxidative phenol coupling with copper(II) chloride gave the corresponding *spiro*-dienone compounds whose deprotection led to (±)-discorhabdins C (**15**) and E (**16**).

5.2.3
(±)-Brevianamides, Paraherquamides, VM55599, and Marcfortines

Prenylated indole alkaloids comprise a large group of secondary metabolites that have been isolated from various fungi (in particular *Penicillium* sp. and *Aspergillus* sp.) (Figure 5.4) [14, 15]. Owing to their molecular complexity and biogenesis these natural products have attracted considerable attention. Paraherquamides (e.g., **52**) most notably display potent anti-parasitic activity.

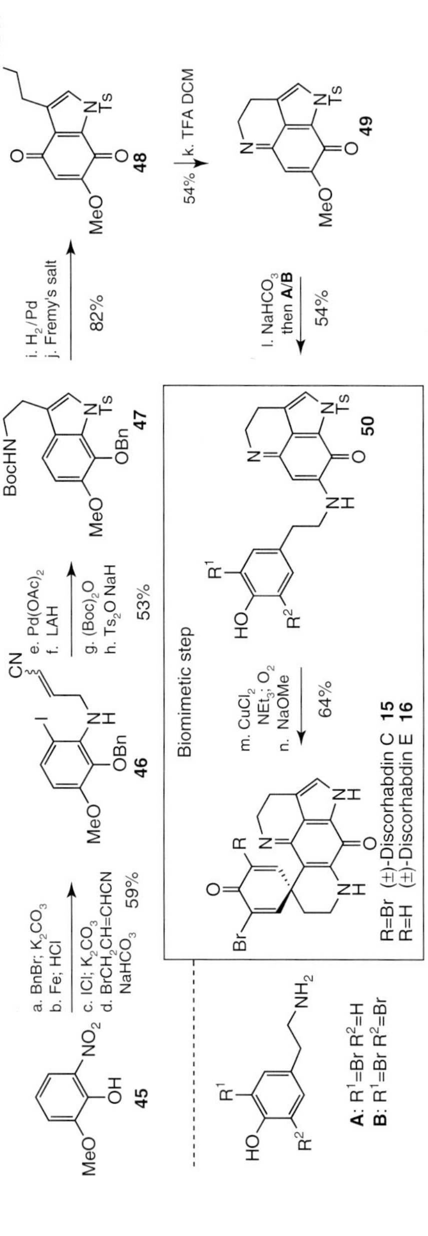

Scheme 5.5 Biomimetic synthesis of discorhabdins C (**15**) and E (**16**) by Heathcock *et al.*

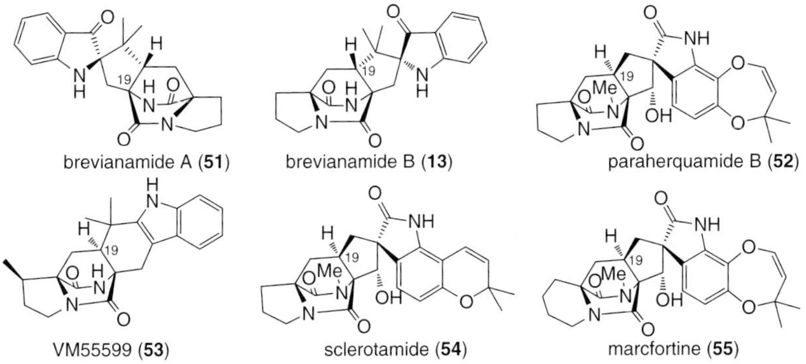

Figure 5.4 Selected members of the family of prenylated indole alkaloids.

The key feature of all these structures is their azabicyclo[2.2.2] ring system that has been proposed to arise from a biogenetic [4+2]-cycloaddition of the isoprene moiety across the diketopiperazine unit (Scheme 5.6) [16]. The relative configuration at C19 (brevianamide numbering) is different within this natural product family. For brevianamides A and B (**51** and **13**, respectively) it is *anti*-, but for all others it is a *syn*-relationship. This difference is explained by the facial divergence in the selectivity of the Diels–Alder reaction, which sets the relative stereochemistry of this carbon center. Although the precursor is an achiral molecule, the brevianamides exhibit optical activity; this example might be a strong

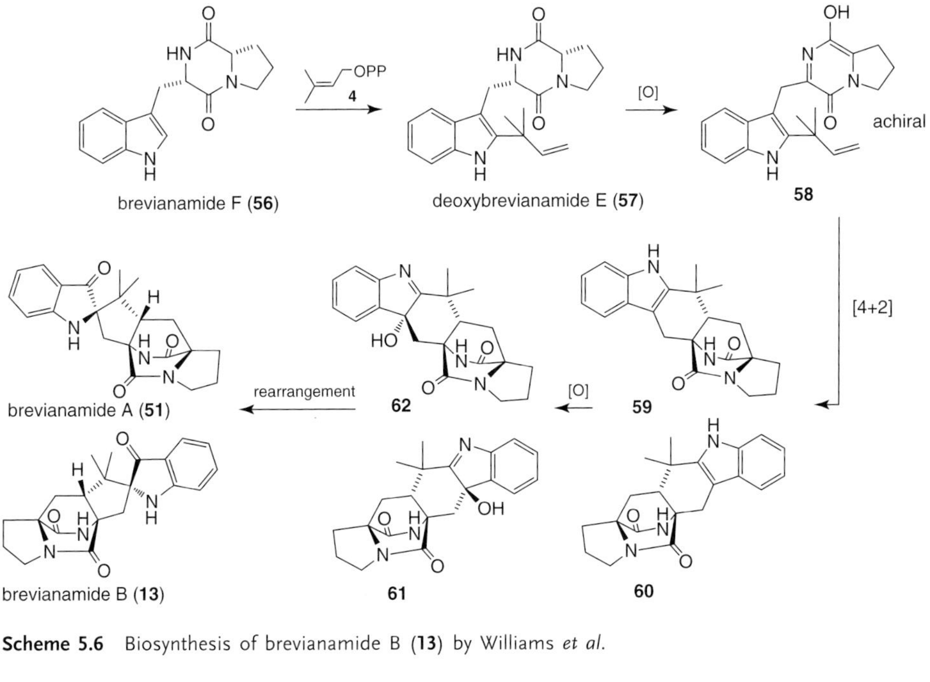

Scheme 5.6 Biosynthesis of brevianamide B (**13**) by Williams *et al.*

hint and important case in the search for the Diels–Alderase enzyme [17]. Except for catalytic antibodies, to date only three reported examples of a Diels–Alderase exist, in addition to ribozymes that are known to catalyze Diels–Alder reactions [18].

The group of R.M. Williams, which is one of the leaders in biosynthetic investigations concerning this class of natural products, has completed numerous total syntheses of several congeners of this class [19], among them biomimetic total syntheses of VM55599 (**53**) [20], and brevianamide B (**13**), with the latter being discussed here (see also Chapter 4) [21]. This synthesis started from compound **57**, which was converted into the lactim ether **62** with Meerwein's salt (Scheme 5.7). Subsequent oxidation of **62** with DDQ (2,3-dichloro-5,6-dicyano-1,4-benzoquinone) gave diene **63**, which was treated with methanolic potassium hydroxide to give *in situ* the desired azadiene **64**, which smoothly underwent an intramolecular Diels–Alder reaction to give the central azabicyclo[2.2.2]ring system that is common to all members of this natural product family. In this case, brevianamide B precursors **65** and **66** were formed in a 2 : 1 ratio in favor of the undesired C19-diastereomer **66**. Oxidation with *m*CPBA (*m*-chloroperbenzoic acid), and treatment with HCl gave indoxyls **67** and **68**. These intermediates under basic conditions underwent ring contraction to form, after acid hydrolysis of the lactim ether, (±)-brevianamide B (**13**) and the 19-epi-compound **69**. This elegant synthesis showcases how predictions about the biosynthetic origin of natural products influence synthetic routes, lead to new insights into chemical reactivity, and give evidence for possible enzymatic transformations of enzymes that remain to be discovered, encouraging those who are in constant search for them.

5.2.4
(+)-Stephacidin A and (−)-Stephacidin B

Prenylated indole alkaloids are an impressive example of how Nature assembles the most complex and diverse structures from a very limited number of rather simple building blocks [15, 22]. In the case of stephacidins these building blocks are tryptophan, proline, and "isoprene units." Stephacidins are fungal metabolites that were isolated from *Aspergillus ochraceus* WC76466 by Bristol Myer Squibb in 2002 and independently in 2001 by Pfizer [23]. They are reported to inhibit the growth of cultured human cancer cells (IC_{50} values 50–100 nM), and are therefore biologically interesting structures. They belong to a large family of natural products together with brevenamides, notoamides, and paraherquamides. Their biosynthesis contains two key steps. The first is an intramolecular Diels–Alder reaction to form their characteristic azabicyclo[2.2.2] ring system [16], in the same fashion as discussed above in the case of brevenamides (Scheme 5.8).

The second key step is an oxidation of stephacidin A (**72**) to avrainvillamide (**73**), which then cyclizes to give the most complex sibling of this natural product family, namely, stephacidin B (**23**). This last and most intriguing oxidation reaction was utilized first by Myers *et al.* [24] and the group of Baran [25] and then in 2007 stephacidin B (**23**) was synthesized by the group of Williams using this last

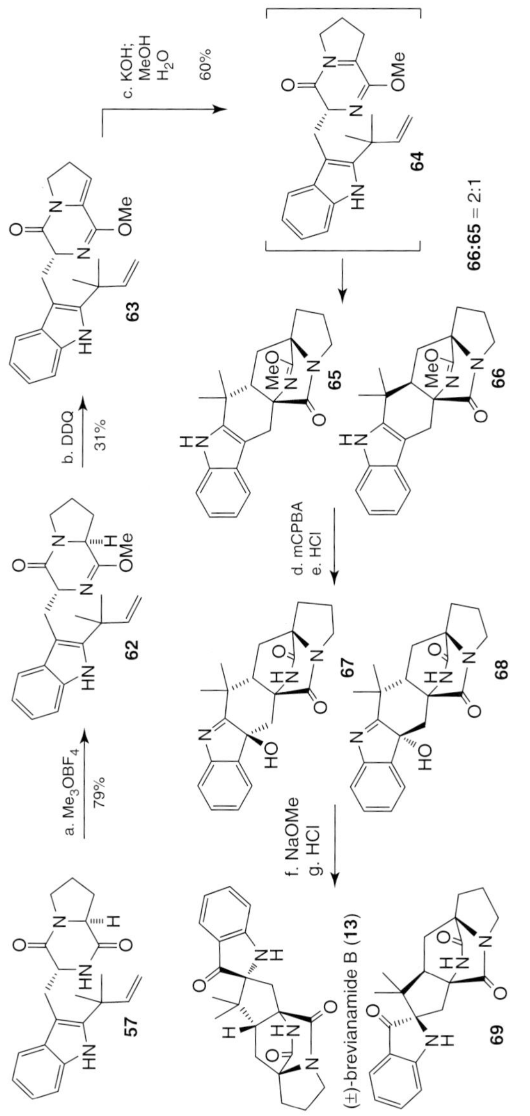

Scheme 5.7 Biomimetic synthesis of brevianamide B (**13**) by Williams *et al.*

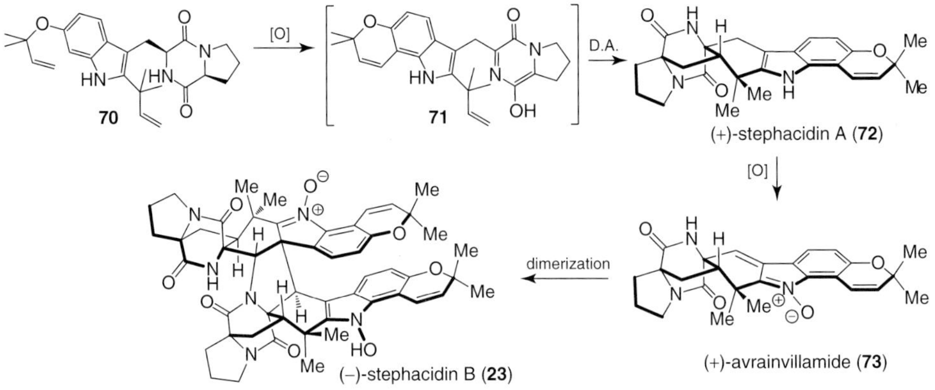

Scheme 5.8 Biosynthesis proposal for stephacidins.

oxidative dimerization reaction [26]. Baran's synthesis is highlighted here. It started with the formation of the annelated pyran ring to compound **74** in three steps, affording **75** in 69% yield (Scheme 5.9). Protection of indole and ester hydrolysis gave carboxylic acid **76**, which was reacted with amine **77** in a peptide coupling reaction to give compound **78**.

The next task was the formation of the azabicyclo[2.2.2] ring system, which was achieved via an oxidative intramolecular enolate coupling of diketopiperazine **79**. This reaction, although very elegant, is not in accord with the proposed biosynthetic transformation for this step. Removal of the MOM protective group and introduction of the *gem*-dimethyl group was followed by dehydration to give isopropylidene compound **81**, which upon heating at 200 °C in sulfolane underwent an ene-reaction that was followed by a 1,2-shift to furnish (+)-stephacidin A (**72**). Oxidation of **72** to (+)-avrainvillamide (**73**) with hydrogen peroxide and selenium dioxide was followed by treatment of **73** with triethylamine, which resulted in a sequential double Michael addition, as outlined in Scheme 5.10, to give (−)-stephacidin B (**23**). Dimerization of **73** also occurred spontaneously in an NMR tube, albeit in lower yields. The biomimetic syntheses of stephacidins show how impressively biosynthetic considerations simplify synthetic routes of mind boggling molecular architectures.

5.2.5
(±)-Chartelline C

Chartelline marine alkaloids were isolated from the bryozoan *Chartella papyracea* by Christophersen and coworkers and characterized in the 1980s [27]. Chartelline C (**14**) is the scarcest member of the chartelline alkaloid family. It consists of an indolenine, an imidazole, and β-lactam heterocycle arrayed in a dense, π-stacking framework that poses numerous challenges for its synthesis [28]. Biosynthetically, chartellines are proposed to stem from one tryptamine, one histidine, and one prenyl unit (Scheme 5.11) [29].

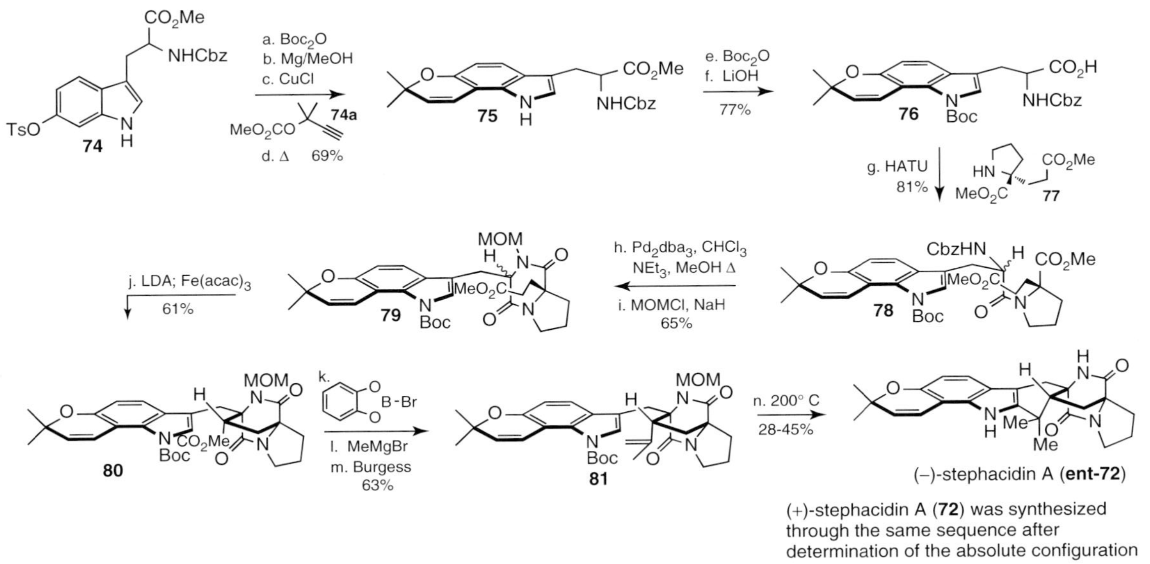

Scheme 5.9 Total synthesis of stephacidin A (**72**) by Baran *et al.*

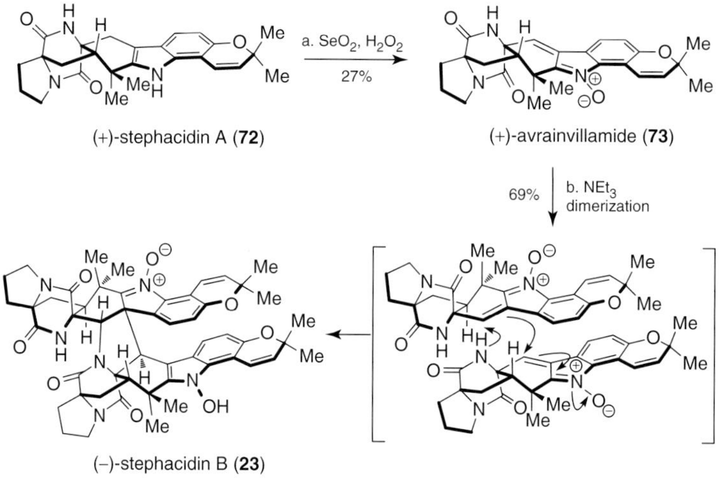

Scheme 5.10 Biomimetic synthesis of stephacidin B (**23**) by Baran *et al.*

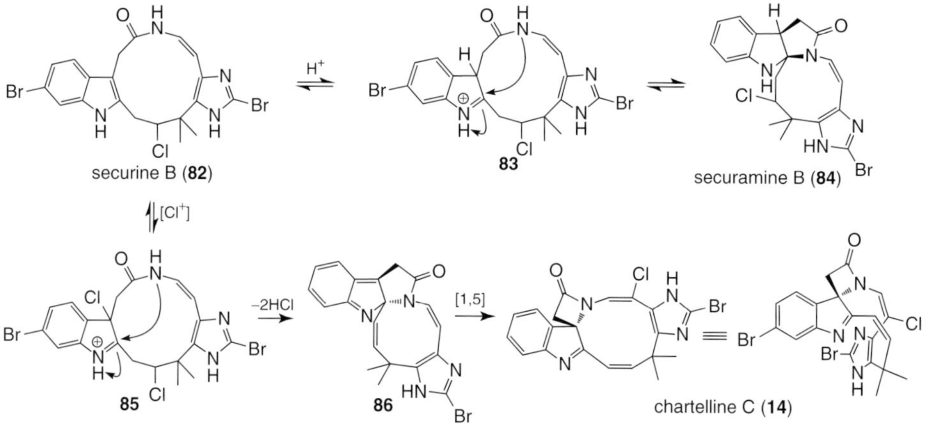

Scheme 5.11 Biosynthesis proposal for chartelline C (**14**).

A related bryozoan, *Securiflustra securifrons*, also produces the securines and securamines, and they are plausible biogenetic precursors to the chartellines. It was reported that solutions of securine B (**82**) in DMSO-d_6 converted into securamine B (**84**), presumably via the protonated intermediate **83** (Scheme 5.11). Redissolving **84** in $CDCl_3$ converted **84** back into **82**. One could imagine that if this isomerization was carried out in the presence of an electrophilic source of chlorine, **82** could be converted into chartelline C (**14**) through the intermediacy of **85** and **86**. The transformation of **86** into **14** can be written as a [1,5]-shift and is a key step in the biomimetic strategy for the synthesis of chartelline C by Baran *et al.* (Scheme 5.12) [30].

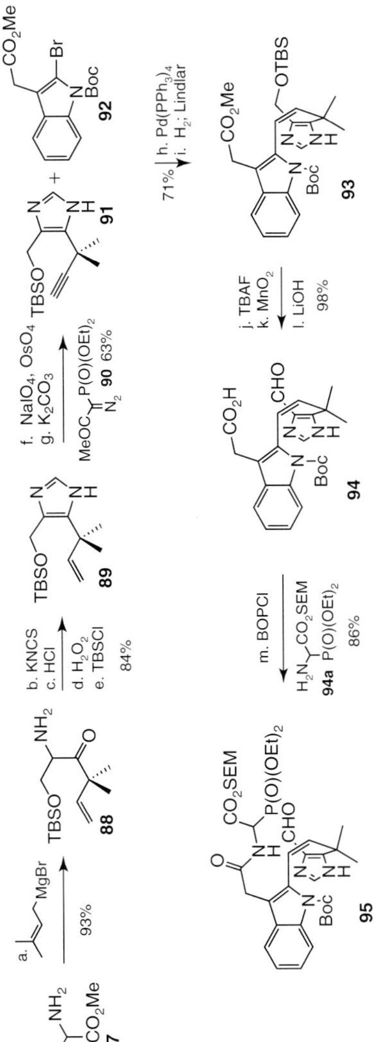

Scheme 5.12 Biomimetic synthetic route towards chartelline C (**14**) by Baran *et al.*

Their synthesis started with the construction of prenylated imidazole **89**, which was achieved in five standard transformations from serine derivative **87**. For the introduction of the *(Z)*-double bond on the macrocycle and the coupling with indole fragment **92** the double bond in **89** was converted into alkyne **91**. Sonogashira coupling of **91** and **92** gave compound **93**, which was deprotected with TBAF, oxidized to give the aldehyde functionality in **94**, and subjected to ester hydrolysis to finally furnish carboxylic acid **94**. This acid was coupled with amine **94a** to yield the substrate for the Horner–Wadsworth–Emmons reaction to close the macrocycle of chartelline C.

Masamune–Roush conditions (Scheme 5.13) afforded the macrocycle in 56% yield, which was brominated to give securine-type compound **96**. The endgame consisted of the same biosynthetic transformation as shown in Scheme 5.11 (**86**→**14**). Baran and coworkers used *N*-bromosuccinimide as an electrophilic bromine source. The reaction was carried out at 180 °C to remove the Boc-protective group, followed by bromination of the C3-position at the indole to give intermediate securamine-type structure **97**. This compound *in situ* underwent a [1,5]-shift to give the desired carboskeleton of chartelline C. Ester hydrolysis mediated by TFA was followed by decarboxylation at 200 °C to yield (±)-chartelline C (**14**). The chartellines with their unique, densely functionalized, and very strained architecture are a very good example of how biosynthetic considerations can lead to the application of new transformations in organic synthesis, if the practitioners keenly follow the biosynthetic proposal, even if it is very difficult to execute in the laboratory.

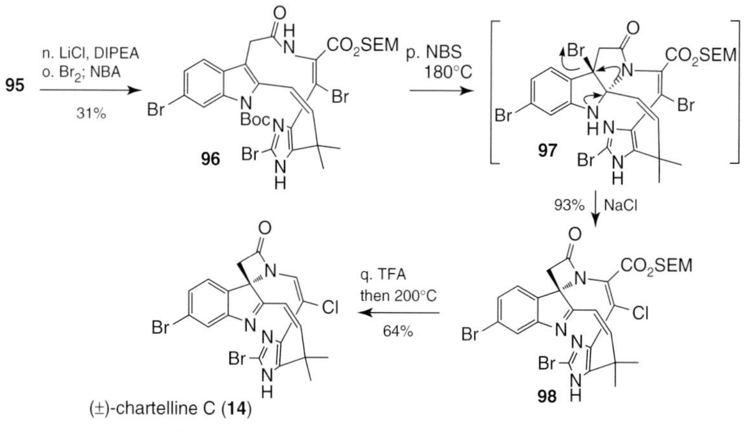

Scheme 5.13 Endgame of the biomimetic synthesis of chartelline C (**14**) by Baran *et al.*

5.2.6
(+)-Welwitindolinone A and (−)-Fischerindole I

The discovery of hapalindole, fischerindole, ambiguine, and welwitindolinone indole alkaloids isolated from marine blue-green algae by Moore and coworkers provided exciting new and unique structures for organic synthesis [31]. Welwitindolinones inspired organic chemists to pursue their synthesis also because of their array of promising biological properties such as insecticidal activities. Welwitindolinone A (**22**) is a densely functionalized oxindole harboring three all-carbon quaternary centers, a neopentyl chlorine atom, and a striking spiro-fused cyclobutane. 12-epi-Fischerindoles I (**103**) and G (**104**) represent the most complex congeners within the fischerindole alkaloid family. A proposal for their biogenetic origin has been put forward by Moore [31], in which a monoterpenoid geranyl-type unit **99** is attacked by an electrophilic chlorine, which again reacts with an indole moiety and thereby links the two different building blocks to give 12-epi-hapalindole E (**101**, Scheme 5.14). Two pathways are now conceivable: The first leads to the formation of oxindole **102**, which cyclizes to form the four-membered ring of welwitindolinone A (**22**). Alternatively, **101** can cyclize under acid catalysis to give 12-epi-fischerindole G (**104**) with its five-membered ring. This in turn is converted into 12-epi-fischerindole I (**103**) via an additional oxidation reaction.

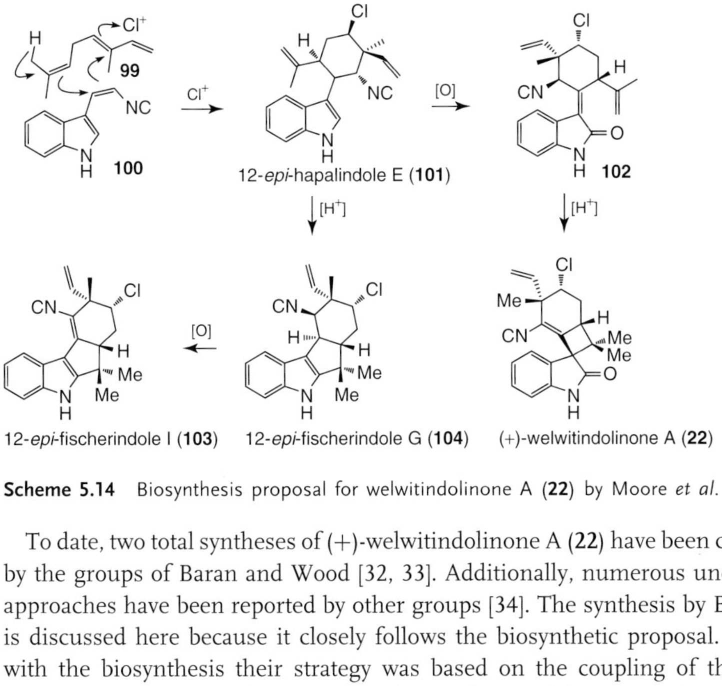

Scheme 5.14 Biosynthesis proposal for welwitindolinone A (**22**) by Moore *et al.*

To date, two total syntheses of (+)-welwitindolinone A (**22**) have been completed, by the groups of Baran and Wood [32, 33]. Additionally, numerous uncompleted approaches have been reported by other groups [34]. The synthesis by Baran *et al.* is discussed here because it closely follows the biosynthetic proposal. In accord with the biosynthesis their strategy was based on the coupling of the terpene moiety with the indole fragment. For the synthesis of the terpene fragment they started from *(S)*-carvone, which was epoxidized in a Scheffer–Weitz reaction, and subsequent vinyl-magnesium bromide addition gave compound **106** with the quaternary carbon center installed. The alcohol functionality of **106** was converted into the chloride via *N*-chlorosuccinimide (NCS). The coupling of the terpene fragment **107** with indole was conducted via a copper(II)-mediated oxidative radical coupling to yield compound **108**. Montmorillonite K-10 (MK-10) enabled cyclization of **108** to give tetracycle **109**, which contained the carboskeleton of fischerindoles G and I. Functional group transformations introduced isonitrile functionality and led to the synthesis of (−)-fischerindole I (**110**). Oxidation of **110** with xenon difluoride gave (+)-welwitindolinone A (**22**) in 44% yield. When comparing the synthetic route with Moore's biosynthetic proposal, strong parallels can be observed. Compound **108** has the same carboskeleton as 12-epi-hapalindole E (**101**) in Scheme 5.14, which is the central precursor in the biosynthetic proposal. The subsequent cyclization reaction that yields tetracycle **109** is strongly related to the conversion of **101** into **104** in Scheme 5.15. In addition, the last step, the oxidative rearrangement, which yields welwitindolinone A (**22**) with its remarkable cyclobutane, can be found in slightly different version in Scheme 5.14 (**102**→**22**).

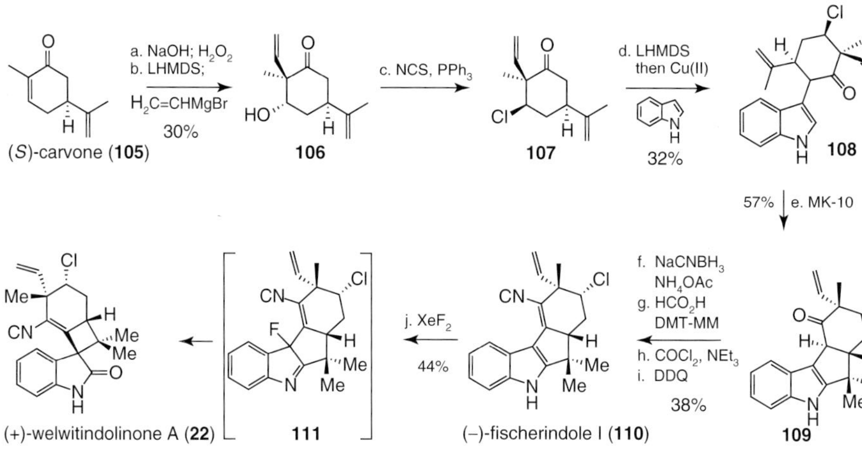

DMT-MM: 4-(4,6-dimethoxy-1,3,5-triazin-2-yl)-4-methylmorpholinium chloride

Scheme 5.15 Biomimetic synthesis of welwitindolinone A (**22**) by Baran *et al.*

5.2.7
(−)-Gelselegine

In recent years, several new indole and oxindole alkaloids have been isolated from *Gelsemium elegans* a plant that is known in Chinese folk medicine as "*Kou-Wen*" or "*Hu-Man-Teng*" [35]. These alkaloids have highly strained polycyclic structures and can be classified into six groups, the sarpagine, humantenine, gelselegine, gelsedine, koumine, and gelsemine types, based on their skeletal types. Sakai and coworkers have considered a possible biogenetic pathway. Their biosynthetic proposal (Scheme 5.16) starts with koumidine (**112**) as central building block [36]. Quaternization of the tertiary amine triggers fragmentation into (19*Z*)-anhydrovobasinediol (**113**). Oxidation of **113** into oxindole rankinidine (**114**) provides access to humantenidine-type alkaloids. To get from humantenidine- to the gelsedine-type alkaloids a ring contraction from a six-membered ring (humantenidine) to a five-membered ring in gelsedine has to take place. The hydroxymethyl

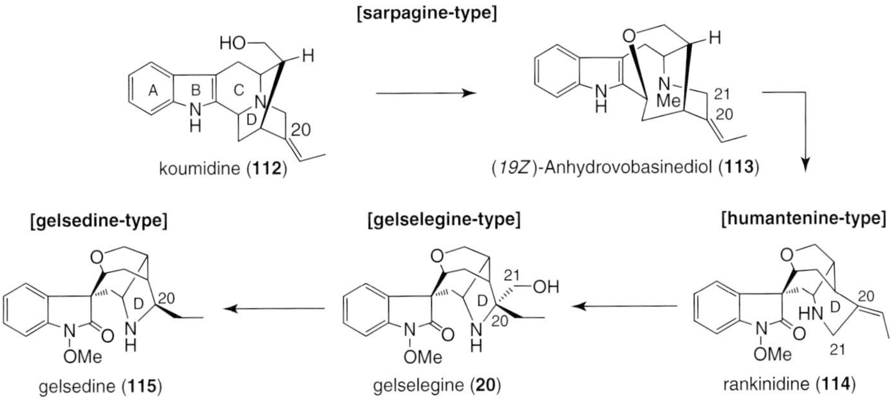

Scheme 5.16 Biosynthesis proposal for gelselegine (**20**) by Sakai *et al.*

group at C20 in gelselegine (**20**) comes from such a rearrangement/ring contraction. Gelsedine, which lacks the hydroxymethyl group, is the biogenetic follow-up product of gelselegine. In this way gelselegine somewhat represents the "missing link" in the biosynthesis between gelsedine (lacking C21), rankinidine, and the humantenidine/sarpagine type alkaloids.

Sakai and coworkers investigated their proposal with a series of chemical transformations, and were able to describe the first synthesis of these chemically and biogenetically unique oxindole alkaloids, in a manner consistent with this biogenetic sequence (Scheme 5.17). They started from koumidine (**112**), which was transformed into (19*Z*)-anhydrovobasinediol-type structure **116**, by reaction

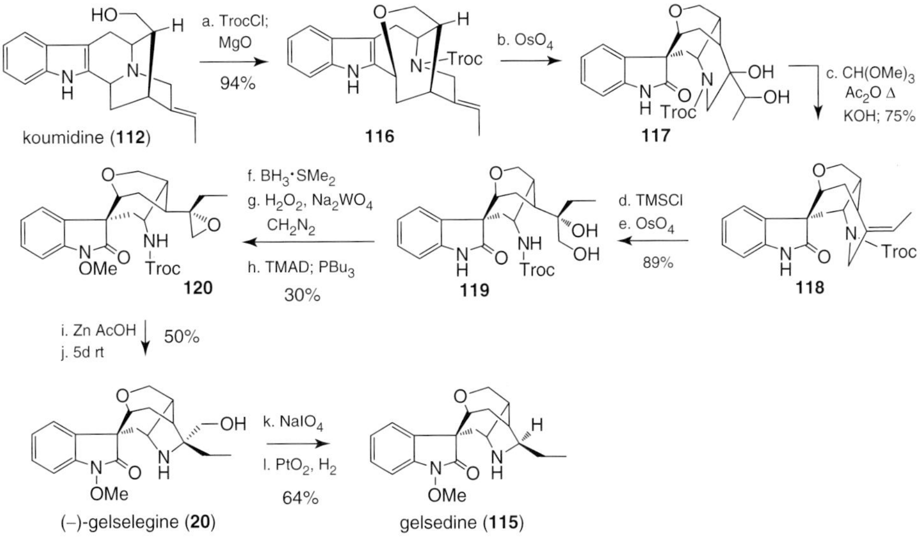

Scheme 5.17 Biomimetic synthesis of gelselegine (**20**) by Sakai *et al.*

with Troc-chloride and subsequent fragmentation reaction. Treatment of **116** with osmium tetroxide not only gave oxidized indolinone system but also afforded diol **117**. This in turn was converted into the cyclic orthoester, which under thermal conditions underwent elimination to give olefin **118** with opposite olefin geometry to that of **116**. Treatment of **118** with trimethylsilyl chloride (TMSCl) shifted the exocyclic double bond into the ring to give an enamide. This was dihydroxylated again with osmium tetroxide to afford diol **119**. This instituted the ring contraction of the former six-membered ring in **118** to the desired five-membered ring of gelselegine (**20**). The following transformations were needed to achieve this goal: continuing from **119**, the indolinone moiety was reduced to the indoline with borane dimethyl sulfide complex. Reoxidation with hydrogen peroxide and sodium tungstate gave the *N*-hydroxylated indolinone moiety. The diol of **119** was treated under Mitsunobu type conditions to afford epoxide **120**. Reductive cleavage of the Troc-protective group was followed by epoxide opening, when left at room temperature for five days, to give the desired (−)-gelselegine (**20**). To gain evidence that gelselegine is the actual precursor or "missing link" in the biosynthesis of gelsedine (**115**), gelselegine was submitted to sodium periodate conditions, giving the glycol cleavage product, which upon reduction with platinum oxide and hydrogen was converted into gelsedine (**115**).

5.2.8
Communesin, Calycanthines, and Chimonanthines

The calycanthaceous alkaloids were first isolated from the plant genus *Calycanthus* [37]. Early isolations of individual members showed a closely related skeleton that differed only in the aminal connectivity for the various natural products. The subtle differences in possible structures made the establishment of the relative and absolute stereochemistry of these compounds very difficult.

The first calycanthoid that was characterized was (+)-calycanthine (**21**). Its structure was chemically established by Robinson and Woodward, and crystallographically by Hamor and Robertson, using the dihydrobromide dehydrated salt [38–40]. Soon after the publication of the crystal structure, the absolute stereochemistry of calycanthine (**21**) was determined using circular dichroism analysis by Mason [41]. The structure of (−)-chimonanthine (**19**), isolated from Chimonanthus fragrans by Hodson *et al.*, was elucidated by Hamor and Robertson through X-ray analysis of the dihydrobromide salt. Chimonanthine bears two indoline units, each with an annulated pyrrolidine, that are 3,3′ – connected. All possible stereoisomers, both the *C*2-symmetric and *meso*-isomers, have been identified. The absolute stereochemistry of the vicinal quaternary carbon centers of (−)-chimonanthine is identical to that of (+)-calycanthine, and these natural products equilibrate under acidic conditions (Scheme 5.18). The total synthesis of *meso*- and racemic chimonanthine and calycanthine via this oxidative coupling was performed by Scott and coworkers (Scheme 5.19) [40c]. They used methyl-magnesium bromide as a base to deprotonate the indole ring, which made it more susceptible to oxidative conditions. Subsequent treatment with iron(III)-chloride furnished dimerization

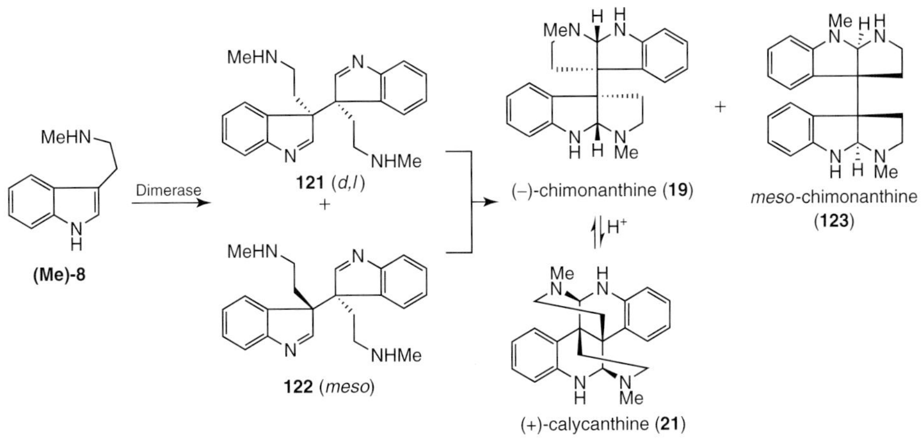

Scheme 5.18 Biosynthesis proposal for calycantheous alkaloids by Woodward and Robinson.

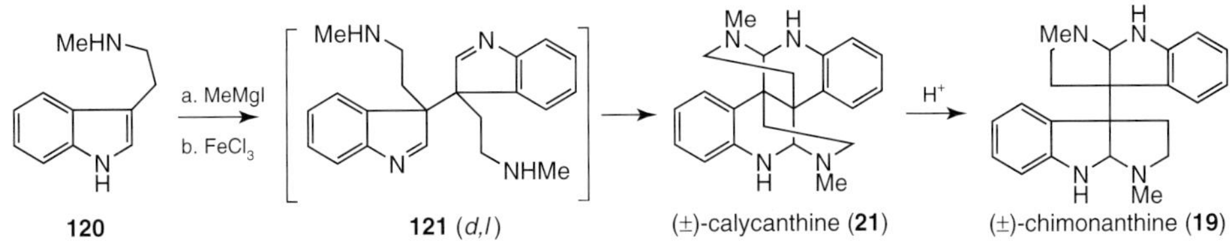

Scheme 5.19 Biomimetic synthesis of calycantheous alkaloids by Scott *et al.*

product **121**, which underwent cyclization to (±)-chimonanthine (**19**). Equilibration indeed gave (±)-calycanthine (**21**). An alternative method of oxidative dimerization of indoles was developed by Takayama and coworkers, which uses hypervalent iodine reagents as oxidants, and provides a rapid synthetic entry into the chimonanthine and calycanthine alkaloids [42]. Woodward and Robinson proposed the biosynthesis of these natural products more than 50 years ago and this proposal was later refined by B. Stoltz. This proposal is shown in Scheme 5.20. It contains an oxidative dimerization of two tryptamine units.

In 2006, Stoltz *et al.* developed a more detailed biosynthetic concept that not only attempted to explain the formation of the calycantheous alkaloids but also of communesins alkaloids from common precursors (Scheme 5.20) [43b]. In analogy to Scheme 5.18, dimer *(R,R)*-**124** could be cyclized and recyclized to form, in a cascade process via intermediates **125** and **126**, the hexacyclic structure **127**, which is N-prenylated to **128** and then oxidatively converted into **12** (Scheme 5.20). An analogous sequence starting form *meso*-**129** would end up with **130**, which is a reasonable precursor to the known alkaloid perophoramidine (**131**).

A few years earlier, in 2003, Stoltz had provided a totally different biosynthesis that contained another natural product–aurantioclavine–as an intermediate in the biosynthesis of communesins [44]. In a model study (Scheme 5.21) some evidence was provided for this concept by generating *ortho*-iminoquinonemethide

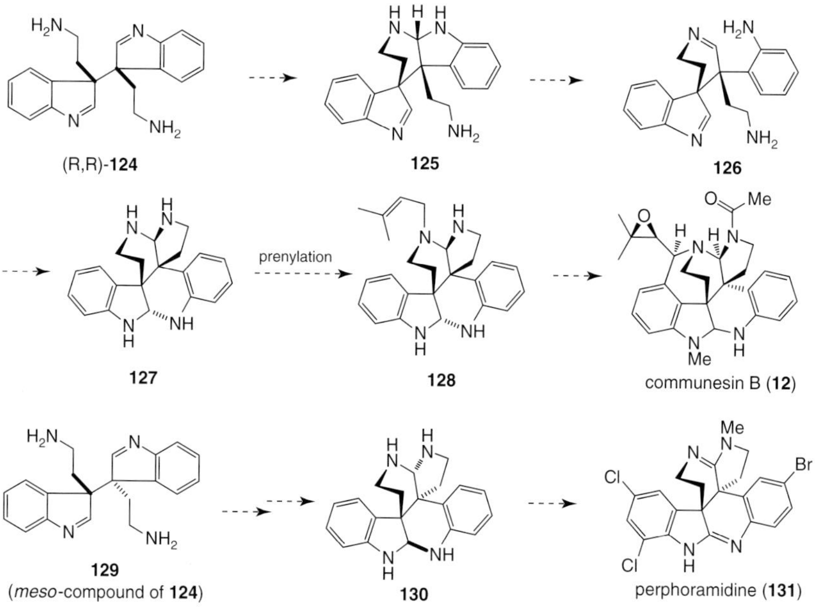

Scheme 5.20 Extended biosynthesis proposal for calycantheous alkaloids and communesins by Stoltz *et al.*

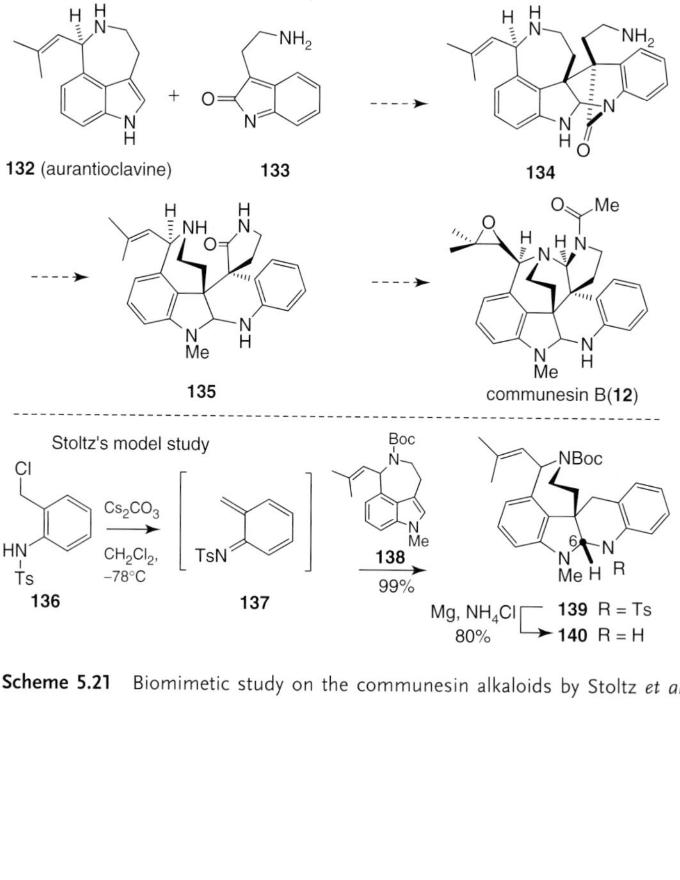

Scheme 5.21 Biomimetic study on the communesin alkaloids by Stoltz *et al.*

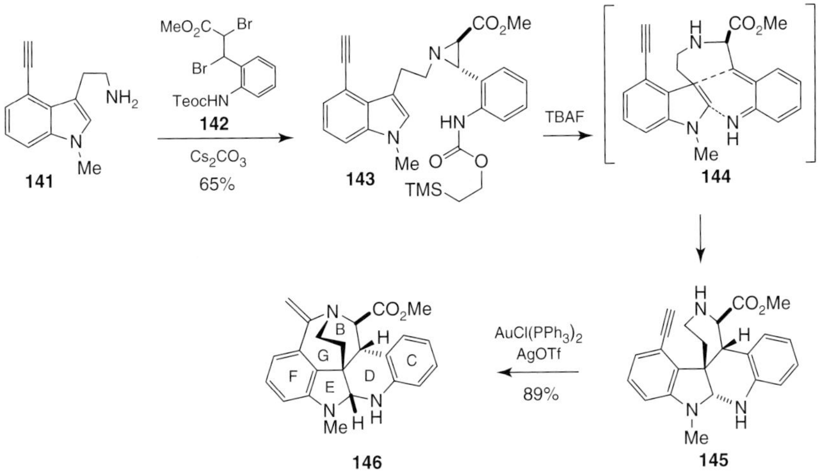

Scheme 5.22 Related study on communesin alkaloids by Funk *et al.*

137 from precursor **136**, and adding it to the aurantioclavine derivative **138**. In fact a diastereomeric mixture of polycycle **139** was obtained containing communesin rings F, E, D, C, and G. The ^{13}C NMR shifts for C6 (84.8 and 83.9 ppm) were in agreement with those of communesin B (82.4 ppm).

Unfortunately, they were never able to complete the total synthesis. The group of Funk also pursued for the synthesis of communesins, and thereby followed a path closely related to Stoltz's older biosynthesis proposal (Scheme 5.22) [45]. Starting from substituted tryptamine **141** and dibromide **142**, aziridine **143** was formed under base catalysis. *ortho*-Quinomethide **144** was formed via TBAF deprotection of the carbamide moiety in compound **143**, which proceeded under formation of ethylene and carbon dioxide, and indeed gave Diels–Alder product **145**. Treatment of **145** with a gold(I)-reagent closed ring G of communesins in **146**, and was the closest that Funk and coworkers could get to the natural product. To date two non-biomimetic total syntheses of communesin F have been completed [46, 47]. The communesin story is one of the rare cases where following a non-biomimetic route has lead to better results. This may be due to the uncertainty in the biosynthesis of these natural products and points out that there is still a lot of research needed to solve this puzzle. Most probably, if the biosynthetic proposal is improved a biomimetic synthesis of communesins will be more efficient than the currently completed ones.

5.2.9
(+)-11,11′-Dideoxyverticillin A

The fungal metabolite (+)-11,11′-dideoxyverticillin A (**10**) is a cytotoxic alkaloid, isolated from a marine *Penicillium* sp., that shows a densely functionalized intricate dimeric epi-dithiodiketopiperazine structure [48]. A hypothesis for the biosynthesis of this molecule envisages the reductive dimerization of monomer **147** followed

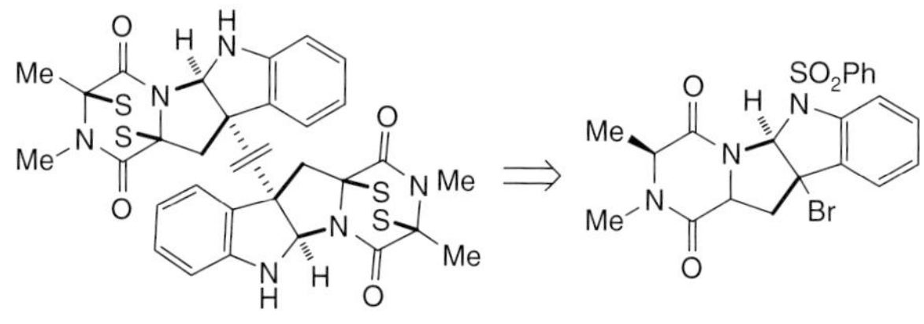

Figure 5.5 Dimerization strategy for the synthesis of (**10**) by Movassaghi *et al.*

by tetrahydroxylation and tetrathiolation reactions (Figure 5.5). Most probably the biomimetic dimerization will not be reductive but oxidative in nature. Brominated compound **147** serves as a surrogate to achieve the same C–C-bond formation under reductive conditions. Therefore, with respect to bond formation it can be regarded as "biomimetic" but not with respect to the reaction conditions.

In the synthesis performed by Movassaghi *et al.* (Scheme 5.23), dipeptide **150** was converted into the diketopiperazine and then the indole nucleus was destroyed by the introduction of bromine to form intermediate **152** which was N-methylated [49]. The crucial dimerization was performed by reduction with a Co(I)-complex to provide dimer **153**, presumably via a radical mechanism. This step is remarkable for its high stereoselectivity, as the 5,5-*cis*-fused ring system is generated exclusively.

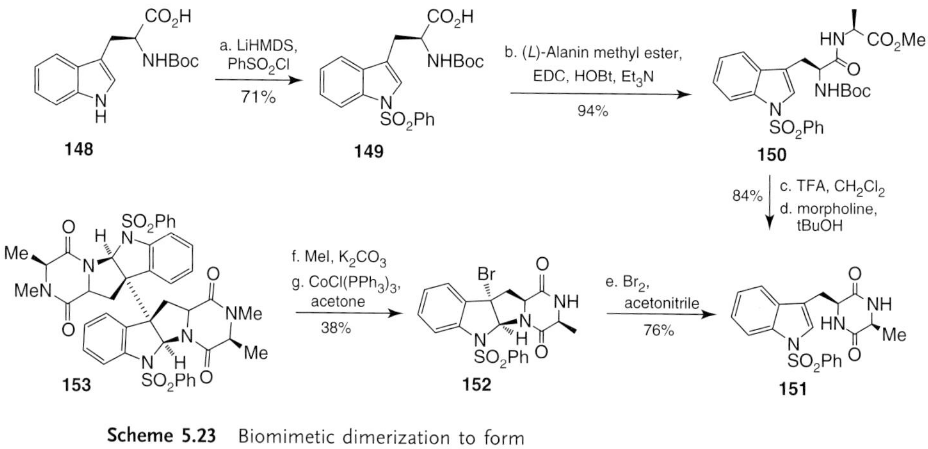

Scheme 5.23 Biomimetic dimerization to form (+)-11,11′-dideoxyverticillin A by Movassaghi *et al.*

Dihydroxylation of **153** with bis(pyridine)-silver(I) permanganate led to the tetracyclic diol **154** as a single diastereomer. TBS-protection of the tertiary OH-groups and removal of the benzenesulfonyl groups with sodium amalgam gave **155**, which on exposure to potassium trithiocarbonate was converted into bis-dithioepanethione **156**. Addition of ethanolamine afforded the unstable proposed biosynthetic precursor, which on oxidation with potassium triiodide finally furnished **10** (Scheme 5.24).

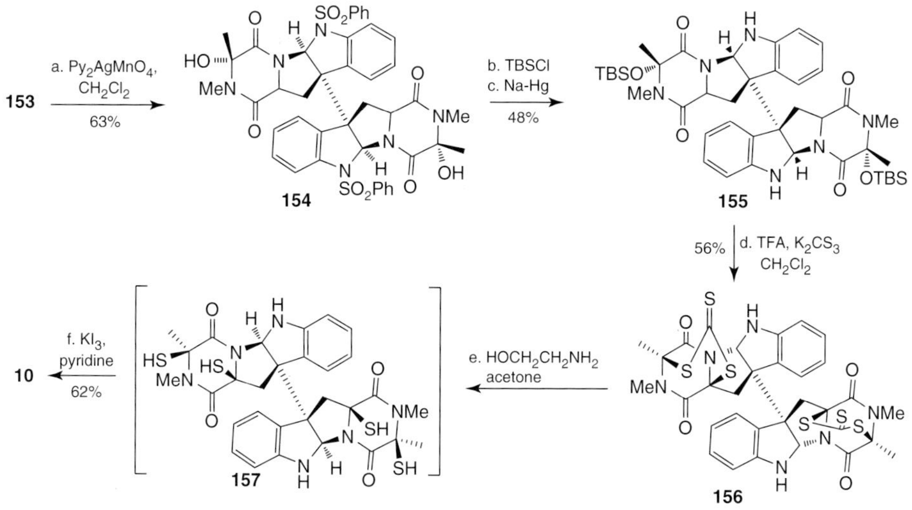

Scheme 5.24 Endgame of the synthesis of (+)-11,11′-dideoxyverticillin A by Movassaghi *et al.*

5.2.10
(±)-Borreverine and (±)-Isoborreverine

(±)-Borrerine (**158**), (±)-borreverine (**18**), and (±)-isoborreverine (**17**) (Scheme 5.25) were originally isolated in 1973 from Borreria verticillata, a plant used in traditional medicine of West Africa for the treatment of skin diseases [50]. Subsequently, in 1991, these natural products were re-isolated from *Flindersia fournieri*, a plant from New Caledonia (West Pacific) [51a].

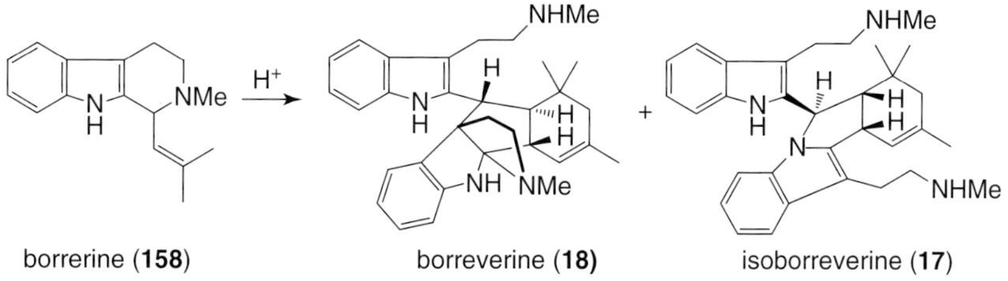

Scheme 5.25 Biosynthesis proposal for borreverine (**18**) and isoborreverine (**17**) by Koch *et al.*

The co-occurrence of these dimeric alkaloids **18** and **17** with (±)-borrerine (**158**) suggested a simple chemical relationship between them. The biosynthetic proposal (Scheme 5.25) contains an acid-catalyzed dimerization of (±)-borrerine (**158**), which leads after the loss of a proton either to (±)-borreverine or (±)-isoborreverine. In fact Koch and coworkers mimicked this reaction and where able to isolate both dimeric indole alkaloids (Scheme 5.26) [52]. The optimum conditions were trifluoroacetic acid in benzene at 65 °C for 30 min. These conditions furnished (±)-borreverine and

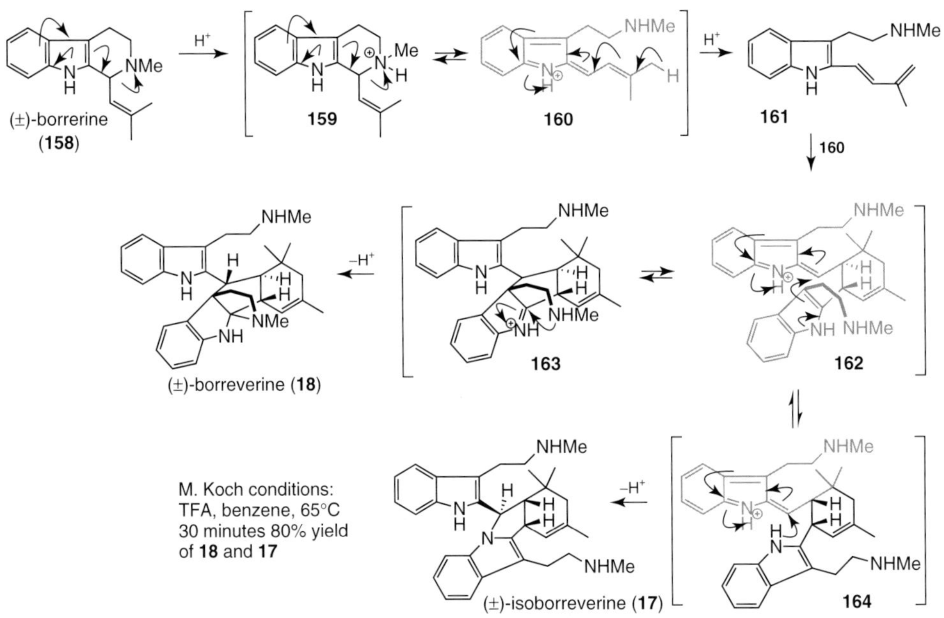

Scheme 5.26 Detailed reaction mechanism of the dimerization reaction.

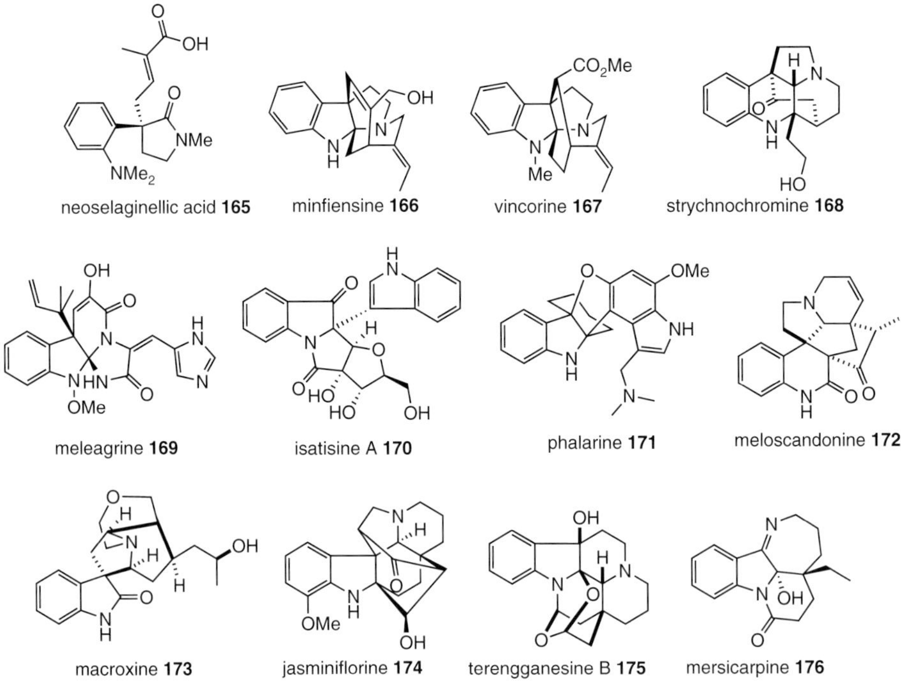

Figure 5.6 Selected indole alkaloids that contain a broken indole nucleus.

(±)-isoborreverine in about equal amounts and a total yield of 80%. Longer reaction times changed the ratio of (±)-borreverine to (±)-isoborreverine in favor of the latter. (±)-Borreverine (**18**) can be quantitatively transformed into (±)-isoborreverine (**17**) after reacting for 12 h under the conditions mentioned above. Additional evidence that the dimerization pathway might definitely be operative in Nature is provided by the fact that all three natural products not only were isolated from the same plant but, so far, were only isolated as racemate, which means that no enzymatic process is involved and the process is spontaneous and thermodynamically favored.

5.3 Conclusion

Indole alkaloids provide a wide variety of different structure types with varying degrees of structural complexity. Some selected structures that were not mentioned in this chapter are given in Figure 5.6. As shown above, this huge architectural diversity is inevitably linked to the characteristic reactivity pattern of the indole nucleus, which can undergo a plethora of skeletal rearrangements that are in most cases initiated by an oxidation reaction. In Nature these oxidations are mostly non-enzymatic processes. This makes this family of natural products amenable to biomimetic synthesis. In this way, the intrinsic reactivity of the molecules allows an increase of molecular complexity in very few steps, if the biosynthetic proposal is followed.

References

1. (a) Wenkert, E. (1962) *J. Am. Chem. Soc.*, **84**, 98–102; (b) Thomas, R. (1961) *Tetrahedron Lett.*, **16**, 544–553; (c) Money, T., Wright, I.G., McCapra, F., and Scott, A.I. (1965) *Proc. Natl. Acad. Sci. U.S.A.*, **53**, 901–903; (d) Battersby, A.R., Burnett, A.R., and Parsons, P.G. (1968) *J. Chem. Soc., Chem. Commun.*, 1280–1281; (e) Battersby, A.R., Burnett, A.R., and Parsons, P.G. (1969) *J. Chem. Soc. C*, 1187–1192; (f) Battersby, A.R., Byrne, J.C., Kapil, R.S., Martin, J.A., Payne, T.G., Arigoni, D., and Loew, P. (1968) *J. Chem. Soc., Chem. Commun.*, 951–953; (g) Eisenreich, W., Bacher, A., Arigoni, D., and Rohdich, F. (2004) *Cell. Mol. Life Sci.*, **61**, 1401–1426; (h) Contin, A., van der Heijden, R., Lefeber, A.W.M., and Verpoorte, R. (1998) *FEBS Lett.*, **434**, 413–416.
2. For syntheses, see: (a) Chavan, S.P., Pathak, A.B., and Kalkote, U.R. (2007) *Synlett*, 2635–2638; (b) Chavan, S.P., Pasupathy, K., Venkatraman, M.S., and Kale, R.R. (2004) *Tetrahedron Lett.*, **45**, 6879–6882; (c) Bennasar, M.-L., Zulaica, E., Juan, C., Alonso, Y., and Bosch, J. (2002) *J. Org. Chem.*, **67**, 7465–7474; (d) Chavan, S.P. and Venkatraman, M.S. (1998) *Tetrahedron Lett.*, **39**, 6745–6748; (e) Ciufolini, M.A. and Roschangar, F. (1996) *Angew. Chem. Int. Ed.*, **35**, 1692–1694; (f) Shen, W., Coburn, C.A., Bornmann, W.G., and Danishefsky, S.J. (1993) *J. Org. Chem.*, **58**, 611–617; (g) Curran, D.P. and Liu, H. (1992) *J. Am. Chem. Soc.*, **114**, 5863–5864; (h) Kametani, T., Ohsawa, T., and Ihara, M. (1980) *Heterocycles*, **14**, 951–953; (i) Tang, C. and Rapoport, H. (1972) *J. Am. Chem. Soc.*, **94**, 8615–8616; (j) Stork, G. and Schultz, A.G. (1971) *J. Am. Chem. Soc.*, **93**, 4074–4075.

3. Wall, M.E., Wani, M.C., Cook, C.E., Palmer, K.H., McPhail, A.I., and Sim, G.A. (1966) *J. Am. Chem. Soc.*, **88**, 3888–3890.
4. (a) Ulukan, H. and Swaan, P.W. (2002) *Drugs*, **62**(2), 2039–2057; (b) Lu, A.J., Zheng, Z.S., Zou, H.J., Luo, X.M., and Jiang, H.L. (2007) *Eur. J. Med. Chem.*, **42**, 307–314; (c) Adams, D.J., Wahl, M.L., Flowers, J.L., Sen, B., Colvin, M., Dewhirst, M.W., Manikumar, G., and Wani, M.C. (2005) *Cancer Chemother. Pharm.*, **57**(2), 145–154.
5. Wenkert, E., Dave, I.G., Lewis, R.G., and Sprague, P.W. (1967) *J. Am. Chem. Soc.*, **89**, 6741–6745.
6. Winterfeldt, E. (1971) *Justus Liebigs Ann. Chem.*, **745**, 23–30.
7. (a) Hutchinson, C.R., Heckendorf, A.H., Daddona, P.E., Hagaman, E., and Wenkert, E. (1974) *J. Am. Chem. Soc.*, **96**, 5609–5611; (b) Hutchinson, C.R. (1981) *Tetrahedron*, **37**, 1047–1065; (c) Takayama, H., Kitajima, M., and Aimi, N. (1999) *J. Synth. Org. Chem. Jpn.*, **57**, 181–193; (d) Carte, B.K., de Brosse, C., Eggleston, D., Hemling, M., Mentzer, M., Poehland, B., Troupe, N., Westley, J.M., and Hecht, S.M. (1990) *Tetrahedron*, **46**, 2747–2760; (e) Aimi, N., Hoshino, H., Nishimura, M., Sakai, S., and Haginawa, J. (1990) *Tetrahedron Lett.*, **31**, 5169–5170; (f) Kitajima, M., Masumoto, S., Takayama, H., and Aimi, N. (1997) *Tetrahedron Lett.*, **38**, 4255–4258.
8. (a) Boch, M., Korth, T., Nelke, J.M., Pike, D., Radunz, H., and Winterfeldt, E. (1972) *Chem. Ber.*, **105**, 2126–2142; (b) Krohn, K. and Winterfeldt, E. (1975) *Chem. Ber.*, **108**, 3030–3042.
9. (a) Anderson, R.J., Raolji, G.B., Kanazawa, A., and Greene, A.E. (2005) *Org. Lett.*, **7**, 2989–2991; (b) Blagg, B.S.J. and Boger, D.L. (2002) *Tetrahedron*, **58**, 6343–6349; (c) Chavan, S.P., Dhawane, A.N., and Kalkote, U.R. (2010) *Tetrahedron Lett.*, **51**, 3099–3101; (d) Chavan, S.P., Pathak, A.B., and Kalkote, U.R. (2007) *Synlett*, 2635–2638; (e) Chavan, S.P., Pathak, A.B., and Kalkote, U.R. (2007) *Tetrahedron Lett.*, **48**, 6561–6563; (f) Chavan, S.P. and Rasapalli, S. (2004) *Tetrahedron Lett.*, **45**, 3113–3115; (g) Chavan, S.P. and Venkatraman, M.S. (2005) *Arkivoc*, 165–169; (h) Kanazawa, A., Muniz, M.N., Baumlova, B., Ljungdahl, N., and Greene, A.E. (2008) *Synlett*, 2275–2278; (i) Liu, G.-S., Dong, Q.-L., Yao, Y.-S., and Yao, Z.-J. (2008) *Org. Lett.*, **10**, 5393–5396; (j) Tagami, K., Nakazawa, N., Sano, S., and Nagao, Y. (2000) *Heterocycles*, **53**, 771–776; (k) Tang, C.-J., Babjak, M., Anderson, R.J., Greene, A.E., and Kanazawa, A. (2006) *Org. Biomol. Chem.*, **4**, 3757–3759; (l) Wu, X.-J., Guo, Q.-C., Zhang, Y., Chen, Y.-Y., and Wang, X.-D. (2008) *Anhui Yiyao*, **12**, 783–785; (m) Yu, S., Luo, Y., Liu, H., Liu, H., and Lu, W. (2010) *Monatsh. Chem.*, **141**, 245–249; (n) Zhang, L., Bao, Y., and Chen, F. (2008) *Zhongguo Yiyao Gongye Zazhi*, **39**, 481–483; (o) Zhou, H.-B., Liu, G.-S., and Yao, Z.-J. (2007) *Org. Lett.*, **9**, 2003–2006; (p) Thomas, O.P., Zaparucha, A., and Husson, H.-P. (2002) *Eur. J. Org. Chem.*, 157–162; (q) Dumas, C., Kan-Fan, C., Royer, J., and Husson, H.-P. (2000) *Eur. J. Org. Chem.*, 3601–3606; (r) For an improved reagent, see: Jiang, W., Zhang, X., and Sui, Z. (2003) *Org. Lett*, **5**, 43–46.
10. For comprehensive reviews see: (a) Kita, Y. (2005) *Curr. Org. Chem.*, **9**, 1567–1588; (b) Makar'eva, T.N., Krasokhin, V.B., Guzii, A.G., and Stonik, V.A. (2010) *Chem. Nat. Comp.*, **46**, 152–153; (c) Jeon, J.-e., Na, Z., Jung, M., Lee, H.-S., Sim, C.J., Nahm, K., Oh, K.-B., and Shin, J. (2010) *J. Nat. Prod.*, **73**, 258–262; (d) Na, M.K., Ding, Y., Wang, B., Tekwani, B.L., Schinazi, R.F., Franzblau, S., Kelly, M., Stone, R., Li, X.-C., Ferreira, D., and Hamann, M.T. (2010) *J. Nat. Prod.*, **73**(3), 383–387; (e) Lang, G., Pinkert, A., Blunt, J.W., and Munro, M.H.G. (2005) *J. Nat. Prod.*, **68**, 1796–1798; (f) Reyes, F., Martin, R., Rueda, A., Fernandez, R., Montalvo, D., Gomez, C., and Sanchez-Puelles, J.M. (2004) *J. Nat. Prod.*, **67**, 463–465; (g) Gunasekera, S.P., Zuleta, I.A., Longley, R.E., Wright, A.E., and Pomponi, S.A. (2003) *J. Nat. Prod.*, **66**, 1615–1617; (h) Dijoux, M.-G., Gamble, W.R., Hallock, Y.F., Cardellina

II, J.H., Van Soest, R., and Boyd, M.R. (1999) *J. Nat. Prod.*, **62**, 636–637; (i) Gunasekera, S.P., McCarthy, P.J., Longley, R.E., Pomponi, S.A., Wright, E., Lobkovsky, E., and Clardy, J. (1999) *J. Nat. Prod.*, **62**, 173–175; (j) Yang, A., Baker, B.J., Grimwade, J., Leonard, A., and McClintock, J.B. (1995) *J. Nat. Prod.*, **58**, 1596–1599; (k) Perry, N.B., Blunt, J.W., Munro, M.H.G., Higa, T., and Sakai, R. (1988) *J. Org. Chem.*, **53**, 4127–4128; (l) Perry, N.B., Blunt, J.W., and Munro, M.H.G. (1988) *Tetrahedron*, **44**, 1727–1734; (m) Perry, N.B., Blunt, J.W., McCombs, J.D., and Munro, M.H.G. (1986) *J. Org. Chem.*, **51**, 5476–5478.

11. Lill, R.E., Major, D.A., Blunt, J.W., Munro, M.H.G., Battershill, C.N., McLean, M.G., and Baxter, R.L. (1995) *J. Nat. Prod.*, **58**, 206–211.
12. (a) Nishiyama, S., Cheng, J.F., Tao, X.L., and Yamamura, S. (1991) *Tetrahedron Lett.*, **32**, 4151–4154; (b) Kita, Y., Tohma, H., Inagaki, M., Hatanaka, K., and Yakura, T. (1992) *J. Am. Chem. Soc.*, **114**(6), 2175–2180; (c) Roberts, D., Joule, J.A., Bros, M.A., and Alvarez, M. (1997) *J. Org. Chem.*, **62**(3), 568–577; (d) White, J.D., Yager, K.M., and Yakura, T. (1994) *J. Am. Chem. Soc.*, **116**(5), 1831–1838; (e) Knoelker, H.J. and Hartmann, K. (1991) *Synlett*, 428–430; (f) Kubiak, G. and Confalone, P. (1990) *Tetrahedron Lett.*, **31**, 3845–3848; (g) Wada, Y., Fujioka, H., and Kita, Y. (2010) *Marine Drugs*, **8**, 1394–1416; (h) Tohma, H., Harayama, Y., Hashizume, M., Iwata, M., Kiyono, Y., Egi, M., and Kita, Y. (2003) *J. Am. Chem. Soc.*, **125**(37), 11235–11240; (i) Tohma, H., Harayama, Y., Hashizume, M., Iwata, M., Egi, M., and Kita, Y. (2002) *Angew. Chem. Int. Ed.*, **41**(2), 348–350.
13. Aubart, K.M. and Heathcock, C.H. (1999) *J. Org. Chem.*, **64**(1), 16–22.
14. (a) Schkeryantz, J.M., Woo, J.C.G., Siliphaivanh, P., Depew, K.M., and Danishefsky, S.J. (1999) *J. Am. Chem. Soc.*, **121**(51), 11964–11975; (b) Ritchie, R. and Saxton, J.E. (1981) *Tetrahedron*, **37**(24), 4295–4303; (c) Kametani, T., Kanaya, N., and Ihara, M. (1981) *J. Chem. Soc., Perkin Trans. 1*, 959–963; (d) Kametani, T., Kanaya, N., and Ihara, M. (1980) *J. Am. Chem. Soc.*, **102**(11), 3974–3975; (e) Lee, B.H., Clothier, M.F., and Pickering, D.A. (1997) *J. Org. Chem.*, **62**(22), 7836–7840; (f) Lee, B.H. and Clothier, M.F. (1997) *J. Org. Chem.*, **62**(6), 1795–1798; (g) Trost, B.M., Cramer, N., and Bernsmann, H. (2007) *J. Am. Chem. Soc.*, **129**(11), 3086–3087.
15. (a) Birch, A.J. and Wright, J.J. (1969) *J. Chem. Soc., Chem. Commun.*, 644–645; (b) Birch, A.J. and Wright, J.J. (1970) *Tetrahedron*, **26**, 2329–2344; (c) Birch, A.J. and Russell, R.A. (1972) *Tetrahedron*, **28**, 2999–3008; (d) Yamazaki, M., Okuyama, E., Kobayashi, M., and Inoue, H. (1981) *Tetrahedron Lett.*, **22**, 135–136; (e) Ondeyka, J.G., Goegelman, R.T., Schaeffer, J.M., Kelemen, L., and Zitano, L. (1990) *J. Antibiot.*, **43**, 1375–1379; (f) Liesch, J.M. and Wichmann, C.F. (1990) *J. Antibiot.*, **43**, 1380–1386; (g) Blanchflower, S.E., Banks, R.M., Everett, J.R., and Manger, B.R. (1991) *J. Antibiot.*, **44**, 492–497; (h) Blanchflower, S.E., Banks, R.M., Everett, J.R., and Reading, C. (1993) *J. Antibiot.*, **46**, 1355–1363; (i) Banks, R.M., Blanchflower, S.E., Everett, J.R., Manger, B.R., and Reading, C. (1997) *J. Antibiot.*, **50**, 840–846; (j) Whyte, A.C., Gloer, J.B., Wicklow, D.T., and Dowd, P.F. (1996) *J. Nat. Prod.*, **59**, 1093–1095; (k) Polonsky, J., Merrien, M.-A., Prangé, T., and Pascard, C. (1980) *J. Chem. Soc., Chem. Commun.*, 601–602; (l) Prangé, T., Billion, M.-A., Vuilhorgne, M., Pascard, C., Polonsky, J., and Moreau, S. (1981) *Tetrahedron Lett.*, **22**, 1977–1980; (m) Sugie, Y., Hirai, H., Inagaki, T., Ishiguro, M., Kim, Y.-J., Kojima, Y., Sakakibara, T., Sakemi, S., Sugiura, A., Suzuki, Y., Brennan, L., Duignan, J., Huang, L.H., Sutcliffe, J., and Kojima, N. (2001) *J. Antibiot.*, **54**, 911–916.
16. (a) Sanz-Cervera, J.F., Williams, R.M., Alberto, M.J., Maria Lopez-Sanchez, J., Gonzalez, F., Eugenia Martinez, M., and Sancenon, F. (2000) *Tetrahedron*, **56**(34), 6345–6358; (b) Williams, R.M., Sanz-Cervera, J.F., Sancenon, F., Marco, J.A., and Halligan, K.M. (1998) *Bioorg. Med. Chem.*, **6**(8), 1233–1241;

(c) Williams, R.M., Sanz-Cervera, J.F., Sancenon, F., Marco, J.A., and Halligan, K. (1998) *J. Am. Chem. Soc.*, **120**(5), 1090–1091.

17. (a) Hilvert, D., Hill, K.W., Nared, K.D., and Auditor, M.T.M. (1989) *J. Am. Chem. Soc.*, **111**, 9261–9262; (b) Heine, A., Stura, E., Yli-Kauhaluoma, J.T., Gao, C., Deng, Q., Beno, B.R., Houk, K.N., Janda, K.D., and Wilson, I.A. (1998) *Science*, **279**, 1934–1940; (c) Kasahara, K., Miyamoto, T., Fujimoto, T., Oguri, H., Tokiwano, T., Oikawa, H., Ebizuka, Y., and Fujii, I. (2010) *ChemBioChem*, **11**, 1245–1252; (d) Katayama, K., Kobayashi, T., Chijimatsu, M., Ichihara, A., and Oikawa, H. (2008) *Biosci. Biotechnol. Biochem.*, **72**, 604–607; (e) Oikawa, H. (2001) *Kagaku Seibutsu*, **39**, 424–426; (f) Oikawa, H. (2005) *Bull. Chem. Soc. Jpn.*, **78**, 537–554; (g) Ose, T., Oikawa, H., and Tanaka, I. (2004) *Seibutsu Butsuri*, **44**, 32–35; (h) Ose, T., Yao, M., Oikawa, H., and Tanaka, I. (2003) *Nippon Kessho Gakkaishi*, **45**, 384–390.
18. Tarasow, T.M. and Eaton, B.E. (1999) *Cell Mol. Life Sci.*, **55**, 1463–1472.
19. (a) Williams, R.M., Glinka, T., Kwast, E., Coffman, H., and Stille, J.K. (1990) *J. Am. Chem. Soc.*, **112**(2), 808–821; (b) Williams, R.M., Cao, J., Tsujishima, H., and Cox, R.J. (2003) *J. Am. Chem. Soc.*, **125**(40), 12172–12178; (c) Williams, R.M. (2002) *Chem. Pharm. Bull.*, **50**(6), 711–740; (d) Williams, R.M. and Cox, R.J. (2003) *Acc. Chem. Res.*, **36**(2), 127–139; (e) Cushing, T.D., Sanz-Cervera, J.F., and Williams, R.M. (1996) *J. Am. Chem. Soc.*, **118**(3), 557–579; (f) Greshock, T.J., Grubbs, A.W., and Williams, R.M. (2007) *Tetrahedron*, **63**(27), 6124–6130.
20. Sanz-Cervera, J.F. and Williams, R.M. (2002) *J. Am. Chem. Soc.*, **124**(11), 2556–2559.
21. Adams, L.A., Valente, M.W.N., and Williams, R.M. (2006) *Tetrahedron*, **62**(22), 5195–5200.
22. Nising, C.F. (2010) *Chem. Soc. Rev.*, **39**(2), 591–599.
23. (a) Qian-Cutrone, J., Huang, S., Shu, Y.-Z., Vyas, D., Fairchild, C., Menendez, A., Krampitz, K., Dalterio, R., Klohr, S.E., and Gao, Q. (2002) *J. Am. Chem. Soc.*, **124**, 14556–14557; (b) Qian-Cutrone, J., Krampitz, K.D., Shu, Y.-Z., Chang, L.-P., Lowe, S.E. (2001) *Patent* US 6291461; (2001) *Chem. Abstr.*, **135**, 236411.
24. (a) Wulff, J.E., Herzon, S.B., Siegrist, R., and Myers, A.G. (2007) *J. Am. Chem. Soc.*, **129**(16), 4898–4899; (b) Herzon, S.B. and Myers, A.G. (2005) *J. Am. Chem. Soc.*, **127**(15), 5342–5344.
25. (a) Baran, P.S., Guerrero, C.A., Ambhaikar, N.B., and Hafensteiner, B.D. (2005) *Angew. Chem. Int. Ed.*, **44**(4), 606–609; (b) Baran, P.S., Hafensteiner, B.D., Ambhaikar, N.B., Guerrero, C.A., and Gallagher, J.D. (2006) *J. Am. Chem. Soc.*, **128**(26), 8678–8693.
26. Greshock, T.J. and Williams, R.M. (2007) *Org. Lett.*, **9**(21), 4255–4258.
27. (a) For the debut of a chartelline alkaloid, see: Chevolot, L., Chevolot, A.-M., Gajhede, M., Larsen, C., Anthoni, U., and Christophersen, C. (1985) *J. Am. Chem. Soc.*, **107**, 4542–4543; (b) For the isolation of chartelline C, see: Anthoni, U., Chevolot, L., Larsen, C., Nielsen, P.H., and Christophersen, C. (1987) *J. Org. Chem.*, **52**, 4709–4712.
28. For studies towards the chartellines and related alkaloids, see: (a) Lin, X. and Weinreb, S.M. (2001) *Tetrahedron Lett.*, **42**, 2631–2633; (b) Pinder, J.L. and Weinreb, S.M. (2003) *Tetrahedron Lett.*, **44**, 4141–4143; (c) Nishikawa, T., Kajii, S., and Isobe, M. (2004) *Chem. Lett.*, **33**, 440–441; (d) Korakas, P., Chaffee, S., Shotwell, J.B., Duque, P., and Wood, J.L. (2004) *Proc. Natl. Acad. Sci. U.S.A.*, **101**, 12054–12057; (e) Nishikawa, T., Kajii, S., and Isobe, M. (2004) *Synlett*, 2025–2027; (f) Sun, C., Lin, X., and Weinreb, S.M. (2006) *J. Org. Chem.*, **71**, 3159–3166; (g) Sun, C., Camp, J.E., and Weinreb, S.M. (2006) *Org. Lett.*, **8**, 1779–1781; (h) Black, P.J., Hecker, E.A., and Magnus, P. (2007) *Tetrahedron Lett.*, **48**(36), 6364–6367; (i) Kajii, S., Nishikawa, T., and Isobe, M. (2008) *Tetrahedron Lett.*, **49**(4), 594–597; (j) Kajii, S., Nishikawa, T., and Isobe, M. (2008) *Chem. Commun.*, **27**, 3121–3123.

29. (a) Baran, P.S., Shenvi, R.A., and Mitsos, C.A. (2005) *Angew. Chem. Int. Ed.*, **44**, 3714–3717; (b) Black, P.J., Hecker, E.A., and Magnus, P. (2007) *Tetrahedron Lett.*, **48**(36), 6364–6367.
30. Baran, P.S. and Shenvi, R.A. (2006) *J. Am. Chem. Soc.*, **128**(43), 14028–14029.
31. Stratmann, K., Moore, R.E., Bonjouklian, R., Deeter, J.B., Patterson, G.M.L., Shaffer, S., Smith, C.D., and Smitka, T.A. (1994) *J. Am. Chem. Soc.*, **116**, 9935–9942.
32. (a) Baran, P.S., Maimone, T.J., and Richter, J.M. (2007) *Nature*, **446**, 404–408; (b) Baran, P.S. and Richter, J.M. (2005) *J. Am. Chem. Soc.*, **127**(44), 15394–15396.
33. Reisman, S.E., Ready, J.M., Weiss, M.M., Hasuoka, A., Hirata, M., Tamaki, K., Ovaska, T.V., Smith, C.J., and Wood, J.L. (2008) *J. Am. Chem. Soc.*, **130**(6), 2087–2100.
34. Studies toward the welwitindolinones: (a) Wood, J.L., Holubec, A.A., Stoltz, B.M., Weiss, M.M., Dixon, J.A., Doan, B.D., Shamji, M.F., Chen, J.M., and Heffron, T.P. (1999) *J. Am. Chem. Soc.*, **121**, 6326–6327; (b) Deng, H. and Konopelski, J.P. (2001) *Org. Lett.*, **3**, 3001–3004; (c) Jung, M.E. and Slowinski, F. (2001) *Tetrahedron Lett.*, **42**, 6835–6838; (d) Lopez-Alvarado, P., Garcia-Granda, S., Alvarez-Rua, C., and Avendano, C. (2002) *Eur. J. Org. Chem.*, 1702–1707; (e) Ready, J.M., Reisman, S.E., Hirata, M., Weiss, M.M., Tamaki, K., Ovaska, T.V., and Wood, J.L. (2004) *Angew. Chem. Int. Ed.*, **43**, 1270–1272; (f) MacKay, J.A., Bishop, R.L., and Rawal, V.H. (2005) *Org. Lett.*, **7**, 3421–3424; (g) Baudoux, J., Blake, A.J., and Simpkins, N.S. (2005) *Org. Lett.*, **7**, 4087–4089; (h) Brailsford, J.A., Lauchli, R., and Shea, K.J. (2009) *Org. Lett.*, **11**(22), 5330–5333; (i) Trost, B.M. and McDougall, P.J. (2009) *Org. Lett.*, **11**(16), 3782–3785; (j) Zheng, P. and Harmata, M. (2007) *Chemtracts*, **20**(1), 20–31.
35. (a) Lin, L., Cordell, G.A., Ni, C., and Clardy, J. (1990) *Phytochemistry*, **29**(9), 3013–3017; (b) Yamada, Y., Kitajima, M., Kogure, N., Wongseripipatana, S., and Takayama, H. (2009) *Tetrahedron Lett.*, **50**(26), 3341–3344.
36. Takayama, H., Kitajima, M., Ogata, K., and Sakai, S. (1992) *J. Org. Chem.*, **57**(17), 4583–4584.
37. Hendrickson, J.B., Rees, R., and Goschke, R. (1962) *Proc. Chem. Soc.*, 383–384.
38. Robinson, R. and Teuber, H. (1954) *J. Chem. Ind.*, 783.
39. Woodward, R.B., Yand, N., and Katz, T.J. (1960) *Proc. Chem. Soc.*, 76–78.
40. (a) Hamor, T.A., Robertson, J.M., Shrivastave, H.N., and Silverton, J.V. (1960) *Proc. Chem. Soc.*, 78–80; (b) Hamor, T.A. and Robertson, J.M. (1962) *J. Chem. Soc.*, 194–205. (c) Scott, A.I., McCapra, F., and Hall, E.S., (1964) *J. Am. Chem. Soc.*, **86**, 302–303.
41. Mason, S.F. (1962) *Proc. Chem. Soc.*, 362.
42. Ishikawa, H., Aimi, N., and Takayama, H. (2002) *Tetrahedron Lett.*, **43**, 5637–5639.
43. (a) Numata, A., Takahashi, C., Ito, Y., Takada, T., Kawai, K., Usami, Y., Matsamura, E., Imachi, M., Ito, T., and Hasegawa, T. (1993) *Tetrahedron Lett.*, **34**, 2355–2358; (b) May, J.A. and Stoltz, B. (2006) *Tetrahedron*, **62**, 5262–5271.
44. May, A., Zeidan, R.K., and Stoltz, B.M. (2003) *Tetrahedron Lett.*, **44**, 1203–1205.
45. (a) Crawley, S.L. and Funk, R.L. (2003) *Org. Lett.*, **5**, 3169–3171; (b) Crawley, S.L. and Funk, R.L. (2006) *Org. Lett.*, **8**, 3995–3998.
46. Yang, J., Wu, H., and Shen Qin, J. (2007) *J. Am. Chem. Soc.*, **129**, 13794–13795.
47. Liu, P., Seo, J.H., and Weinreb, S.M. (2010) *Angew. Chem. Int. Ed.*, **49**(11), 2000–2003.
48. (a) Son, B.W., Jensen, P.R., Kauffman, C.A., and Fenical, W. (1999) *Nat. Prod. Res.*, **13**, 213–222; (b) Gardiner, D.M., Waring, P., and Howlett, B.J. (2005) *Microbiology*, **151**, 1021–1032.
49. Kim, J., Ashenhurst, J.A., and Movassaghi, M. (2009) *Science*, **324**(5924), 238–241.
50. Pousset, J.L., Kerharo, J., Maynart, G., Monseur, X., Cavé, A., and Goutarel, R. (1973) *Phytochemistry*, **12**(9), 2308–2310.

51. (a) Balde, A.M., Pieters, L.A., Gergely, A., Wray, V., Claeys, M., and Vlietinck, A.J. (1991) *Phytochemistry*, **30**(3), 997–1000; (b) Tillequin, F. and Koch, M. (1979) *Phytochemistry*, **18**(9), 1559–1561; (c) Tillequin, F., Rousselet, R., Koch, M., Bert, M., and Sévenet, T. (1979) *Ann. Pharm. Fr.*, **37**(11–12), 543–548.

52. Tillequin, F., Koch, M., Pousset, J.L., and Cavé, A. (1978) *J. Chem. Soc., Chem. Commun.*, (19), 826–828.

6
Biomimetic Synthesis of Manzamine Alkaloids*

Romain Duval and Erwan Poupon

6.1
Introduction

Since the isolation of manzamine A (**1**) in 1986 [1] the manzamine group of alkaloids[1)] has been enriched continuously by the discovery of novel marine compounds with unprecedented molecular architectures (Figures 6.1 and 6.2). Nearly 100 of alkaloids have been isolated to date from sponges of the order Haplosclerida and Dictyoceratida. This apparently heterogeneous family of alkaloids[2)] encompasses:

1) 3-Alkylpyridines and 3-alkylpyridinium salts (see the examples of monomeric theonelladin A (**2**) [2], oligomeric cyclostellettamine A (**3**) [3], and niphatoxin B (**4**) [4], and also polymeric structures such as halitoxins (**5**) [5]–Figure 6.1);
2) Elaborated and sometimes highly complex structures; see the examples of manzamine A (**1**) [1], sarain A (**6**) [6], keramaphidin B (**7**) [7], halicyclamine A (**8**) [8], manadomanzamine A (**9**) [9], nakadomarine A (**10**) [10], madangamine C (**11**) [11], misenine (**12**) [12], and upenamide A (**13**) [13] (Figure 6.2).

Despite their high structural diversity and variable sponge origin, the manzamine alkaloids exhibit common structural features, such as polycyclic bis-nitrogenated cores and macrocyclic alkyl loops, suggesting a common biosynthetic origin. This led to the proposal of "universal" biogenetic hypotheses for these complex secondary compounds, and motivated their biomimetic synthesis by several research groups.[3)]

* In memory of the late Dr Christian Marazano whose creativity and humanity will always inspire us.

1) The term "*manzamine alkaloids*" will be used throughout the chapter, as recommended by Marazano and colleagues, instead of less specific "3-alkylpiperidine alkaloids."

2) Although identical alkaloids were sometimes isolated from different sponges, only the one organism from which the molecule was first characterized is given in this chapter.

3) Of particular interest and with important consequences among sponge-derived secondary metabolites is the real origin of the molecules. Whether they are produced by the sponge itself or by associated symbionts raises exciting questions (who possess the genes?).

Biomimetic Organic Synthesis, First Edition. Edited by Erwan Poupon and Bastien Nay.

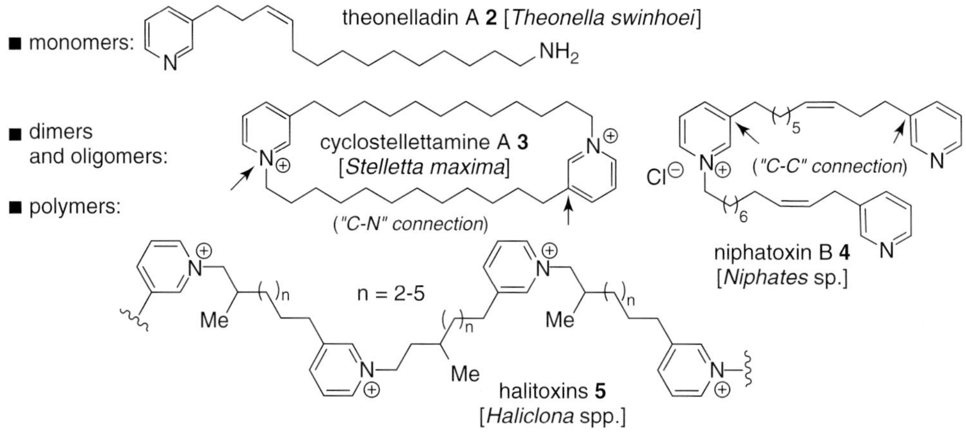

Figure 6.1 Examples of simple "manzamine alkaloids": 3-alkylpyridines and pyridinium salts.

While numerous articles have already reviewed these fascinating alkaloids and their total syntheses [14], to the best of our knowledge none has yet concentrated on comparing their perceived biogenesis and biomimetic synthesis. This chapter reviews several possible biogenetic relationships (mapped in Schemes 6.1, 6.24, 6.34, 6.36, and 6.38 and, mainly, Scheme 6.42) between the most representative members of the manzamine alkaloids. This tree-like, "from simple to complex," description integrates biomimetic chemistry studies to illustrate how, and to what extent, some hypotheses were validated or invalidated by experimental synthesis. Comprehensive reviews on the structure and sources of manzamine alkaloids as well as their total synthesis and biological activities will be found elsewhere [14]. From the discovery of manzamine A to the brilliant intuitions of Baldwin and Marazano and to the latest development in total synthesis we will embark on a journey that covers 25 years of discoveries of fascinating natural substances and biosynthetically driven chemical endeavors (see timescale on Figure 6.3).

6.2 Two Complementary Hypotheses: An "Acrolein Scenario" and a "Malondialdehyde Scenario"

6.2.1 From Fatty Aldehydes Precursors to Simple 3-Alkyl-Pyridine Alkaloids

The manzamine alkaloids, which feature large alkyl chains or loops, must at least partly originate from polyacetate metabolism. In a memorable paper entitled "On the biosynthesis of manzamines" published in 1992, this observation led Baldwin and Whitehead from the University of Oxford to propose fatty dialdehydes

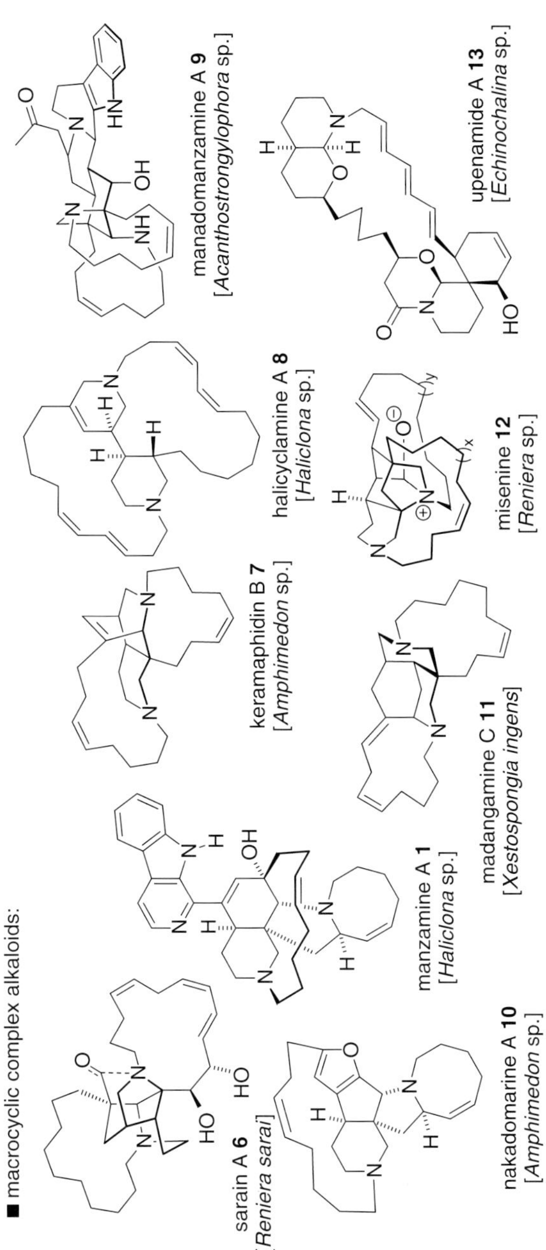

Figure 6.2 Examples of complex manzamines: macrocyclic polycyclic alkaloids.

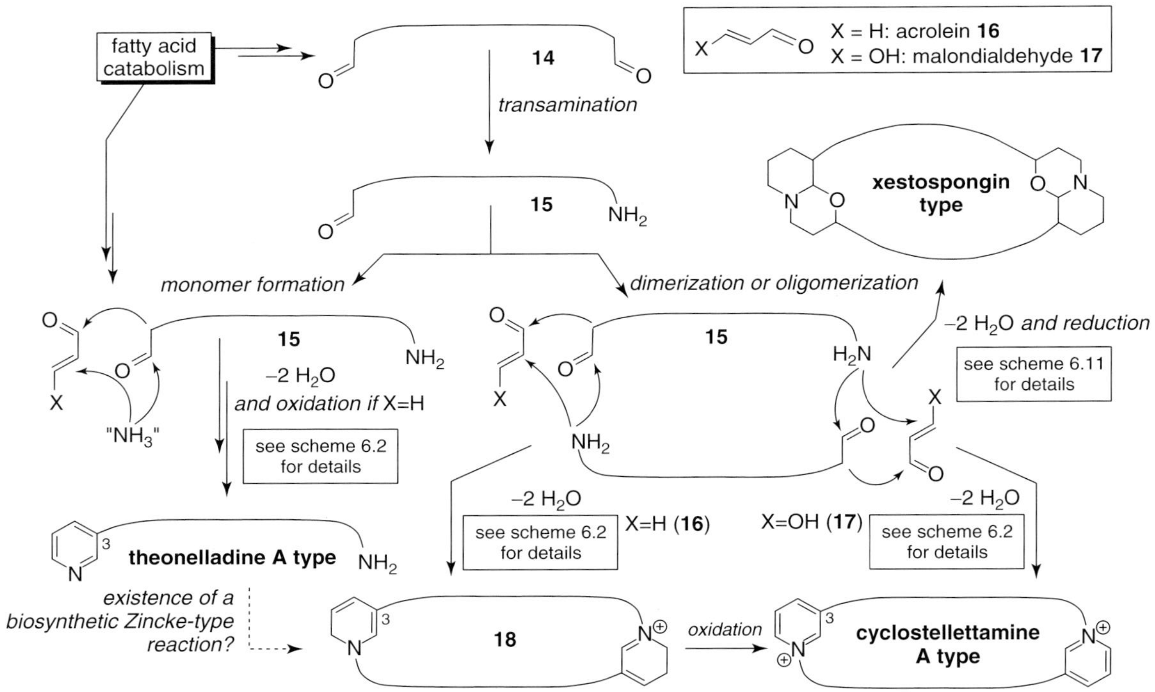

Scheme 6.1 Proposed biosynthesis of 3-alkylpyridine alkaloids: (i) *monomers*: need for an exogenous nitrogen source (ammonia equivalents) and (ii) *dimers (and oligomers)*: self-aminating process, that is, one amino-aldehyde partner is the nitrogen source for the other.
Lipophilic chains and macrocycles of the natural alkaloids are depicted by loops in Schemes 6.1, 6.11, 6.25 and 6.30, allowing a vision of the possible relationships between heterocyclic cores of representative alkaloids, irrespective of specific structural differences (such as chain length, unsaturations, etc.).

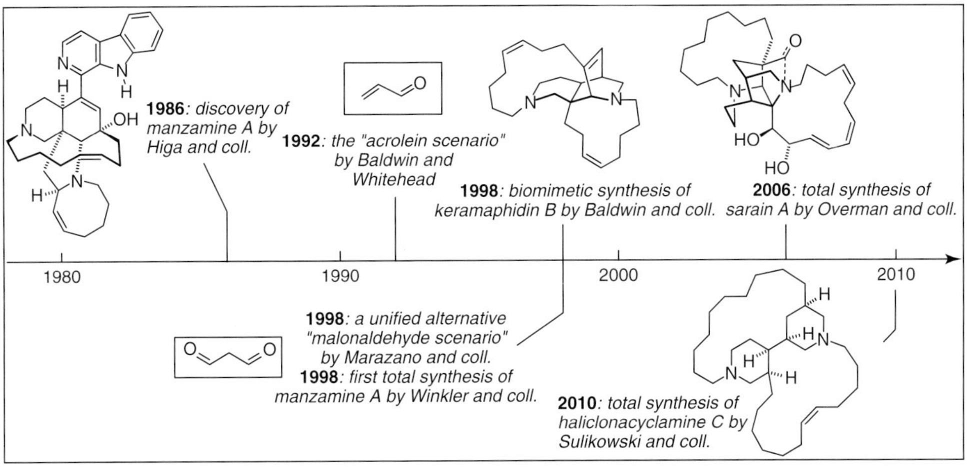

Figure 6.3 Selected milestones in 25 years of manzamine alkaloid chemistry.

(possibly coming from fatty acid catabolism) as universal, bifunctional precursors of what would become the "manzamine family" [15]. According to this seminal hypothesis, dialdehydes in C_8–C_{16} such as **14** would be monoaminated with either pyridoxamine (via transamination) or ammonia (via reductive amination) to yield amino-aldehydes **15** (Scheme 6.1). Also plausibly produced by fatty acid degradation (see below), acrolein **16** (the original C_3 species hypothesized by Baldwin) or malondialdehyde **17** (alternative C_3 species proposed by Marazano in 1998 at the Institut de Chimie des Substances Naturelles in Gif-sur-Yvette [16]) would react with **15** and a source of ammonia, furnishing pyridine alkaloids of the theonelladin A type. On the other hand, dimerization of amino-aldehydes **15** in the presence of two equivalents of acrolein **16** or malondialdehyde **17** would furnish alkaloids of the cyclostellettamine type via dihydropyridiniums **18**. Formation of alkaloids of the xestospongin type would occur in the case of alkyl chains β-hydroxylated relative to the nitrogen (Section 6.3.2). Scheme 6.2 presents detailed mechanisms for pyridine ring formation according to both scenarios. Importantly, late oxidation of dihydropyridine species would be required to yield the final pyridine/pyridinium skeletons when acrolein **16** is incorporated, contrary to malondialdehyde **17**. Also important and central to the modified hypothesis involving **17** is the postulate of the existence of two types of C_5 reactive units: glutaconaldehydes and aminopentadienals.

In vivo, malondialdehyde (**17**) results mainly from the catabolism and peroxidation of polyunsaturated fatty acids such as arachidonic or linolenic acid (Scheme 6.3). Different mechanisms have been proposed, some involving radical reactions with reactive oxygen species (ROS) [17]. The existence of an acrolein radical (**19**) has been postulated, which could react with a hydroxyl radical to give malondialdehyde **17**. From this hypothesis, a simple reduction of **19** would explain the formation of acrolein **16**.

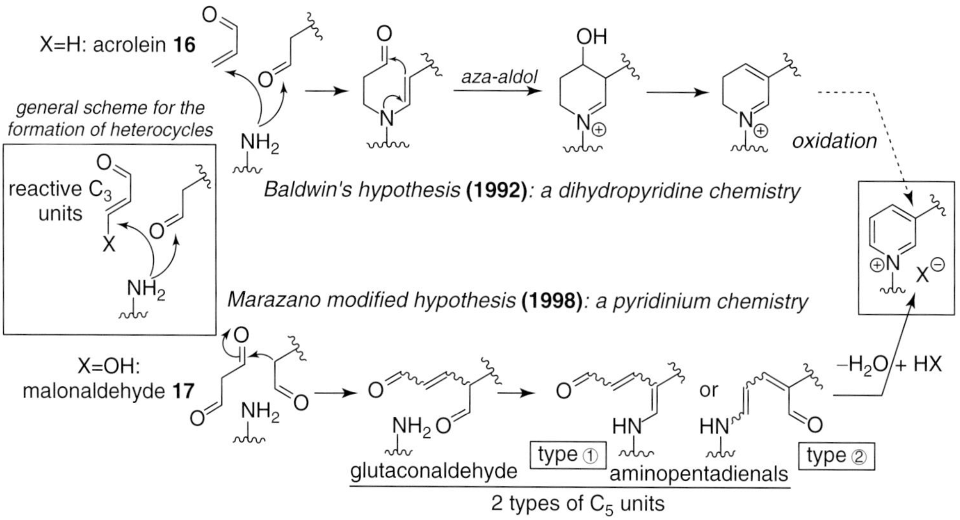

Scheme 6.2 Detailed pyridine ring formation according to both scenarios.

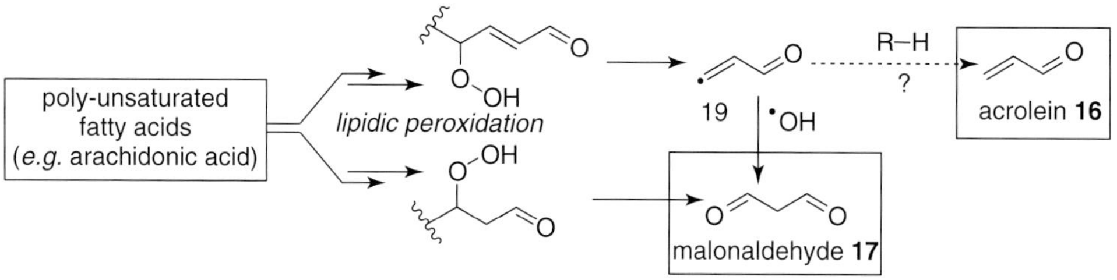

Scheme 6.3 Plausible origin of C_3-reactive units from lipidic peroxidation.

Alternative biosynthetic pathways have been proposed. To date, they have not been corroborated by biomimetic chemical synthesis but they merit attention. Amade, Thomas, and colleagues founded their hypothesis for the biosynthesis of 3-alkylpyridiniums when they isolated pachychalines A (**20**) and B (**21**) from a Caribbean *Pachychalina* species [18] and pachychaline D (**22**) from a *Callyspongia* species [19]. Given the presence of a homospermidine fragment on pachychaline B (**21**) and D (**22**), the authors proposed this diamine as a possible C_3 unit provider and put together a unified scenario for both C-N and C-C connection patterns necessitating oxidation steps of primary amines into the corresponding imines (Scheme 6.4).

Imines/enamines cascades could be responsible for the formation of pyridiniums with a C_3 diamine acting as a leaving group. Whatever the biosynthetic scheme devised by chemists, it is striking that it is impossible to avoid some kind of C_3 unit.

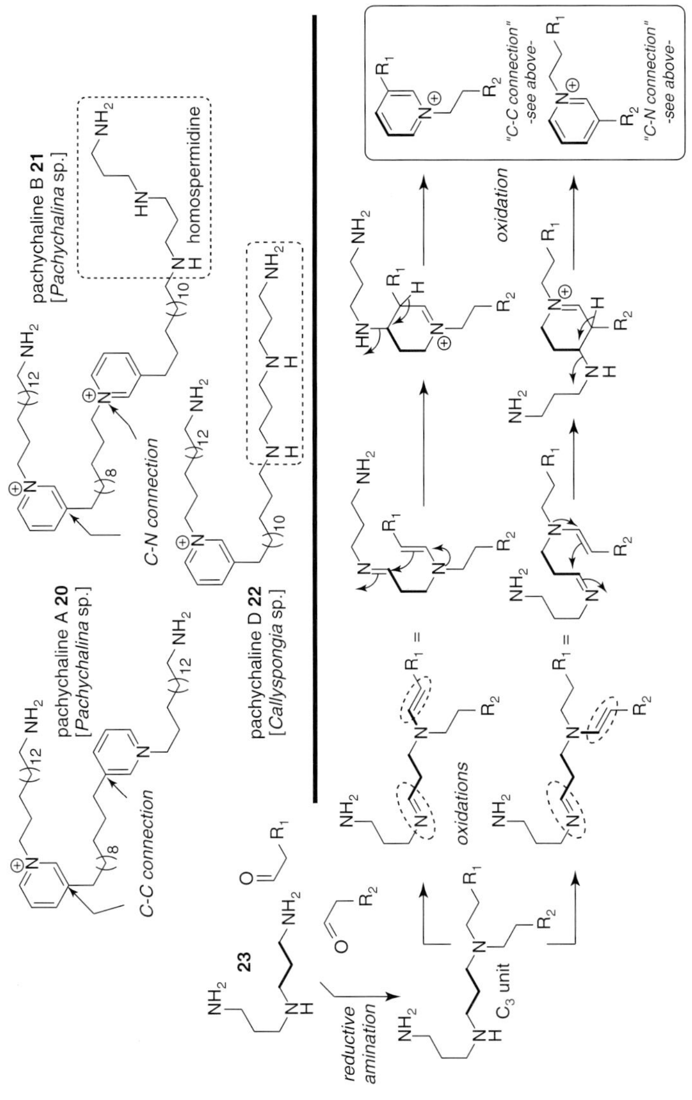

Scheme 6.4 Alternative biosynthetic hypotheses for 3-alkylpyridiniums based on the pachychaline series.

6.2.2
Biomimetic Synthesis of Dihydropyridines and Dihydropyridinium Salts

The order of events in this postulated biosynthesis of pyridine rings, involving imine/enamine formation, Michael reaction, and aldol/aza-aldol reaction, is unknown. To our knowledge, its closest synthetic equivalent is the Chichibabin synthesis of pyridines, where ammonia or primary amines and aliphatic aldehydes in excess react at elevated temperatures to yield 3,4,5-trisubstituted dihydropyridines/dihydropyridinium salts that spontaneously oxidize to pyridines/pyridinium salts (Scheme 6.5) [20]. This reaction cannot be exploited when pyridines substituted with different groups, or dihydropyridine intermediates, are desirable. However, Marazano and colleagues developed a versatile strategy related to the Chichibabin synthesis to by-pass these limitations and access Baldwin's intermediates, based on the coupling of Strecker (intermediate **24**), Michael (intermediate **25** formed with **16**), and aza-aldol reactions. This "one-pot" procedure capitalizes on the particular reactivity of zinc triflate and directly furnishes 1,3-disubstituted dihydropyridiniums masked under the form of α-aminonitriles **26** (Scheme 6.5) [21]. Those stable equivalents favorably compare to dihydropyridinium **27** obtained from pyridinium salts **28**, following the classical reduction-modified Polonovski reaction sequence developed by Husson and colleagues (via tetrahydropyridine **29** and *N*-oxide **30**) [22]. Dihydropyridinium **27** can in turn be formed from **25** upon

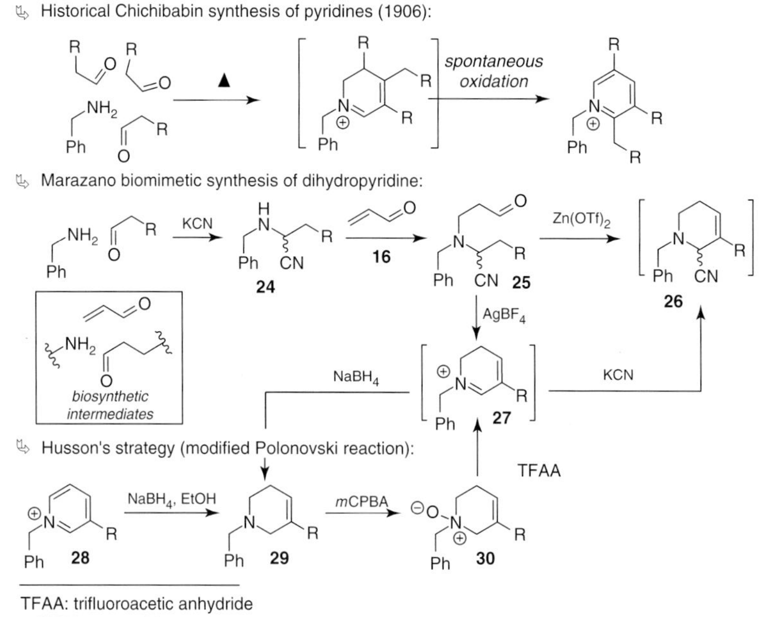

TFAA: trifluoroacetic anhydride
*m*CPBA: *meta*-chloroperbenzoic acid

Scheme 6.5 Syntheses of biomimetic equivalents of dihydropyridines.

treatment with a silver salt and be reduced to **29** with sodium borohydride, usually with high selectivity, or trapped with cyanide ions to yield **26**.

Conditions for deprotonating stable salts such as **31**, prepared *in situ* by heating masked dihydropyridinium **32**, to give the corresponding dihydropyridine **33** (that was unstable but could be trapped), were disclosed by the Marazano group (Scheme 6.6) [23]. The possibility of favoring one isomer over the other (dihydropyridinium **34**) soon appeared to be a challenging problem in gaining chemoselectivity (see the following sections).

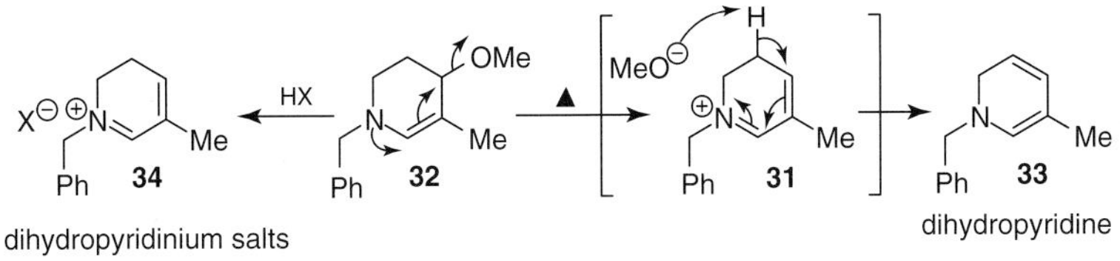

Scheme 6.6 Isomerization of dihydropyridinium salts to dihydropyridine.

6.2.3
A Tool Box of Biomimetic C_5 Reactive Units from the "Old" Zincke Reaction

Marazano *et al.* revisited the century-old Zincke reaction, a nucleophilic ring opening of electron-deficient pyridinium ("Zincke salts," easily prepared from pyridines and electrophiles such as cyanogen bromide or 2,4-dinitrochlorobenzene) [24].

Scheme 6.7 presents the mechanism generally admitted for the Zincke reaction [25]. Ring opening of pyridinium **35** occurs with the first equivalent of amine (usually accompanied by a dark red coloration of the reaction mixture). A second equivalent of amine then reacts with **36** to provide **37** with extrusion of a 2,4-dinitroaniline moiety. In solution, **37** is in equilibrium with two aminopentadienimines (**38** and **39**). Upon heating, intermediate **37** may cyclize into pyridinium salts **40** with elimination of one amine moiety [26].

Aminopentadienimines of type **38** and **39** are of particular interest for accessing biomimetic equivalents of postulated intermediates, namely aminopentadienals and glutacondialdehydes (Scheme 6.2) [27]. Convenient and scalable accesses to substituted glutacondialdehyde salts from the corresponding pyridinium Zincke salts were recently disclosed[4] [28]. Specifically, in these cases, a secondary amine such as dimethylamine is employed for the ring-opening to afford firstly salts **42**, then biomimetic equivalents **43** of biosynthetic aminopentadienals. These latter can be hydrolyzed into glutacondialdehyde salts **44** when treated with potassium hydroxide (Scheme 6.8).

4) Overcoming thereby some drawbacks of the "classical" Zincke reaction, such as the use of two equivalents of amine and the propensity of aminopentadienals to form pyridinium salts.

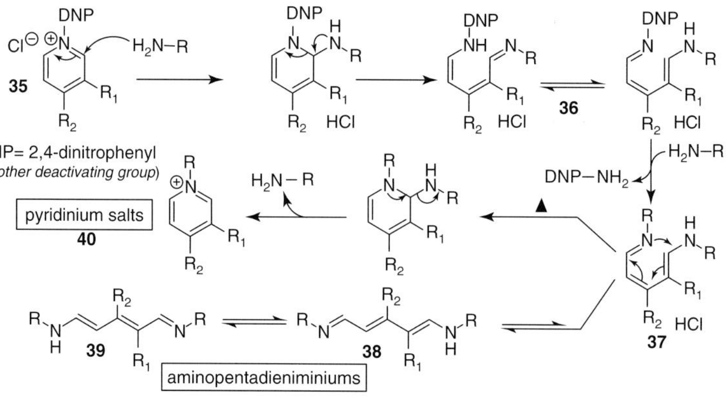

Scheme 6.7 Mechanism of the Zincke reaction.

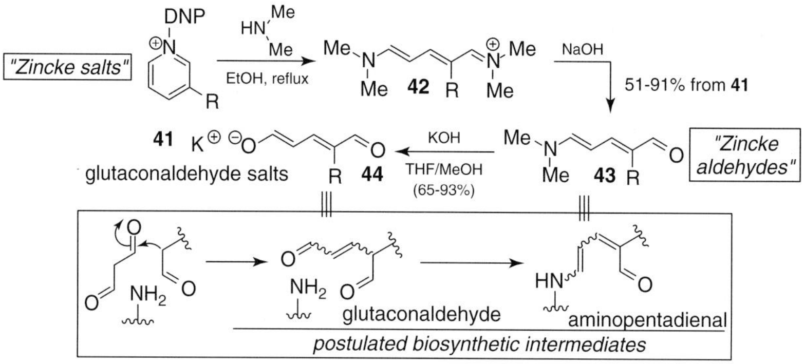

Scheme 6.8 Glutacondialdehydes and aminopentadienals as biosynthetic intermediates and biomimetic equivalents.

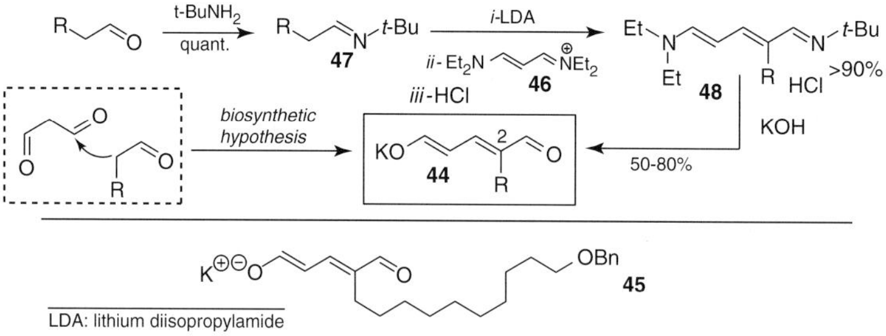

Scheme 6.9 Alternative synthetic pathway towards substituted glutacondialdehyde salts, using vinamidinium salts as biomimetic equivalents of malonodialdehyde.

More elaborated glutaconaldehyde salts of type **44** (as exemplified by compound **45**, Scheme 6.9) can also be prepared starting from various aldehydes and vinamidinium salts **46** (via **47** and **48**) [29]. Interestingly, this strategy is reminiscent of the first fundamental step in Marazano's hypothesis of pyridine ring formation, that is, the reaction of a fatty aldehyde with malondialdehyde **17** (see Scheme 6.2 for details).

6.3
Biomimetic Synthesis of Pyridinium Marine Sponge Alkaloids

6.3.1
Biomimetic Total Synthesis of Cyclostellettamine B and Related 3-Alkylpyridiniums

To test their "malondialdehyde" scenario and to demonstrate the suitability of Zincke chemistry toward this end, the access to cyclostellettamine B (**49**) [3] was studied by Marazano and colleagues [16a]. They performed a pseudo-dimerization of two 3-aminoalkylpyridines (**50**, **51**) of different chain lengths,[5] using sequential pyridinium *N*-activation (via **52** and **53**). Cyclostellettamine B (**49**) was thus elegantly obtained in a biomimetic "domino-Zincke" reaction (Scheme 6.10). A similar philosophy permitted the total synthesis of haliclamine A (**54**) [30, 31] and niphatoxin B (**4**) [32], and also that of two original molecules isolated and synthesized by the Köck group, that is, viscosamine (**55**) (a trimeric 3-alkylpyridinium) [33] and a monomeric but cyclic 3-alkylpyridinium alkaloid (**56**) (Scheme 6.10) [34].

6.3.2
Biomimetic Synthesis of Xestospongins and Related Structures

Xestospongins are macrocyclic bis-1-oxaquinolizidine alkaloids isolated from *Xestospongia exigua* (syn. *Neopetrosia exigua*) [35]. Many other structures are closely related to xestospongins such as araguspongins, and the interested reader is referred to general review articles [14]. We focus herein on the biomimetic synthesis of xestospongins A (**57**) (Scheme 6.11) and C (**58**) [35] as well as (+)-araguspongin B (**59**) (Scheme 6.12) [36] by the Baldwin group in 1998, which also permitted the establishment of their correct absolute configurations [37].

Biosynthetically, starting from bis-hydroxypyridinium dimer **60** or the corresponding dihydropyridinium salt **61**, intramolecular trapping of the iminiums would explain the formation of the oxaquinolizidine ring systems and the natural substances after a reduction step on **62** (Scheme 6.11).

Conformational and configurational differences and/or equilibria between natural substances in this series can be seen to occur via iminium/enamine epimerizing equilibrium, involving ring opening/reclosure to cyclic aminals. The biomimetic

5) The incorporation of chains of various lengths is of course a critical point in the total synthesis of such molecules.

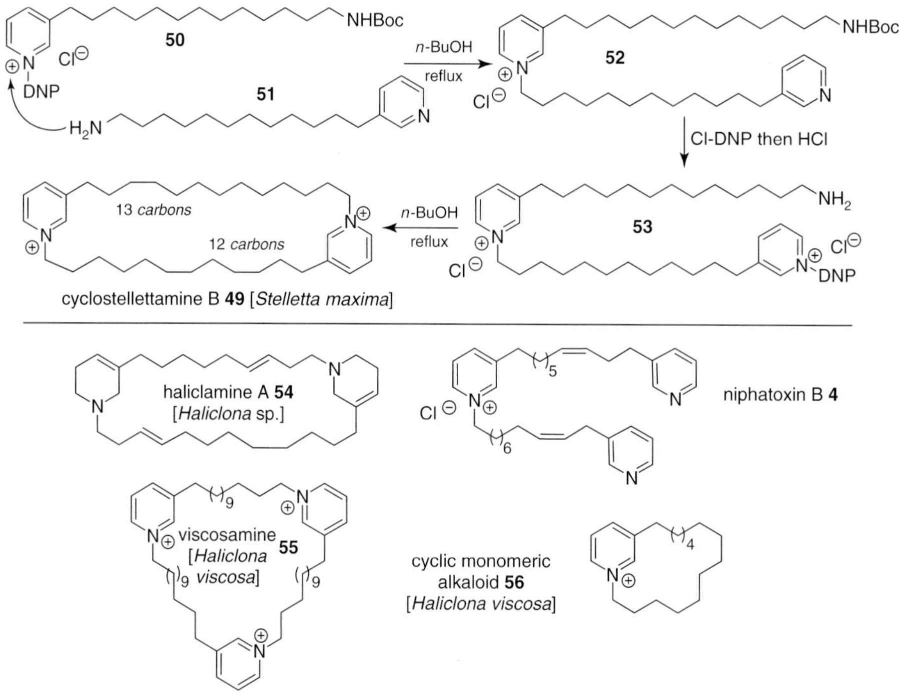

Scheme 6.10 Biomimetic synthesis of cyclostellettamine B and the structure of alkaloids synthesized using a similar philosophy.

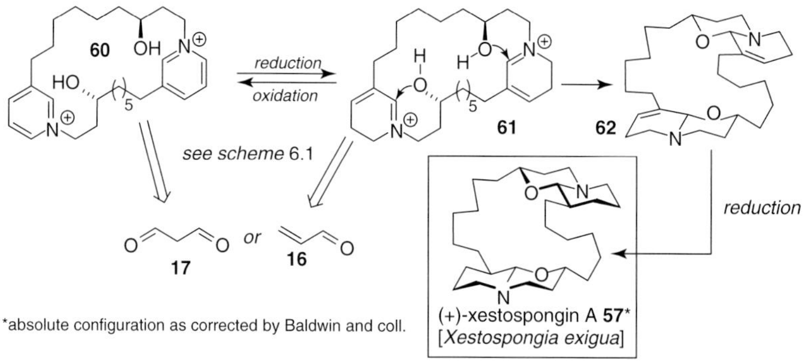

Scheme 6.11 Biosynthetic proposal for xestospongin A and related structures.

synthesis of *ent*-xestospongins A (*ent*-**57**) and C (*ent*-**58**) and *ent*-araguspongin B (*ent*-**59**) from *ent*-**60** and *ent*-**62** as depicted in Scheme 6.12 probably proceeds via this pathway, and implicates intermediate *ent*-**61**. Two distinct reaction conditions permitting a reduction of the unsaturated piperidine without reduction of the masked iminium were studied (i.e., hydrogenation with catalytic rhodium or Raney nickel), and gave different ratios of the three natural substances *ent*-**57** and *ent*-**58**

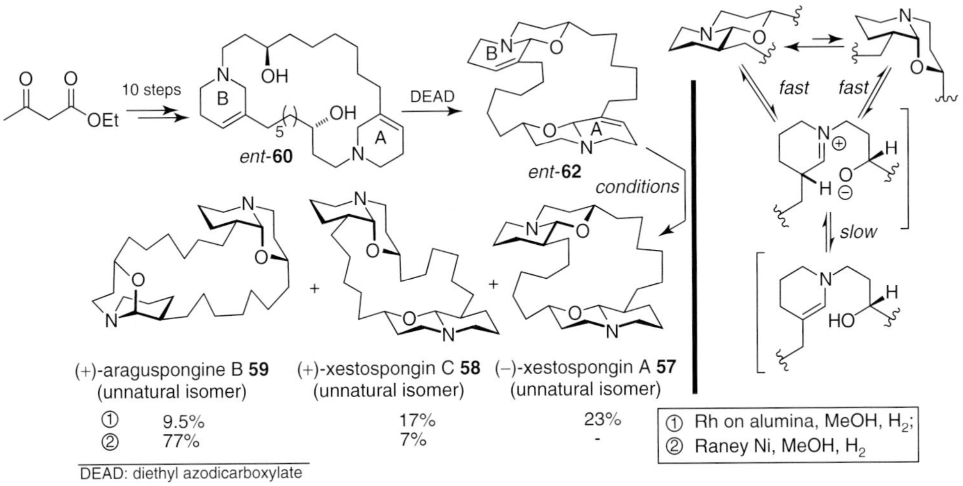

Scheme 6.12 Biomimetic synthesis by the Baldwin group.

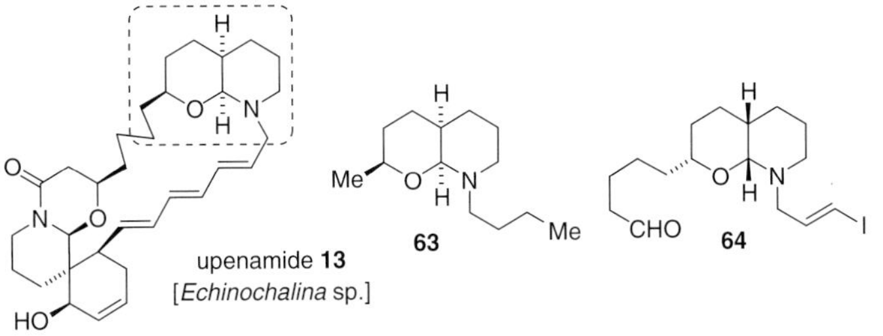

Figure 6.4 Selected synthetic approaches to upenamide.

and *ent*-**59**. Clear establishment of the absolute configurations of (+)-xestospongin A (**57**) and (−)-xestospongin C (**58**) and questions concerning those of araguspongin (**59**) alkaloids were also discussed by Baldwin and colleagues [37].

Similar stereoelectronic outcomes, which we will not discuss here, were studied during the synthesis of the octahydropyrano-pyridine ring system of upenamide (**13**) by the Marazano [38] and Sulikowski groups [39]. The fragments (**63** and **64**) prepared by each group are presented in Figure 6.4. To date, this fascinating alkaloid has resisted total synthesis.

6.3.3
Is the Zincke-Type Pyridine Ring-Opening Biomimetic?

To the best of our knowledge, Zincke-type pyridine or pyridinium ring formations are poorly documented in biochemistry. As an example (Scheme 6.13), the biosynthesis of quinolinic acid (**65**)–a direct precursor of nicotinamide adenine dinucleotide (NAD)–from tryptophan-derived **66** was proposed to take place via acyclic **67** followed by a 6π-electrocyclization (demonstrated with model systems)

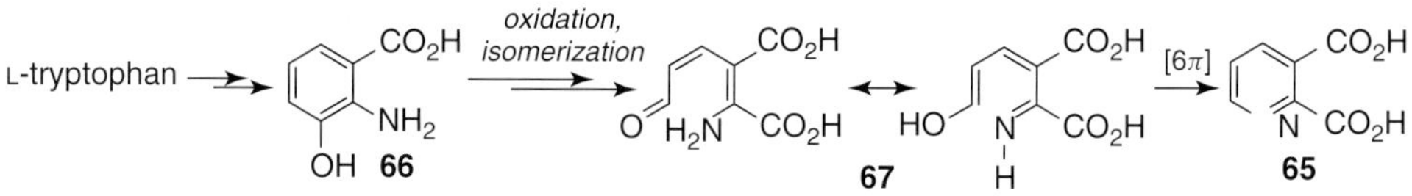

Scheme 6.13 Biosynthesis of quinolinic acid.

(Scheme 6.13) [40]. If discovered in secondary metabolism, such a transformation would putatively connect theonelladine-type and cyclostellettamine-type alkaloids via a biosynthetic Zincke-type reaction (cf. Scheme 6.1).

6.3.4
Alkylpyridines with Unusual Linking Patterns

6.3.4.1 Biomimetic Synthesis of Pyrinodemin A

With a *cis*-cyclopent[*c*]isoxazolidine ring system linking two alkylpyridine chains, pyrinodemin A (**68**) (Scheme 6.14), the first representative of a small group of four alkaloids [41], has been the subject of several publications [42]. In fact, only the total synthesis in combination with degradation experiments of **68** permitted establishment of the correct structure of this intriguing natural product as far as the position of the side-chain double bond is concerned. In 2005 [43], the Kobayashi group put forward clear conclusions establishing the position of the double bond and the racemic character of the central core–despite a (−) reported optical rotation in the original paper [41a]. We will, in this section, primarily deal with the biomimetic access to the central bicyclic system of **68**, which was logically proposed to biosynthetically arise from a [3 + 2] cycloaddition between a nitrone and a *(Z)*-alkene, which in turn arise from two precursors (aldehyde **69** and amine **70** sharing the same number of carbons and a similarly positioned cis-double bond). The key cycloaddition was exploited in most total syntheses of **68** and resulted in a stereocontrolled formation of the bicyclic system in good yields.

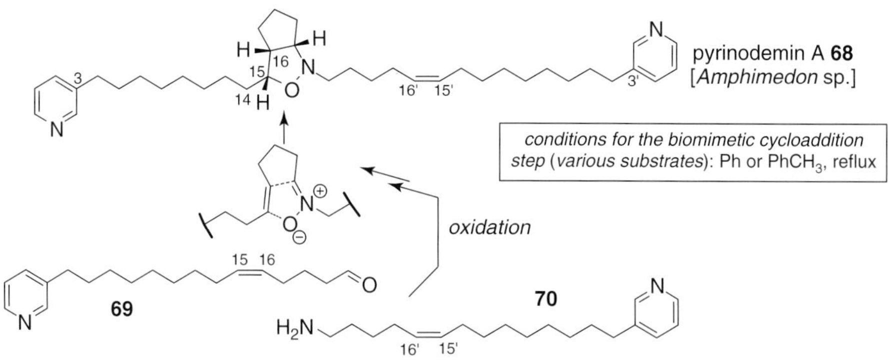

Scheme 6.14 Biosynthetic considerations for pyrinodemin A.

6.3.4.2 Biomimetic Synthesis of Pyrinadine A

Another intriguing linking pattern is the one observed in pyrinadine A (**71**) isolated from a Cribochalina sp. by Kobayashi and colleagues (Scheme 6.15) [44]. It consists of an uncommon in Nature diazoxy group, presumably resulting from the oxidative dimerization of a hydroxylamine such as **72** (obtained from the oxidation of the corresponding amine **73**).

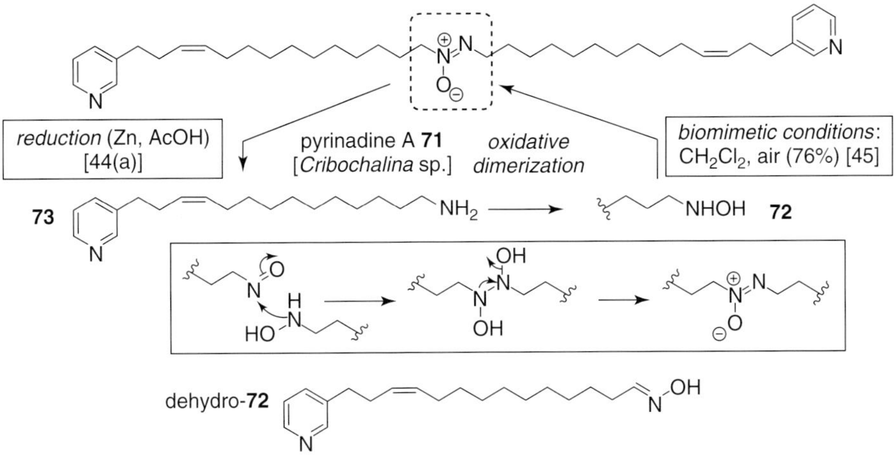

Scheme 6.15 Pyrinadine A: plausible biosynthetic origin and biomimetic access.

In 2009, Lee and colleagues [45] successfully mimicked the process in the laboratory with a clean and spontaneous conversion of synthetic precursor **72** into **71** under simple aerial conditions according to the mechanism proposed in the box in Scheme 6.15. Keeping in mind that plausible precursor dehydro-**72** is a known natural product [41b], the exact role of enzymes is clearly questioned in such cases and an artifactual origin cannot be ruled out.[6)]

6.4 Development of Baldwin's Hypothesis: From Cyclostellettamines to Keramaphidin-Type Alkaloids

6.4.1 Linking Pyridinium Alkaloids and Manzamine A-Type Alkaloids

Competitive to the redox processes that may take place between cyclostellettamine-related alkaloids **74** and postulated dihydropyridinium salts counterparts such as **75**, intramolecular Diels–Alder cycloaddition of bis-dihydropyridinium **75** might occur as represented in Scheme 6.16, yielding a bridged pentacyclic

6) The existence of unsymmetrical molecules such as pyrinadines C–G, see Reference [44], adds further credence to this statement.

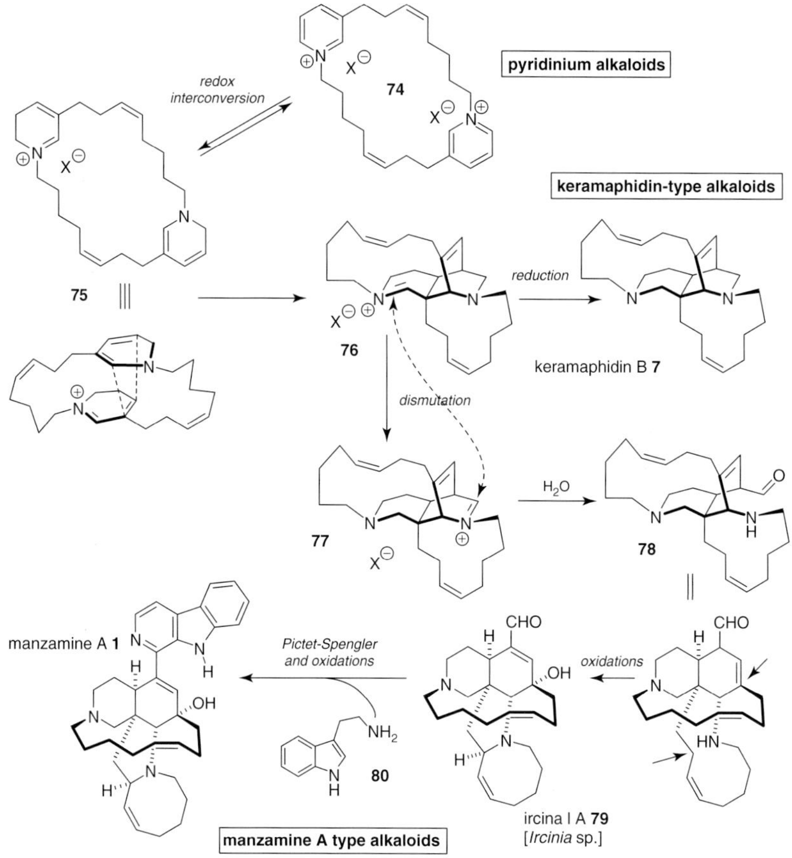

Scheme 6.16 Baldwin's hypothesis: the missing link between pyridinium alkaloids and manzamine A.

intermediate (**76**). Dismutation[7] reaction would give rise to a new iminium (**77**), which upon hydrolysis would provide **78** as a direct precursor of ircinal A (**79**), a natural substance isolated for the first time in 1992 from Ircinia sp. (notably, after Baldwin's proposals) [46]. Formation of manzamine A (**1**) through a Pictet–Spengler reaction with tryptamine (**80**) followed by oxidation into the final β-carboline would then be easily explained.

The pertinence of Baldwin's proposal comes from the fact that key pentacyclic intermediate **76** was postulated in 1992 for the biogenesis of manzamine A (**1**), before its natural occurrence became apparent some time later with the isolation of keramaphidin B (**7**) (1994) and related analogs. In fact, simple iminium reduction of **76** can explain the biosynthesis of keramaphidin B-type alkaloids, placing thereby

7) This term will be used cautiously in the present chapter, regarding the absence of knowledge of the precise redox mechanisms involved biosynthetically.

76 as a cornerstone in the general biosynthetic mapping of manzamine alkaloids. This example is probably one of the most brilliant demonstrations of the power of "retrobiosynthesis" as it beautifully paved the way to rich biomimetic endeavors.

6.4.2 Biomimetic Total Synthesis of Keramaphidin B

6.4.2.1 Model Studies (1994)

To validate their hypothesis, Baldwin and colleagues successfully carried out model reactions that permitted the synthesis of the central core of keramaphidin B (**7**) by *intermolecular* Diels–Alder reaction between two molecules of dihydropyridinium salts **81**, prepared from picoline via *N*-oxide **82** (Scheme 6.17) [47]. Incubation of **81** in an aqueous buffer at pH 8.3 for 18 h followed by treatment with sodium borohydride yielded mainly unsaturated piperidine **83** but also **84** (10% yield based on *N*-oxide **82**), the awaited polycyclic core reminiscent of that of keramaphidin B (**7**), probably formed via iminium **85**.[8)]

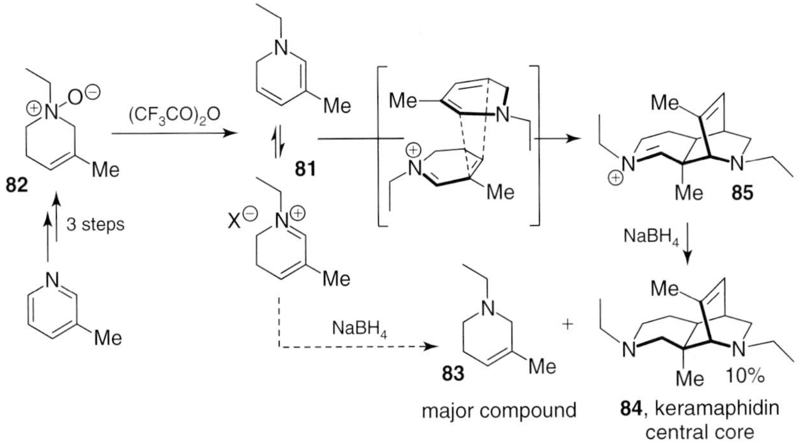

Scheme 6.17 Baldwin's hypothesis: model studies toward keramaphidin B.

6.4.2.2 Total Synthesis of Keramaphidin B (1998)

Four years later, the Baldwin group made a significant contribution to the art of total (biomimetic) synthesis (Scheme 6.18). They first obtained cyclostellettamine-type **74** by dimerization of 3-tosyloxyalkenyl pyridine **86**, which was converted into postulated biosynthetic intermediate **75**, via intermediate **87**, using the classical reduction/modified Polonovski reaction sequence (*vide supra* Schemes 6.5 and 6.17). Following equilibrium of bis-dihydropyridinium **75** in aqueous medium

8) The publication ended with:

"investigations into an intramolecular variant of the cycloaddition [...] are in progress with the aim of accomplishing the total biomimetic syntheses of manzamines and ingenamine via keramaphidin B."

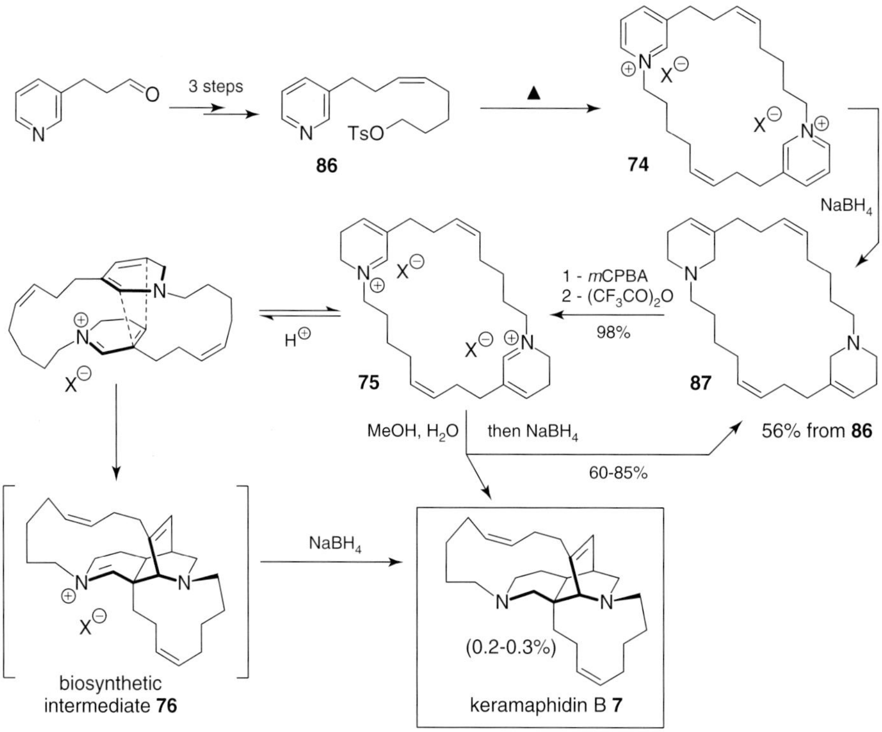

Scheme 6.18 Validation of Baldwin's hypothesis: total synthesis of keramaphidin B.

and reduction, the authors were able to isolate keramaphidin B (7) in 0.2–0.3% overall yield [48]. As already pointed out in the model studies (Scheme 6.17), the major compound resulting from this last reaction was recyclable bis-tetrahydropyridine **87**.

6.4.3
Drawbacks of the "Acrolein" Scenario

While this total synthesis achievement clearly demonstrated the validity of the model, the extremely low yield of keramaphidin B (7) became the subject of puzzling investigations. This result was explained by important side-reactions (mainly dismutation of dihydropyridines via intermolecular hydride transfer), and by the high energy barrier of the macrocyclic cycloaddition. The experimental drawbacks observed with this pioneering model are detailed below and were part of the reason why a modified scenario was concomitantly proposed by the Marazano group.

6.4.3.1 Very Low Yield of the *Endo*-Intramolecular Diels–Alder Reaction

This is obviously the major drawback of Baldwin's total synthesis of keramaphidin B (Scheme 6.19). This key step could benefit *in vivo* from the intervention of a

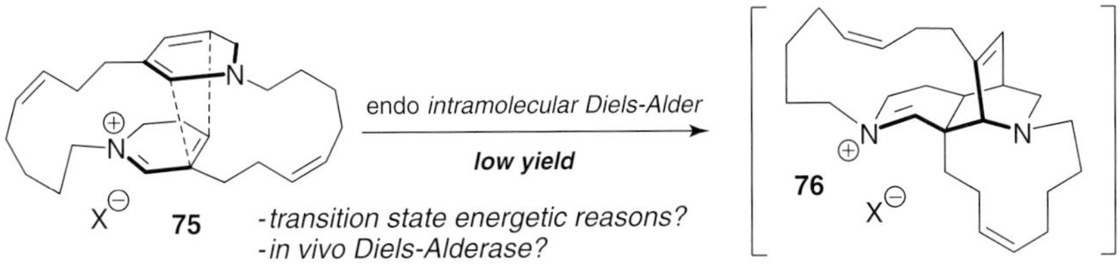

Scheme 6.19 Low yield of the Diels–Alder reaction.

"Diels–Alderase" that could limit the conformational mobility of the transition state, thereby minimizing the entropic factor.

However, molecular modeling studies conducted on intermediate **75** revealed the existence of conformations close to the required transition state [48b]. The kinetic preference of **75** to disproportionate was thereby put forward as the main reason for the low yield of the synthesis.

6.4.3.2 Undesirable Transannular Hydride Transfers

One of the reasons why the above Diels–Alder reaction was not as efficient as expected is probably the existence of a favorable transannular dismutation, leading after reduction to Δ^3-piperidines. This disproportionation phenomenon, which was observed on model systems with dihydropyridinium salt **81** [47], occurred to a greater extent with **75** in the natural product synthesis (Scheme 6.20). Several experimental evidences suggested that the isolation of reduced **87** after treatment of the reaction mixture with sodium borohydride resulted from the reduction of **88** (arising from spontaneous dismutation), and not from the reduction of biosynthetic intermediate **75**. The intervention of a putative "Diels–Alderase" could also exclude or limit this disproportionation reaction.

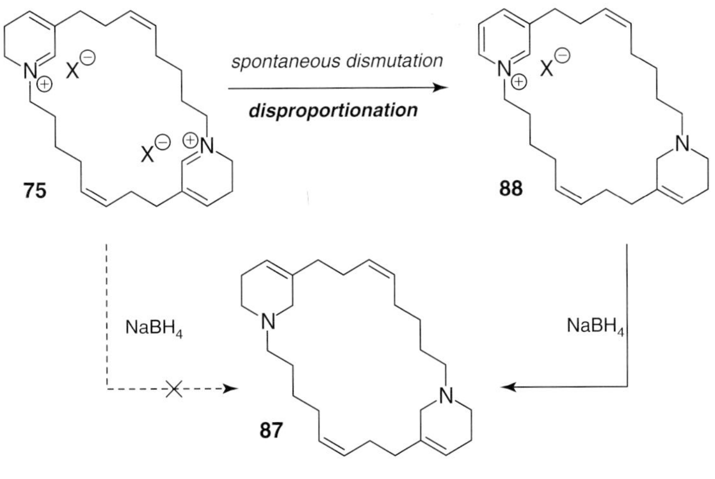

Scheme 6.20 The disproportionation issue.

6.4.3.3 Conversion of a "Keramaphidin" Skeleton into an "Ircinal/Manzamine" Skeleton Was Not Experimentally Possible

According to Baldwin's biogenetic hypothesis, alkaloids of the keramaphidin B type are the immediate precursors of ircinal type and manzamine A type alkaloids, following regioselective oxidation and iminium hydrolysis (cf. Scheme 6.16). To test this hypothesis, the Marazano group synthesized aminonitrile **89**, which was submitted to decyanation–hydrolysis using tetrafluoroboric acid in aqueous medium. Although the N_2-iminium **90** was clearly observed by NMR, aldehyde **91** could never be obtained even under various forcing hydrolytic conditions (Scheme 6.21) [49]. This result disfavors the perception that manzamine aldehydes originate from the hydrolysis of keramaphidin-type iminium as proposed by J.E. Baldwin [15]. However, one should keep in mind that this hydrolysis equilibrium could be driven biosynthetically by a connected equilibrium (e.g., proton shift to $\alpha\beta$-unsaturated aldehyde), or any irreversible transformation (e.g., Pictet–Spengler reaction).

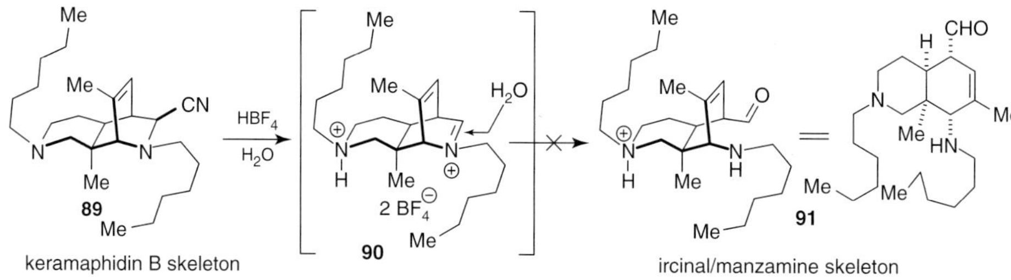

Scheme 6.21 Failure to convert a keramaphidin skeleton into a ircinal/manzamine A skeleton on a model system.

6.5 "Malondialdehyde Scenario:" A Modified Hypothesis Placing Aminopentadienals as Possible Precursors of Manzamine Alkaloids

6.5.1 Keramaphidin/Ircinal Connection

The modified hypothesis, based on the intervention of malondialdehyde C_3-reactive units to explain the formation of pyridinium salts, sets the stage for a universal model of biosynthesis for the manzamine alkaloids. Armed with a long experience in the pyridinium chemistry, the Marazano group turned its attention to the manzamine alkaloids in the mid-1990s. In their 1998 and 1999 papers [16], for the first time the pyridinium/aminopentadienal chemistry was put forward to explain the divergent formation of ircinal/manzamine and halicyclamine alkaloids.

Whereas polycyclic intermediate **76** (resulting from the intramolecular Diels–Alder reaction between dihydropyridinium salts, cf. Scheme 6.16) was central in Baldwin's hypothesis, intermediate **92** featuring (i) a dihydropyridinium moiety

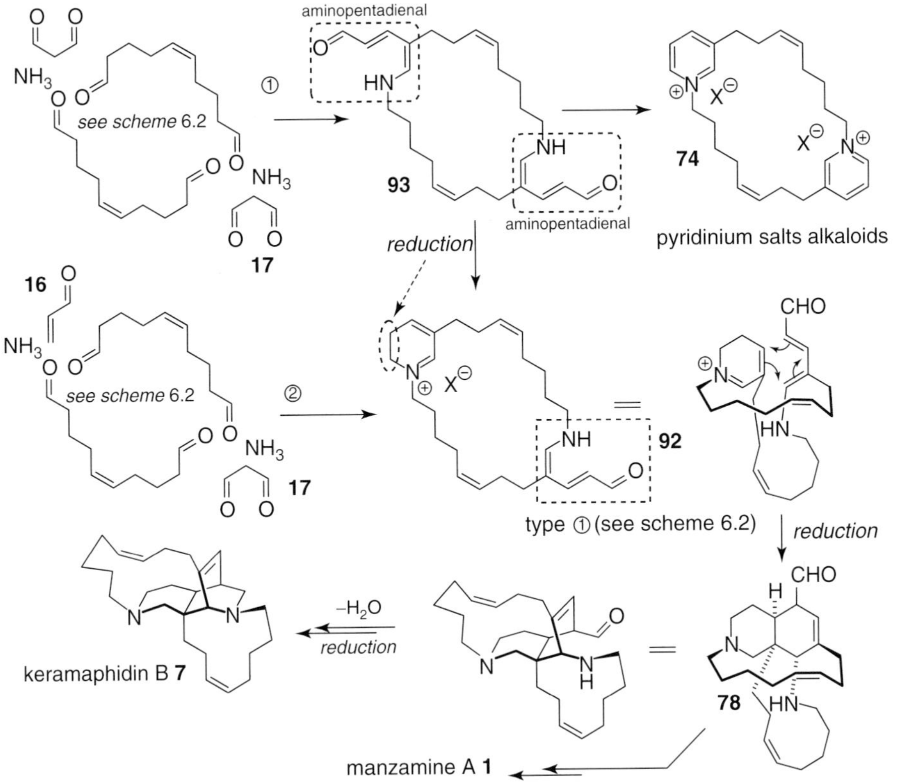

Scheme 6.22 Biosynthetic scheme towards manzamine A and keramaphidin B according to Marazano's hypothesis.

(resulting from either the "acrolein" or "malondialdehyde" pathways ① and ②, respectively, on Scheme 6.22) and (ii) an open-chain aminopentadienal, is obviously the key element of the modified hypothesis (Scheme 6.22). It should be mentioned here that intermediate **78** closely related to ircinal and resulting formally from an intramolecular Diels–Alder reaction of **92** is therefore a precursor of the keramaphidin (**7**) skeleton according to this model. Additionally, an intermediate such as **93** could also explain the biosynthesis of cyclostellettamine-related pyridinium **74**.

6.5.2
Halicyclamine Connection

The aim of this model was also to explain the formation of halicyclamine-type alkaloids (Scheme 6.23) from precursor **94**, that is, similar to precursor **92** with the difference that it integrates a type-2 aminopentadienal moiety (according to Scheme 6.2). Halicyclamine-type alkaloids would this time be generated by an intramolecular 1,4-addition of the aminopentadienal moiety reacting not as a diene

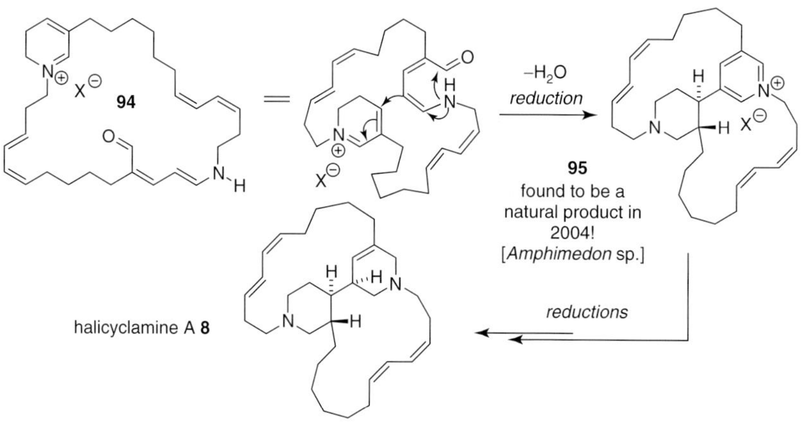

Scheme 6.23 Biosynthetic scheme towards halicyclamine A according to the modified scenario.

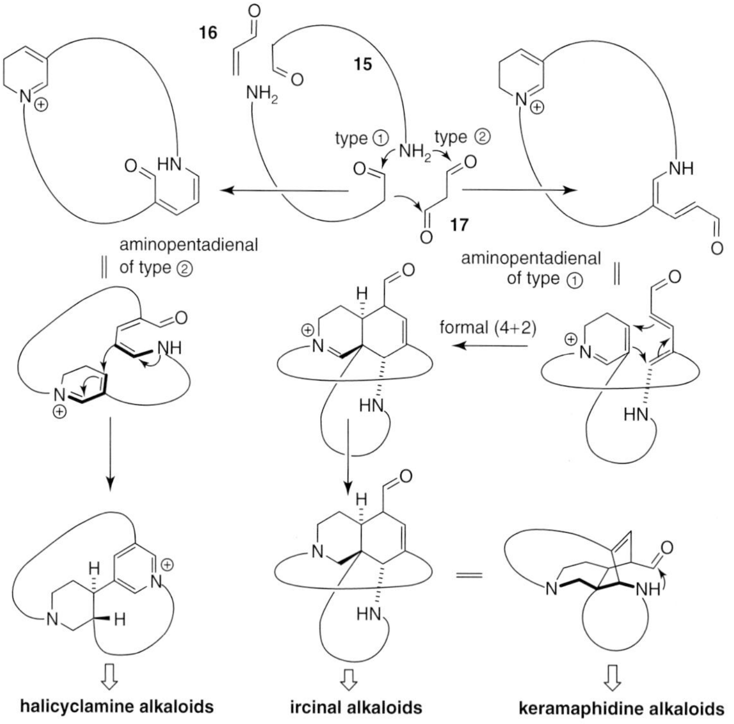

Scheme 6.24 The aminopentadienal connection between representative manzamine alkaloids.

(Scheme 6.22) but as an enamine. Illustrated in Scheme 6.23 with the biosynthetic proposal for halicyclamine A (**8**) the scenario was consolidated some years later with the isolation of postulated pyridinium **95** as a natural substance from *Amphimedon* sp. [50].[9)]

When considering only the biosynthesis of the central cores of the different sub-classes of alkaloids evoked up to now, the homogeneity of the model is even more striking, as represented in Scheme 6.24. Starting from intermediates of type **15**, access to the halicyclamine, keramaphidin, and ircinal series is explained via the reactivity of the aminopentadienal system (acting as either diene or enamine).

6.6 Testing the Modified Hypothesis in the Laboratory

6.6.1 Biomimetic Models toward Manzamine A

According to Marazano's hypothesis, which involves malondialdehyde as the crucial C_3 building-block, aminopentadienal-dihydropyridinium species like **92** would undergo intramolecular (4 + 2) cycloaddition to yield an iminium that already possesses the main structural features of ircinal alkaloids (Scheme 6.22). To probe this hypothesis, Marazano and colleagues reacted dihydropyridinium **34** and aminopentadienoate **96**, which unexpectedly furnished amine **97** (Scheme 6.25). This behavior was explained by intramolecular hydride transfer to the formed iminium **98**, followed by loss of butanal upon hydrolysis of **99** [16b].

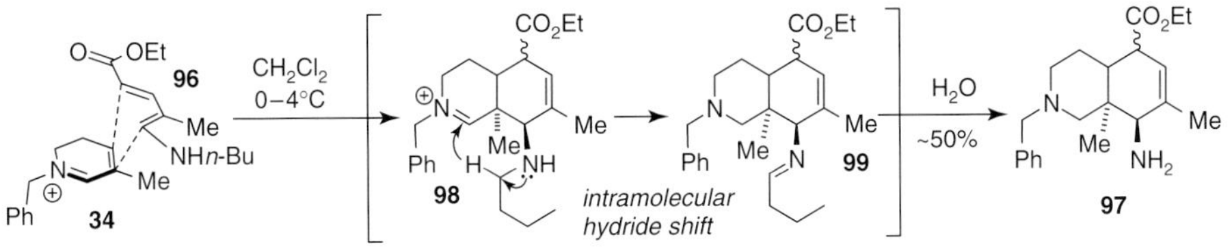

Scheme 6.25 A (4 + 2) cycloaddition strategy towards an ircinal model.

While many strategies relied on the use of the Diels–Alder reaction in the numerous synthetic approaches to manzamine A [51], this one seems to be among the most efficient since all crucial functionalities are brought together in a single step. Indeed, the construction of the ABC-ring system of manzamine A was published soon after by the same team (Scheme 6.26) [52]. The choice of a correct substitution pattern on the starting dihydropyridinium salt **100** permitted

9) The Marazano group extended the Baldwin hypotheses (based on the dihydropyridinium chemistry) to explain the biosynthesis of halicyclamine A (and also sarain A) in a 1995 publication [53] before developing their own modified model.

the construction of the C-ring with a (4 + 2) cycloaddition from **101**, furnishing ircinal analog **102** in good overall yield.

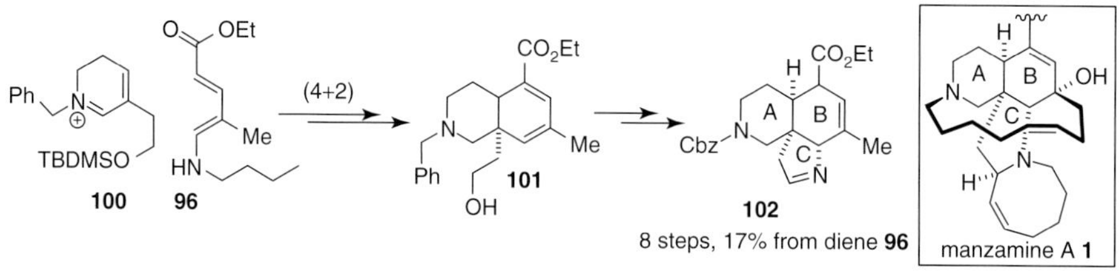

Scheme 6.26 Biomimetic synthesis of the ABC-ring system of manzamine A.

No further developments concerning the biomimetic approach to ircinal/manzamine A series were published in the following years by the group of Marazano, until 2008 and the publication of a general approach validated by unprecedented results [21]. The authors combined Zincke-type chemistry, involving biomimetic species such as aminopentadienals and glutacondialdehydes, and their "Chichibabin-like" synthesis of dihydropyridinium depicted in Scheme 6.5. Thus, Strecker–Michael adduct **103** was reacted with aminopentadienal **104** in Lewis acid conditions, yielding in moderate yield bicyclic iminopentadienal **105** (Scheme 6.27). This stable compound, when treated with acetic anhydride followed by reduction and final hydrolysis, furnished dienal **106** as a biomimetic model of ircinal alkaloids, following an impressive cascade sequence of rearrangements of the aminopentadienal system (Scheme 6.27).

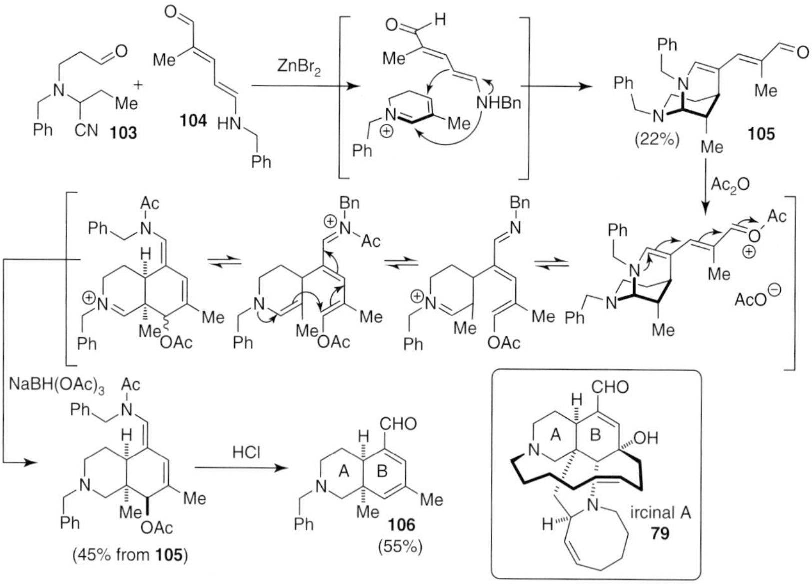

Scheme 6.27 Aminopentadienal scenario towards the AB-ring system of manzamine A.

6.6.2
Biomimetic Models toward Halicyclamines

The first clear link between keramaphidin B (**7**) and halicyclamine alkaloids was experimentally demonstrated in 1995 by the Marazano team [53]. When performing a similar reaction to the one described in 1994 by Baldwin and colleagues [47] (cf. Scheme 6.17), consisting of studying the behavior of dihydropyridinium salt **108** in solution (generated *in situ* from **107**), a significant amount of **109** was isolated along with awaited **110** and **111** (Scheme 6.28).

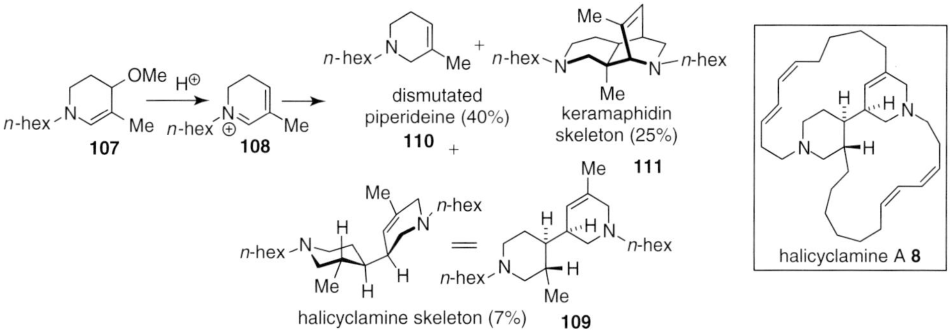

Scheme 6.28 First biomimetic synthesis of halicyclamine-related structures.

Similarly, keramaphidin model **112**, when submitted to regioselective photo-oxidative α-cyanation, was turned into halicyclamine model **113** via retro-aza-Mannich fragmentation of iminium **114** and cyanide retrapping (Scheme 6.29) [49]. This result suggests that a regioselective N^1-oxidation of keramaphidin B type alkaloids might be implicated in the biogenesis of the halicyclamines.

Further, Marazano and colleagues observed that treating polycyclic aminal adduct **115** resulting from nucleophilic reaction of aminopentadienal **116** with dihydropyridinium salt **34** [16b] in acidic conditions gave pyridinium **117**, which was directly reduced with sodium borohydride to give a 3 : 1 ratio of compounds

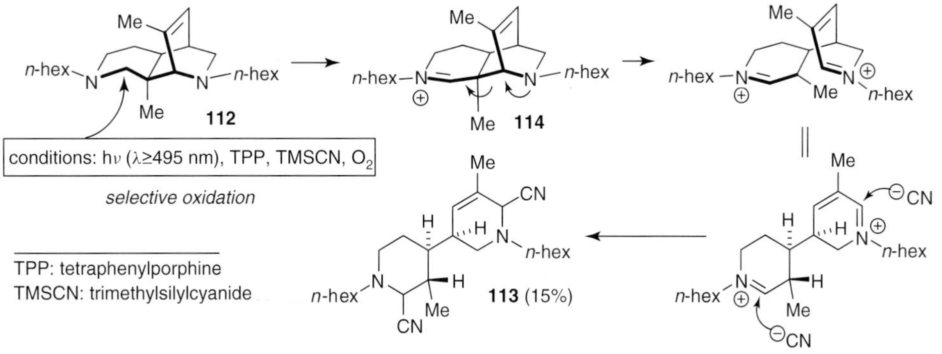

Scheme 6.29 Selective oxidation of a keramaphidin model.

118 and **119** (Scheme 6.30a). The main compound (**118**) was found to possess the halicyclamine A (**8**) central core whereas **119**, the minor adduct, can be considered as a halicyclamine B (**120**) model [54] (despite the opposite stereochemistry of one stereocenter on the tetrahydropyridine ring). In its recent development of the halicyclamine chemistry [21], the Marazano group observed that reaction of Strecker adduct **103** and aminopentadienal **116** in the presence of zinc triflate directly furnished pyridinium **121** (Scheme 6.30b). Pyridinium **121** was further reduced to the above-described mixture of bis-piperidines **118** and **119** in good overall yield. Taken together, these results suggest that dehydrohalicyclamines such as **95** (i.e., pyridinium species) could be the actual biogenetic precursors of halicyclamine-type alkaloids after reduction, and not the opposite.

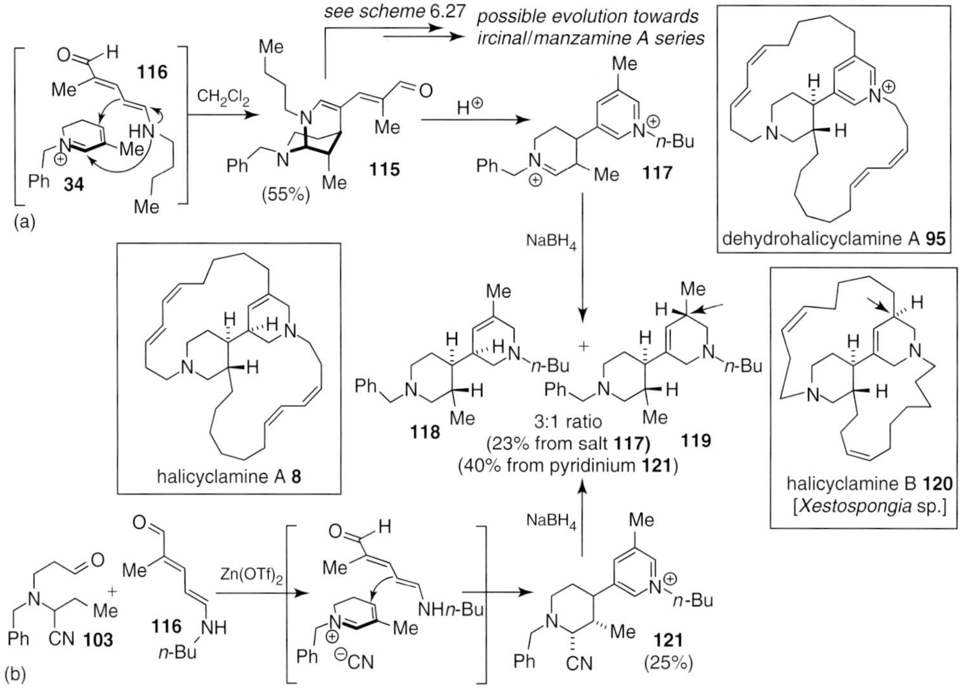

Scheme 6.30 First (a) and second (b) generation approaches to halicyclamines.

The latest developments to tackle the total synthesis of halicyclamine A (**8**) include the synthesis of a macrocyclic target molecule, demonstrating thereby the feasibility of intramolecular reaction (Scheme 6.31) [55]. Precursor **123** was prepared in eleven steps from tetradecandioic acid (**122**). Intermediate **124** was effectively reached with, now classical in this series, zinc triflate. Compound **124** then collapsed in acidic conditions to provide α-aminonitrile/pyridinium **125**, which was finally converted into halicyclamine A model **126** during a reduction step that appeared to be both regio- and diastereoselective (to be compared to the obtaining of a mixture of regioisomers in the case of intermolecular reaction,

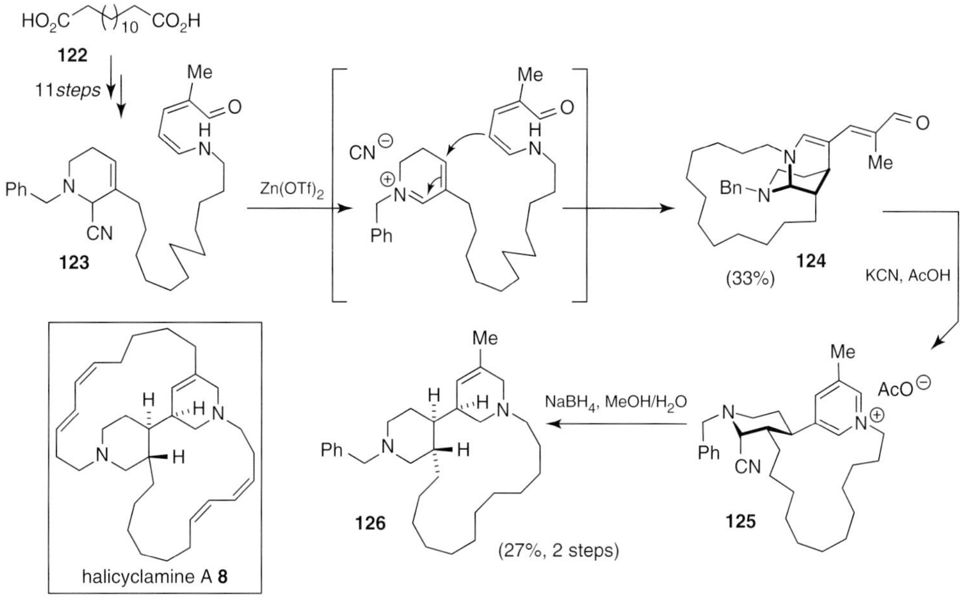

Scheme 6.31 Latest developments toward a biomimetic synthesis of halicyclamine A.

see Scheme 6.30). In view of this achievement, a biomimetic total synthesis of halicyclamine A (**8**) is reasonably within reach.

An alternative pathway, involving a late introduction of the nitrogen atoms, was communicated in 2003 (Scheme 6.32) [56]. Compound **127** was prepared as a stabilized equivalent of biomimetic intermediate **128**, itself reminiscent of postulated biosynthetic intermediate **129**. The strategy towards **127** firmly exploited the biosynthetic proposals as it consisted of sequential condensation of aldehyde and malonaldehyde equivalents. Its reactivity towards primary amines was then studied and led to the formation–via (among others) pyridinium **130**–of four diastereomers, including compound **131**, which display the same relative stereochemistry as halicyclamine A. Notably, even if conceivable in principle, no manzamine A type compounds were formed during these investigations, probably because of the irreversibility of pyridinium formation.

Overall, these short and convergent biomimetic syntheses, which rely on identical reactants brought to distinct reaction fates, are based on the fusion of Baldwin's seminal hypothesis (dihydropyridinium-based) and Marazano's modified theory (aminopentadienal-based) (Scheme 6.33). In one of the most exciting recent achievements in natural product biomimetic synthesis, application of the Baldwin–Marazano concepts delivered core analogs of ircinal/manzamine A and the halicyclamines, constituting a strong presumption of how those alkaloids relate biosynthetically. Biomimetically speaking, the most impressive fact is that the entire sequences of reactions are promoted in cascades depending on the Lewis acid; it is therefore difficult to imagine more straightforward ways to access these complex families of molecules.

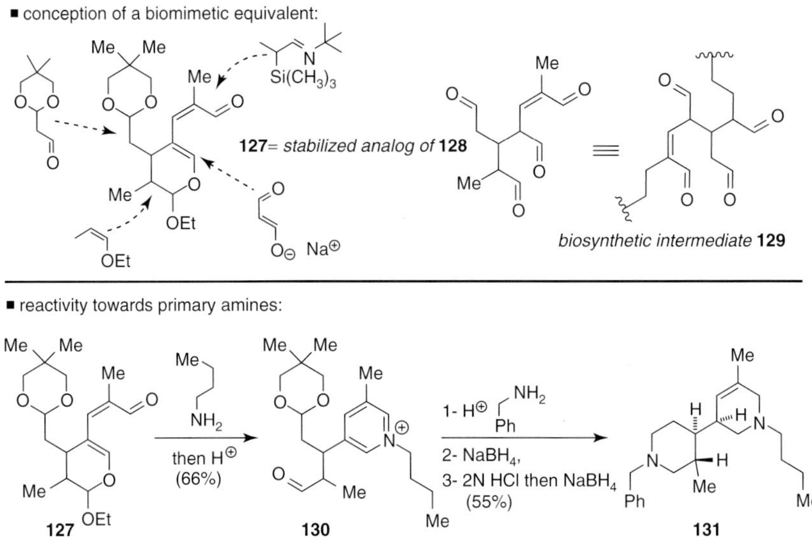

Scheme 6.32 Alternative strategy towards halicyclamines with the late introduction of nitrogens.

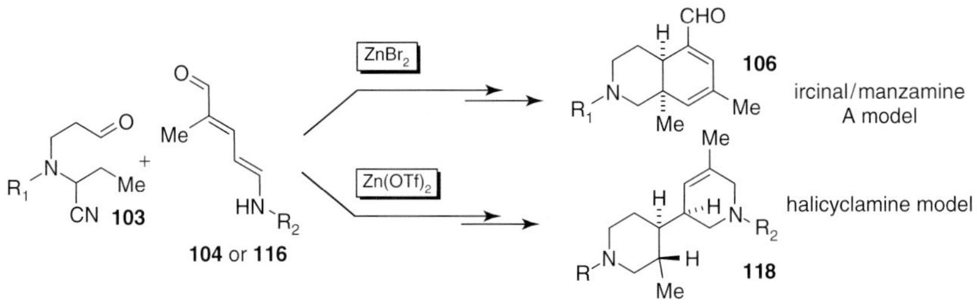

Scheme 6.33 Marazano divergent route to either ircinal/manzamine A or halicyclamine-type alkaloids.

6.7
Biomimetic Approaches toward Other Manzamine Alkaloids

6.7.1
Biomimetic Models of Madangamine Alkaloids

Pro-ircinal-type alkaloids could undergo core fragmentation via intracyclic (path A) or pericyclic (path B) vinylogous retro-Mannich reactions (Scheme 6.34). Following redox transformations, a sequence of enamination and vinylogous aza-Mannich reaction would eventually produce madangamine type alkaloids.

Path (B) was pioneered biomimetically by Marazano and colleagues (Scheme 6.35) in an oxidized version, based on (carboxyethyl)acetoacetate dianion

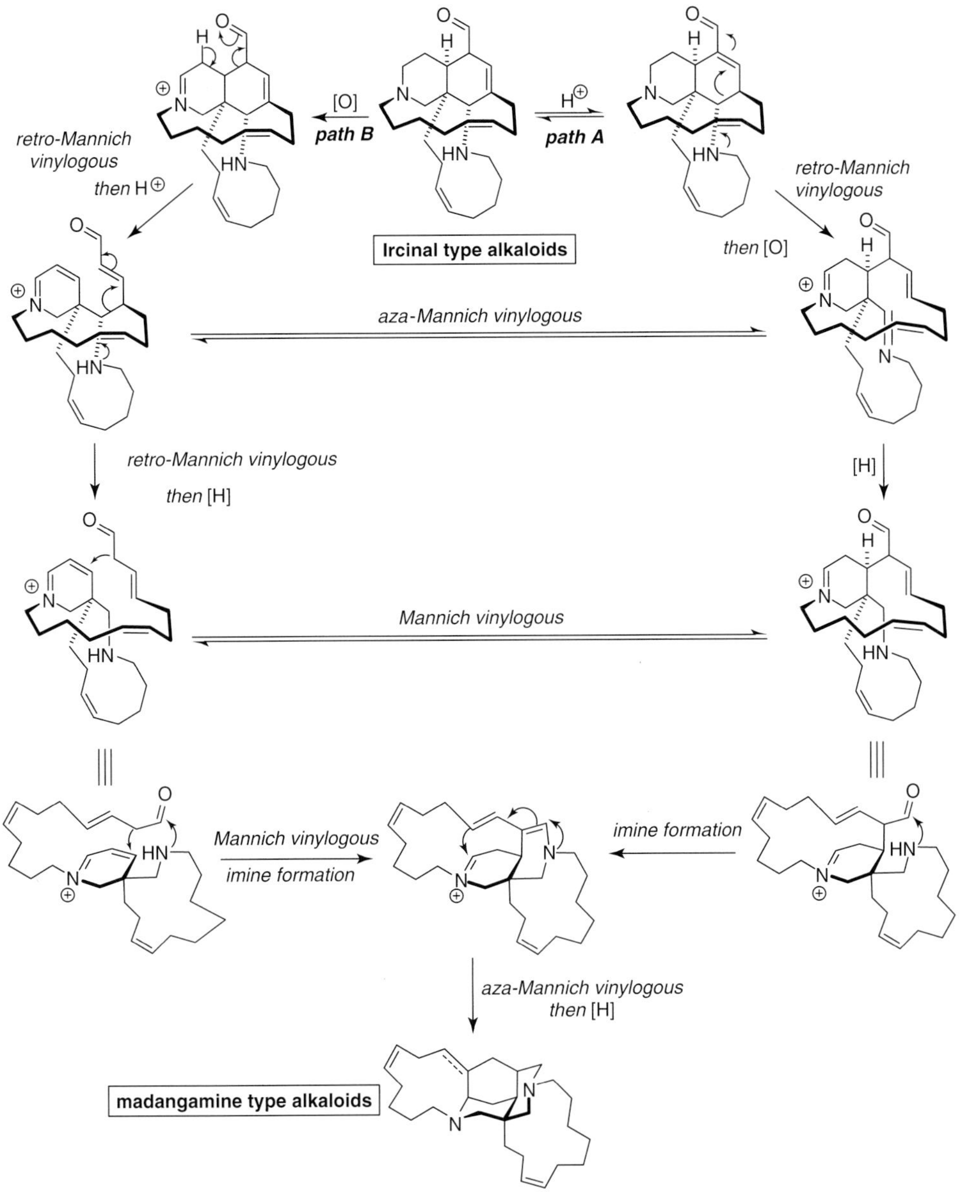

Scheme 6.34 Possible biogenesis of madangamine C type alkaloids.

and quaternarized dihydropyridinium **133** as a biomimetic equivalent of postulated intermediate **132** (see Scheme 6.34) [57]. Following a double Mannich addition, tricycle **134** was obtained in close analogy with the core of madangamine C-type alkaloids. In 2011, the total synthesis of madangamine type alkaloids remains a mountain to climb [58].

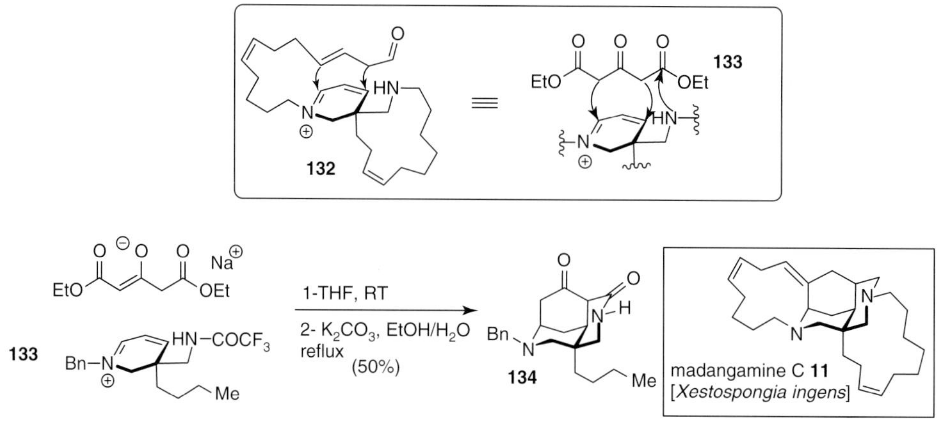

Scheme 6.35 Biomimetic synthesis of the madangamine C core.

6.7.2 Biomimetic Model of Nakadomarine A

Alkaloids of the ircinal A type could also undergo intracyclic fragmentation via vinylogous retro-Mannich reaction to give **135** (Scheme 6.36). Subsequently, a vinylogous Mannich reaction would enable ring closure of **135**, yielding fused tetracycle **136**. Final cyclization to furan would produce alkaloids of the nakadomarine A type. Alternatively, furan formation could occur from diketonic **135** to **137**, enabling a furan-Mannich intramolecular cyclization to nakadomarine alkaloids.

This last hypothesis was validated in the laboratory by Nishida and colleagues on models [59], before their publication of the first total synthesis of (+)-nakadomarine

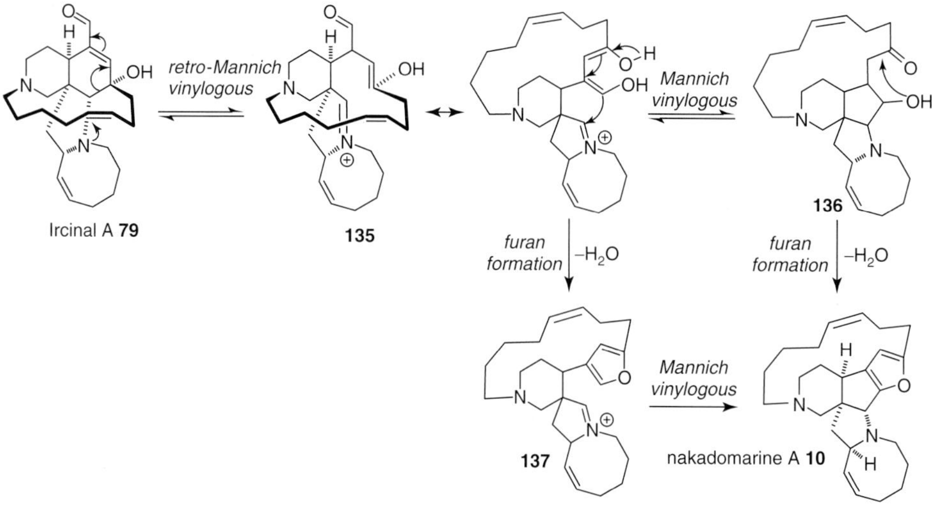

Scheme 6.36 Possible biogenesis of nakadomarine A from ircinal A.

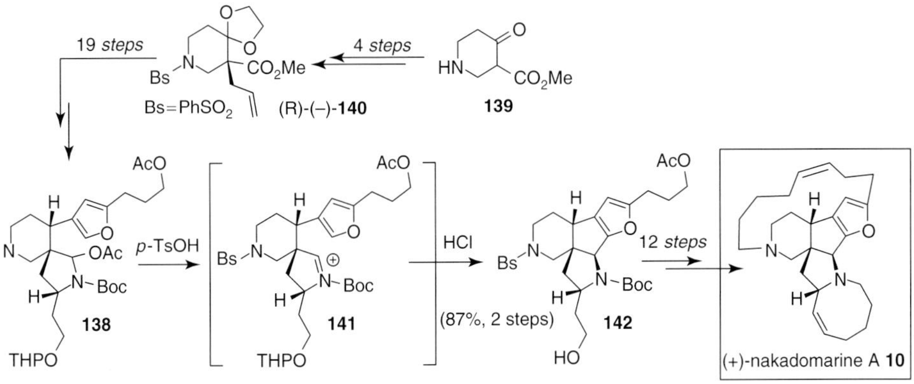

Scheme 6.37 Focus on the biomimetic key step in the first total synthesis of nakadomarine A.

A (**10**) (see also Section 6.9.8) in 2003 (Scheme 6.37) [60], featured by a key biomimetic step. Advanced intermediate **138** was prepared in 23 steps from racemic **139** (via optically active intermediate **140**). The authors made successful an intramolecular furan-iminium cyclization of spiropyrrolidine **141** into **142**, an elegant way of mimicking the presumed core biosynthesis of nakadomarine A type alkaloids.

6.7.3
Biomimetic Models of Sarains: A Side Branch of the Manzamine Tree

With its highly intricate diazatricyclic central core and two macrocyclic side chains sarain A (**6**), isolated from Reniera sarai in 1986 and fully characterized in 1989 (X-ray analysis of a diacetate derivative), is one of the most complex manzamine alkaloid (Scheme 6.38), featuring an unprecedented pentacyclic, box-like heterocyclic architecture. Inspection of this alkaloid reveals a polyenic 1,2,3-aminodiol, sphingolipid-like moiety, suggesting a distinct biogenetic origin relative to the manzamines previously described in this chapter. Indeed, retro-biosynthetic analysis of sarain A type alkaloids[10] using iminium-based disconnections provides with key amino-aldehyde **15** and two C_3 synthons [postulated as malondialdehyde (**17**) according to Marazano's model], along with cyclic amino acid **144** (that may result from the catabolism of sphingolipid **143**) (Scheme 6.38) [61]. From these biogenetic elements, but with a philosophy identical to what we have seen up to now, it thus appears that sarain A type alkaloids are not directly connected to the elaborated manzamines presented before, with which they only share simple amino-aldehyde precursors **15**. Alkaloids related to sarain A should thus be regarded as branching

10) So far, sarain A is the only isolated manzamine alkaloid to possess this unprecedented polycyclic core.

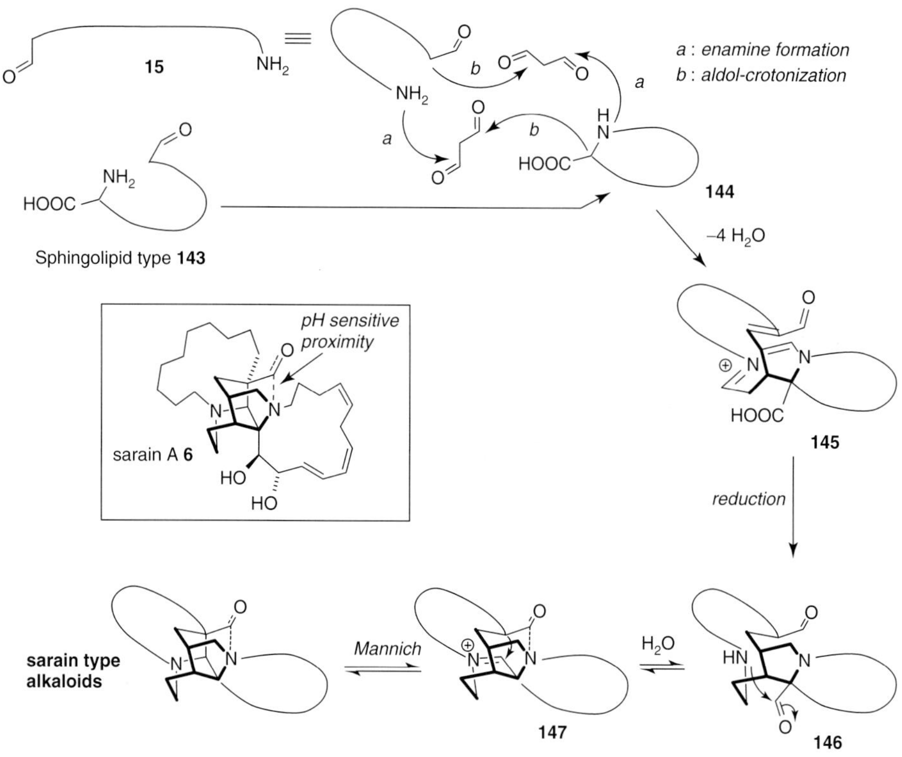

Scheme 6.38 Proposed biogenesis of sarain A-type alkaloids.

off early on in the manzamine metabolism, in a case of divergent biosynthesis, with postulated intermediates **145–147** represented on Scheme 6.38.

Marazano *et al.* pioneered the biomimetic synthesis of the heterocyclic core of sarain A. In their 1999 paper, a first model, albeit incomplete, gave the first experimental evidence that manzamines and sarains are biosynthetically related (Scheme 6.39). A thermodynamic mixture of two aldehydes (**148** and **149**) was obtained when reacting dihydropyridinium salt **34** and glutaconaldehyde salt **150**. The sequence occurred via the rearrangement of aminal **151** into iminium **152** on contact with alumina. Compound **148** contains a bicyclic system reminiscent of that of sarain A **6**.

A few years later [61b], using β-bromoacrylamide **153** as aminated malondialdehyde equivalent (Scheme 6.40), and benzylidene **154** as an amino acid surrogate, another model study was disclosed. Those reactants were coupled under basic conditions to yield glutarimide **155** after benzylidene hydrolysis. Next, **155** was coupled with malondialdehyde (**17**) sodium salt to furnish aminopropenal **156**, which was *N*-alkylated to yield **157**. After a series of tedious reductions of the glutarimide moiety, the authors were able to obtain aminal **158**, which underwent Sakurai-type cyclization under the described conditions to afford tricyclic sarain A model **159**.

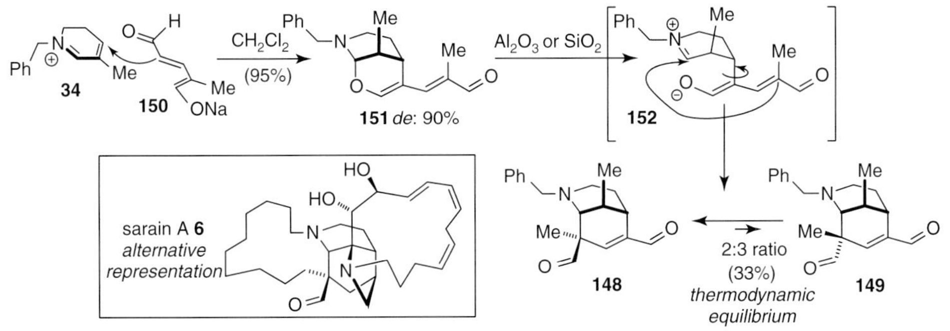

Scheme 6.39 Biomimetic synthesis of a first sarain A model.

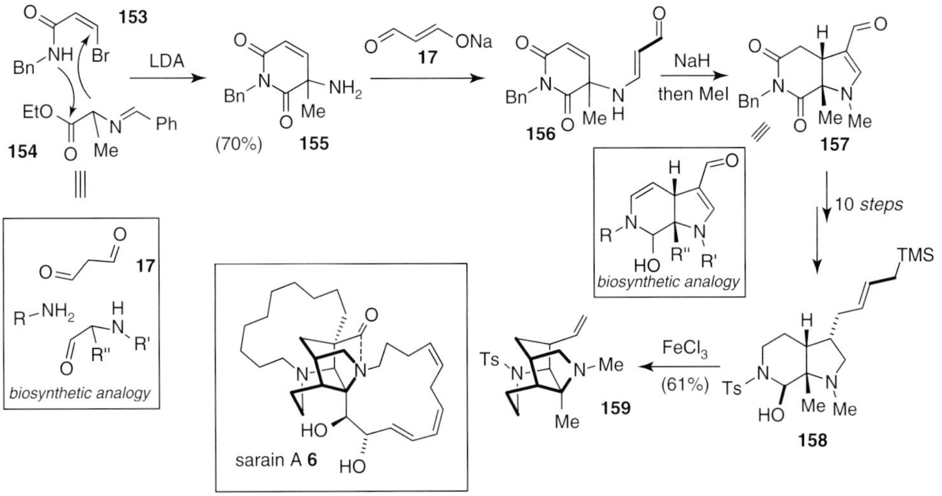

Scheme 6.40 Biomimetic synthesis of a second sarain A model.

6.8
A Biomimetic Tool-Box for the Synthesis of Manzamine Alkaloids: Glutaconaldehydes and Aminopentadienals

As seen above, simple aminopentadienals constitute C_5 biomimetic equivalents of long-chain aminopentadienals (types 1 and 2), regarded as key precursors of the manzamine alkaloids. Glutacondialdehydes are hydrolyzed counterparts of aminopentadienals and can also be seen as equivalents of postulated aldehyde intermediates (refer to Scheme 6.2). Such biomimetic C_5 nucleophiles can be obtained in a few steps from simple starting materials, using Mannich addition of imine anions onto vinamidinium salts (cf. Scheme 6.9), or, in a more versatile manner, using the Zincke opening of *N*-activated pyridinium salts (cf. Scheme 6.8). In their first model reactions (Schemes 6.25 and 6.26), the Marazano group exploited the reactivity of more stable aminopentadienoates (but presenting inappropriate

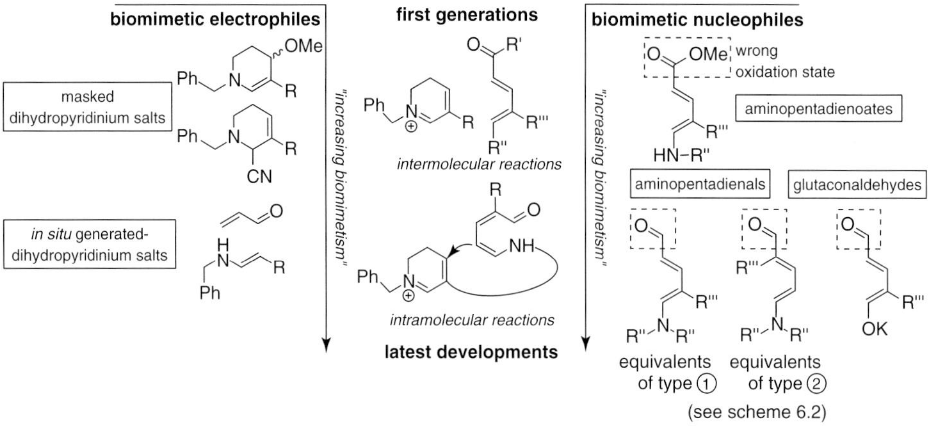

Figure 6.5 Evolution of the biomimetic chemical tool-box used for accessing manzamine alkaloids.

oxidation relative to aldehydes), though those species were rapidly abandoned in favor of more versatile aminopentadienals (Figure 6.5).

On the borderline of the chemistry of manzamine alkaloids, several experimental studies were conducted to delimit the scope of reactivity of glutaconaldehydes. More or less biosynthetically related structures were obtained. One of the simplest reactions reported is the self-dimerization of glutaconaldehyde **150** formed *in situ* from the corresponding sodium salt under acidic conditions (Scheme 6.41) [62]. In fact, according to the mechanism depicted, cinnamaldehyde **160** could be obtained in very good yield. A similar outcome was observed when a monoprotected malonaldehyde unit **161** was engaged in the cascade [63]. Formation of the resulting adduct **162**, which displays the same aromatic pattern as the one encountered in **160**, was explained by the loss of a carbon in the form of a formic acid molecule.

The chemistry of glutaconaldehydes was recently explored beyond biomimetic considerations, especially by the Vanderwal group. Intramolecular ring openings

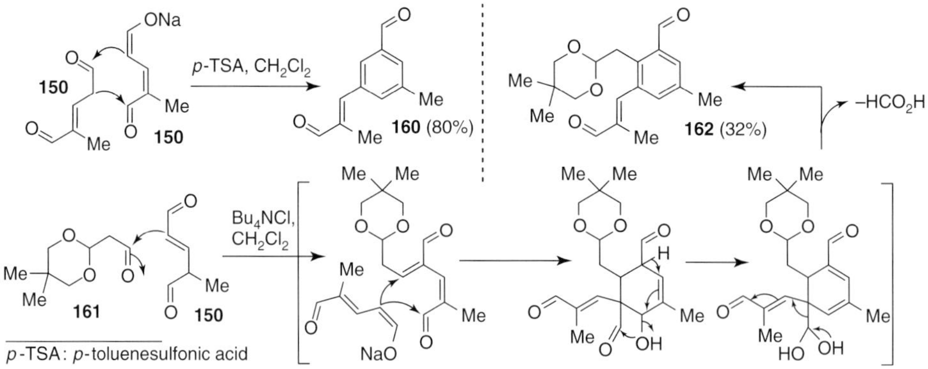

Scheme 6.41 Selected examples of reactivity of glutaconaldehydes.

of pyridinium salts (including "Zincke salts") gave easy access to multiple heterocycles [64]. Aminopentadienals ("Zincke aldehydes") were also shown to undergo pericyclic cascades to provide synthetically useful *(Z)*-$\alpha, \beta, \gamma, \delta$,-unsaturated amides [65] or served for the preparation of δ-tributylstannyl-$\alpha, \beta, \gamma, \delta$,-unsaturated aldehydes [66]. Applications to the synthesis of natural products or natural product-like analogs include the access to indolomonoterpenic alkaloid cores of the strychnane, aspidospermane, or ibogane types (with a total synthesis of norfluorocurarine), a formal synthesis of porothramycins or the total synthesis of nicotine and analogs [67]. Nuhant, Delpech, and colleagues disclosed a methodology driven study dealing with the activation of aminopentadienals towards nucleophiles which also enabled completion of the synthesis of protoemetinol [68].

6.9 Biosynthesis of Manzamine Alkaloids: Towards a Universal Scenario

In the following, we propose a general mapping of the biosynthesis of typical manzamine alkaloids, as emerging from presumed biogenetic relationships as well as biomimetic chemistry evidence. This mapping finds its introduction in Schemes 6.1, 6.24, and 6.38 and is extended in Scheme 6.42. The reader will find additional details in terms of intermediates and alternative biosynthetic connections in the corresponding isolated schemes. To provide the simplest yet broadest vision of the possible relationships between the manzamine alkaloids, we again depict macrocycles by large loops (see footnote 3).

6.9.1 From Fatty Acids to Long-Chain Aminoaldehydes and Sarain Alkaloids

As depicted in Scheme 6.1, fatty acid degradation would produce two kinds of biosynthetic reactants: (i) dialdehydes **14** (C_8-C_{16}) that would become monoaminated to yield amino-aldehydes **15**; (ii) acrolein (**16**) (the original C_3 species hypothesized by Baldwin) or malondialdehyde (**17**) (alternative C_3 species proposed by Marazano in 1998). Incorporation of a sphingolipid moiety to this metabolism would open the path towards sarain A-type alkaloids (Scheme 6.38), which appear to branch off very early from the biogenetic trunk of the manzamine family.

6.9.2 Pyridine Alkaloids: Theonelladine, Cyclostellettamine, and Xestospongin-Type Alkaloids

Condensation of amino-aldehydes **15** with a source of ammonia would furnish theonelladine-type alkaloids. Alternatively, dimerization of **15** in the presence of two equivalents of acrolein or malondialdehyde would give rise not only to alkaloids of the cyclostellettamine type but also of the xestospongin-type when the alkyl chains are β-hydroxylated (Schemes 6.1 and 6.11).

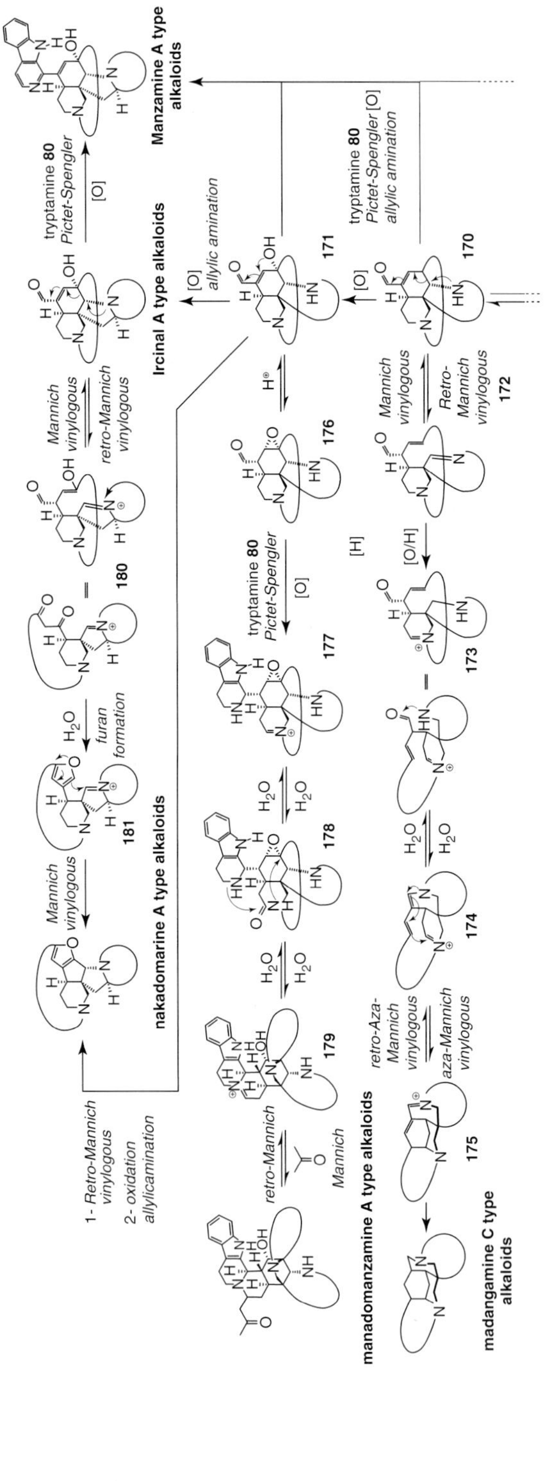

Scheme 6.42 Mapping of the presumed biogenetic relationships between representative manzamine alkaloids.

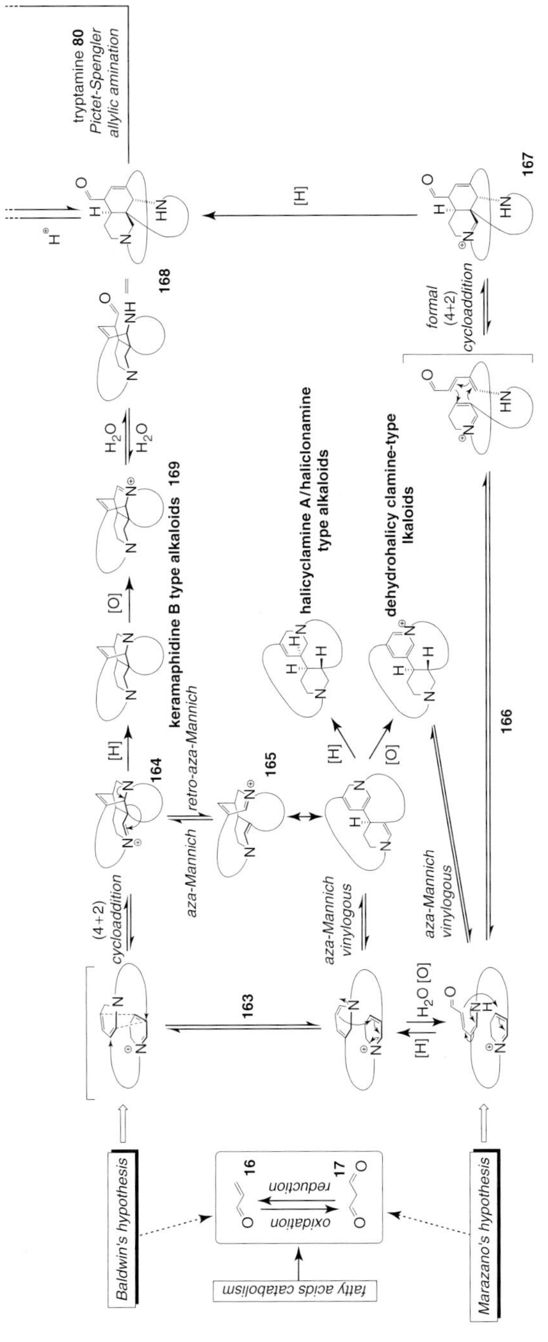

Scheme 6.42 (Continued)

6.9.3
From Cyclostellettamines to Keramaphidin and Halicyclamine/Haliclonamine Alkaloids

Incorporation of acrolein as C_3 precursor according to Baldwin's model implies late, possibly spontaneous, oxidation of dihydropyridine/dihydropyridinium intermediates to pyridine/pyridinium. Competitive with these oxidation processes, two kinds of reaction might occur: (i) intramolecular Diels–Alder cycloaddition of bis-dihydropyridiniums **163** to bridged intermediates **164**, which could either be reduced to keramaphidin B type alkaloids or (ii) undergo retro-aza-Mannich fragmentation to afford bis-piperidine skeletons **165** typical of halicyclamine/haliclonamine alkaloids following redox transformation. These alkaloids could also arise directly from bis-dihydropyridinium **163** upon intramolecular vinylogous aza-Mannich cyclization, and also from aminopentadienal-dihydropyridinium **166** if malondialdehyde (**17**) is involved as key C_3 unit (Marazano's model).

6.9.4
Spinal Cord of Manzamine Metabolism: The Ircinal Pathway

Aminopentadienal-dihydropyridiniums **166** are the direct precursors of isoquinoline aldehydes **167** following intramolecular [4 + 2] cycloaddition. Such pro-ircinals represent the earliest entry point in the biosynthesis of ircinal alkaloids (**168** to ircinal A type alkaloids), of which they possess most structural features, except for their pyrrolidine ring. Pro-ircinals **168** could also be accessed by regioselective oxidation of keramaphidin type alkaloids to **169** followed by iminium hydrolysis, although this theory has not been supported experimentally. All pro-ircinals and ircinal intermediates should be considered central to the "manzamine metabolism," and seem to constitute branching points toward the majority of structurally representative alkaloids (Scheme 6.24).

6.9.5
From Ircinal and Pro-ircinals to Manzamine A Alkaloids

Although ircinal-type molecules are seen as immediate precursors for manzamine A type alkaloids, it should be kept in mind that β-carboline formation is not the obligate last step toward these alkaloids. Indeed, pro-ircinals **170** and **171** are candidates for oxidative allylic amination/β-carboline formation, making them all potentially direct precursors of manzamine A-type alkaloids.

6.9.6
From Pro-ircinals to Madangamine Alkaloids

Pro-ircinal alkaloids **170** could undergo a vinylogous retro-Mannich fragmentation, giving rise to spiro-piperidine **172**. Following intramolecular redox transfer or sequential oxidoreduction, spiranic tetrahydropyridinium **173** would be produced. Madangamine-type alkaloids would eventually be yielded upon cyclizing enamination (**174**) and vinylogous aza-Mannich reaction with a final reduction of **175**.

6.9.7
From Pro-ircinals to Manadomanzamine Alkaloids

Intramolecular epoxidation of pro-ircinals **171** would produce isoquinoline aldehydes **176**, prone to Pictet–Spengler reaction with tryptamine (**80**). Following regioselective oxidation, iminium **177** would be hydrolyzed to propionaldehyde derivative **178**, which is amenable to iminium formation with the neo-formed tetrahydroisoquinoline system. Mannich addition of an acetone equivalent onto the residual iminium of **179** would eventually produce manadomanzamine-type alkaloids.

6.9.8
From Ircinals and Pro-ircinals to Nakadomarine Alkaloids

Ircinal A-type alkaloids could undergo vinylogous retro-Mannich fragmentation, identical to the one undergone by **170** in the supposed biosynthesis of madangamine alkaloids (cf. Scheme 6.34). Subsequently, cyclization of diketone **180** to furan **181** would enable intramolecular vinylogous Mannich addition to effect ring closure, yielding the fused tetracyclic system typical of nakadomarine A-type alkaloids. The implication of furan nucleophiles in this biosynthesis is directly suggested from biomimetic experiments (cf. Scheme 6.37).

Approaching the end of this marine story, let us finally emphasize how the chemist's intuitions proved to be right and were corroborated (sometimes afterwards) by the isolation and characterization of new informative structures of natural products. Bear in mind that the structure of ircinal A (**79**) and keramaphidin (**7**) were not yet known when Whitehead and Baldwin proposed their pioneering biosynthetic model, and that pyridinium-piperidine intermediates such as **95** were postulated before being discovered in sponges (cf. Scheme 6.23). As the latest nod from Nature to the chemical community, the manzamine-type alkaloids zamamidines [69] [see the structure of zamamidine C (**182**), Scheme 6.43] recently gave striking presumptive evidence for the C_3 (acrolein/malonaldehyde) scenario. In fact, a C_3 link such as **16** clearly unifies a tetrahydromanzamine A molecule with a β-carboline moiety; a plausible biosynthesis from ircinal A (**79**) is therefore easily conceivable with the intervention of two molecules of tryptamine (**80**).

6.10
Total Syntheses of Manzamine-Type Alkaloids

The Baldwin–Whitehead and Marazano's biosynthetic hypotheses have provided a useful framework to develop synthetic approaches to the manzamines alkaloids.

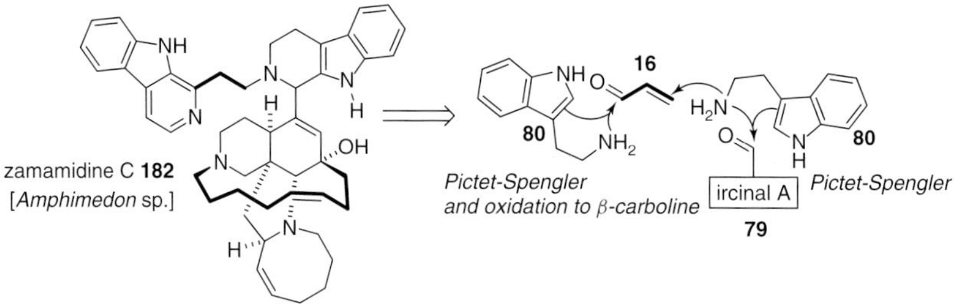

Scheme 6.43 Retrobiosynthesis of zamamidine C.

Nevertheless, stepwise strategies have been initiated worldwide over the last 20 years toward these remarkable alkaloids. To date, the total synthesis of manzamine A (**1**) [70], nakadomarine A (**10**) [60, 71], haliclonacyclamine C [72], and sarain A (**6**) [73] have been achieved and beautifully illustrate the state-of-the art in methods for highly complex molecule construction. These total syntheses as well as the numerous chemical approaches are not included in this chapter. Natural ircinals and manzamines have been subjected to semi-synthetic transformations, especially in the Hamann group, providing a wide range of derivatives for diverse biological screenings and studies [74]. In addition, total synthesis has also enabled the preparation of various simplified analogs that are unreachable from natural material [75].

6.11
Conclusion

From a biosynthetic standpoint, the most striking observation is that the great majority of manzamine alkaloids can be connected by means of reversible reactions (such as Mannich, aza-Mannich, Michael, and aldol), and thus be potentially interconvertible at a biochemical level. According to this hypothesis, only a limited number of alkaloids should be considered structurally "terminal," that is, those few formed by enzymatically irreversible steps (such as Pictet–Spengler cyclization to β-carboline systems). In this general biogenetic proposal, reductions would have the role of freezing reactive intermediates (e.g., iminium species) into stable alkaloids. However, it must be realized that several alternative reoxidations (i.e., at other levels of the molecule) probably remain feasible biochemically, due to similar redox potentials of related iminium alkaloids. Moreover, conformation-induced intramolecular electron transfers and dismutations have the potential to occur spontaneously, as suggested by several observations in the laboratory (cf. Schemes 6.18 and 6.25). Overall, the conception of "manzamines in equilibrium" for this rich metabolism of marine alkaloids (that might be driven by subtle ecological changes) yields a particularly striking picture of dynamic chemical evolution and diversity-oriented biogenesis. These latter suggestions have been, at least partly, magnificently demonstrated by track records of

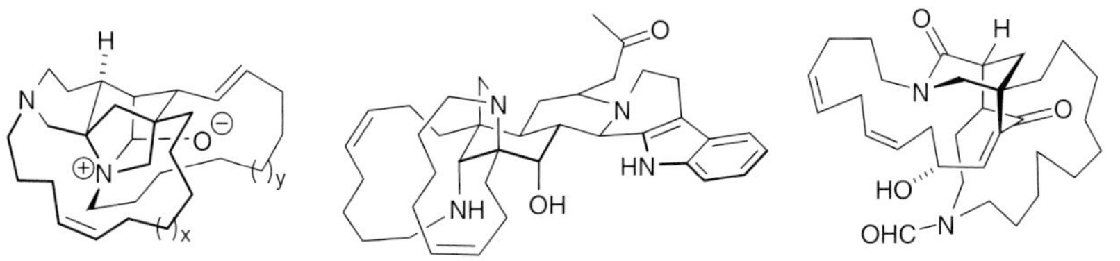

misenine **12** manadomanzamine A **9** haliclonine A **183** [*Haliclona* sp.]

Figure 6.6 Challenging manzamines as future synthetic targets.

successful applications at the biomimetic chemistry level (cf. Schemes 6.18, 6.28, and 6.33).

Synthetic effort toward the manzamine alkaloids will certainly continue in the future as new exciting structures periodically appear in the literature. Molecules such as misenine (**12**) (for which no logical biosynthetic route can yet be proposed), complex, highly rearranged haliclonine A (**183**) [76], or indolic manadomanzamine (**9**), will surely keep stimulating chemists because of their intrinsic beauty and not only because of interesting biological properties (Figure 6.6). Future synthetic endeavors will most probably be guided by the body of growing biosynthetic studies currently performed worldwide with marine organisms.

References

1. Sakai, R., Higa, T., Jefford, C.W., and Benardinelli, G. (1986) *J. Am. Chem. Soc.*, **108**, 6404–6405.
2. Kobayashi, J., Murayama, T., Ohizumi, Y., Sasaki, T., Ohta, T., and Nozoe, S. (1989) *Tetrahedron Lett.*, **20**, 4833–4835.
3. Fusetani, N., Asai, N., Matsunaga, S., Honda, K., and Yasumuro, K. (1994) *Tetrahedron Lett.*, **35**, 3967–3970.
4. Talpir, R., Rudi, A., Ilan, M., and Kashman, Y. (1992) *Tetrahedron Lett.*, **33**, 3033–3034, see Reference [16] for an important comment concerning the structure of niphatoxin A.
5. Schmitz, F.J., Hollenbeak, K.H., and Campbell, D.C. (1978) *J. Org. Chem.*, **43**, 3916–3922.
6. Cimino, G., Mattia, C.A., Mazarella, L., Puliti, R., Scognamiglio, G., Spinella, A., and Trivellone, E. (1989) *Tetrahedron*, **45**, 3863–3872.
7. Kobayashi, J., Tsuda, M., Kawasaki, N., Matsumoto, K., and Adachi, T. (1994) *Tetrahedron Lett.*, **35**, 4383–4386.
8. Jaspers, M., Pasupathy, V., and Crews, P. (1994) *J. Org. Chem.*, **59**, 3253–3255.
9. Peng, J., Hu, J.-F., Kazi, A.B., Li, Z., Avery, M., Péraud, O., Hill, R.T., Franzblau, S.G., Zhang, F., Schinazi, R.F., Wirtz, S.S., Tharnish, P., Kelly, M., Wahyuono, S., and Hamann, M.T. (2003) *J. Am. Chem. Soc.*, **125**, 13382–13386.
10. Isolation and structure determination: Kobayashi, J., Watanabe, D., Kawasaki, N., and Tsuda, M. (1997) *J. Org. Chem.*, **62**, 9236–9239.
11. (a) Kong, F., Andersen, R.J., and Allen, T.M. (1994) *J. Am. Chem. Soc.*, **116**, 6007–6008; (b) Kong, F., Graziani, E.I., and Andersen, R.J. (1998) *J. Nat. Prod.*, **61**, 267–271.
12. Guo, Y., Trivellone, E., Sconamiglio, G., and Cimino, G. (1998) *Tetrahedron*, **54**, 541–550.
13. Jimenez, J.I., Goetz, G., Mau, C.M.S., Yoshida, W.Y., Sheuer, P.J., Williamson, R.T., and Kelly, M. (2000) *J. Org. Chem.*, **65**, 8465–8469.
14. See among others: (a) Peng, J., Rao, K.V., Choo, Y.-M., and Hamann, M.T. (2008) in *Modern Alkaloids, Structure,*

Isolation, Synthesis and Biology (eds E. Fattorusso and O. Tagliatela-Scafati), Wiley-VCH Verlag GmbH, Weinheim, pp. 189–232; (b) Hu, J.-F., Hamann, M.T., Hill, R., and Kelly, M. (2003) in *The Alkaloids, Chemistry and Biology*, vol. 60 (ed. G.A. Cordell), Academic Press, San Diego, pp. 207–285.

15. Baldwin, J.E. and Whitehead, R.C. (1992) *Tetrahedron Lett.*, **33**, 2059–2062.
16. (a) Kaiser, A., Billot, X., Gateau-Olesker, A., Marazano, C., and Das, B.C. (1998) *J. Am. Chem. Soc.*, **120**, 8026–8034; (b) Jakubowicz, K., Ben Abdeljelil, K., Herdemann, M., Martin, M.-T., Gateau-Olesker, A., Al Mourabit, A., Marazano, C., and Das, B.C. (1999) *J. Org. Chem.*, **64**, 7381–7387.
17. Many mechanisms have been proposed involving among others the formation of endoperoxides or enzymatic reactions in the course of the biosynthesis of inflammatory mediators such as prostaglandins and thromboxanes and so on. See for example: Esterbauer, H., Schaur, R.J., and Zollner, H. (1991) *Free Radic. Biol. Med.*, **11**, 81–128.
18. Laville, R., Thomas, O.P., Berrué, F., Reyes, F., and Amade, P. (2008) *Eur. J. Org. Chem.*, 121–125.
19. Laville, R., Thomas, O.P., and Amade, P. (2009) *Pure Appl. Chem.*, **81**, 1033–1040.
20. Chichibabin, A.E. (1906) *J. Russ. Phys. Chem. Soc.*, **37**, 1229.
21. Wypych, J.-C., Nguyen, T.M., Nuhant, P., Bénéchie, M., and Marazano, C. (2008) *Angew. Chem. Int. Ed.*, **47**, 5418–5421.
22. See for example: Husson, H.-P., Grierson, D., and Harris, M. (1980) *J. Am. Chem. Soc.*, **102**, 1064–1082.
23. Gil, L., Gateau-Olesker, A., Marazano, C., and Das, B.C. (1995) *Tetrahedron Lett.*, **36**, 707–710.
24. (a) Zincke, T. (1903) *Justus Liebigs Ann. Chem.*, **330**, 361–374; (b) Zincke, T. (1904) *Justus Liebigs Ann. Chem.*, **333**, 296–345; (c) Zincke, T. and Wurker, W. (1905) *Justus Liebigs Ann. Chem.*, **338**, 107–141.
25. Review article: Cheng, W.-C. and Kurth, M.J. (2002) *Org. Prep. Proc. Int.*, **34**, 585–608.
26. Especially useful for the synthesis of chiral non-racemic pyridinium salts, see among many applications: (a) Compère, D., Marazano, C., and Das, B.C. (1999) *J. Org. Chem.*, **64**, 4528–4532; (b) Guilloteau-Bertin, B., Compère, D., Gil, L., Marazano, C., and Das, B.C. (2000) *Eur. J. Org. Chem.*, 1391–1399.
27. Review article: Becher, J. (1980) *Synthesis*, 589–612.
28. Nguyen, T.M., Peixoto, S., Ouairy, C., Nguyen, T.D., Bénéchie, M., Marazano, C., and Michel, P. (2010) *Synthesis*, 103–109.
29. Wypych, J.-C., Nguyen, T.M., Bénéchie, M., and Marazano, C. (2008) *J. Org. Chem.*, **73**, 1169–1172.
30. Isolation: Fusetani, N., Yasumuro, K., and Matsunaga, S. (1989) *Tetrahedron Lett.*, **30**, 6891–6894.
31. Michelliza, S., Al Mourabit, A., Gateau-Olesker, A., and Marazano, C. (2002) *J. Org. Chem.*, **67**, 6474–6478; (b) See also, the total synthesis of cyclostellettamines A–F by the Baldwin group exploiting a *N*-oxide strategy: Baldwin, J.E., Spring, D.R., Atkinson, C.E., and Lee, V. (1998) *Tetrahedron*, **54**, 13655–13680.
32. Kaiser, A., Marazano, C., and Maier, M. (1999) *J. Org. Chem.*, **64**, 3778–3782.
33. (a) Isolation of viscosamine: Volk, C.A. and Köck, M. (2003) *Org. Lett.*, **5**, 3567–3569; (b) total synthesis of viscosamine: Timm, C. and Köck, M. (2006) *Synthesis*, 2580–2584.
34. (a) Isolation and total synthesis: Timm, C., Volk, C.A., Sasse, F., and Köck, M. (2008) *Org. Biomol. Chem.*, **6**, 4036–4040; (b) See also: Timm, C., Mordhorst, T., and Köck, M. (2010) *Mar. Drugs*, **8**, 483–497.
35. Nakagawa, M., Endo, M., Tanaka, N., and Gen-Pei, L. (1984) *Tetrahedron Lett.*, **25**, 3227–3230.
36. Kobayashi, M., Kawazoe, K., and Kitagawa, I. (1989) *Chem. Pharm. Bull.*, **37**, 1676–1678.
37. Baldwin, J.E., Melman, A., Lee, V., Firkin, C.R., and Whitehead, R.C. (1998) *J. Am. Chem. Soc.*, **120**, 8559–8560.
38. Maia, A.A., Mons, S., Pereira de Freitas Gil R., and Marazano, C. (2004) *Eur. J. Org. Chem.*, 1057–1062.

39. Kiewel, K., Luo, Z., and Sulikowski, G.A. (2007) *Org. Lett.*, **9**, 5051–5054.
40. Colabroy, K.L. and Begley, T.P. (2005) *J. Am. Chem. Soc.*, **127**, 840–841.
41. (a) Tsuda, M., Hirano, K., Kubota, T., and Kobayashi, J. (1999) *Tetrahedron Lett.*, **40**, 4819–4820; (b) Hirano, K., Kubota, T., Tsuda, M., Mikami, Y., and Kobayashi, J. (2000) *Chem. Pharm. Bull.*, **48**, 974–977
42. (a) Baldwin, J.E., Romeril, S.P., Lee, V., and Claridge, T.D.W. (2001) *Org. Lett.*, **3**, 1145–1148; (b) Snider, B.B. and Shi, B. (2001) *Tetrahedron Lett.*, **42**, 1639–1642; (c) Romeril, S.P., Lee, V., Claridge, T.D.W., and Baldwin, J.E. (2002) *Tetrahedron Lett.*, **43**, 327–329; (d) Romeril, S.P., Lee, V., Baldwin, J.E., Claridge, T.D.W., and Odell, B. (2003) *Tetrahedron Lett.*, **44**, 7757–7761; (e) Morimoto, Y., Kitao, S., Okita, T., and Shoji, T. (2003) *Org. Lett.*, **5**, 2611–2614; (f) Pouilhès, A., Amado, A.F., Vidal, A., Langlois, Y., and Kouklovsky, C. (2008) *Org. Biomol. Chem.*, **6**, 1502–1510
43. (a) Ishiyama, H., Tsuda, M., Endo, T., and Kobayashi, J. (2005) *Molecules*, **10**, 312–316.
44. (a) Pyrinadine A: Kariya, Y., Kubota, T., Fromont, J., and Kobayashi, J. (2006) *Tetrahedron Lett.*, **47**, 997–998; (b) pyrinadines B–G: Kariya, Y., Kubota, T., Fromont, J., and Kobayashi, J. (2006) *Bioorg. Med. Chem.*, **14**, 8415–8419
45. Anwar, M. and Lee, V. (2009) *Tetrahedron Lett.*, **65**, 5834–5837.
46. Kondo, K., Shigemori, H., Kikuchi, Y., Ishibashi, M., Sasaki, T., and Kobayashi, J. (1992) *J. Org. Chem.*, **57**, 2480–2483.
47. Baldwin, J.E., Claridge, T.D.W., Heupel, F.A., and Whitehead, R.C. (1994) *Tetrahedron Lett.*, **35**, 7829–7832.
48. (a) Baldwin, J.E., Claridge, T.D.W., Culshaw, A.J., Heupel, F.A., Lee, V., Spring, D.R., Whitehead, R.C., Boughtflower, R.J., Mutton, I.M., and Upton, R.J. (1998) *Angew. Chem. Int. Ed.*, **37**, 2661–2663; (b) Baldwin, J.E., Claridge, T.D.W., Culshaw, A.J., Heupel, F.A., Lee, V., Spring, D.R., and Whitehead, R.C. (1999) *Chem. Eur. J.*, **5**, 3154–3161.
49. Gomez, J.-M., Gil, L., Ferroud, C., Gateau-Olesker, A., Martin, M.-T., and Marazano, C. (2001) *J. Org. Chem.*, **66**, 4898–4903.
50. Matsunaga, S., Miyata, Y., van Soest, R.W.M., and Fusetani, N. (2004) *J. Nat. Prod.*, **67**, 1758–1760.
51. Among many others, see for example: Kita, Y., Toma, T., Kan, T., and Fukuyama, T. (2008) *Org. Lett.*, **10**, 3251–3253, and references cited therein.
52. Herdemann, M., Al Mourabit, A., Martin, M.-T., and Marazano, C. (2002) *J. Org. Chem.*, **67**, 1890–1897.
53. Gil, L., Baucherel, X., Martin, M.-T., Marazano, C., and Das, B.C. (1995) *Tetrahedron Lett.*, **36**, 6231–6234.
54. Isolation: Harrison, B., Talapatra, S., Lobkovsky, E., Clardy, J., and Crews, P. (1996) *Tetrahedron Lett.*, **37**, 9151–9154.
55. Sinigaglia, I., Nguyen, T.M., Wypych, J.-C., Delpech, B., and Marazano, C. (2010) *Chem. Eur. J.*, **16**, 3594–3597.
56. Sanchez-Salvatori, M.R. and Marazano, C. (2003) *J. Org. Chem.*, **68**, 8883–8889.
57. Tong, H.M., Martin, M.-T., Chiaroni, A., Bénéchie, M., and Marazano, C. (2005) *Org. Lett.*, **7**, 2437–2440.
58. For recent approaches, see: Amat, M., Pérez, M., Proto, S., Gatti, T., and Bosch, J. (2010) *Chem. Eur. J.*, **16**, 9438–9441, and references cited therein.
59. Nagata, T., Nishida, A., and Nakagawa, M. (2001) *Tetrahedron Lett.*, **42**, 8345–8349.
60. Nagata, T., Nakagawa, M., and Nishida, A. (2003) *J. Am. Chem. Soc.*, **125**, 7484–7485.
61. (a) Hourcade, S., Ferdenzi, A., Retailleau, P., Mons, S., and Marazano, C. (2005) *Eur. J. Org. Chem.*, 1302–1310; (b) Ge, C.S., Hourcade, S., Ferdenzi, A., Chiaroni, A., Mons, S., Delpech, B., and Marazano, C. (2006) *Eur. J. Org. Chem.*, 4106–4114.
62. Sanchez-Salvatori, M.R., Lopez-Giral, A., Ben Abdeljelil, K., and Marazano, C. (2006) *Tetrahedron Lett.*, **47**, 5503–5506.
63. Lopez-Giral, A., Mahuteau-Betzer, F., Gateau-Olesker, A., and Marazano, C. (2003) *Eur. J. Org. Chem.*, 1859–1867.
64. Kearney, A.M. and Vanderwal, C.D. (2006) *Angew. Chem. Int. Ed.*, **45**, 7803–7806.

65. Steinhardt, S.E., Silverston, J.S., and Vanderwal, C.D. (2008) *J. Am. Chem. Soc.*, **130**, 7560–7561.

66. Michels, T.D., Rhee, J.U., and Vanderwal, C.D. (2006) *Org. Lett.*, **10**, 4787–4790.

67. (a) Martin, D.B.C. and Vanderwal, C.D. (2009) *J. Am. Chem. Soc.*, **131**, 3472–3473; (b) Michels, T.D., Kier, Kearney, A.M., Vanderwal, C.D. (2010) *Org. Lett.*, **12**, 3093–3095; (c) Peixoto, S., Nguyen, T.M., Crich, D., Delpech, B., Marazano, C. (2010) *Org. Lett.*, **12**, 4760–4763.

68. Nuhant, P., Raikar, S.B., Wypych, J.-C., Delpech, B., and Marazano, C. (2009) *J. Org. Chem.*, **74**, 9413–9421.

69. (a) Takahashi, Y., Kubota, T., Fromont, J., and Kobayashi, J. (2009) *Org. Lett.*, **11**, 21–24; (b) Yamada, M., Takahashi, Y., Kubota, T., Fromont, J., Ishiyama, A., Otoguro, K., Yamada, H., Omura, S., and Kobayashi, J. (2009) *Tetrahedron*, **65**, 2313–1317

70. (a) Winkler, J.D. and Axten, J.M. (1998) *J. Am. Chem. Soc.*, **120**, 6425–6426; (b) Martin, S.F., Humphrey, J.M., Ali, A., and Hillier, M.C. (1999) *J. Am. Chem. Soc.*, **121**, 866–867; (c) Humphrey, J.M., Liao, Y., Ali, A., Rein, T., Wong, Y.-L., Chen, H.-J., Courtney, A.K., and Martin, S.F. (2002) *J. Am. Chem. Soc.*, **124**, 8584–8592; (d) Toma, T., Kita, Y., and Fukuyama, T. (2010) *J. Am. Chem. Soc.*, **132**, 10233–10235.

71. (a) Ono, K., Nakagawa, M., and Nishida, A. (2004) *Angew. Chem. Int. Ed.*, **43**, 2020–2023; (b) Young, I.S. and Kerr, M.A. (2007) *J. Am. Chem. Soc.*, **129**, 1465–1469; (c) Jakubec, P., Cockfield, D.M., and Dixon, D.J. (2009) *J. Am. Chem. Soc.*, **131**, 16632–16633; (d) C. D. Vanderwal recently wrote a review article with a fine analysis of the different strategies, and the interested reader can advantageously refer to this article: Vanderwal, C.D. (2010) *Angew. Chem. Int. Ed.*, **49**, 2830–2832

72. (a) Smith, B.J. and Sulikowski, G. (2010) *Angew. Chem. Int. Ed.*, **49**, 1599–1602; (b) Smith, J.B., Qu, T., Mulder, M., Noetzel, M.J., Lindsley, C.W., and Sulikowski, G.A. (2010) *Tetrahedron*, **66**, 4805–4810

73. (a) Garg, N.K., Hiebert, S., and Overman, L.E. (2006) *Angew. Chem. Int. Ed.*, **45**, 2912–2915; (b) Becker, M.H., Chua, P., Downham, R., Douglas, C.J., Garg, N.K., Hiebert, S., Jaroch, S., Matsuoka, R.T., Middleton, J.A., Ng, F.W., and Overman, L.E. (2007) *J. Am. Chem. Soc.*, **129**, 11987–12002.

74. See among other recent publications: (a) Peng, J., Kudrimoti, S., Prasanna, S., Odde, S., Doerksen, R.J., Pennaka, H.K., Choo, Y.-M., Rao, K.V., Tekwani, B.L., Madgula, V., Khan, S.I., Wang, B., Mayer, A.M.S., Jacob, M.R., Tu, L.C., Gertsch, J., and Hamann, M.T. (2010) *J. Med. Chem.*, **53**, 61–76; (b) Wahba, A.E., Peng, J., Kudrimoti, S., Tekwani, B.L., and Hamann, M.T. (2009) *Bioorg. Med. Chem.*, **17**, 7775–7782

75. See among others: (a) Winkler, J.D., Londregan, A.T., and Hamann, M.T. (2006) *Org. Lett.*, **8**, 2591–2594; (b) Winkler, J.D., Londregan, J.R., Ragains, A.T., and Hamann, M.T. (2006) *Org. Lett.*, **8**, 3407–3409

76. Jang, K.H., Kang, G.W., Jeon, J., Lim, C., Lee, H.-S., Sim, C.J., Oh, K.-B., and Shin, J. (2009) *Org. Lett.*, **11**, 1713–1716.

7
Biomimetic Synthesis of Marine Pyrrole-2-Aminoimidazole and Guanidinium Alkaloids

Jérôme Appenzeller and Ali Al-Mourabit

"la chimie est à la biologie ce que le solfège est à la musique"
(i.e., "chemistry is to biology what musical notation is to music.")
Pierre Potier

7.1
Introduction

Targeted and exhaustive metabolomic studies have become crucial for the comprehension of the origin and reactivity of natural substances. The latter investigations have become extremely important for any biogenetic proposal and even for conceiving efficient biosynthetic experiments using labeled early precursors. Prior to the association of structures to build families of biogenetically related molecules, the bioprospecting and collecting of natural sources for molecule isolation have to focus on phylogenetically related species. Although the fingerprints of secondary metabolites generated by biogenetic key-steps in living systems are essentially highly controlled by the genome, it is important to take into account that enzymatic key events could be followed by intrinsic spontaneous reactions toward each final molecule. Some highly chemo- and stereoselective transformations could be due exclusively to structural pre-organization of the molecules and to the highly chiral environment in cells. Isolation and reactivity analysis of classes of molecules might contribute to efficient synthetic plans, in particular for molecules that could be synthesized by cascade processes. The objective of this chapter is to compile some relevant biomimetic reactions and synthetic achievements based on biomechanistic analyses in the pyrrole-2-aminoimidazole and guanidinium alkaloid series.

To distinguish biomimetic from non-biomimetic syntheses, we propose some (not absolute) criteria:

- synthetic plan inspired by biosynthetic hypotheses;
- the conversion of a natural product into another natural product following biomechanistic thoughts;
- the use of a minimum number of protecting groups;

Biomimetic Organic Synthesis, First Edition. Edited by Erwan Poupon and Bastien Nay.

- the use of original biomimetic reactions (often one-pot reactions) with mild conditions in a biomimetic medium (water as solvent, use of acids or bases, photochemistry, . . .);
- the design of straightforward and highly chemoselective reactions.

7.1.1
Introduction to Pyrrole-2-Aminoimidazole (P-2-AI) Marine Alkaloids

Pyrrole-2-aminoimidazole alkaloids (P-2-AIs) constitute a family of alkaloids that to date have been isolated exclusively from marine sponges, particularly from Agelasidae, Halichondridae, and Axinellidae families. The extraordinary molecular diversity can be generated from the central precursor clathrodin (**1**) and its brominated derivatives, that is, hymenidin (**2**) and oroidin (**3**). These $C_{11}N_5$ molecules contain a pyrrole carboxamide moiety linked to a 2-aminoimidazole ring by a propenamine chain. The highly diverse family of P-2-AIs metabolites can be conceived of as derivatives of these $C_{11}N_5$ monomers with regards to oxidation, reduction, hydration, or methylation of both 2-amino-4(5)-vinylimidazole and pyrrole-2-carboxamide units. Polycyclization events or dimerization could take place, involving tautomerism of the 2-amino-4(5)-vinylimidazole system or the nucleophilic properties of the pyrrole (Scheme 7.1, path a) or the tautomerically controlled C–C and C–N bond formation between two monomers (path b), respectively. Complex transformations of clathrodin into dimeric P-2-AIs involving tautomerism, C–C bonds formation, C–N bonds formation, oxidation, and cascade reactions could lead to interesting molecules like palau'amine (**11**) congeners (path c). Finally, further fragmentation events of the previous compounds would afford degraded members, clinching this entire class of marine natural products (path d).

The present chapter completes our previous comprehensive review analyzing the biomechanistic aspects of P-2-AIs [1], and is dedicated to biomimetic achievements. Many of the recent synthetic studies based on the biogenetic proposals culminated in the total synthesis of several P-2-AIs members. Importantly, the most intricate steps and retrosynthetic strategies for P-2-AIs alkaloids were directly inspired by the biochemical pathways crossed back and forth by the numerous biogenetic proposals put forward for their biosynthesis after isolation. The P-2-AIs metabolites will be classified in terms of linear monomers, cyclic monomers, simple dimers, complex dimers, and tetramers. Linear monomers such as dispacamide A (**4**) are directly related to oroidin while cyclic monomers such as dibromoagelaspongin (**6**), dibromophakellin (**7**), and agelastatin A (**8**) contain new rings in addition to 2-aminoimidazole and pyrrole. In simple dimers ($2 \times C_{11}N_5$), the two subunits are linked by one or two bonds, such as in sceptrin (**9**), whereas in complex dimers more than two bonds appear between the two $C_{11}N_5$ subunits, such as in benzosceptrin A (**10**). Palau'amine (**11**) and its close congeners require several reactions, including chlorination, cascade reaction, oxidations, and hydrolysis. Only a few tetramers such as stylissadine A (**12**) have been reported; their presence indicates the promising molecular diversity of P-2-AIs marine alkaloids. Not surprisingly, some degradation compounds such as **13–17** were also reported.

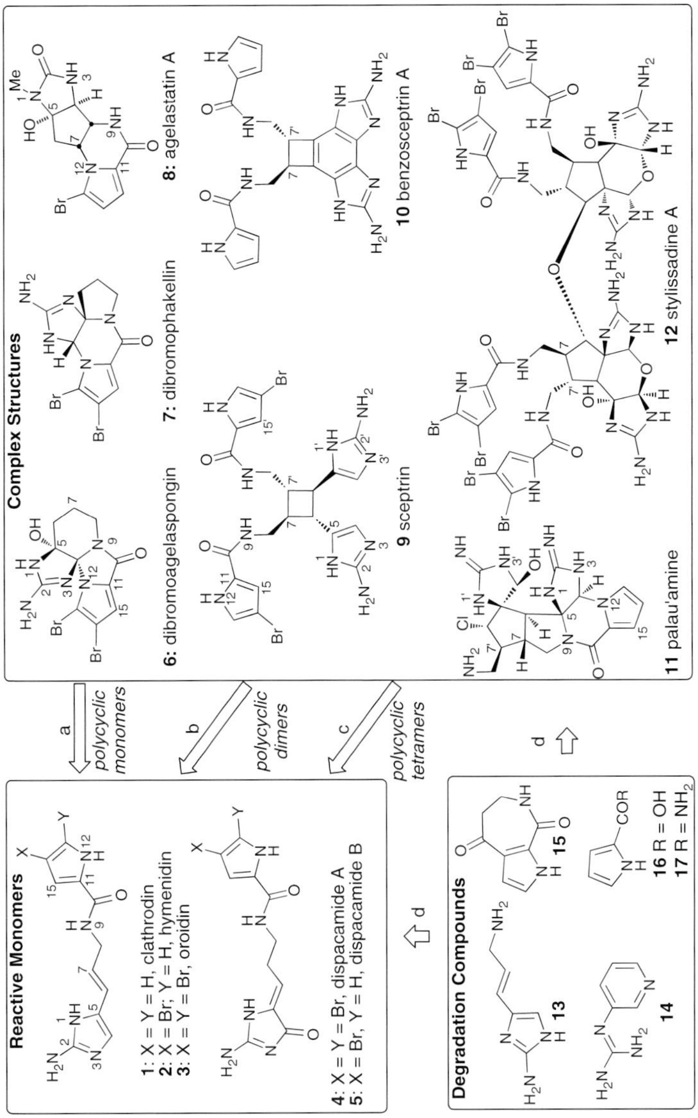

Scheme 7.1 Clathrodin (**1**) and some examples of related P-2-AIs.

The case of the misleading propenamine **13** in regards to biogenesis is interesting. While it was considered to be an early precursor of oroidin through acylation by 4,5-dibromopyrrole-2-carboxylic acid, its formation is very likely the result of the hydrolysis of oroidin or similar derivatives.

Exhaustive reviews dealing with P-2-AIs structures, activities, and syntheses have been published recently [2]. We focus here on the biomimetic syntheses of P-2-AIs published since 2005 and also on recent insights in terms of biogenetic hypotheses. The biogenesis of P-2-AIs can be studied at two levels of complexity: (i) from the pool of amino acids to clathrodin (**1**) and dispacamide (**4**, **5**) skeletons or (ii) from the latter to more complex structures via polycyclizations or dimerizations (Scheme 7.2). In 2001, our group postulated that the tautomeric equilibria of the aminopropenylimidazole portion of oroidin (**3**) were the key processes involved in the formation of both polycyclic monomers and cyclized dimers [1]. This is certainly a consequence of the corresponding ambivalent nucleophilic/electrophilic behavior of oroidin derivatives. The proposed universal chemical pathway for the formation of P-2-AIs members, known up to 2000, is still valid. All complex dimeric P-2-AIs members isolated during the last decade [2f, 3] fit in with the general chemical pathway that we have previously suggested for their formation and relative stereochemical relationships. The correction of palau'amine relative configuration (**11**), and consequently those of congeners [4], supported the hypothesis. The universal biochemical pathway is also in accordance with the conserved stereochemical relationships within the dimer subclass. Some additional and particular biogenetic hypotheses have been suggested by Baran and Köck starting from sceptrin (**9**) [2c, 5]. In fact, the most fundamental question regarding the biosynthesis of P-2-AI alkaloids concerns the first event connecting monomers like hymenidin (**2**) to dimers like sceptrin (**9**).

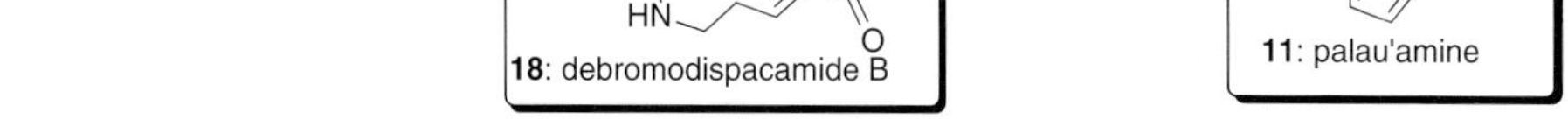

Scheme 7.2 Pre- and post-clathrodin (**1**) events involved in the biogenesis of P-2-AIs.

The biogenetic hypothesis defined the basic $C_{11}N_5$ structure of P-2-AIs alkaloids and introduced the concept of the dual reactivity originated by the 2-aminoimidazole section (Figure 7.1).

The ambivalent effect that propagates along the vinylogous chain defines five possible tautomers (I–V) with various modes of cyclization and dimerization. Nucleophilic and electrophilic reacting positions can vary depending on the tautomer engaged in the dimerization step [1]. The investigations, including molecular

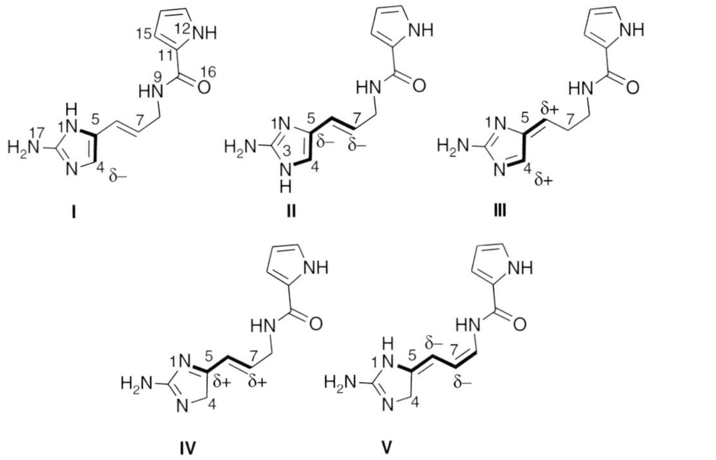

Tautomers of type I, II, III, IV and V will be extensively used in the following section

Figure 7.1 Tautomerism in the building blocks of the P-2-AI monomer clathrodin (**1**); ambivalent reactivity of vinylogous 2-aminoimidazole.

calculations, concluded that the tautomeric interconversion of I–V is more easily explained in acidic conditions [6]. The process leading to the selection of the reacting tautomers in Nature raises an interesting question. The dynamic hydrogen-bonding interaction of monomers like clathrodin (**1**) with the host enzyme is probably one of the most intriguing biochemical catalytic systems. The proposed isomerization mechanism is probably operative with precise modulation dictated by the sponge genome. The consequence appears in the composition of the P-2-AI targeted metabolome, which changes with the sponge species. From a synthetic point of view, the apparently obvious dimerization step proved difficult to realize in the laboratory. The biogenesis process of P-2-AI metabolites can be divided into two major events (Scheme 7.2): (i) pre-clathrodin events linking early precursors, presumably amino acids, to monomers like clathrodin (**1**) or debromodispacamide B (**18**) (Scheme 7.2) and their brominated derivatives (e.g., oroidin) and (ii) chemical transformations of the crucial precursors clathrodin or debromodispacamide B into higher order P-2-AI members.

7.1.2
Proposed Biogenetic Hypothesis for Clathrodin (1) and Related Monomers Starting from α-Amino Acids

Regarding the pre-clathrodin events several more or less intuitive biosynthetic pathways have been proposed by different groups starting from various amino acids. The postulated key steps rely on some presumed key precursors isolated from phylogenetically related sponges or co-isolated together with other P-2-AIs members. In some cases, synthetic results and/or radioactive labeling experiments provided

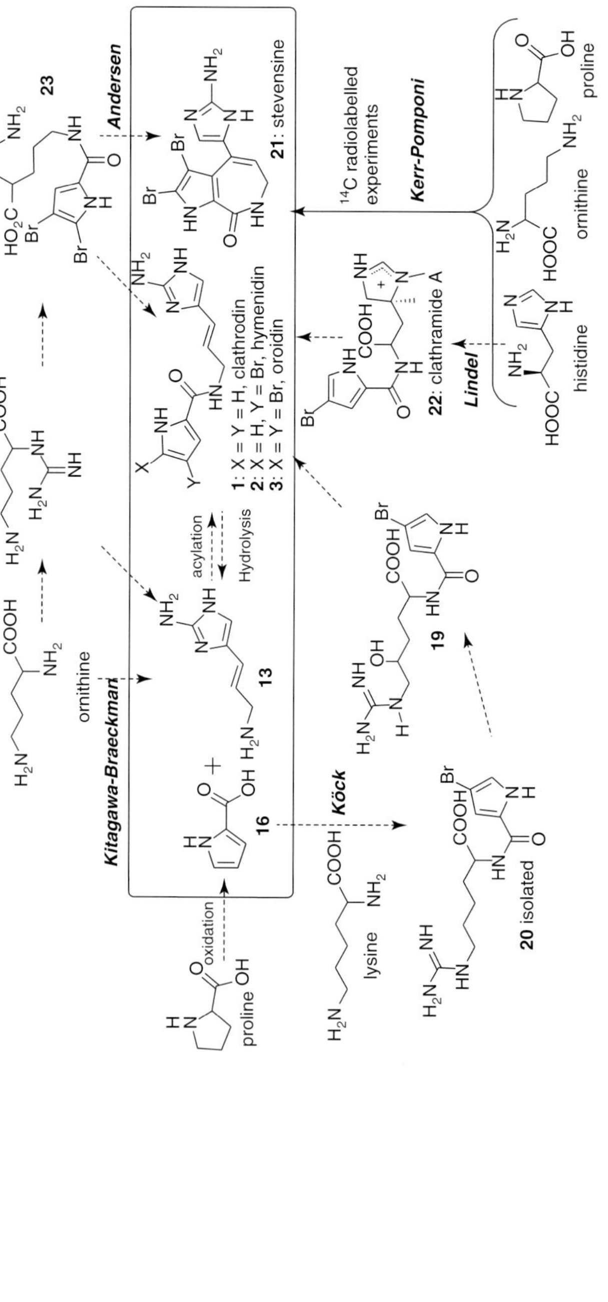

Scheme 7.3 Biosynthetic proposals for clathrodin and its brominated derivative oroidin (3).

additional support. Failing any biosynthetic demonstration for the real early precursors involved in the formation of the 2-aminoimidazole and 2-aminoimidazolone parts, hypotheses on the pre-clathrodin events remain purely speculative at present. The most commonly cited route as deduced from the isolated metabolites would start from proline, ornithine, histidine, arginine, or lysine amino acids. Scheme 7.3 gives an overview of the different proposals.

Pyrrole-2-carboxylic acid (**16**) and 2-amino-5-(3-amino)propylimidazole (**13**) have been isolated from many sponges. This "misleading" result suggests that clathrodin (**1**) could be formed by the acylation of **13** by **16** (Scheme 7.3). If the latter suggestion is right, the biosynthesis of oroidin could be inferred from understanding the biogenesis of compounds **13** and **16** independently. Proline was assumed by all authors to be the precursor of the pyrrole-2-carboxamide part by double oxidation, while several possibilities have been suggested for the origin of the 2-amino-5-(3-amino)propylimidazole unit.

Many hypotheses based on primary amino acids have been proposed for **13** while proline was exclusively considered for pyrrole-2-carboxylic acid (**16**). According to us, the origin of the 2-aminoimidazole moiety could actually be much more subtle. In fact, 2-amino-5-(3-amino)propylimidazole (**13**), supposed to be the precursor of oroidin, might rather be the product of its hydrolysis.

For Kitagawa [7] and Braeckman [8], ornithine and its guanidic analog are the precursors of the vinyl-2-aminoimidazole (**13**) (see Scheme 7.3). More recently Andersen [9] and coworkers completed this proposal by an alternative route for the incorporation of ornithine into oroidin alkaloids via a guanidine-containing fragment (**23**) related to stevensine (**21**).

Kerr *et al.* reported the incorporation of the ^{14}C-labeled proline, ornithine, and histidine into the cyclized oroidin derivative stevensine (**21**) in a cell culture of the sponge *Teichaxinella morchella* [10]. The mechanistic incorporation of histidine into oroidin remains unresolved. Unfortunately, only low levels of radioactivity and specific incorporation levels were observed with amino acids in the Kerr experiments ([U-^{14}C]His, 1460 dpm (0.026%); [U-^{14}C]ornithine, 1300 dpm (0.024%)), and these results along with the lack of *in vivo* biosynthetic studies leave the question unanswered [11]. Lindel [12] complemented the Kerr's proposal regarding the incorporation of histidine. This could proceed via the clathramide A like precursor **22** (Scheme 7.3). The quaternary methyl group in **22** would provide the additional carbon, leading to the $C_{11}N_5$ skeleton. The inclusion of this carbon was thought to proceed through a cyclopropane intermediate. However, this step is based on yet unobserved amination of histidine at position 2 of the imidazole nucleus.

Köck isolated [13] the guanidine-containing bromopyrrole compound **20** from the sponge *Agelas wiedenmayeri*. This compound, which corresponds to the guanidylated lysine 4-bromopyrrole-2-carboxamide, was also highlighted as a precursor of P-2-AI early intermediates. Further hydroxylation, cyclization, and decarboxylation reactions could afford the oroidin skeleton from compound **19** (see Scheme 7.3).

More recently Al-Mourabit and coworkers proposed a biogenetic pathway [14] for P-2-AI early precursors from proline and related arginine. This proposal (Scheme 7.4) was based on the isolation of verpacamides A (**24**) and C (**25**) from the

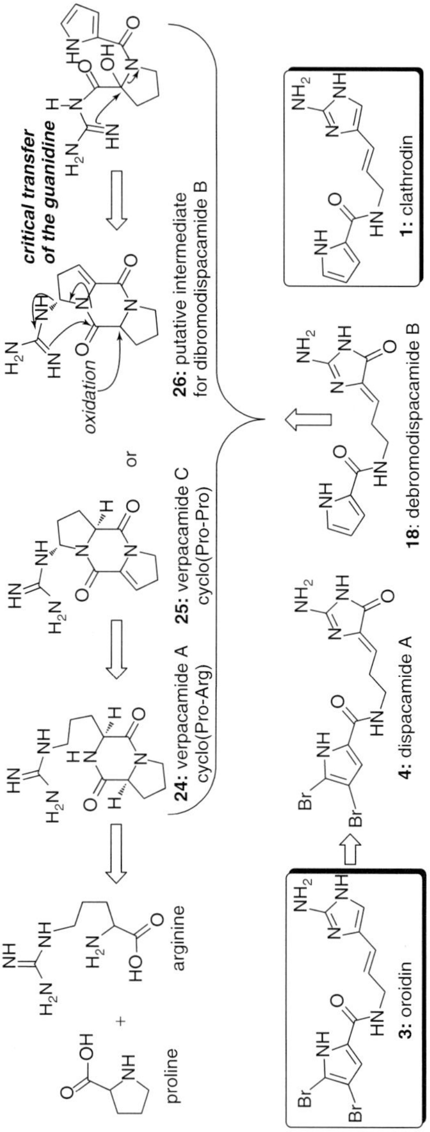

Scheme 7.4 Al-Mourabit's retro-biogenetic proposal for P-2-AIs via cyclo(Pro-Arg) and further cyclo(Pro-Pro) diketopiperazines as precursors.

sponge *Axinella vaceletti*. The consideration of cyclo(Pro-Arg) and cyclo(Pro-Pro) diketopiperazines distinguishes this hypothesis from others. This family of proline and arginine diketopiperazines could be the precursor of dispacamide A via dioxygen-mediated oxidation. Notably, two prolines and one guanidine (or one proline and one arginine) yield the $C_{11}N_5$ skeleton of P-2-AIs. Importantly, the co-isolated verpacamides A (**24**) and C (**25**) show an increasing degree of oxidation *en route* to P-2-AIs through arginine cyclization into a proline residue bearing the required guanidine. The biomimetic syntheses of dispacamides following a similar pathway have been achieved recently in the group of Al-Mourabit (Section 7.3) [15, 16].

7.2 Ground Work of George Büchi: Dibromophakellin (7) Synthesis from Dihydrooroidin (31)

Among the many contributions of George Büchi (1921–1998) in the field of organic chemistry, his ground work on the synthesis of natural products over 50 years reveals his elegant style. Several creative and original syntheses of natural products were indeed achieved in Büchi's laboratory. His synthetic conceptions were often associated with biomimetic synthesis. The synthesis of parazoanthoxanthin (**27**) [17] and dibromophakellin (**7**) [18], following some biomimetic thoughts, can be seen as a landmark in the biomimetic chemistry of P-2-AIs and complex guanidinium alkaloids in general. The first total synthesis of dibromophakellin (**7**) will be depicted and discussed in details. Büchi first investigated the oxidation of 2-aminoimidazoles, which was applied to the syntheses of parazoanthoxanthin (**27**) and dibromophakellin (**7**) (Scheme 7.5). The key steps involved the acidic dimerization of 2-aminoimidazolylethanol (**28**) to give **27** via vinylimidazole **29**, and the oxidative intramolecular cyclization of dihydrooroidin (**30**) to give **7**.

Dibromophakellin (**7**) [19] is a complex and compact tetracyclic structure associated with mild biological activities. Dibromophakellin exhibits surprising stability to hydrolysis despite the presence of two aminal groups. Its unusual architecture made it a very attractive target for total synthesis. Dibromophakellin has been isolated from several sponges of the families Agelasidae and Axinellidae in its two enantiomeric forms. Another interesting facet of this structure is that

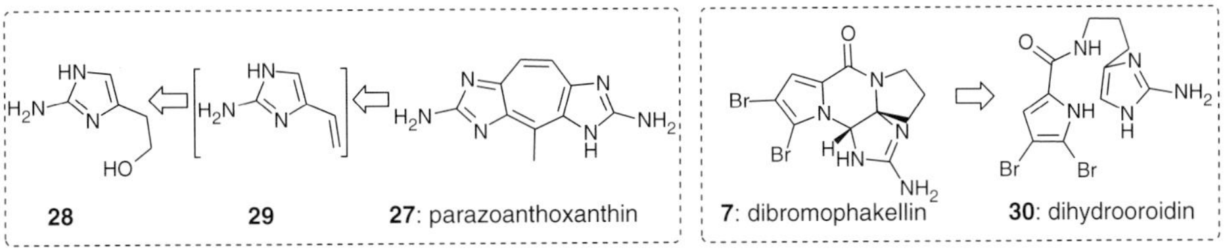

Scheme 7.5 Büchi's foundation of oxidative and ambivalent reactivity of 2-aminoimidazole (2-AI) toward marine metabolites.

dibromophakellin is a subunit met in one of the most challenging complex marine metabolites: palau'amine (**11**). The construction of the dibromophakellin skeleton *en route* to palau'amine is still a hot topic. Dibromophakellin is biogenetically related to oroidin through formal N9–C5 and N12–C4 cyclizations. Büchi proposed dihydrooroidin (**30**) as a common precursor to both oroidin (**3**) and dibromophakellin (**7**) (Scheme 7.6).

Dihydrooroidin could lead to oroidin through oxidation at C6–C7 or to dibromophakellin under the same oxidative conditions through N9–C5 and N12–C4 cyclizations. This strategy was remarkable at the time in the sense that very few P-2-AI members had been isolated from sponges in 1980.

The synthesis of dihydrooroidin (**30**) was achieved in five steps and 35% overall yield starting from L-citrulline. It began with the preparation of 2-aminoimidazole-4-propanamine (**31**) from citrulline. The key step was the Lancini condensation [20] of the corresponding aminoaldehyde with cyanamide leading to the 2-aminoimidazole nucleus. The aminoaldehyde was prepared by Akabori reduction [21]. The pyrrole subunit was introduced at a late stage of the synthesis using the acylation of **31** in the presence of 4,5-dibromo-2-trichloroacetylpyrrole. The dihydrooroidin (**30**) was then oxidized by elemental bromine to yield racemic dibromophakellin (**7**) after treatment with *t*BuOK. The precise mechanism involved in the latter reaction cascade remains unclear. This ground-breaking synthesis was straightforward and easily scalable, and still provides a rapid access to racemic dibromophakellin. A similar strategy was used 25 years later by Harran and coworkers in a synthetic study of dibromophakellin [22]. The principles illustrated in this synthesis were developed by Horne and coworkers in the synthesis of oroidin, dispacamides, dibromoisophakellin, dibromophakellstatin, and even dimeric mauritiamine [23]. The 2-aminoimidazole part, which is thought to be very sensitive to oxidative reactions, could be installed using preparative synthetic chemistry. Although the polycyclization of the dihydrooroidin (**30**) intermediate was seemingly biomimetic, the biosynthetic pathway leading to the 2-aminoimidazole (2-AI) from citrulline remains obscure. The principal findings of Büchi's work are certainly the high propensity of 2-aminoimidazole to be oxidized and its ambivalent reactivity. These properties are clearly reminiscent of those employed in the synthesis of parazoanthoxanthin (**27**) by Büchi himself, and also of many subsequent developments of P-2-AIs chemistry.

7.3 Biomimetic Synthesis of P-2-AI Linear and Polycyclic Monomers

The pre-clathrodin synthetic details linking amino acids toward monomers like clathrodin or debromodispacamide and their brominated derivatives are unknown. Only the biomechanistic analyses of isolated P-2-AIs and their synthetic achievements based on speculative biosynthesis have contributed to progress in the understanding of these early steps. The isolation of an increasing number of

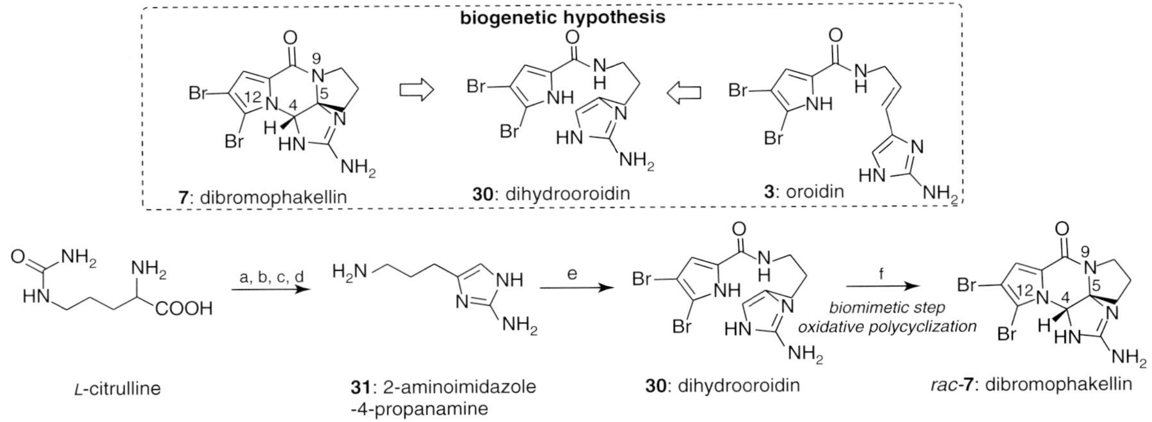

Scheme 7.6 Büchi's biogenetic proposal for oroidin (**3**) and dibromophakellin (**7**), and biomimetic synthesis of *rac*-**7** from L-citrulline. *Reagents and conditions*: (a) EtOH, HCl; (b) Na/Hg; (c) NH_2CN, HCl; (d) NaOH, heat, 71% overall yield from L-citrulline; (e) 4,5-dibromo-2-trichloroacetylpyrrole, Na_2CO_3, DMF, 50%; and (f) (i) Br_2, AcOH, (ii) *tert*-BuOK, 2-butanol; quantitative.

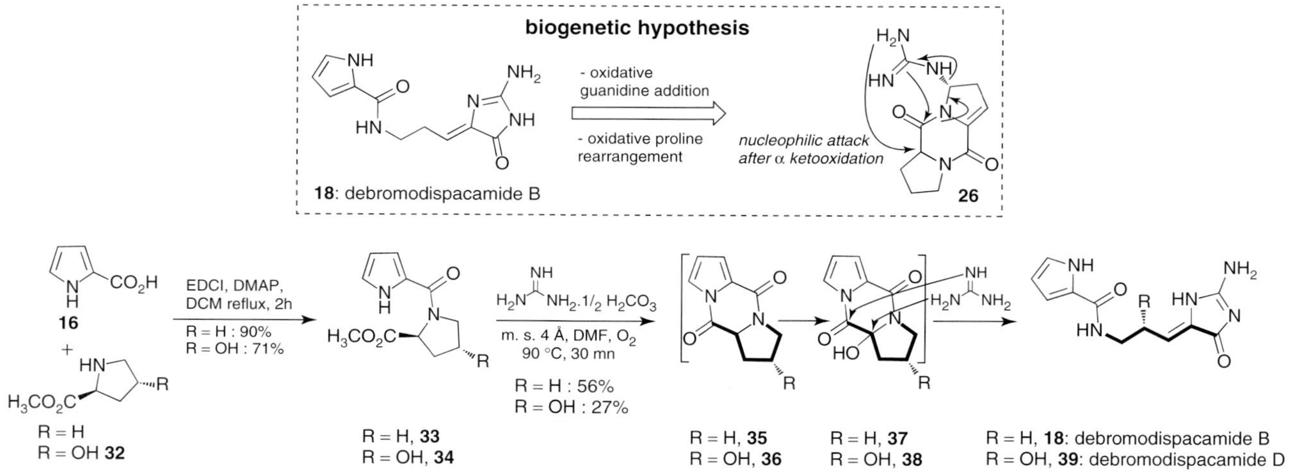

Scheme 7.7 Al-Mourabit's biomimetic synthesis of debromodispacamides B and D.

putative biosynthetic intermediates was extremely important in developing hypotheses regarding the pre-clathrodin or pre-dispacamide biochemical events. The logic constructed through intermediate isolation from natural sources, combined with synthetic achievements, is of high value for the formulation of biosynthetic scenarios. The next section is based mainly on this reasoning.

7.3.1 Biomimetic Synthesis of Linear Monomers

In contrast with complex P-2-AI members, very little has been made in terms of the biomimetic synthesis of simple monomers. Considering that these early transformations are very important for the elaboration of chemical diversity, our group has focused over the past years on biogenetically inspired syntheses of oroidin (**3**), dispacamides, and other linear monomers.

7.3.1.1 Debromodispacamides B (18) and D (39) and Dispacamide A (4)

The suggested hypothesis starts from the putative key diketopiperazine of type **26**, combining proline and arginine (Schemes 7.4 and 7.7).

The synthesis of debromodispacamides B (**18**) and D (**39**) was completed using the same strategy [16] in one step from the pseudodipeptides **33** or **34** in the presence of guanidine and air dioxygen. When the reaction with guanidine was run in degassed solvent under argon, no rearrangement of proline into 2-aminoimidazolone was observed. The one-pot reaction of the pseudopeptide pyrrole-proline methyl esters **33** and **34** in the presence of guanidine carbonate and air oxygen, in DMF at 90 °C, was found to afford stereoselectively the desired debromodispacamides B (**18**) and D (**39**). The natural products were isolated in 44% yield from proline and 19% from 4-hydroxyproline, respectively, in two steps only and without protecting groups. Notably, the synthesis of *(R)*-debromodispacamide D from L-*trans*-4-hydroxyproline methyl ester (**32**) was completely stereoselective and stereospecific whereas the natural product was isolated racemic. The bromine-containing dispacamide A (**4**) was prepared from **33** using the same strategy (Scheme 7.8) [15].

7.3.1.2 Clathrodin (1) and Its Brominated Derivative Oroidin (3)

The biomimetic synthesis of oroidin and related monomers [24] was inspired by the biogenetic proposals involving pyraxinine (**14**) [25] and dibromoagelaspongin (**6**) [26]. The oroidin motif can be identified in **6**, an intricate tetracyclic P-2-AI member with two adjacent quaternary carbons. The pyridine containing pyraxinine **14** could be obtained from the vinylaminoimidazole (**13**) isolated from sponges. The proximity between oroidin (**3**) and dibromoagelaspongin (**6**) led to the formulation of a common chemical approach using the oxidative addition of guanidine derivatives to 1,2-dihydropyridine (**44**) (Scheme 7.9). The mechanistic rationalization for the formation of the pyridine derivative **14** from the 2-aminoimidazole (**13**) was based on the cyclization of tautomer **43** into **42** followed by the its ring opening when exposed to acidic or basic conditions. Based on this biogenetic proposal, a retrosynthetic scheme for oroidin was planed. The conversion of **43** into the

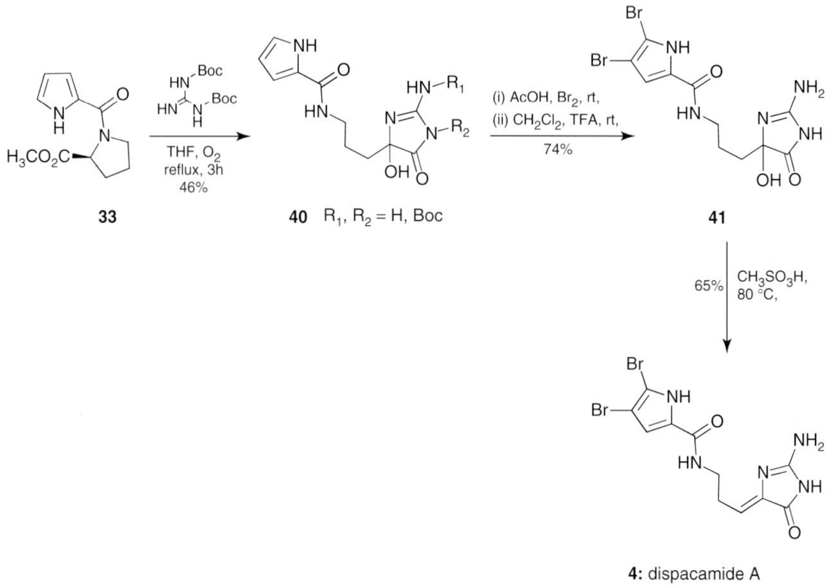

Scheme 7.8 Al-Mourabit's synthesis of dispacamide A (**4**).

guanidic pyridine intermediate was assumed to be reversible through the putative intermediate **42**. Mechanistic evidence for this process was reported by Al-Mourabit *et al.* [27].

The key step of this sequence was the oxidative addition of 2-aminopyrimidine (2-AP) mediated by bromine on the 1,2-dihydropyridine derivative (**44**) obtained by reductive acylation of pyridine. The masked 2-aminoimidazoline (**45**) was then deprotected to yield the expected oroidin derivative **46** and *(Z)*-oroidin (**47**). A final isomerization in acidic medium yielded oroidin (**3**). This four-step synthesis proved to be short and afforded a reasonable overall yield. Similar access to clathrodin (**1**) was proposed using non-brominated reagents [28].

7.3.2
Biomimetic Synthesis of Cyclized Monomers

7.3.2.1 Cyclooroidin (48)

An efficient conversion of oroidin into racemic cyclooroidin was reported by Lindel and coworkers by heating oroidin in protic medium [29]. The N12–C7 bond formation is believed to proceed via the azafulvene type IV tautomer (Scheme 7.10).

7.3.2.2 Dibromoagelaspongin (6)

The polycyclized monomer dibromoagelaspongin (**6**) first isolated by Struchkov *et al.* [26] exhibits an intriguing tetracyclic structure with a quaternary carbon

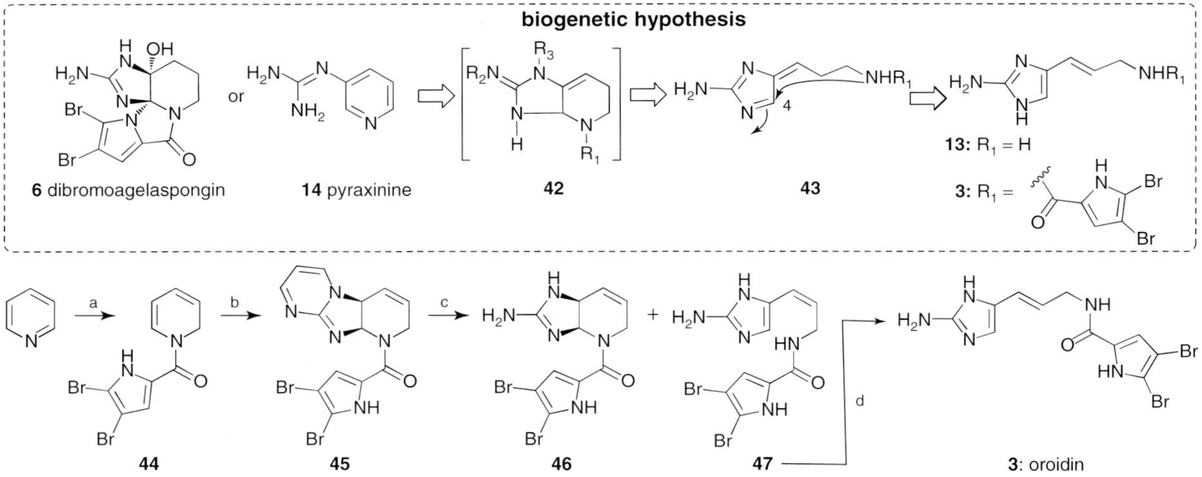

Scheme 7.9 Al-Mourabit's synthesis of oroidin (**3**) based on the biogenetic proposal for pyraxinine (**14**) and dibromoagelaspongine (**6**). *Reagents and conditions*: (a) 4,5-dibromo-2-carbonyl chloride, $NaBH_4$, −78 °C, 20%; (b) 2-aminopyrimidine, Br_2, 42%; (c) NH_2OH, TEA, EtOH, heat, 47%; and (d) TFA-DCM, 50 °C, 71%.

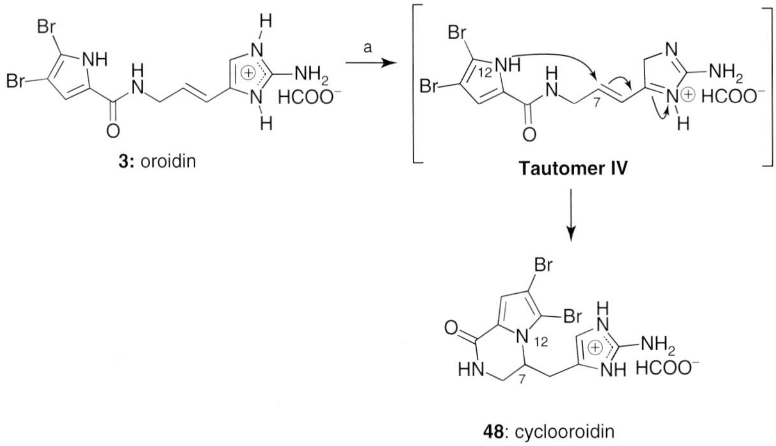

Scheme 7.10 Lindel's conversion of oroidin into *rac*-cyclooroidin (**48**). *Reagents and conditions*: (a) H_2O/EtOH (4 : 1), 95 °C, sealed tube, 45 h, 93%.

connected to three different nitrogens. Struchkov *et al.* proposed that the putative imidazolinone **49** would cyclize to afford the skeleton of **6** (Scheme 7.11). Al-Mourabit and Potier proposed a biomimetic pathway from oroidin (**3**) through the dual tautomer of type III. After cyclization into aminals **50** (path a) [1] or **51** (path b) [27] and tautomerization into more stable intermediates **52** and **53**, respectively, similar N9–C4 oxidative cyclizations could yield the hemiaminal dibromoagelaspongin (**6**). The notable macrocycle **50** (path a) was proposed as the common intermediate for both dibromophakellin and dibromoagelaspongin [1].

The first N–C bond proceeded through tautomer III of the natural product oroidin itself (Scheme 7.11). Although this proposal has not been tested experimentally, the suggested transannular cyclization of the nine-membered intermediate **50** was used in the recently reported total synthesis of palau'amine [30]. An alternative avenue based on Büchi's oxidation of dihydrooroidin (**30**) for the formation of **6** (path c) [18] was proposed by Feldman [31]. The reaction was thought to occur via **54** and **55**, followed by an oxidative closure into the dibromoagelaspongin (**6**). Direct C4–N9 bond formation is also plausible (path d) to yield **56**.

Inspired by paths b–d, Feldman and coworkers achieved the first total synthesis [31] of dibromoagelaspongin (**6**) in 16 steps and 4.7% overall yield from an advanced imidazole sulfoxide (**57**, Scheme 7.12).

After acylation with *N*-SEM protected dibromopyrrole derivative, the dihydrooroidin intermediate **58** was formed. A Pummerer oxidative cyclization yielded the N9–C4 bond in compound **59**. The second biomimetic oxidative cyclization was realized in presence of NCS to yield **60** after N12–C4 bond formation. Regioselective oxidation and cyclization allowed for the formation of the tetracyclic core of agelaspongin with a bis-aminal at C4. Amination of **60** through the azide **61** and methoxy/hydroxy exchange led to *rac*-dibromoagelaspongin (**6**).

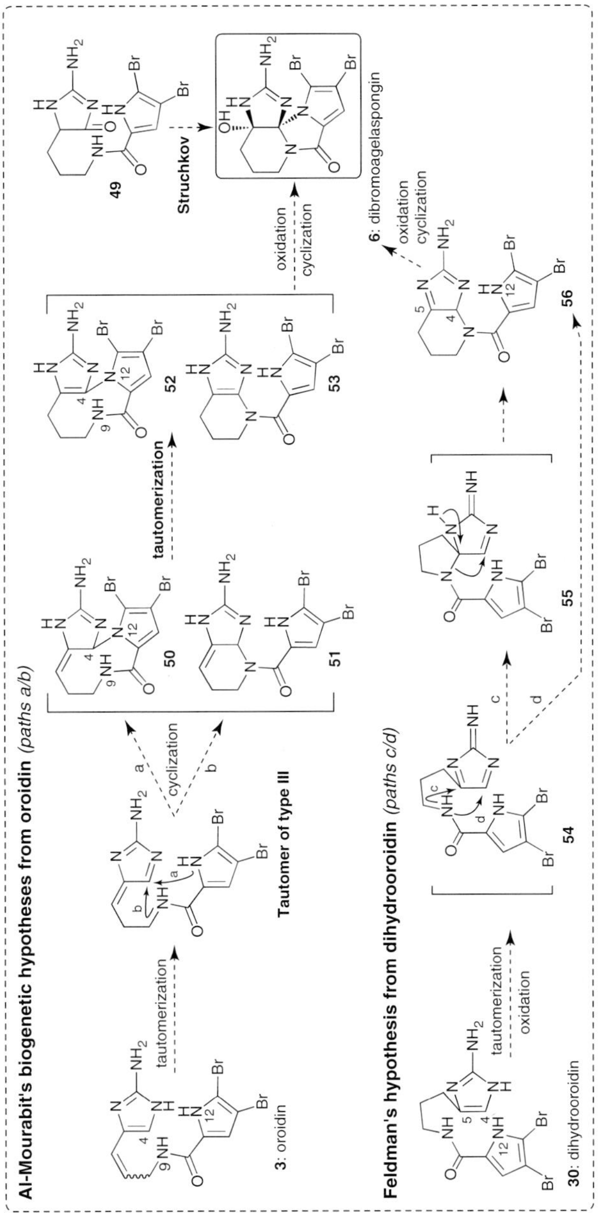

Scheme 7.11 Biomimetic hypotheses for dibromoagelaspongin (6).

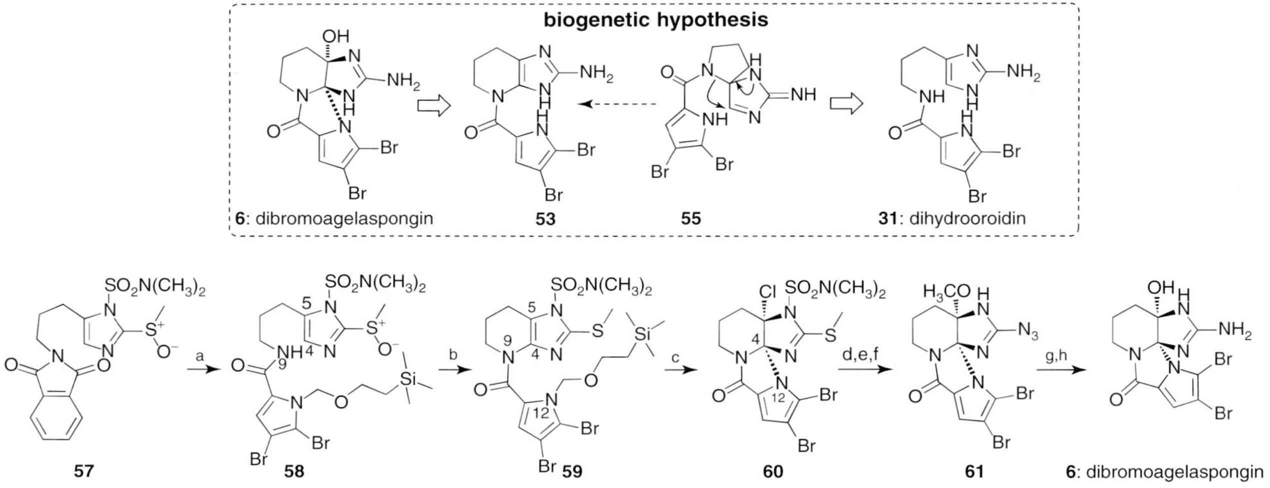

Scheme 7.12 Feldman's total synthesis of *rac*-dibromoagelaspongin (**6**). *Reagents and conditions*: (a) (i) $(NH_2)_2$, (ii) 2-trichloroacetyl-4,5-dibromo-*N*-SEM-pyrrole, Na_2CO_3, and (iii) HCl 1, 5M, 45%; (b) Tf_2O, 2,6-lutidine, CH_2Cl_2, $-78\,^{\circ}C$, 45%; (c) (i) TFA then Bu_4NF and (ii) NCS, CH_2Cl_2, 65%; (d) HCl, CH_3OH; (e) mCPBA; (f) $TMSN_3$, ZnI_2, 36%; (g) H_2, Pd, TFA; and (h) H_2O, 90%.

Al-Mourabit *et al.* tested the biomimetic approach considering the intermediate **53** (Scheme 7.13) [32]. The strategy was based on a six-step sequence using 2-aminopyrimidine, which has the notable advantage of being less polar than guanidine. The tetrahydropyridine (**62**) was prepared from 5-aminopentanol and 2-trichloroacetylpyrrole on a multi-gram scale. It reacted with 2-AP in the presence of NIS to afford **63** in 52% yield. Bromination of the pyrrole ring by Br_2 yielded **64**.

Compared to other oxidative reagents, the use of $BaMnO_4$ to promote the dehydrogenation of intermediate **64** proved more successful, affording **65** in 34% yield. The protected cyclic oroidin **65** was used to investigate the biomimetic oxidative cyclization into the fused tetracyclic core of **6**. The protected dibromoagelaspongin **66** was obtained using DMDO as oxidant. Although deprotection attempts of **66** to give **6** failed, validation of this biomimetic approach was achieved by the preparation of the intricate tetracyclic compound **66**.

7.3.2.3 **Dibromophakellin (7) and Dibromophakellstatin (69)**

Dibromophakellin is the representative member of the cyclized monomers, previously synthesized by Büchi in a pioneering work. In a biomimetic strategy, the N12–C4 and N9–C5 bonds should be formed in one step starting from open-chain precursors related to oroidin. Other non-biomimetic syntheses have proved to be interesting alternative routes, as for the related (−)-dibromophakellstatin (**69**) [33] reported by Lindel and coworkers. Büchi's synthesis of phakellin was improved by Horne [34] and extended to phakellstatin by using a NBS-mediated oxidation. Feldman reported a biomimetic synthesis of phakellstatin [35] based on a Pummerer extended reaction towards the aminoimidazole ring (Scheme 7.14). The sulfide **67** was obtained in five steps and converted into the oroidin-like intermediate **68** after deprotection and coupling with 4,5-dibromo-2-trichloroacetylpyrrole. Oxidative tetracyclization starting from compound **68** was successful using the Stang reagent, PhI(CN)OTf, with formation of the key N12–C4 and N9–C5 bonds. Further oxidation mediated by CAN led straightforwardly to dibromophakellstatin (**69**). The conversion of **69** into dibromophakellin (**7**) was finally achieved by enol ether formation with Meerwein's reagent and amination using ammonium propionate.

While the following syntheses are not biomimetic, they are deliberately included in our discussion to illustrate biological inspiration from proline and cyclo(Pro-Pro) diketopiperazine. In 2008, Romo [36], and later Nagasawa [37], reported the enantioselective syntheses of (+)-phakellin starting from L-proline to construct the pyrrole-proline ketopiperazine moiety of dibromophakellin (**7**). The same strategy was employed for the first asymmetric synthesis of the close (+)-phakellin derivative (+)-phakellstatin by Romo in 2003 [38]. The synthesis confirmed the absolute configuration of the natural product, which is (−)-phakellstatin (hydrolyzed phakellin). The synthesis of the natural (−)-enantiomer by Lindel [39] confirmed the same configuration. The main features of Romo's synthesis of the unnatural enantiomer (+)-dibromophakellstatin are outlined here. A pivotal step in the approach was the diastereoselective desymmetrization of the optically active cyclo (L-Pro-L-Pro) diketopiperazine **70** (Scheme 7.15). Stereocontrolled C-acylation of the corresponding monoenolate with benzyl chloroformate produced **71** as the

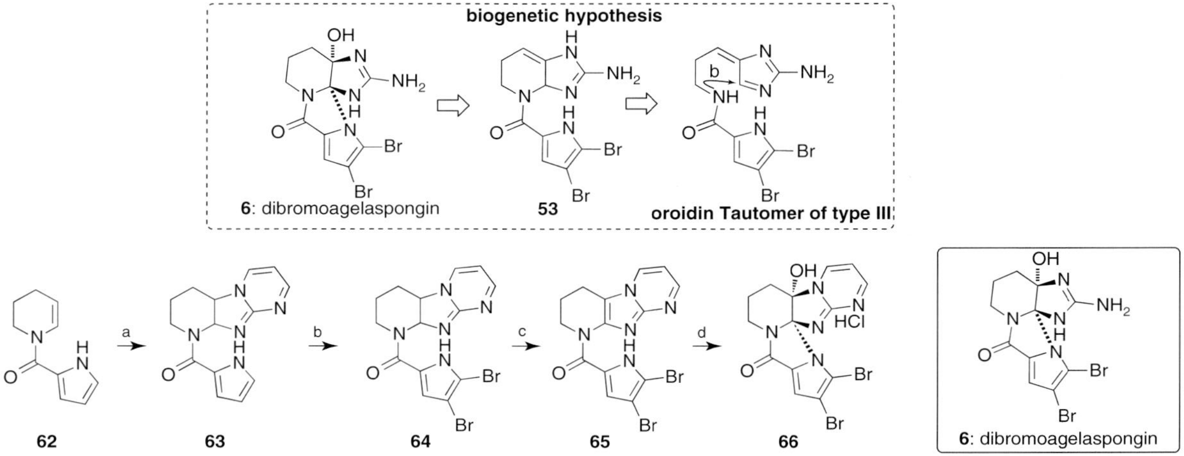

Scheme 7.13 Al-Mourabit's approach to *rac*-dibromoagelaspongin (**6**). *Reagents and conditions*: (a) NIS, 2-aminopyrimidine (2-AP), DMF/CH_3CN, 52%; (b) Br_2, CH_2Cl_2; (c) $BaMnO_4$, 34% overall yield; and (d) DMDO, acetone then HCl, Et_2O, quantitative.

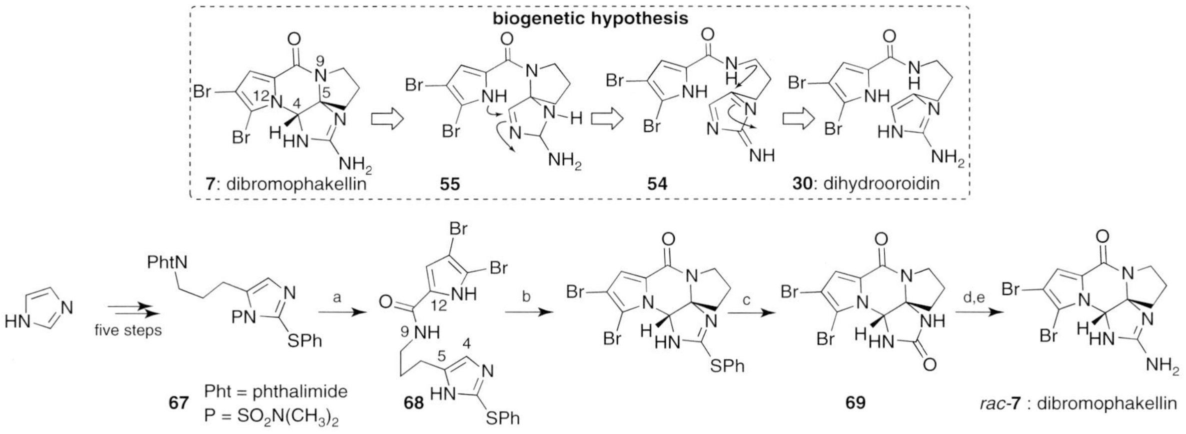

Scheme 7.14 Feldman's synthesis of dibromophakellstatin (**69**) and dibromophakellin (**7**) using a Pummerer extended reaction. *Reagents and conditions*: (a) (i) $(NH_2)_2$ and (ii) 4,5-dibromo-2-trichloroacetylpyrrole, 64%; (b) PhI(CN)OTf, iPr_2NEt, CH_2Cl_2/CH_3OH, 60–73%; (c) CAN, CH_3CN/H_2O, 80–93%; (d) Et_3OBF_4, $NaHCO_3$, CH_2Cl_2, 45%; and (e) $EtCO_2-$, NH_4+, heat, 59%.

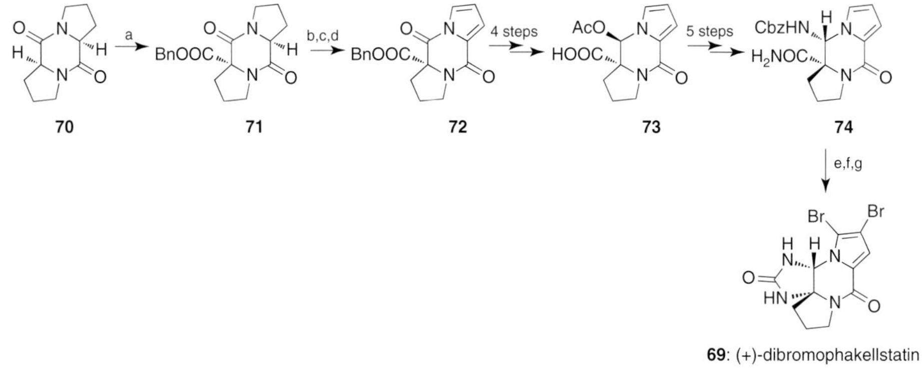

Scheme 7.15 Romo's synthesis of (+)-dibromophakellstatin (**69**). *Reagents and conditions*: (a) KHMDS, ClCOOBn, THF, 70%; (b) PhSeBr, KHMDS, THF; (c) DMDO, CH_2Cl_2; (d) SeO_2, dioxane, 44% overall yield; (e) $PhI(TFA)_2$, pyridine, CH_3CN; (f) H_2, Pd/C, MeOH; and (g) NBS, THF, 35% overall yield. KHMDS = potassium 1,1,1,3,3,3-hexamethyldisilazide.

major product. A finely tuned sequence of oxidation then yielded the pyrrole derivative **72** in three steps. Acylcarbinol (**73**) could be constructed using four additional steps of functional group transformations. Compound **74** with the two nitrogens in place required five more steps. The next key step was a Hoffmann rearrangement under oxidative conditions leading to the cyclic urea moiety of the targeted natural product. The synthesis of (+)-dibromophakellstatin (**69**) was achieved after Cbz-hydrogenolysis and pyrrole dibromination.

Lindel and coworkers proposed a concise synthesis of the natural (−)-dibromophakellstatin (**69**) using the same methodology they employed for the racemic mixture [40]. Scheme 7.16 shows the key steps of the synthesis.

The diastereoselective synthesis of compound **75** was achieved in four steps starting from 4,5-dibromo-2-carboxypyrrole and L-*trans*-4-hydroxyproline methyl ester. Debromination of the pyrrole followed by dehydration led to the bis-enamine **76**. This compound was the substrate of the key diamination step using TsONHCOOEt. Hydroxyproline was the source of stereogenic information to control selectively

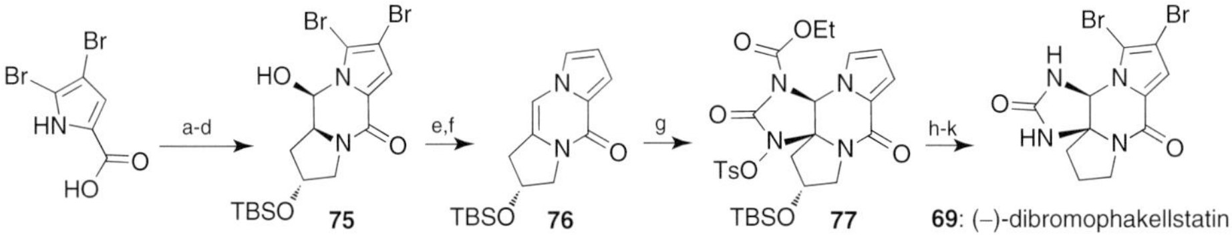

Scheme 7.16 Lindel's synthesis of (−)-dibromophakellstatin (**69**). *Reagents and conditions*: (a) $SOCl_2$, MeCN, reflux; (b) HO-Pro-OMe, Na_2CO_3, room temp. (rt); (c) TBSCl, imidazole, DMF, rt; (d) DIBAL-H, CH_2Cl_2, −78 °C, 49% overall yield; (e) H_2, Pd/C, NEt_3, MeOH, CH_2Cl_2, rt; (f) MsCl, DBU, 0 °C, 81% overall yield; (g) TsONHCOOEt, CaO, CH_2Cl_2, 15 °C, 50%; (h) $NEt_3 \cdot 3HF$, THF, rt, 97%; (i) CBr_4, PPh_3, rt, 73%; (j) SmI_2, THF–MeOH, rt, 77%; and (k) NBS, THF–MeCN, rt, 47%.

the stereochemistry of the annulation during the formation of tetracycle **77**. The reductive removal of the hydroxyl group was achieved via alkyl bromide formation and its reduction using SmI_2. Final dibromination of the pyrrole afforded the natural product (−)-dibromophakellstatin (**69**).

The remarkable synthesis of (+)-phakellin (**83**) by Romo is depicted here (Scheme 7.17). The concise phakellin annulation approach is reminiscent of the approach developed for palau'amine by the same group. The hemiaminal **78**, prepared from L-proline in three steps, was aminated into diastereoisomers **79** and **80** using diphenylphosphoryl azide followed by hydrogenolysis (Scheme 7.17). The diastereoisomer **79** was converted into the thermodynamically stable diastereoisomer **80** in basic medium. After the amidination of **80**, the following step consisted in the crucial oxidative cyclization of compound **81**, leading to the *N*-Tces phakellin (**82**) and performed using iodonium benzene-diacetate and magnesium oxide. Final deprotection yielded the (+)-phakellin (**83**). The mechanism proposed by Romo for the oxidative phakellin cyclization parallels the plausible biomimetic interconversion of the natural ugibohlin (**84**) and dibromoisophakellin (**85**) (box in Scheme 7.17) [41].

The enantiopure pyrrole-hydroxyproline-diketopiperazine **36** (Scheme 7.18) used in the synthesis of debromodispacamides (Section 7.3.1) was the starting material of the synthesis of (+)-phakellin hydrochloride (**90**) reported by Nagasawa and coworkers [37]. After selective oxidation and dehydration, the resultant α,β-unsaturated amide **86** was reduced under Luche conditions to yield allylic alcohol **87** after acetylation and desilylation of hydroxyl groups. The key step of the synthesis was the Overman trichloroacetimidate rearrangement [42], which provided the tertiary amide **88** with complete transfer of chirality on the quaternary carbon. The acetyl group in **88** was then replaced by a silyl group, through the thermodynamically favored isomerization at the carbinolamine position. *N,N*-Boc protected guanidine was then introduced to afford compound **89** after three additional steps. The final cyclization took place after silyl group removal, formation of a mesylate ester, followed by cyclization and *N*-Boc deprotection in acidic conditions. The synthesis of (+)-phakellin hydrochloride (**90**) was achieved in 18 steps. This synthesis provided the first example of the non-intuitive Overman [3,3] sigmatropic rearrangement applied to an enamide (**87**).

7.3.2.4 **Hymenialdisines (91)**

Hymenialdisines are another example of cyclized oroidin monomers isolated from numerous sponges and containing a pyrroloazepinone system [2i]. The two *(Z)*-**91** and *(E)*-**91** diastereoisomers of hymenialdisine interconvert in a pH- and concentration-dependent equilibrium, with *(Z)*-**91** being the most thermodynamically stable isomer.

(Z)-Hymenialdisine [*(Z)*-**91**] shows a nanomolar kinase inhibitory activity against a wide panel of kinases [43], making synthetic analogs with this type of scaffold very attractive from a medicinal chemistry point of view. Nguyen and Tepe recently reviewed the preparation of analogs and their evaluation as kinase inhibitors [44].

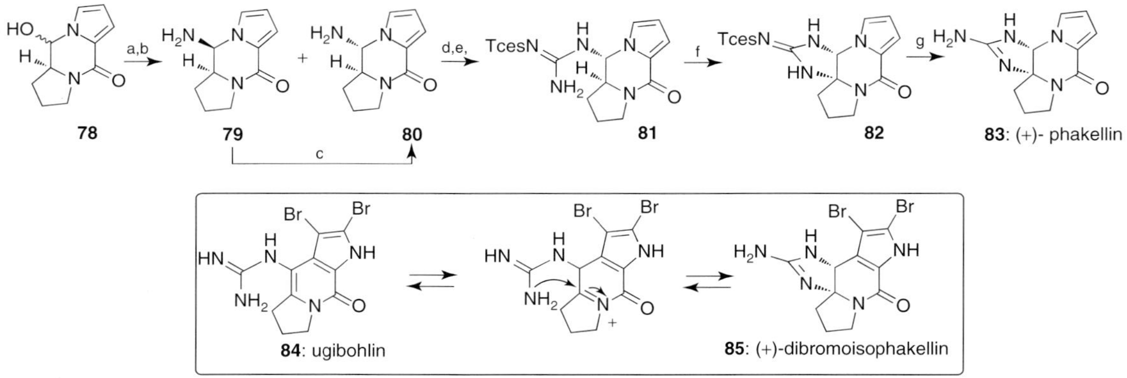

Scheme 7.17 Romo's enantioselective synthesis of phakellin (**83**) and biomimetic interconversion of dibromoisophakellin (**85**) into ugibohlin (**84**). *Reagents and conditions*: (a) diphenylphosphoryl azide, DBU, THF, rt, 63%; (b) H_2, Pd/C, MeOH, rt, 70% overall yield for the two epimers; (c) K_2CO_3, CH_3OH, 60 °C, 94%; (d) TcesN = C(Cl)SMe, $NEt_3CH_2Cl_2$, rt, 93%; (e) $HgCl_2$, HMDS, CH_3CN, rt, 77%; (f) $PhI(OAc)_2$, MgO, CH_2Cl_2, microwaves, 30–38%; and (g) Zn, AcOH, CH_3OH, 40 °C, 80%. DBU = 1,8-diazabicycloundec-7-ene, HMDS = 1,1,1,3,3,3-hexamethyldisilazane, and Tces = 2,2,2-trichloroethoxysulfonyl.

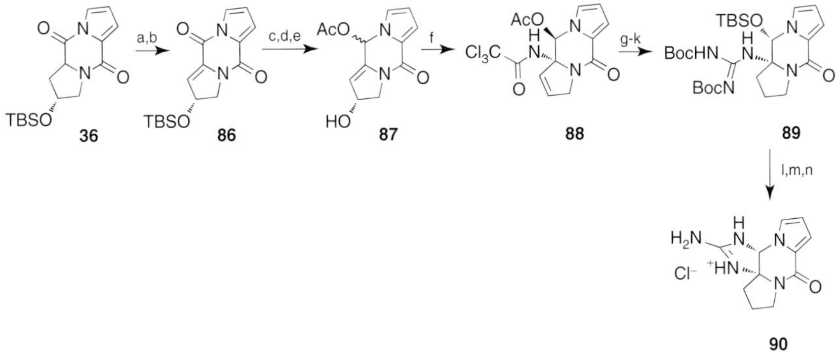

Scheme 7.18 Nagasawa's enantioselective synthesis of (+)-phakellin hydrochloride (**90**). *Reagents and conditions*: (a) IBX, trimethylamine *N*-oxide, DMSO, rt; (b) MsCl, NEt_3, CH_2Cl_2, rt, 52% overall yield; (c) $NaBH_4$, $CeCl_3$, EtOH/THF (1 : 1), 0 °C; (d) Ac_2O, pyridine, CH_2Cl_2, rt, 75% overall yield; (e) HF · NEt_3, THF, rt, 92%; (f) CCl_3CN, DBU, CH_2Cl_2, 0 °C to rt, 48%; (g) K_2CO_3, MeOH, rt, 96%; (h) TBSCl, NaH, THF, 0 °C, 82%; (i) DIBAL-H, toluene, −80 °C, 97%; (j) H_2, Raney Ni, EtOH, rt, 99%; (k) NBoc = C(SMe)NHBoc, AgOTf, NEt_3, MeCN, rt, 95%; (l) TBAF, THF, 0 °C, 99%; (m) MsCl, NEt_3, CH_2Cl_2, reflux, 91%; and (n) aq. HCl, MeOH, rt, 99%. DBU = 1,8-diazabicycloundec-7-ene.

Biosynthetically, hymenialdisines could arise from the related hymenin (**92**) by an oxidative process. Hymenin could be in turn formed from the corresponding tautomer III of oroidin (**3**), allowing for the C15–C6 bond formation [1]. This strategy has been applied by Horne and coworkers [45] in the synthesis of hydantoin analogs axinohydanthoins (**95**). The linear precursor **93** was obtained in four steps from ornithine methyl ester, using Büchi's and Horne's methodologies outlined in the synthesis of dibromophakellin (Scheme 7.19). The biomimetic key cyclization was performed using TFA to yield the pyrroloazepinone **94**. A final oxidation with three equivalents of bromine yielded axinohydantoins **95** as a 7/9 mixture of the *(E/Z)* isomers.

While the biomimetic approaches are inspired from biosynthetic considerations, some synthetic strategies for the synthesis of *(Z)*-hymenialdisine rely on its degradation into bromoaldisine **96** [46]. The construction of the pyrroloazepinone system in **91** is intuitive. The introduction of the 2-aminoimidazolone could be performed using a large variety of reactants [2i]. A similar synthetic plan was used by Horne [47] for the synthesis of hymenin (**92**) (Scheme 7.20). By finely tuned oxidations, hymenin was then converted into *(Z)*-hymenialdisine (**91**).

The pyrroloazepinone **98** was obtained from the alcohol **97** in two steps and the aminoimidazole part was then introduced in acidic medium to yield hymenin (**92**). Bromination of the aminoimidazole yielded **99**, from which **100** was obtained after hydrolysis by heating in acetic acid. An additional oxidation in acidic medium yielded the α,β-unsaturated 2-aminoimidazolone ring of hymenialdisine with partial protodebromination. The natural products *(Z)*-hymenialdisine (**91**) and *(Z)*-bromohymenialdisine (**101**) were isolated.

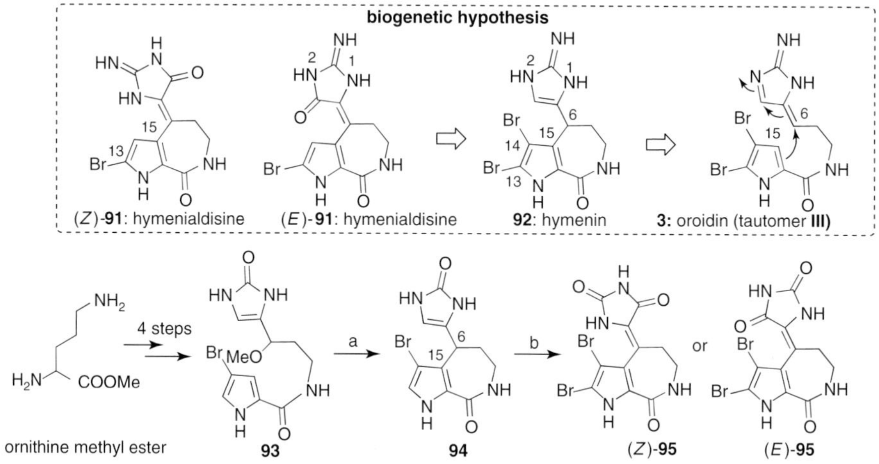

Scheme 7.19 Biogenetic hypothesis for hymenialdisines and Horne's synthesis of axinohydantoins. *Reagents and conditions*: (a) TFA, rt, 30% and (b) Br_2, 3 equiv, AcOH, AcONa, 80% for the two diastereoisomers.

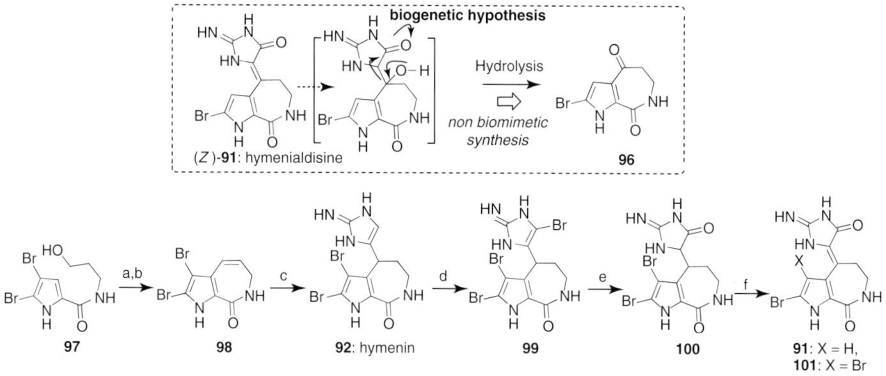

Scheme 7.20 Hydrolysis of hymenialdisine [(*Z*)-**91**] into bromoaldisine **96** could be foreseen in Horne's syntheses of hymenin (**92**) and (*Z*)-hymenialdisine (**91**). *Reagents and conditions*: (a) Swern oxidation; (b) CH_3SO_3H, rt, 80%; (c) 2-aminoimidazole, CH_3SO_3H, rt, 65%; (d) Br_2, TFA, rt, 95%; (e) AcOH, H_2O, reflux, 72%; and (f) CH_3SO_3H, HBr cat., 90 °C, sealed tube, 33% for **101** and 27% for **91**.

7.3.2.5 Agelastatins

Agelastatins A–F (**8**, **105–109**) belong to the class of tetracyclic P-2-AI monomers that were isolated from *Agelas dendromorpha* [48] and *Cymbastela* sp. [49]. Agelastatins E and F were isolated recently [50] in our group from *Agelas dendromorpha* as well. These natural products exhibit a densely functionalized tetracyclic core (A–D) with a central cyclopentane C ring bearing four contiguous stereocenters (Scheme 7.21). The reported derivatives resulted from the variation of the bromination degree in addition to the N1 methylation and C4/C5 hydroxylation/methoxylation.

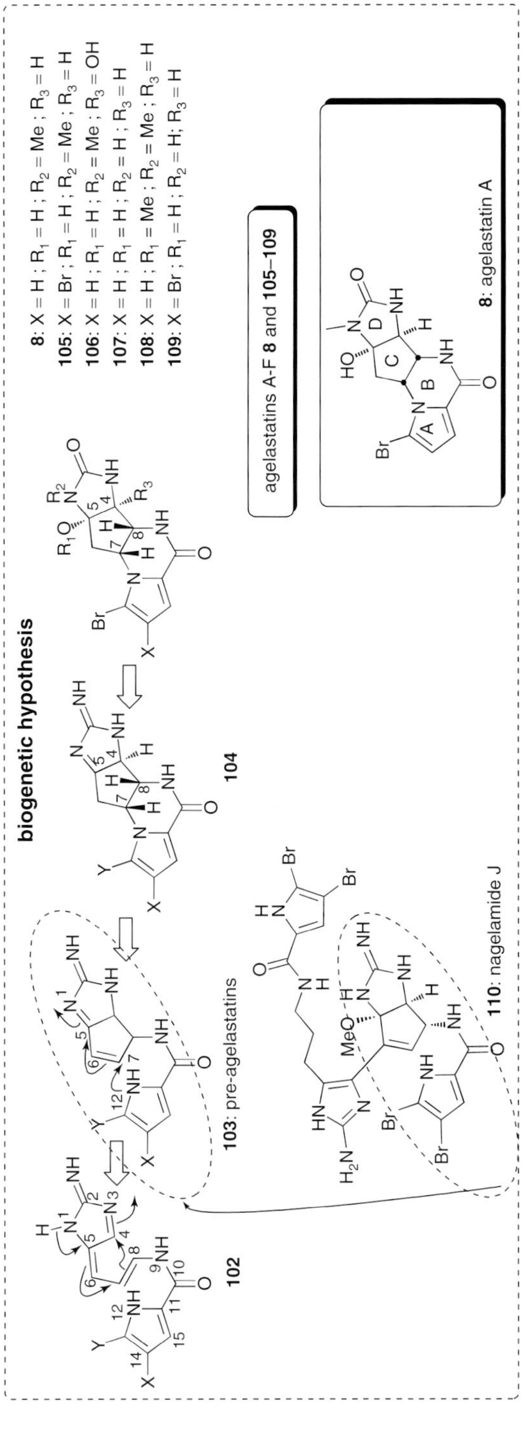

Scheme 7.21 Biogenetic hypothesis for agelastatins and related derivatives.

Agelastatin A (**8**) has been reported to be a powerful antitumoral agent [51]. These synthetically challenging bioactive molecules have been targeted by many groups in the recent past. The first synthesis of racemic agelastatin A has been reported by Weinreb and coworkers [52].

The biogenetic hypothesis formulated in our group is depicted in Scheme 7.21. Agelastatins could be formed starting from the precursor pre-agelastatin **103** by N12–C7 bond formation via a conjugated addition of the nitrogen from pyrrole on the conjugated imine N1–C5–C6–C7 leading to the tetracycle skeleton **104**. A pre-agelastatin motif is clearly shown by the dimeric natural product nagelamide J (**110**) recently isolated by Kobayashi and coworkers [53]. The pre-agelastatin could be readily formed from oxidized oroidin (**102**) by C4–C8 cyclopentane formation. Importantly, the bromination pattern (on C13) in agelastatin A is very seldom found in the P-2-AIs family and could provide a particular reactivity favoring the agelastatin pathway.

From a synthetic point of view, the stereocontrolled installation of four stereocenters on the cyclopentane remains the most challenging task. Non-biomimetic syntheses of the cyclopentane have been completed using various synthetic tools and strategies reviewed elsewhere [2d, i]: *N*-sulfinyl Diels–Alder cycloaddition (Weinreb), ring-closing metathesis from a linear precursor (Davis, Hale, Ichikawa, Chida), or functionalization of a cyclopentane mainly by allylic substitution (Trost, Tanaka, Wardrop) have been proposed. We focus now on the later stage of these syntheses.

Most agelastatin syntheses reported to date [54] made use, for the late stages, of a similar biogenetically related aza-Michael conjugate addition to form the crucial N12–C7 bond. The subsequent addition of methylurea to the remaining ketone led to ring D closure with hemiaminal formation at C5. This sequence is illustrated in Scheme 7.22 for Wardrop's synthesis of agelastatin A (**8**). Treatment of cyclopentene

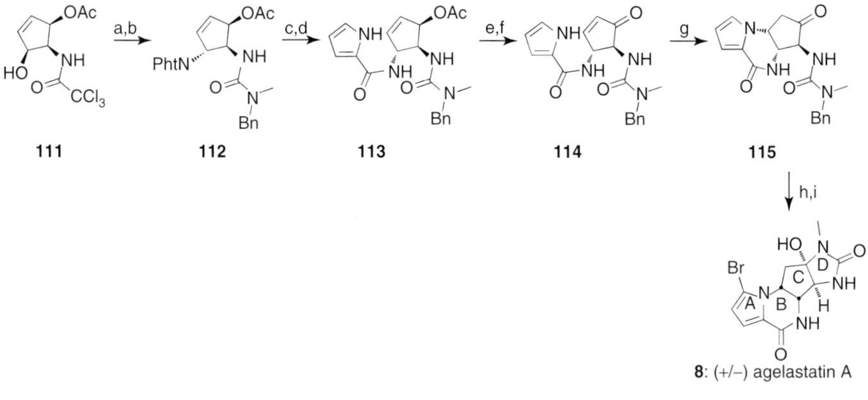

Scheme 7.22 Wardrop's completion of the synthesis of (±)-agelastatin A (**8**). *Reagents and conditions*: (a) DEAD, phthalamide, PPh_3, THF, 0 °C to rt; (b) Bn(Me)NH, $NaHCO_3$, DMF, 100 °C, 62% for two steps; (c) NH_2 NH_2, THF, rt; (d) 2-pyrrolecarboxylic acid, EDC, CH_2Cl_2, rt, 85% for two steps; (e) K_2CO_3, MeOH, CH_2Cl_2, rt; (f) IBX, DMSO, rt, 91% for two steps; (g) K_2CO_3, DMSO, 100 °C, 48%; and (h) H_2, $Pd(OH)_2$, THF, rt, 61%; (i) NBS, MeOH, THF, rt, 75%.

111 under Mitsunobu conditions with phthalamide followed by the trichloromethyl substitution with N-methyl-N-benzylamine provided the derivative **112** with N9 and the urea moiety in position. Phthalamide removal with hydrazine followed by acylation with 2-pyrrolecarboxylic acid led to compound **113**. After methanolysis of the acetate ester, oxidation of the resulting allylic alcohol with *o*-iodoxybenzoic acid (IBX) yielded the cyclopentenone **114**. The next biogenetically inspired step was the key N12–C7 bond formation by conjugate aza-Michael addition of the nitrogen from pyrrole onto the cyclopentenone ring, performed with potassium carbonate in DMSO. Hydrogenolysis of the N-benzyl group in ketone **115** using Pearlman's catalyst afforded spontaneously the annulation of the *N,N*′-disubstituted urea. Further regioselective bromination by *N*-bromosuccinimide yields racemic agelastatin A (**8**). Notably, Feldman proposed a similar synthesis in 2002 [55].

7.4
Biomimetic Synthesis of P-2-AIs Simple Dimers

7.4.1
Mauritiamine

Mauritiamine (**116**) isolated by Fusetani [56] from *Agelas mauritiana* is an oxidative dimerization product of oroidin. Fusetani and coworkers proposed an oxidation–dehydration process prior to the oroidin dimerization connecting C4–C5′. The formation of mauritiamine fits into the general chemical pathway involving tautomers I–V (Figure 7.1) suggested by Al-Mourabit and Potier [1]. The first total synthesis of mauritiamine reported by Horne seems in accordance with the biomimetic suggestions [23b]. The strategy involving heterodimerization of intermediates **117** and **118** was applied (Scheme 7.23).

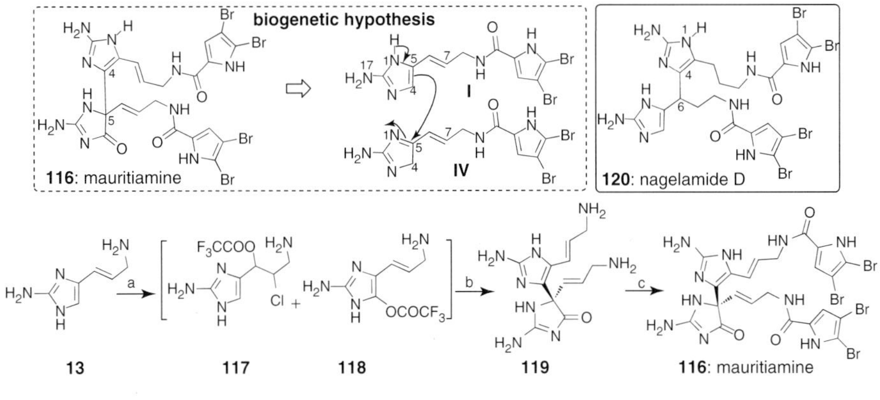

Scheme 7.23 Horne's synthesis of C4–C5 connected mauritiamine (**116**) and the structure of the C4–C6 connected nagelamide D (**120**). *Reagents and conditions*: (a) NCS, TFA, rt; (b) MeOH–xylene, 135 °C, 23% for two steps; and (c) 4,5-dibromo-2-(trichloroacetylpyrrole), DMF, rt, 65%.

The bis-imidazole core **119** of mauritiamine (**116**) was obtained in 23% yield from aminovinyl-2-aminoimidazole **13** and was acylated twice with 4,5-dibromo-2-(trichloroacetyl)pyrrole to obtain the natural product in 65% yield.

Although several natural P-2-AI dimers with one C–C connection were isolated [57], their synthesis does not seem experimentally obvious. It is worth mentioning here the total synthesis of the C4–C6 connected nagelamide D (**120**) reported by Lovely [58]. The authors indicated that the NMR of the synthetic compound did not match that of the natural product. The question of the existence of nagelamide D as a natural product is thus still open.

7.4.2
Sceptrins, Ageliferins, and Oxysceptrins

Sceptrin (**9**) [59] and ageliferin (**121**) are formally made up of two hymenidin (**2**) subunits. The biogenetic hypothesis formulating the formation of sceptrin from hymenidin (**2**) via [2 + 2] cycloaddition was proposed by Faulkner. The authors mentioned, though, some unfruitful [2 + 2] cycloaddition attempts from oroidin. A non-concerted and unified mechanism was later proposed by Rinehart for sceptrin (**9**) and ageliferin (**121**) [60] instead of independent [2 + 2] or [4 + 2] cycloaddition, respectively.

A general and common chemical pathway was put forward for sceptrin (**9**), ageliferin (**121**), and palau'amine (**11**) congeners in 2001 by Al-Mourabit and Potier [1]. The dual reactivity introduced above, involving tautomers I–V, was proposed to govern the corresponding nucleophilic/electrophilic behavior of the monomers leading to multifaceted intermediates (Scheme 7.24). The first step engaging tautomers II and IV would lead to the key connection between the nucleophilic C7 of II and the electrophilic analogous carbon C7 of IV. The multifaceted intermediate **122** displayed the appropriate structure for conversion into sceptrin (**9**) or ageliferin (**121**) through two modes of cyclizations.

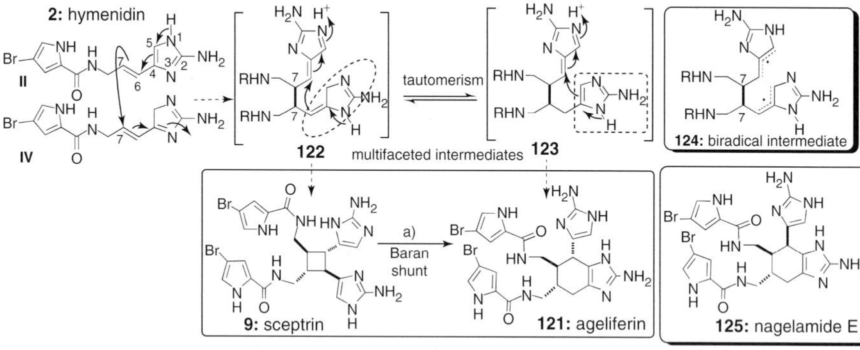

Scheme 7.24 Common biosynthetic pathway proposed for sceptrin/ageliferin and the alternative Baran's radical shunt. *Reagents and conditions*: (a) H_2O, 200 °C, microwave, 40%.

The analysis shows that the scheme is extremely dynamic and that the interconversions of sceptrin and ageliferin are conceivable. The conversion of sceptrin (**9**) into ageliferin (**121**) has been realized by Baran [5]. The rearrangement under microwave conditions was compatible with a radical or ionic mechanism. The authors proposed a radical shunt that transformed the vinyl cyclobutane of sceptrin into the cyclohexane ring of ageliferin. Biradical intermediate **124** [61] was put forward to explain the formation of nagelamide E (**125**) (epi-ageliferin). In contrast to the universal pathway starting from the same linear precursor via multifaceted intermediates, the [1,3]-sigmatropic rearrangement of sceptrin eludes the question of what is the real link between monomers like hymenidin (**2**) and dimers. In other words, is the first dimerization event dissociable from the ionic **122** or the radical **124** intermediates? The drastic conditions used by Baran to turn back from sceptrin to the putative intermediate **124** indicate a high activation barrier required for the reverse transformation. The direct access to the multifaceted intermediate **122** through the formation of the first C7–C7 bond indicates the likely central role of the first dimerization event for the subsequent transformations.

A series of oxysceptrins (**127**) could also derive from sceptrin by simple oxidation [60]. Exposure of sceptrin (**9**) to peracetic acid followed by an acidic treatment provided **127** [62], while the over oxidized natural product nakamuric acid (**128**) was obtained by oxidative cleavage of the dihydroxy intermediate **126** (Scheme 7.25).

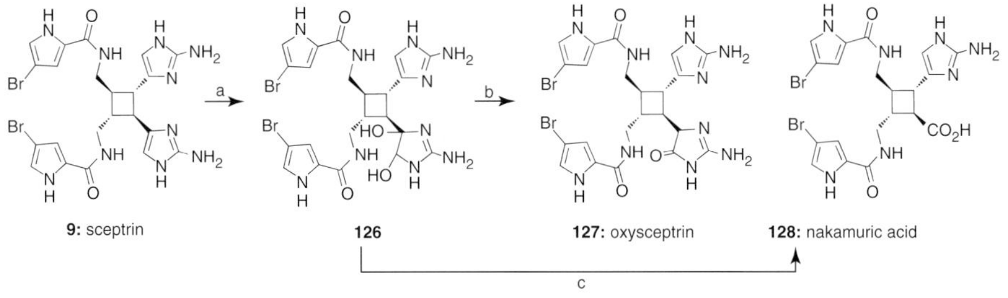

Scheme 7.25 Baran's synthesis of oxysceptrin (**127**) and nakamuric acid (**128**). *Reagents and conditions*: (a) AcO_3H, H_2O, 50%; (b) AcOH, 140 °C, 65%; and (c) $NaIO_4$, AcOH–NaOAc buffer, 60%.

7.5
Biomimetic Synthesis of Complex Dimers: Palau'amine and Related Congeners

Palau'amine (**11**) is the most attractive member of this family of compounds since it displays an extremely complex architecture with six fused heterocycles and eight contiguous stereocenters. Furthermore, this compound shows an impressive nanomolar immunosuppressive activity [63]. Its synthesis recently by the group of Baran [30], partially based on previous biosynthetic proposals, proved that biomechanistic speculations can provide valuable approaches not only for biosynthetic experiments but also for synthetic strategies.

7.5.1
Common Chemical Pathway for P-2-AI Biosynthesis

Unlike other classes of alkaloids, the P-2-AI alkaloids show an unprecedented use of multifaceted intermediates that provide a high degree of complexity and molecular diversity. Complex P-2-AI dimers like palau'amine (**11**) can be mapped back to two $C_{11}N_5$ clathrodin (**1**) subunits linked by several C–C and C–N bonds. The initial proposal for the structure (**11**' Figure 7.2) of palau'amine was made in 1993 by Scheuer and coworkers. Its structure was revised in 2007 by Köck [4a] and Quinn [64] by NMR spectral means. This revision deals with the relative configuration of C6 and C7′ stereocenters and made uniform the relative configuration around the cyclopentane core of complex P-2-AI dimers. The homogenization of the stereochemistry extended the list of arguments to a universal biochemical pathway for the entire family [1]. The correction into the highly strained *trans*-azabicyclo[3.3.0]octane subunit was not obvious, as *trans*-fused bicyclo[3.3.0]octanes are energetically disfavored.

The extreme chemical complexity of palau'amine (**11**) and congeners axinellamines A (**133**) and B (**134**) and massadine (**135**) makes biomimetic approaches very attractive to design a common synthetic plan for these related alkaloids and to minimize the number of steps. As shown below, the fully substituted cyclopentane core E of palau'amine (with C5 bearing the spiroaminoimidazole ring F) (Figure 7.3) is a common feature of this family of compounds. The A-B-C subunit corresponds to phakellin. Thus, the reactivity generating the phakellin subunit should proceed along the same lines. From an oroidin-like precursor, the key bonds between the two oroidin subunits are C7–C7′ and C5–C6′.

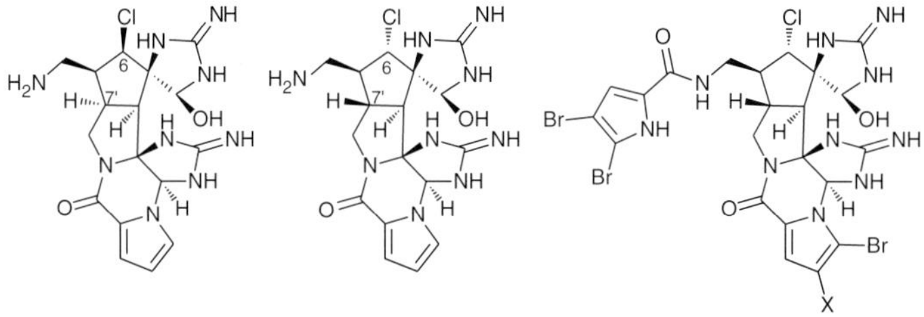

11′: palau'amine initial structure   **11**: palau'amine revised structure   **129**: X = H konbu'acidin A; **130**: X = Br konbu'acidin B

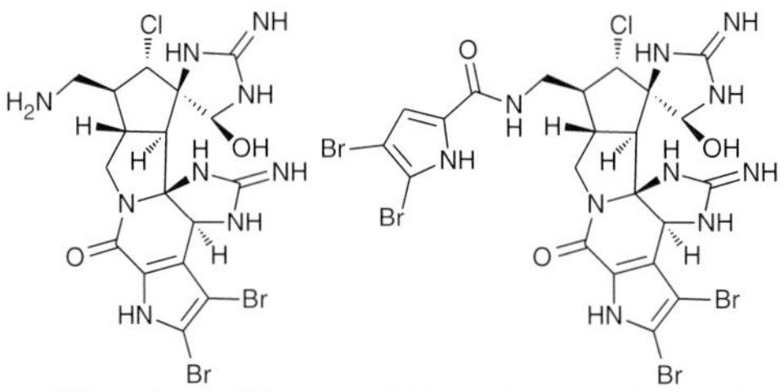

131: styloguanidine   **132**: tetrabromostyloguanidine

Figure 7.2 Palau'amine: previous (**11**′) and revised (**11**) structures and congeners (**129**–**132**).

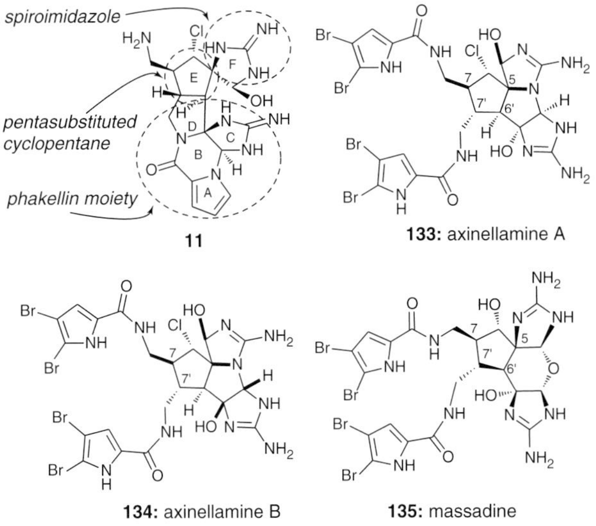

Figure 7.3 Structure of palau'amine (**11**) and related dimers.

Interestingly, the C7–C7′ bond is conserved in all complex dimers. We will now discuss different biogenetic hypotheses proposed for palau'amine and congeners.

7.5.2
First Proposal Based on a Diels–Alder Key Step

Scheuer and Kinnel's biosynthetic proposal for palau'amine (Scheme 7.26) was based on the previous structure of palau'amine (**11**): the key step proposed by these authors was a Diels–Alder cycloaddition with aminovinylimidazole (**13**) as dienophile and a dehydrophakellin of type **137** as diene [65]. The cycloadduct (**138**) with H6′/H7′ in cis relation would be obtained. An additional oxidative ring contraction mediated by chlorohydroxylation could then lead to palau'amine (**11**). This hypothesis was weakened by the structural revision since in the revised structure H6′/H7′ are trans and ring contractions usually proceed with retention of configuration. The isolation of konbu'acidins (**129**) as precursors of palau'amine has suggested another route with clathrodin (**1**) as the dienophile. Palau'amine would then be obtained through amide hydrolysis of the konbu'acidin like compound **142**. Although the dehydrophakellin **137** was not isolated from sponges, the isolation of the phakellin oxidized product **136** was reported quite recently [66].

7.5.3
Universal Chemical Pathway

A universal chemical pathway for the formation of more complex, dimeric, polycyclic P-2-AI members known up to 2000, involving the dual reactivity of the 2-aminoimidazole was proposed by Al-Mourabit and Potier [1]. The presence of

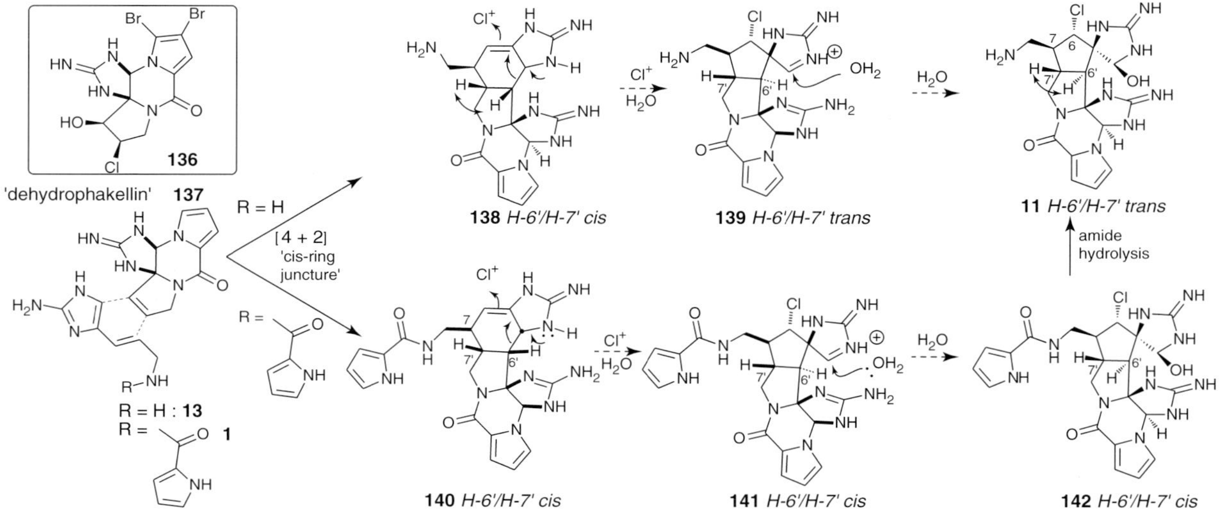

Scheme 7.26 Scheuer and Kinnel's biogenetic proposal for palau'amine.

similar structural motifs and functionalities in several P-2-AIs regardless of their monomeric or dimeric nature suggests common biosynthetic transformations leading to their formation. Variations of Al-Mourabit's postulated biosynthetic intermediates were further deepened by Baran and Köck [2c]. The C7–C7′ multifaceted intermediate A **143** could follow many plausible modes of cyclization, including the formation of the second multifaceted intermediate B **145** (Scheme 7.27). The putative common precursor **145** towards several higher order P-2-AIs would provide the observed trans relative stereochemistry in the [3.3.0] bicyclic core common to all higher order P-2-AIs, including palau'amine (**11**), axinellamines (**148**), massadine chloride (**149**), and styloguanidine (**150**).

Interestingly, the macrophakellin (**151**) and macroisophakellin (**152**) subunits were found in the palau'amines and the regioisomeric styloguanidines that could arise from macrocyclic fleeting intermediates "macropalau'amine" **146** and "macrostyloguanidine" **147**, respectively (see also Scheme 7.33 below). The macrocyclic intermediates could then collapse in a transannular fashion, generating the strained *trans*-azabicyclo[3.3.0]octane core of these alkaloids. Variations of Al-Mourabit's postulated biosynthetic intermediate **145** were exploited by Baran for the synthesis of axinellamine A and B [67], massadine [68], and palau'amine [30].

7.5.4
Intramolecular Aziridinium Mediated Mechanism for the Formation of Massadine (141) from Massadine Chloride (155)

Displacement of the chlorine by a hydroxyl group at C6 through an aziridinium intermediate was investigated by Romo and coworkers [69]. The aziridinium mechanism is an interesting way to explain the conversion of massadine chloride (**149**) into massadine (**135**) with retention of configuration (Scheme 7.28). During synthetic studies toward palau'amine and congeners, the Romo group indeed reported an unexpected conversion of cyclopentyl chloride **154** into its derivative **155** in the presence of azide, which surprisingly proceeded with retention of configuration (Scheme 7.28). This group concluded that a mechanism involving neighboring group participation of the spiroaminoimidazole might be in operation through the aziridinium species **153**. Furthermore, it was proposed that a similar process could account for the biosynthesis of massadine **135** from an earlier chlorinated precursor (**149**).

7.5.5
Aziridinium Mechanism for the Formation of the Tetramer Stylissadine A

In 2007 Baran and Köck [2c] provided support for the pathway proposed by Romo, by isolating the unprecedented massadine chloride (**149**), and reported its rapid conversion into massadine in an aqueous medium at 60 °C. This behavior was thought to be due to the favorable geometrical disposition between the aminoimidazole nitrogen and the chlorine atom. Such a mechanism was proposed for the dimerization of massadine chloride (**149**) into stylissadine A (**2**) (Scheme 7.29).

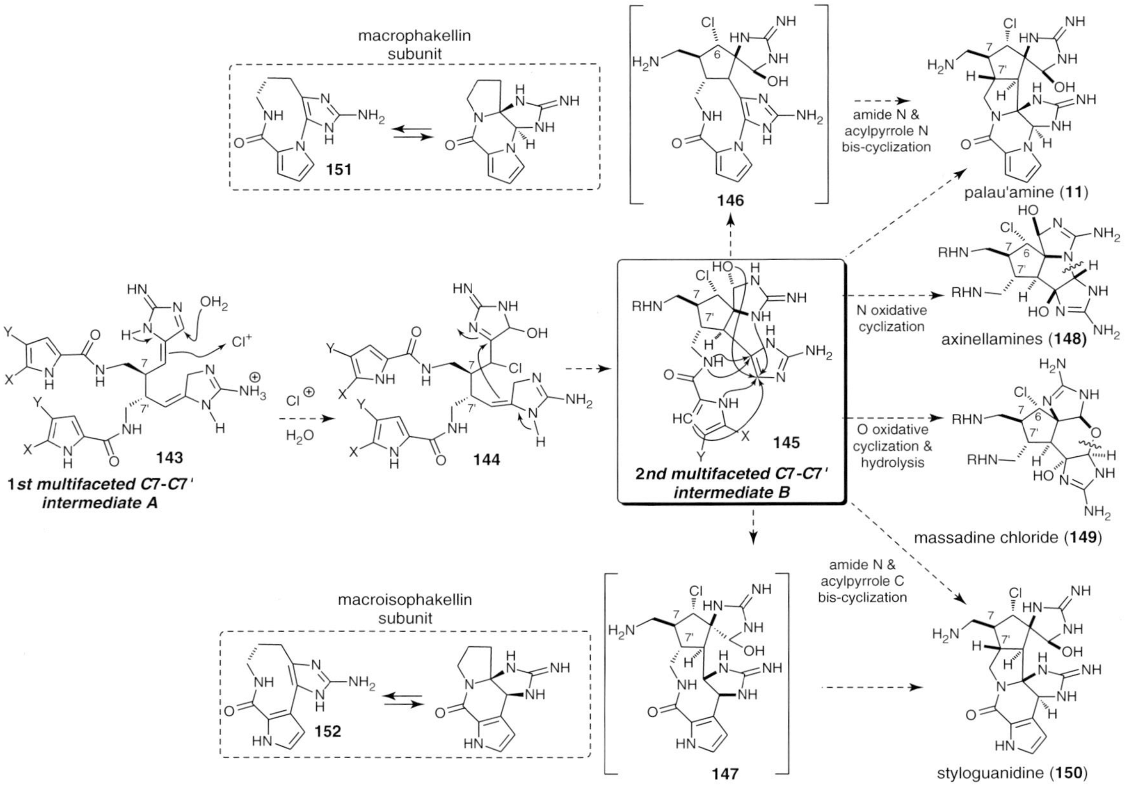

Scheme 7.27 Al-Mourabit's unified biogenetic proposal for palau'amine and congeners [1] and further developments of Baran and Köck [2c].

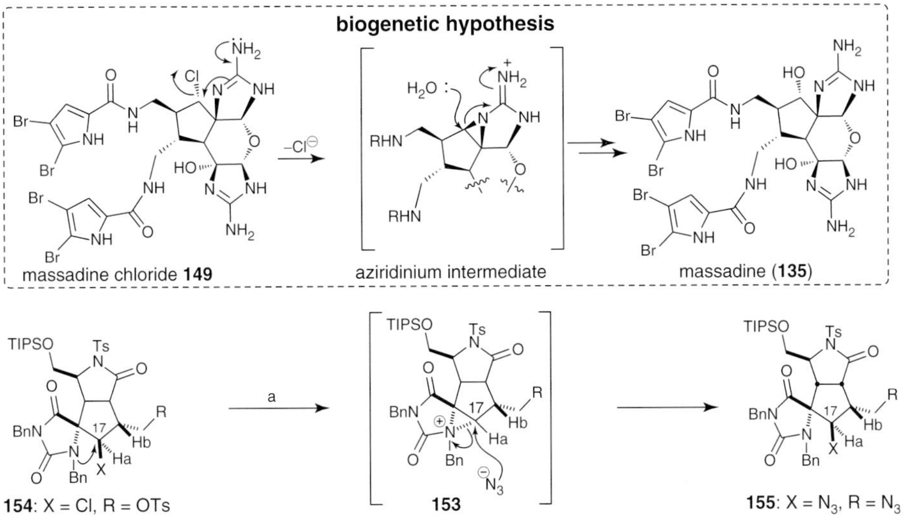

Scheme 7.28 Biosynthesis proposal for massadine: Romo's aziridinium intermediate suggested for the displacement of chloride with retention of configuration. *Reagents and conditions*: (a) NaN_3, DMF, 105–120 °C.

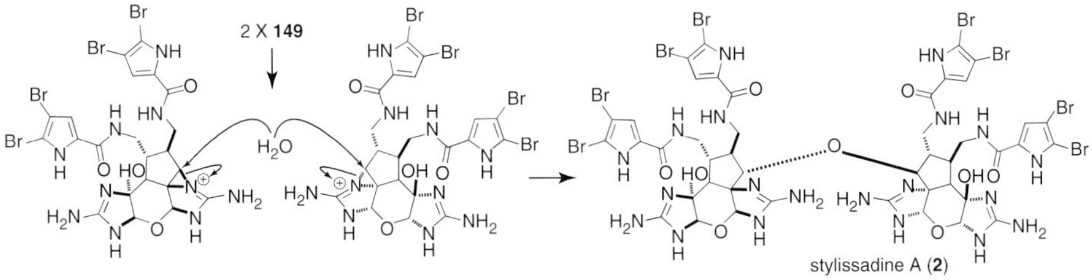

Scheme 7.29 Baran and Köck's proposed formation of stylissadine A from massadine.

7.5.6 Synthetic Achievements

Inspired by the reactivities showed in these biogenetic proposals, several groups have proposed synthetic approaches for complex P-2AI dimers. Obviously, the most accomplished synthetic work that took advantage from these hypotheses was reported by Baran and coworkers. Using very effective preparative organic chemistry, including new and chemoselective reaction discoveries, this group completed the first total synthesis of complex dimers using late biomimetic steps [30, 68]. Scheme 7.30 depicts the synthesis of the common precursor.

The Diels–Alder cycloadduct **159** was converted into the diazide **160** in six steps; the azide groups played the role of the masked amines needed for the introduction of pyrrolecarboxamides. Five additional steps, involving a key intramolecular aldol

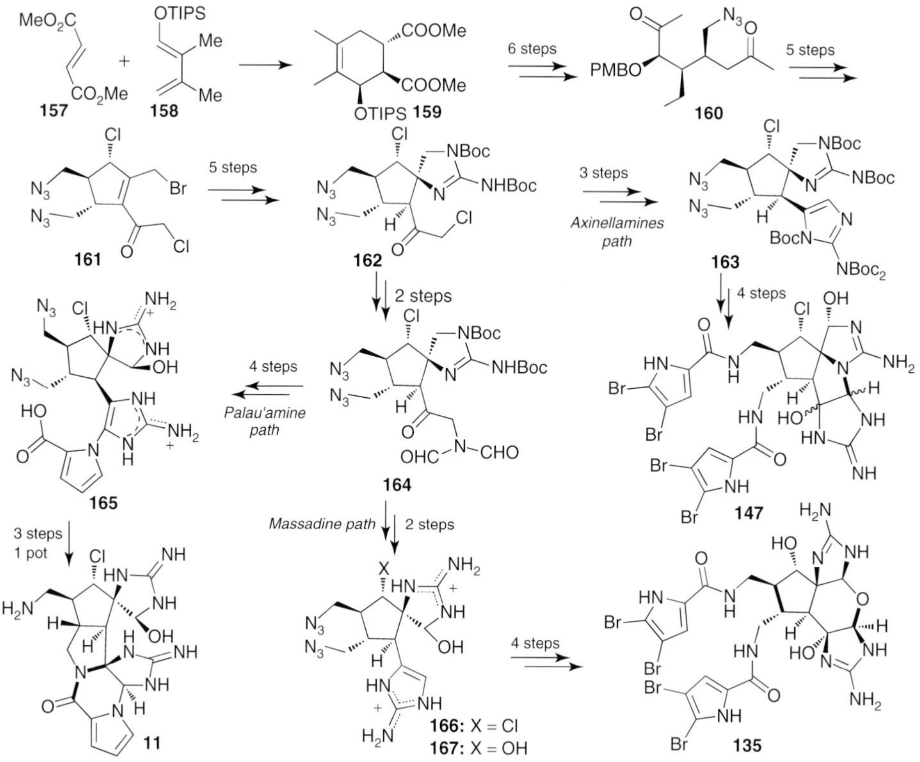

Scheme 7.30 Baran's syntheses of precursors for axinellamines A/B, massadine (**135**) and palau'amine (**11**).

reaction, were needed to build the cyclopentene **161**, which was functionalized to obtain the Boc-protected spiroimidazole **162** in five steps via a key stereoselective oxidation by IBX. Compound **162** shows the fully substituted cyclopentane core of palau'amine (**11**) and congeners with the right relative configurations of the five stereocenters. This compound – obtained in multi-gram quantity via scalable procedures at a relative early stage of the synthesis – was used as a common intermediate toward the synthesis of axinellamines, massadine, and palau'amine. In the axinellamines path, the introduction of a second aminoimidazole yielded bis-aminoimidazole **163** in three steps while in the palau'amine and massadine paths the bis-formamide **164** was first prepared. The biomimetic reactions of these syntheses are located in the last steps, which are detailed in the next sections.

7.5.6.1 **Axinellamines A/B**

The crucial step in the synthesis of axinellamines was inspired from biomechanistic analysis (Scheme 7.31). *N*-Boc removal starting from **163** yielded bis-aminoimidazole **166**. The oxidation of the C4′=C5′ imidazolic double bond by DMDO led to the N1–C4′ linkage and to the introduction of the hydroxyl group on

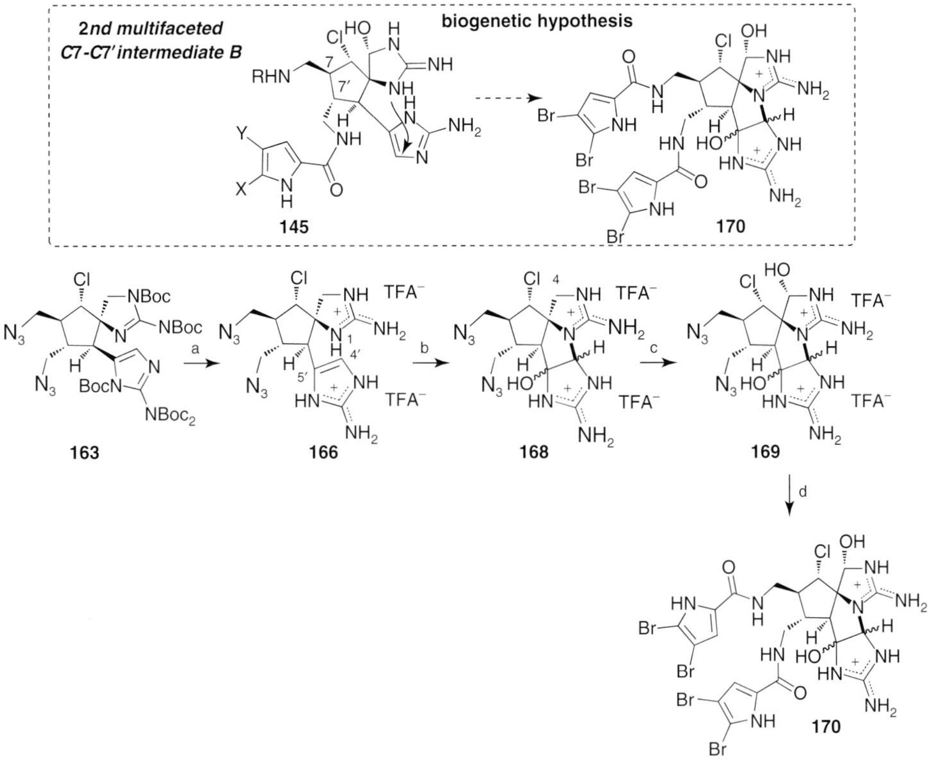

Scheme 7.31 Biogenetic hypothesis and Baran's final biomimetic steps for the synthesis of axinellamines (**170**). *Reagents and conditions*: (a) TFA–CH_2Cl_2 (2 : 1), 100%; (b) DMDO, H_2O then TFA-CH_2Cl_2(2 : 1); (c) silver(II) picolinate, H_2O; 40% for the two diastereoisomers over two steps; and (d) 1,3-propanedithiol, triethylamine, MeOH then 4,5-dibromo-2-trichloroacetylpyrrole, diisopropylamine, DMF, 69%.

C5′ of **168**. Two diastereoisomers with cis fusion were obtained. The highly selective silver(II) picolinate-mediated oxidation at the C4 position of the unprotected spirocyclic core (**168**) yielded the hemiaminal **169**. Azide reduction in presence of 1,3-propanedithiol followed by acylation with 4,5-dibromo-2-trichloroacetylpyrrole yielded the two diastereoisomers axinellamines A and B (**170**) as TFA salts.

7.5.6.2 **Massadine Chloride (149) and Massadine (135)**

The total syntheses of massadine chloride (**149**) and massadine (**135**) were achieved by the Baran's group using the same methodology as for the constitutional isomeric axinellamines. The bis-formamide **164** (Scheme 7.32) was hydroxylated on C4 by silver(II) picolinate in the presence of TFA to deliver deformylated ammonium **171**. The installation of the second 2-aminoimidazole ring was made by the Lancini's cyanamide methodology [20]. This proceeded with partial chlorine displacement by a hydroxyl group on C6 with retention of configuration, as

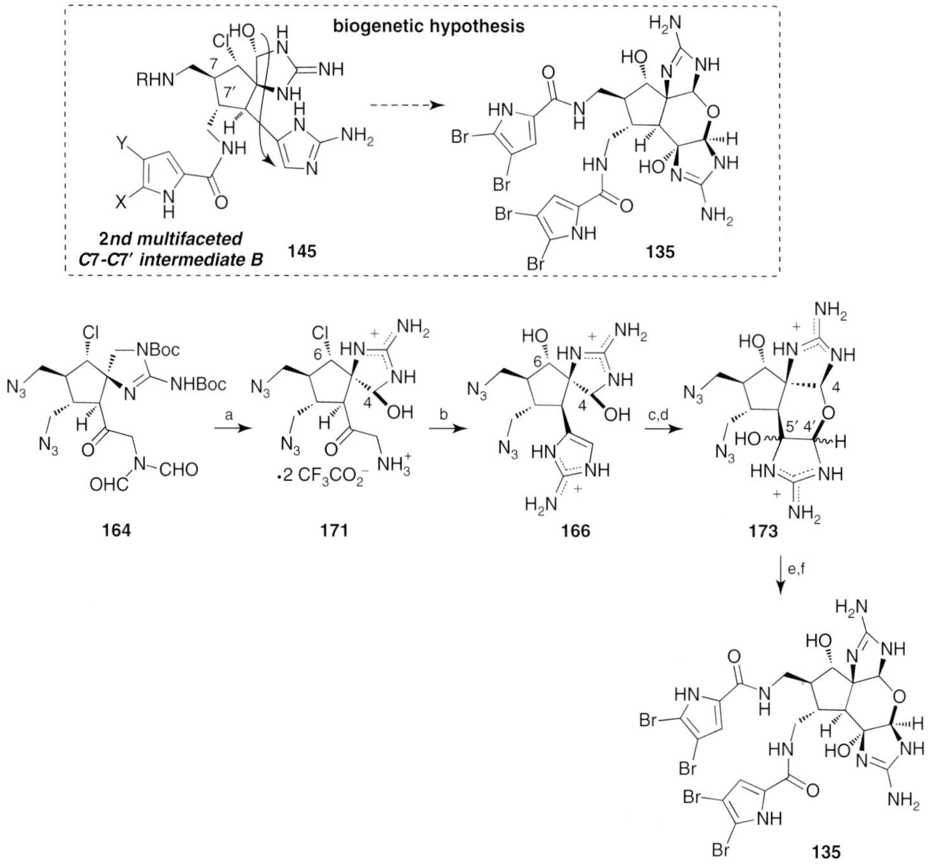

Scheme 7.32 Biogenetic hypothesis and Baran's final biomimetic steps for the synthesis of massadine (**135**). *Reagents and conditions*: (a) silver(II) picolinate, H_2O–TFA then TFA, 84%; (b) cyanamide, NaOH to pH 5, 24%; (c) DMDO, H_2O–TFA; (d) TFA, 65% over two steps; (e) PtO_2, H_2; and (f) 4,5-dibromo-2-trichloroacetylpyrrole, diisopropylamine, DMF, 40% over two steps.

proposed by Romo [69]. The diol **166** was then transformed into compound **173** with the ether bond C4–O–C4′ of massadine through oxidation of the C4′=C5′ double bond by DMDO followed by addition of TFA. Control of the pH was required to favor the regioselective cyclization versus the axinellamine cyclization mode. Surprisingly, the configuration of the major diastereomer **173** at C4′–C5′ corresponds to that of epi-massadine. The authors have postulated that epi-massadine should exist in nature, as for axinellamines A and B isomers. Starting from **173**, final azide hydrogenations and acylation of the resulting amines with 4,5-dibromo-2-trichloroacetylpyrrole yielded massadine (**135**). The synthesis of massadine chloride was completed starting from the C6 chloro-analog of **166** by the same procedures.

7.5.6.3 **Palau'amine (11)**

Based on the latter strategy, Baran and coworkers reported the first total synthesis of palau'amine (**11**) in 2010 [30]. Scheme 7.33 describes the final biomimetic steps. The strained *trans*-fused 5,5-ring system was obtained from the macrocyclic intermediate **146**, following the transannular attack that was put forward by Al-Mourabit and Potier for the formation of the phakellin subunit. The strategy provided evidence for the chemical pathway in which the "phakellin" substructure was reached through a transannular cyclization of a "macropalau'amine" precursor **146**. After the elegant preparation of the intermediate **175** from **164**, the ambivalent reactivity of the 2-aminoimidazole part allowed the access to palau'amine (**11**).

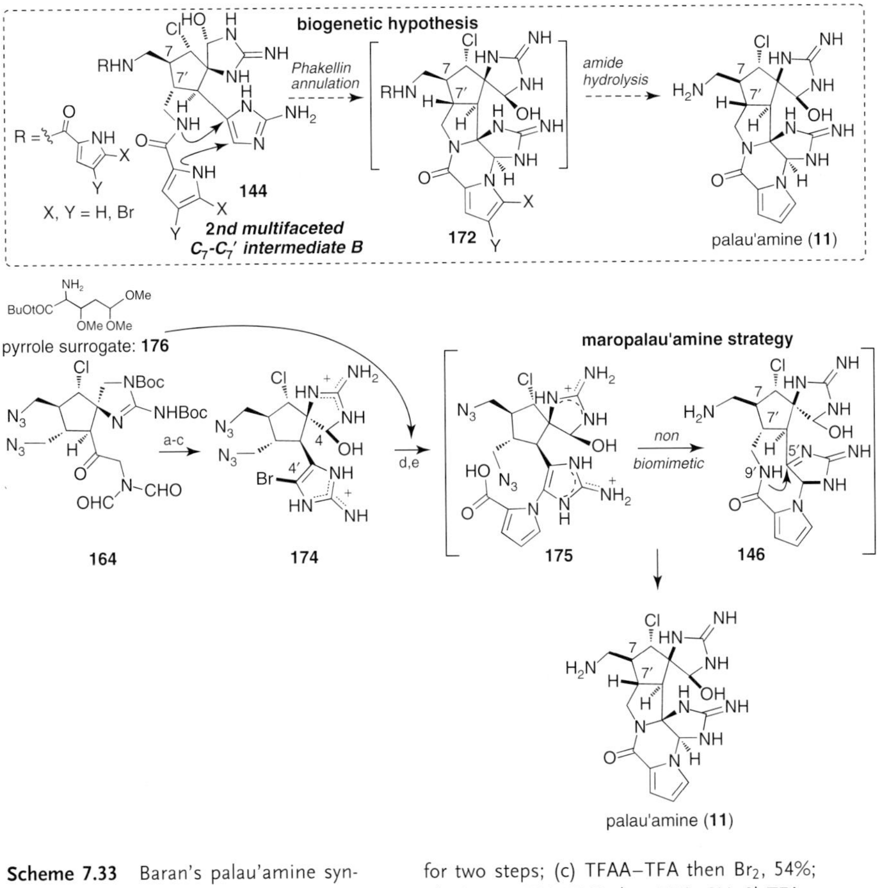

Scheme 7.33 Baran's palau'amine synthesis with the final likely biomimetic macropalau'amine formation. *Reagents and conditions*: (a) H_2O–TFA then silver(II) picolinate; (b) cyanamide, brine, pH 5, 64% for two steps; (c) TFAA–TFA then Br_2, 54%; (d) AcOH, **176**, THF then TFA-CH_2Cl_2TFA, 44%; and (e) $Pt(OAc)_2$, H_2, TFA–H_2O then EDC–HOBt–DMF then TFA, 17% one pot.

7.6
New Challenging P-2-AI Synthetic Targets and Perspectives

Many efforts were made to clarify the chemical pathways toward complex P-2-AI metabolites, but the factors controlling the fleeting reactivity of P-2-AI alkaloids are still rather obscure. Several interesting and challenging molecules are currently under chemical and biological studies.

Recent isolation of novel P-2-AI dimeric members, benzosceptrin A (**10**) [3] and stylissazole C (**177**) [70], demonstrates the incredible potential of P-2-AI metabolites to generate original molecules and rare organic architectures. While benzosceptrin A shows a highly strained benzocyclobutane, stylissazole C is one of the first example of dimerization involving exclusively N–C bond formations (Scheme 7.34). These members add another dimension to the molecular diversity of P-2-AI metabolites and further highlight the unique dual reactivity of the vinylogous 2-aminoimidazole clathrodin (**1**), hymenidin (**2**), and oroidin (**3**) precursors.

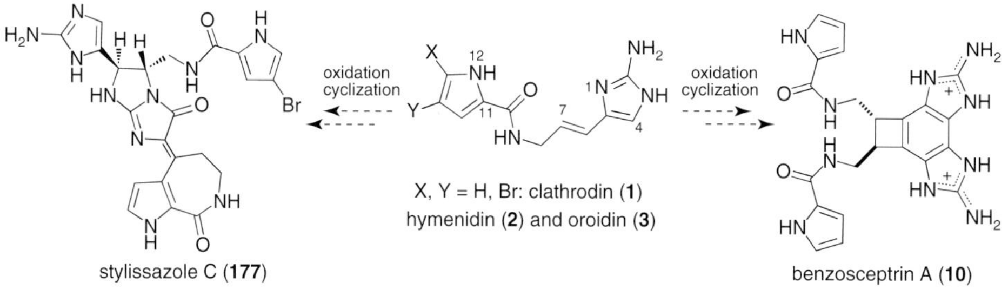

Scheme 7.34 New challenging structures: benzosceptrin A and stylissazole C.

Biomimetic synthesis of P-2-AI metabolites by various groups led to better understanding of the reactivity and the potential biosynthetic early precursors. The lack of biosynthetic studies by labeled precursors underpins the crucial role of research using the combination of isolation and synthesis to construct the chemical pathway puzzle. The isolation of new intermediates helps to explore the ability of the corresponding phylogenetically related sponges to create new molecules. The utility of biosynthetic speculations inspired from the reactivity of the molecules for understanding both synthetic and living systems seems obvious. The molecules identified in living matter result from complex processes that became understandable through the discovery of the impressive molecular diversity and reasoned methods of organic chemistry. In future, essential insights into the actual P-2-AI biosynthetic machinery are expected to give decisive information on the enzymes involved. The generation of the presumed universal precursors like clathrodin (**1**) and their conversion into higher order P-2-AIs like palau'amine (**11**) or benzosceptrin (**10**) seems very subtle. The dynamic hydrogen-bonding interaction induced by the duality of the 2-aminoimidazole is probably an interesting biochemical catalytic system still to be discovered. Synthetic plans inspired by biosynthetic hypothesis need straightforward and highly chemoselective reactions. In several cases of natural product families, the chemical correlations/transformations following biomechanistic inspiration have demonstrated their high value for effective biomimetic and total synthesis.

References

1. Al-Mourabit, A. and Potier, P. (2001) *Eur. J. Org. Chem.*, 237–243.
2. (a) Hoffmann, H. and Lindel, T. (2003) *Synthesis*, 1753–1783; (b) Jacquot, D.E.N. and Lindel, T. (2005) *Curr. Org. Chem.*, **9**, 1551–1565; (c) Köck, M., Grube, A., Seiple, I.B., and Baran, P.S. (2007) *Angew. Chem. Int. Ed.*, **46**, 6586–6594; (d) Weinreb, S.M. (2007) *Nat. Prod. Rep.*, **24**, 931–948; (e) Arndt, H.D. and Riedrich, M. (2008) *Angew. Chem. Int. Ed.*, **47**, 4785–4788; (f) Aiello, A., Fattorusso, E., Menna, M., and Taglialatela–Scafati, O. (2008) in *Modern Alkaloids: Structure, Isolation, Synthesis and Biology* (eds E. Fatorusso and O. Taglialatela–Scafati), Wiley-VCH Verlag GmbH, Weinheim, pp. 271–304; (g) Feldman, K.S. and Fodor, M.D. (2009) *Synthesis*, 3162–3173; (h) Heasley, B. (2009) *Eur. J. Org. Chem.*, 1477–1489; (i) Forte, B., Malgesini, B., Piutti, C., Quartieri, F., Scolaro, A., and Papeo, G. (2009) *Mar. Drugs*, **7**, 705–753.
3. See Appenzeller, J., Tilvi, S., El Bitar, H., Martin, M.T., Gallard, J.F., Tran Huu Dau, E., Debitus, C., Laurent, D., Moriou, C., and Al-Mourabit, A. (2009) *Org. Lett.*, **11**, 4874–4877, and references therein.
4. (a) Grube, A. and Köck, M. (2007) *Angew. Chem. Int. Ed.*, **46**, 2320–2324; (b) Kobayashi, H., Kitamura, K., Nagai, K., Nakao, Y., Fusetani, N., van Soest, R.W.M., and Matsunaga, S. (2007) *Tetrahedron*, **48**, 2127–2129; (c) Buchanan, M.S., Carroll, A.R., and Quinn, R.J. (2007) *Tetrahedron Lett.*, **48**, 4573–4574.
5. Baran, P.S., Zografos, A.L., and O'Malley, D.P. (2004) *Angew. Chem. Int. Ed.*, **43**, 2674–2677.
6. Wei, Y. and Zipse, H. (2008) *Eur. J. Org. Chem.*, **47**, 3811–3816, and references cited therein.
7. Kitagawa, I., Kobayashi, M., Kitanaka, K., Kido, K., and Kyogoku, Y. (1983) *Chem. Pharm. Bull.*, **31**, 2321–2328.
8. Braeckman, J.-C., Daloze, D., Stoller, C., and van Soest, R.W.M. (1992) *Biochem. Syst. Ecol.*, **20**, 417–431.
9. Linington, R.G., Williams, D.E., Tahir, A., van Soest, R., and Andersen, R.J. (2003) *Org. Lett.*, **5**, 2735–2738.
10. Andrade, P., Willoughby, R., Pomponi, S.A., and Kerr, R.G. (1999) *Tetrahedron Lett.*, **40**, 4775–4778.
11. Wang, Y.G., Morinaka, B.I., Reyes, J.C.P., Wolff, J.J., and Molinski, T.F. (2010) *J. Nat. Prod.*, **73**, 428–735.
12. Lindel, T., Hochgürtel, M., Assmann, M., and Köck, M. (2000) *J. Nat. Prod.*, **63**, 1566–1569.
13. Assmann, M., Lichte, E., van Soest, R.W.M., and Köck, M. (1999) *Org. Lett.*, **1**, 455–457.
14. Vergne, C., Boury-Esnault, N., Perez, T., Martin, M.-T., Adeline, M.-T., Tran Huu Dau, E., and Al-Mourabit, A. (2006) *Org. Lett.*, **8**, 2421–2424.
15. Travert, N. and Al-Mourabit, A. (2004) *J. Am. Chem. Soc.*, **126**, 10252–10253.
16. Vergne, C., Appenzeller, J., Ratinaud, C., Martin, M.T., Debitus, C., Zaparucha, A., and Al-Mourabit, A. (2008) *Org. Lett.*, **10**, 493–496.
17. Braun, M., Büchi, G., and Bushey, D.F. (1978) *J. Am. Chem. Soc.*, **100**, 4208–4213.
18. Foley, L.H. and Büchi, G.H. (1981) *J. Am. Soc. Chem.*, **104**, 1776–1777.
19. (a) Sharma, G.M. and Burholder, P.R. (1971) *J. Chem. Soc., Chem. Commun.*, 151–152; (b) Sharma, G.M. and Magdoff-Fairshild, B. (1977) *J. Org. Chem.*, **42**, 4118–4124.
20. (a) Lancini, G.C. and Lazzari, E. (1966) *J. Antibiot.*, **3**, 152–154; (b) Lancini, G.C., Lazzari, E., Arioli, V., and Bellani, P. (1969) *J. Med. Chem.*, **12**, 775–780.

21. Akabori, S. (1933) *Chem. Ber.*, **66**, 151–158.
22. Nakadai, M. and Harran, P.G. (2006) *Tetrahedron Lett.*, **47**, 3933–3935.
23. (a) Wiese, K.J., Yakushijin, K., and Horne, D.A. (2002) *Tetrahedron Lett.*, **43**, 5135–5136; (b) Olofson, A., Yakushijin, K., and Horne, D.A. (1997) *J. Org. Chem.*, **62**, 7918–7919.
24. Shroif-Grégoire, C., Travert, N., Zaparucha, A., and Al Mourabit, A. (2006) *Org. Lett.*, **8**, 2961–2964.
25. Al-Mourabit, A., Pusset, M., Chtourou, M., Gaigne, C., Ahond, A., Poupat, C., and Potier, P. (1997) *J. Nat. Prod.*, **60**, 290–291.
26. Fedoreyev, S.A., Ilyin, S.G., Utkina, N.K., Maximov, O.B., Reshetnyak, M.V., Antipin, M.Y., and Struchkov, Y.T. (1989) *Tetrahedron*, **45**, 3487–3492.
27. Abou-Jneid, R., Ghoulami, S., Martin, M.-T., Tran Huu Dau, E., Travert, N., and Al-Mourabit, A. (2004) *Org. Lett.*, **6**, 3933–3936.
28. Schroif-Grégoire, C., Travert, N., Zaparucha, A., and Al-Mourabit, A. (2006) *Org. Lett.*, **8**, 2961–2964.
29. Pöverlein, C., Breckle, G., and Lindel, T. (2006) *Org. Lett.*, **8**, 819–821.
30. Seiple, I.B., Su, S., Young, I.S., Lewis, C.A., Yamaguchi, J., and Baran, P.S. (2010) *Angew. Chem. Int. Ed.*, **49**, 1095–1098.
31. Feldman, K.S. and Fodor, M. (2008) *J. Am. Chem. Soc.*, **130**, 14964–14965.
32. Picon, S., Tran Huu Dau, E., Martin, M.-T., Retailleau, P., Zaparucha, A., and Al Mourabit, A. (2009) *Org. Lett.*, **11**, 2525–2528.
33. Zöllinger, M., Mayer, P., and Lindel, T. (2006) *J. Org. Chem.*, **71**, 9431–9439.
34. Wiese, K.J., Yakhushijin, K., and Horne, D.A. (2002) *Tetrahedron Lett.*, **43**, 5135–5136.
35. Feldman, K.S., Skoumbourdis, A.P., and Fodor, M.D. (2007) *J. Org. Chem.*, **72**, 8076–8086.
36. Wang, S. and Romo, D. (2008) *Angew. Chem. Int. Ed.*, **46**, 1304–1306.
37. Imaoka, T., Iwamoto, O., Noguchi, K., and Nagasawa, K. (2009) *Angew. Chem. Int. Ed.*, **47**, 3799–3801.
38. Poullennec, K.G. and Romo, D. (2003) *J. Am. Chem. Soc.*, **125**, 6344–6345.
39. Zoellinger, M., Mayer, P., and Lindel, T. (2007) *Synlett*, 2756–2758.
40. Jacquot, D.E.N., Zoellinger, M., and Lindel, T. (2005) *Angew. Chem. Int. Ed.*, **44**, 2295–2298.
41. Travert, N., Martin, M.-T., Bourguet-Kondracki, M.-L., and Al-Mourabit, A. (2005) *Tetrahedron Lett.*, **46**, 249–252.
42. Overman, L.E. (1980) *Acc. Chem. Res.*, **13**, 218–224.
43. Meijer, L., Thunnissen, A.M., White, A.W., Garnier, M., Nikolic, M., Tsai, L.H., Walter, J., Cleverley, K.E., Salinas, P.C., Wu, Y.Z., Biernat, J., Mandelkow, E.M., Kim, S.H., and Pettit, G.R. (2000) *Chem. Biol.*, **7**, 51–63.
44. Nguyen, T.N.T. and Tepe, J.J. (2009) *Curr. Med. Chem.*, **16**, 3122–3143.
45. Barrios Sosa, A.C., Yaskushijin, K., and Horne, D.A. (2002) *J. Org. Chem.*, **67**, 4498–4500.
46. Schmitz, F.J., Gunasekara, S.P., Lakshmi, V., and Tillekeratne, L.M.V. (1985) *J. Nat. Prod.*, **48**, 47–53.
47. Xu, Y., Yakushijin, K., and Horne, D.A. (1997) *J. Org. Chem.*, **62**, 456–464.
48. D'ambrosio, M., Guerriero, A., Debitus, C., Ribes, O., Pusset, J., Leroy, S., and Pietra, F. (1993) *J. Chem. Soc., Chem. Commun.*, 1305–1306.
49. Hong, T.W., Jimenez, D.R., and Molinski, T.F. (1998) *J. Nat. Prod.*, **61**, 158–161.
50. Tilvi, S., Moriou, C., Martin, M.-T., Gallard, J.-F., Sorres, J., Patel, K., Petek, S., Debitus, C., Ermolenko, L., and Al-Mourabit, A. (2010) *J. Nat. Prod.*, **73**, 720–721.
51. Hale, K.J., Domostoj, M.M., El-Tanani, M., Campbell, F.C., and Mason, C.K. (2005) in *Strategies and Tactics in Organic Synthesis*(ed. M. Harmata), Academic Press, London.
52. (a) Anderson, G.T., Chase, C.E., Koh, Y.-H., Stien, D., and Weinreb, S.M. (1998) *J. Org. Chem.*, **63**, 7594–7595; (b) Weinreb, S.M. (1999) *J. Am. Chem. Soc.*, **121**, 9574–9579.
53. Araki, A., Tsuda, M., Kubota, T., Mikami, Y., Fromont, J., and Kobayashi, J. (2007) *Org. Lett.*, **9**, 2369–2371.
54. Dickson, D.P. and Wardrop, D.J. (2009) *Org. Lett.*, **11**, 1341–1344.

55. (a) Feldman, K.S. and Saunders, J.C. (2002) *J. Am. Chem. Soc.*, **124**, 9060–9061; (b) Feldman, K.S., Saunders, J.C., and Wrobleski, M.L. (2002) *J. Org. Chem.*, **67**, 7096–7109.

56. Tsukamoto, S., Kato, H., Hirota, H., and Fusetani, N. (1996) *J. Nat. Prod.*, **59**, 501–503.

57. Endo, T., Tsuda, M., Okada, T., Mitsuhashi, S., Shima, H., Kikuchi, K., Mikami, Y., Fromont, J., and Kobayashi, J. (2004) *J. Nat. Prod.*, **67**, 1262–1267.

58. Bhandari, M.R., Sivappa, R., and Lovely, C.J. (2009) *Org. Lett.*, **11**, 1535–1538.

59. Walker, R.P., Faulkner, D.J., van Engen, D., and Clardy, J. (1981) *J. Am. Chem. Soc.*, **103**, 6772–6773.

60. Keifer, P.A., Schwartz, R.E., Koker, M.E.S., Hugues, R.G. Jr., Rittschof, D., and Rinehart, K.L. (1991) *J. Org. Chem.*, **56**, 2965–2975.

61. Northrop, B.H., O'Malley, D.P., Zographos, A.L., Baran, P.S., and Houk, K.N. (2006) *Angew. Chem. Int. Ed.*, **45**, 4126–4130.

62. O'Malley, D.P., Li, K., Maue, M., Zographos, A.L., and Baran, P.S. (2007) *J. Am. Chem. Soc.*, **129**, 4762–4775.

63. Kinnel, R.B., Gehrken, H.P., and Scheurer, P.J. (1993) *J. Am. Chem. Soc.*, **115**, 3376–3377.

64. Buchanan, M.S., Carroll, R., and Quinn, R.J. (2007) *Tetrahedron Lett.*, **48**, 4573–4574.

65. (a) Kinnel, R.B., Gehrken, H.P. and Scheuer, P.J. (1993) *J. Am. Chem. Soc.*, **115**, 3376–3377; (b) Kinnel, R.B., Gehrken, H.-P., Swali, R., Skoropowski, G., and Scheuer, P.J. (1998) *J. Org. Chem.*, **63**, 3281–3286.

66. Tsukamoto, S., Tane, K., Ohta, T., Matsunaga, S., Fusetani, N., and van Soest, R.W.M. (2001) *J. Nat. Prod.*, **64**, 1576–1578.

67. O'Malley, D.P., Yamaguchi, J., Young, I.S., Seiple, I.B., and Baran, P.S. (2008) *Angew. Chem. Int. Ed.*, **47**, 3581–3583.

68. Su, S., Seiple, I.B., Young, I.S., and Baran, P.S. (2008) *J. Am. Chem. Soc.*, **130**, 16490–16491.

69. Wang, S., Dilley, A.S., Poullennec, K.G., and Romo, D. (2006) *Tetrahedron*, **62**, 7155–7161.

70. Patel, K., Laville, R., Martin, M.-T., Tilvi, S., Moriou, C., Gallard, J.-F., Ermolenko, L., Debitus, C., and Al-Mourabit, A. (2010) *Angew. Chem. Int. Ed.*, **48**, 4775–4779.

8
Biomimetic Syntheses of Alkaloids with a Non-Amino Acid Origin

Edmond Gravel

8.1
Introduction

As discussed in other chapters, a vast majority of the alkaloids known to date are biosynthetically derived from amino acids such as tryptophan, tyrosine, lysine, and many others. Nevertheless, a certain number of alkaloids stand out as exceptions and find their origin in polyketide or terpene precursors instead. The present chapter is intended to be a survey of some relevant biomimetic synthetic efforts toward these particular natural products. Alkaloids of polyketide origin will be treated first and will constitute the major part of this contribution, following which terpenoid alkaloids will be discussed.

8.2
Galbulimima Alkaloids

In the late 1950s, Ritchie *et al.* discovered a new family of alkaloids [1] that were isolated throughout the following decade from the bark of relic trees native to Papua New Guinea and northern Australia belonging to the *Galbulimima* (also known as *Himantandra*) genus (Himantandraceae) [2]. These natural products have been divided into three classes [3] based on structural features [4] (Figure 8.1) and the number of isolated molecules in this family is still increasing [5]. While all three classes share the same *trans*-decalin ring system, the lactone derivatives of Class I seem rather simple compared to complex alkaloids of Class II or Class III.

In the late 1960s, the Ritchie group had already pointed out a possible biogenetic precursor common to all alkaloids of the family [2d] without disclosing much detail about the plausible required transformations. A more detailed biosynthetic hypothesis for alkaloids of Class I was later proposed by the Baldwin group [6], followed by a unified biosynthetic route to alkaloids of Class II and Class III suggested by Movassaghi *et al.* [7].

Biomimetic Organic Synthesis, First Edition. Edited by Erwan Poupon and Bastien Nay.

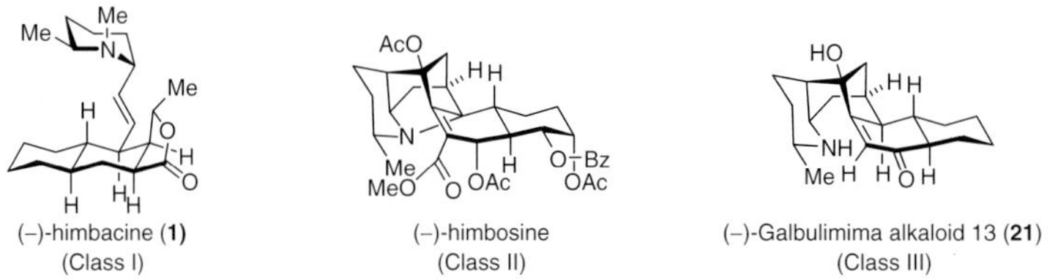

Figure 8.1 Representative structures of the different classes of *Galbulimima* alkaloids.

8.2.1 Alkaloids of Class I

Himbacine (**1**), belonging to Class I, was the first alkaloid of the family to be fully described [2a] and was shown to be a selective muscarinic antagonist, hence potentially opening the way to a new class of drug molecules for the treatment of Alzheimer's disease [8]. Since then, **1** has motivated many synthetic efforts [9], among which Baldwin *et al.* proposed the first biomimetic total synthesis [6a] and later extended their approach to other alkaloids of Class I such as himbeline (**2**) and himandravine (**3**) [6b].

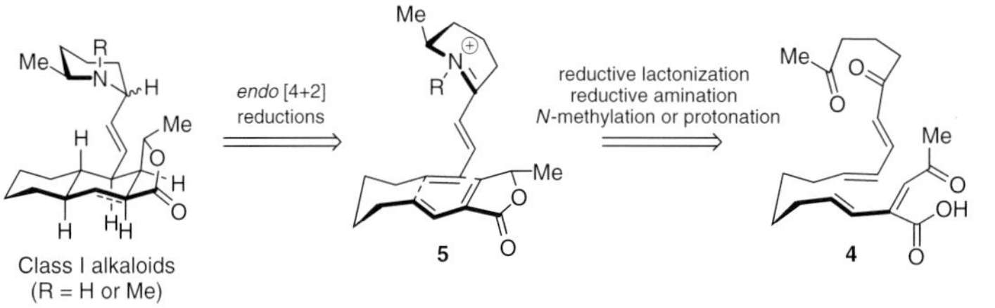

Scheme 8.1 Baldwin's biosynthetic hypothesis for the origin of Class I *Galbulimima* alkaloids.

Their strategy relied on their own biosynthetic hypothesis (summarized in Scheme 8.1), according to which polyketide derivative **4** would first undergo reductive lactonization followed by reductive amination and *N*-methylation or protonation to give rise to intermediate **5**. The tetrahydropyridinium ring of the latter would serve as activator for a biological Diels–Alder reaction via an *endo* transition state that would lead, after hydride reduction of the iminium ion to form a *cis*- or *trans*-piperidine ring, to the general scaffold of Class I alkaloids. As a synthetic precursor of postulated intermediate **5**, the authors decided to prepare dienone **6**. They predicted that the latter, upon removal of the Boc group, would cyclize into iminium **5′** which would then spontaneously undergo the cycloaddition process (Scheme 8.2) [6a].

Key biomimetic intermediate **6** was assembled from chiral phosphonate **7** (obtained in 9% yield over three simple steps) and aldehyde **8** (obtained in 29% yield over seven steps) by a Horner–Emmons reaction that proceeded in 50%

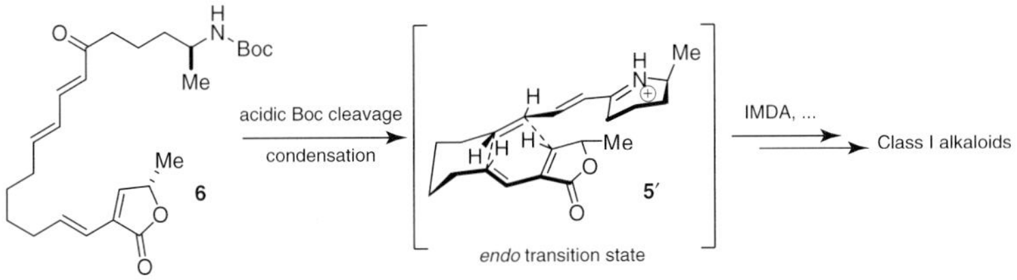

Scheme 8.2 Conversion of dienone **6** into intermediate **5′**.

yield. The Boc group was cleaved by treatment of **6** with trifluoroacetic acid at 0 °C (very likely leading to the formation of **5′** after condensation); the mixture was then allowed to slowly warm up to room temperature to enable the cycloaddition process, and the reaction was finally quenched by addition of sodium cyanoborohydride followed by aqueous sodium bicarbonate. This key biomimetic sequence yielded **9** as a 1 : 1 mixture of isomers (epimers at C6) that could not be separated. Boc protection was thus performed on the crude reaction mixture followed by selective reduction of the trisubstituted double bond with Adam's catalyst and, after purification by flash chromatography, compound **10** and compound **10′** were obtained in 11% yield each from **6**. TFA treatment to remove the Boc groups of **10** and **10′** permitted the formation of himbeline (**2**) and himandravine (**3**), respectively, in high yields. *N*-methylation of **2** finally afforded himbacine (**1**) in 74% yield (Scheme 8.3) [6b].

Interestingly, by their approach, Baldwin *et al.* have accessed three of the four Class I *Galbulimima* alkaloids. The fourth Class I alkaloid, himgravine (**11**), was not synthesized but might have been obtained by *N*-methylation of the isomer of **9** with C6 in the (*R*)-configuration, had the latter been separable from its epimer.

8.2.2
Alkaloids of Class II and Class III

Different research groups have studied the synthesis of Class II [10, 11] and Class III [7, 12] alkaloids and the group of Movassaghi has proposed a unified biosynthetic pathway for these molecules (Scheme 8.4). They postulated that polyketide precursor **12** (analog of **4**, oxidized at C16) can undergo reductive amination followed by condensation, giving rise to tetrahydropyridinium **13**, which will then cyclize via an intramolecular Diels–Alder (IMDA) cycloaddition. On the obtained cycloadduct **14**, the ketone at C20 is free to react (contrary to the Class I pathway in which it is engaged in the lactone cycle) in its enol form by conjugate addition to permit the formation of the C21–C8 bond and yield intermediate **15**. After tautomerization of the resulting imine into the corresponding endocyclic enamine, attack of the latter on the C20 ketone would lead to the formation of the C5–C20 bond, and upon reduction of the residual imine give rise to intermediate **16**. Enone **17**, resulting from the oxidation of **16**, can be seen as the

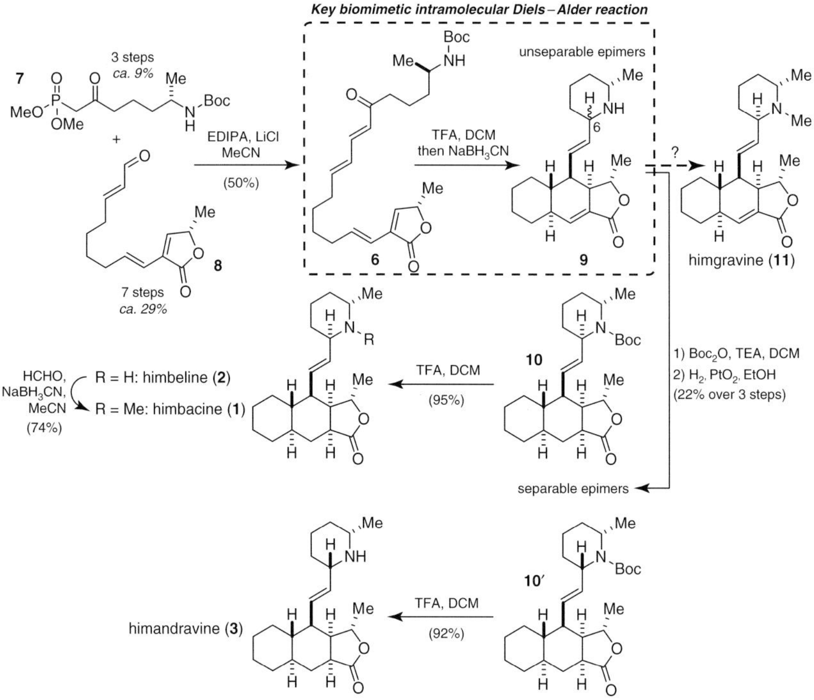

Scheme 8.3 Baldwin's biomimetic approach to Class I *Galbulimima* alkaloids [6b].

pivotal intermediate that will lead either to **18** (precursor of Class III alkaloids) by conjugate addition (forming the N–C19 bond) and decarboxylation or to **19** and then **20** (precursor of Class II alkaloids) by oxidation [7].

Inspired by this putative biosynthetic pathway, the authors achieved the total synthesis of both alkaloids of Class II (himandrine) [10] and Class III (Galbulimima alkaloid 13 or GB 13) [7]. During these investigations, they obtained exciting results that support their hypothesis.

In 2006, with the synthesis of GB 13 (**21**) [7], they demonstrated the introduction of the C5–C20 bond from synthetic intermediate **22** (obtained in 15% yield over eleven steps). Deprotection of the silyl enol ether of **22** gave transient compound **23** (biomimetic equivalent of **15**), which readily cyclized by formation of the C5–C20 bond and upon treatment with sodium borohydride afforded **24** in circa 70% yield[1] (Scheme 8.5). In this one-pot process, three contiguous stereocenters were formed with a high level of diastereoselectivity, and target alkaloid **21** was obtained after three additional synthetic steps. This synthesis also provided the first evidence that revision of the absolute stereochemistry of Class II/III alkaloids was needed.

1) The yield here is given based on (−)-**22** even though the isomers were separated at a later stage in the synthesis.

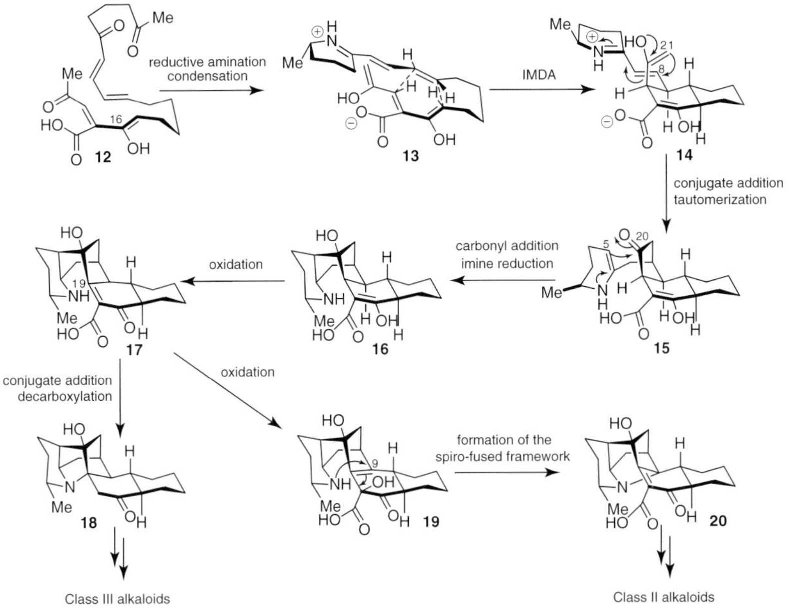

Scheme 8.4 Movassaghi's biosynthetic hypothesis for Class II and Class III alkaloids.

More recently [10], while completing the first total synthesis of a Class II alkaloid, the authors demonstrated that the formation of the N–C9 bond, leading to the spirofused polycyclic framework, could result from the oxidative spirocyclization of intermediate **25** (synthetic equivalent of **17**, obtained in circa 1% yield over 20 steps). The latter was treated with *N*-chlorosuccinimide to supposedly afford α-chloroester **26** (biomimetic equivalent of **19**) and permit the formation of **27**, by intramolecular allylic displacement, in 89% yield (Scheme 8.5). Only two further synthetic steps (reduction of the enone into enol and benzoylation) were required to complete the first synthesis of himandrine (**28**).

8.3 Cyclic Imine Marine Alkaloids

In the past two decades, Uemura *et al.* have been implicated in many studies regarding the chemistry of marine alkaloids. In the course of their work, it occurred to them that several of these natural products bore a common structural feature: a cyclic imine either spiro- or side-fused to a cyclohexene. Most of the alkaloids concerned were suspected to originate from dinoflagellates and the authors proposed that a general scenario, based on an IMDA reaction involving an iminium-activated dienophile (*vide infra*), might account for their biosynthesis.

Members of this family of natural compounds include symbioimine, pinnatoxins, pteriatoxins, and gymnodimines that will be discussed in the following sections,

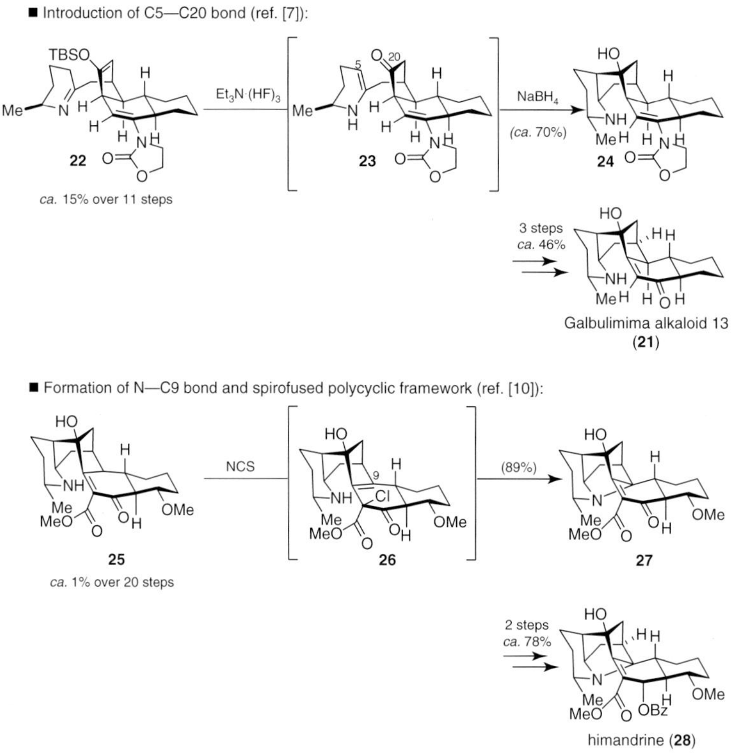

Scheme 8.5 Biomimetic processes in Movassaghi's synthetic efforts toward Class II and Class III alkaloids.

but also spirolides [13], prorocentrolides [14], and spiroprorocentrimine [15], which are presented in Figure 8.2 but will not be detailed any further.

8.3.1
Symbioimine and Neosymbioimine

In 2004 the research group of Uemura reported the isolation of symbioimine (**29**) from *Symbiodinium* sp. (Symbiodiniaceae), a dinoflagellate cultivated from its symbiotic partner, the marine flatworm *Amphiscolops* sp. (Convolutidae) [16]. The molecule is composed of an uncommon 6,6,6-tricyclic iminium ring system and an aryl sulfate moiety (Figure 8.3).

In terms of biological activity, this singular tetracyclic iminium sulfate alkaloid inhibits the differentiation of RAW264 cells into osteoclasts ($EC_{50} = 44\ \mu g\ ml^{-1}$) and can thus be considered a potential candidate for the prevention and treatment of osteoporosis. Symbioimine also shows a significant anti-inflammatory potential as it inhibits cyclooxygenase-2 activity (32%) at 10 μM without affecting cyclooxygenase-1 activity. The structure and biological activity of this new natural

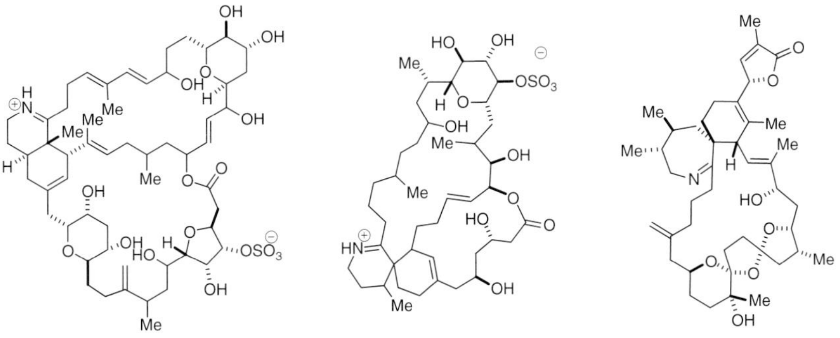

prorocentrolide B (*Prorocentrum lima*)
spiro-prorocentrimine (*Prorocentrum* sp.)
spirolide C (*Alexandrium ostenfeldii*)

Figure 8.2 Examples of marine alkaloids featuring a cyclic imine fused to a cyclohexene.

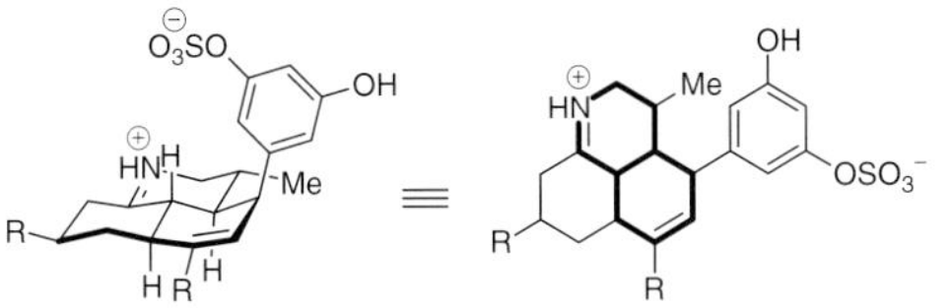

R = H: symbioimine (**29**)
R = Me: neosymbioimine (**30**)
cyclic iminium side-fused to cyclohexene

Figure 8.3 Structure of symbioimine and neosymbioimine.

product have stimulated several efforts toward its synthesis [17]. One year after the discovery of symbioimine, Uemura *et al.* reported the isolation of neosymbioimine (**30**), a compound closely related to the former [18].

In their first paper, the authors proposed that the biosynthesis of symbioimine involved the cyclization of polyketide precursor **31** by an IMDA cycloaddition proceeding via an *exo* transition state followed by condensation to form the iminium. This proposal appeared quite unusual considering that most literature precedents concerning IMDA on such substrates were in favor of an *endo* transition state [19]. An alternative biosynthetic pathway was proposed along with the neosymbioimine isolation report and suggested that the IMDA occurred through an *endo* transition state after formation of dihydropyridinium precursor **32**. The obtained cycloadduct **33** could then undergo epimerization at C4 to yield symbioimine (Scheme 8.6).

This latter hypothesis was supported, in the following years, by studies toward the biomimetic synthesis of symbioimine (Scheme 8.7) from the groups of Snider [17c,d] and Thomson [17e].

Snider *et al.* started with the synthesis of biomimetic intermediate **34** from simple starting material in nine synthetic steps with 55% overall yield. Key compound **34** was obtained as a mixture of isomers with a *cis*/*trans* ratio of 3 : 1 at C2–C3 and an (*E*/*Z*) ratio of 6 : 1 for the C9=C10 olefin. Upon treatment with $BF_3 \cdot Et_2O$, tetrahydropyridinol **34** was converted into dihydropyridinium **35**, which directly

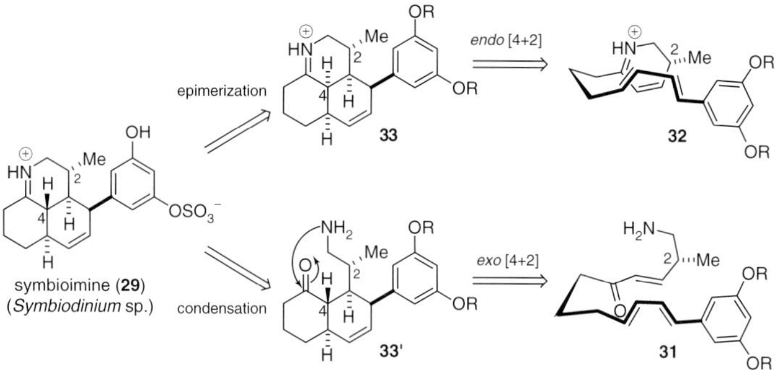

Scheme 8.6 Proposed biosynthetic origin of symbioimine.

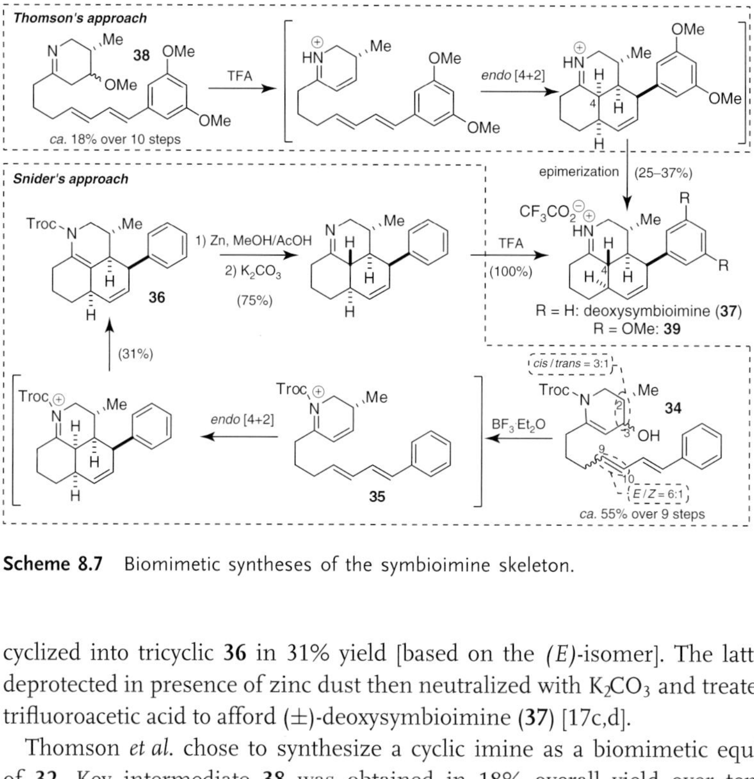

Scheme 8.7 Biomimetic syntheses of the symbioimine skeleton.

cyclized into tricyclic **36** in 31% yield [based on the *(E)*-isomer]. The latter was deprotected in presence of zinc dust then neutralized with K_2CO_3 and treated with trifluoroacetic acid to afford (±)-deoxysymbioimine (**37**) [17c,d].

Thomson *et al.* chose to synthesize a cyclic imine as a biomimetic equivalent of **32**. Key intermediate **38** was obtained in 18% overall yield over ten steps as a mixture of isomers at the C3 position but with control over the absolute configuration at C2. Treatment of **38** with TFA permitted the generation of a transient dihydropyridinium that readily cyclized and epimerized into tricyclic adduct **39** in 25–37% yield. Optically active symbioimine (**29**) was then obtained upon phenol deprotection and selective sulfation (22% yield for the two steps), thus completing the first enantioselective total synthesis of this alkaloid [17e].

Both routes share the same key biomimetic IMDA during which two new rings and four stereocenters are diastereoselectively formed in a single step (with comparable yields in the two teams), which is quite impressive in terms of molecular complexity. Notably, the presence of a single methyl substituent at C2 is responsible for the π-facial selectivity of the cycloaddition.

Apart from the enantioselectivity, the two approaches mainly differ from the fact that Thomson's synthesis proceeds via an iminium ion intermediate rather than an acyl iminium ion (which requires further deprotection after cycloaddition) as proposed in Snider's work.

Very recently, the group of Chruma disclosed a model study with further arguments in favor of the revised biosynthetic hypothesis involving the *endo*-IMDA [20]. In this report, the authors showed that when all-*(E)* intermediate **40** (closely related to **31**) underwent an IMDA reaction it was via an *endo* transition state, as depicted in Scheme 8.8, and not via an *exo* transition state as originally proposed. The reaction yielded a single diastereomer (**41**, epimeric at C3 and C4 compared to the postulated biosynthetic intermediate **33′**), confirming that the methyl substituent at C2 is sufficient to induce the π-facial selectivity of the IMDA. Moreover, this study demonstrated that *(Z)*-enones, such as **42**, isomerized into the all-*(E)* form before undergoing *endo*-IMDA, thus highlighting the importance of constraining these precursors in the form of dihydropyridinium salts to prevent such isomerization and increase the dienophile's reactivity.

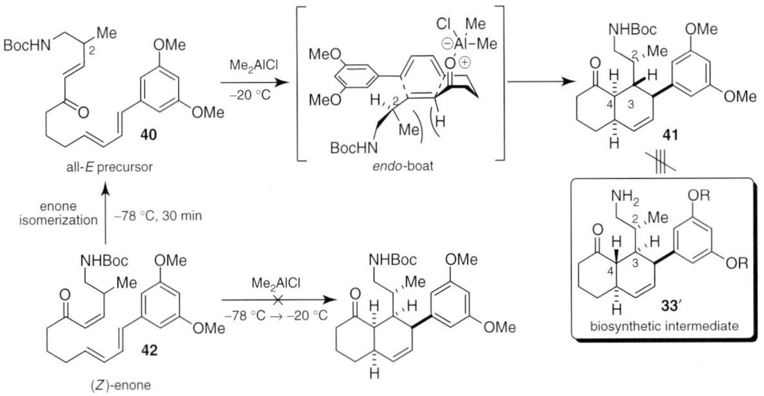

Scheme 8.8 Chruma's contribution to our understanding of the biosynthesis of symbioimine.

8.3.2
Pinnatoxins and Pteriatoxins

Described by the group of Uemura in 1995 [21], pinnatoxin A (**43**) was the first representative isolated in this macrocyclic toxin family that was later enlarged with the report of pinnatoxins B, C [22], D [23], E, F, and G [24] as well as

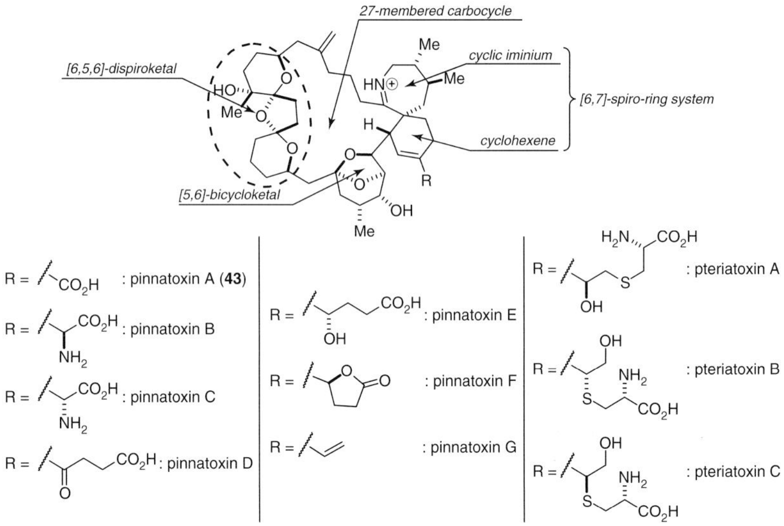

Figure 8.4 General structure of pinnatoxins and pteriatoxins.

pteriatoxins A–C (Figure 8.4) [25]. Extracted from *Pinna muricata* (Pinnidae),[2] and also present in other Okinawan bivalves such as *P. attenuata, P. atropurpurea*, or *Atrina pectinata*, pinnatoxins are suspected of being responsible for the frequent food poisoning symptoms (paralysis, diarrhea, convulsion, etc.) following the ingestion of this commonly eaten shellfish, and have been shown to be Ca^{2+} channel activators.

Structurally, all these molecules share the same highly complex macrocyclic skeleton featuring a [6,5,6]-dispiroketal moiety, a [6,7]-spiro-ring system composed of cyclic iminium attached to a cyclohexene and a [5,6]-bicycloketal (tetrasubstituted dioxabicyclo[3.2.1]octane). The difference between the toxins lies in the nature of the side-chain attached to the cyclohexene's olefin (Figure 8.4).

The intriguing molecular architecture of these natural products combined with their pronounced biological activity motivated several chemists to work on their synthesis, especially on that of pinnatoxin A (**43**) [26]. The first reported total synthesis of **43** was achieved by the group of Kishi [26a] and was based on Uemura's biosynthetic hypothesis [21a] that proposed an IMDA reaction followed by condensation to cyclic imine for the formation of the [6,7]-spiro-ring system (Scheme 8.9 depicts the possible biosynthetic origin of pinnatoxin A).

In this work, the authors first synthesized precursor diketone **44**, which, upon treatment with camphorsulfonic acid, yielded the tricyclic dispiroketal core (formation of three new cycles and two new stereocenters) bearing a free hydroxyl at C24, as a mixture of epimers at C19. Fortunately, under classical silylation conditions,

2) Pteriatoxins were extracted from *Pteria penguin* (Pteriidae).

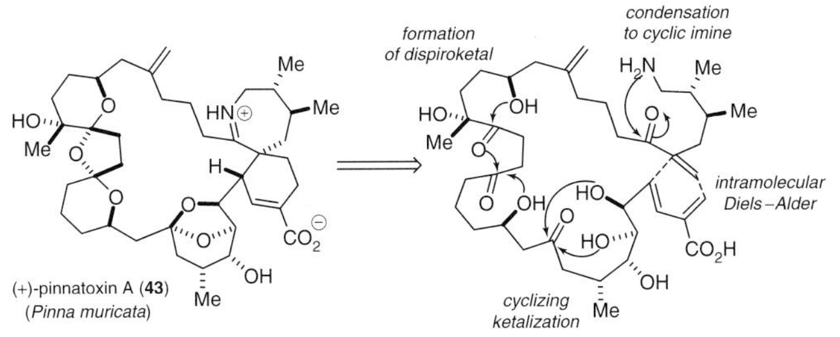

Scheme 8.9 Plausible biosynthetic origin of pinnatoxin A.

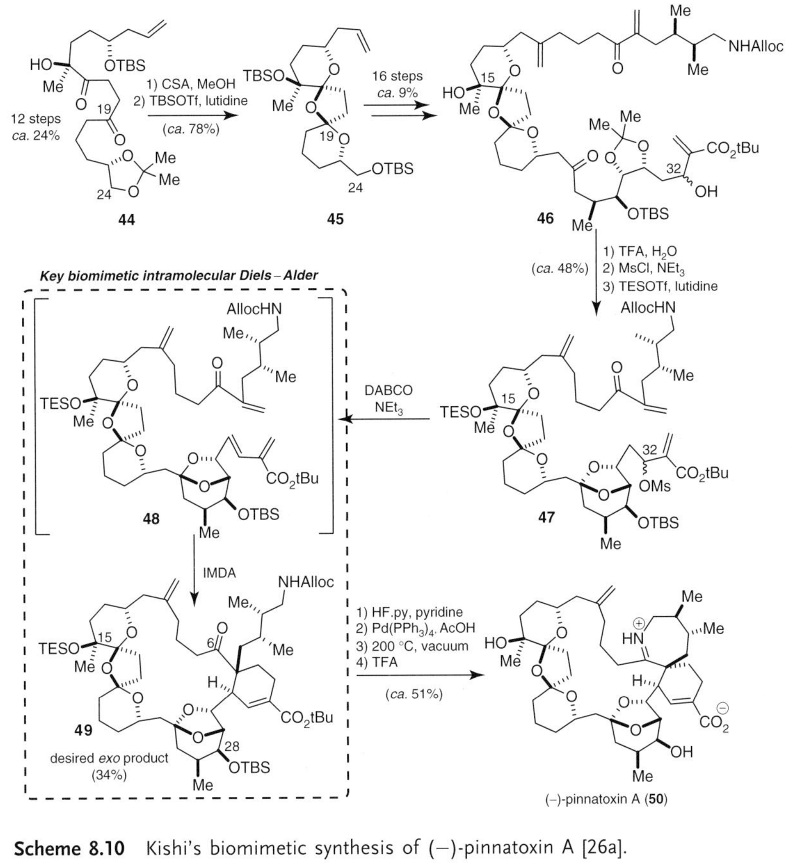

Scheme 8.10 Kishi's biomimetic synthesis of (−)-pinnatoxin A [26a].

the undesired epimer isomerized to the natural series and afforded intermediate **45** in 78% yield for the two steps (Scheme 8.10). Acetonide **46** was obtained after 16 synthetic steps (circa 9% overall yield) and was converted into bicycloketal **47** in 48% yield over three steps (acetonide cleavage, mesylation of C32 secondary hydroxyl, protection of C15 tertiary hydroxyl). The S_N2' displacement of the C32 mesylate with DABCO permitted the transformation of **47** into transient diene **48**, which then readily cyclized via an IMDA reaction to give rise to *exo* cycloadduct **49** as the major product (34%). Deprotection of C15 and C28 hydroxyls as well as terminal amine were easily carried out under classical conditions, but the cyclization into imine by condensation of the amine and C6 ketone required drastic conditions (heating at 200 °C under vacuum for 1 h). Finally, (−)-pinnatoxin A (**50**) was obtained (after cleavage of the *tert*-butyl ester) in 51% yield over the four last steps (Scheme 8.10) [26a].

Notably, this impressive synthesis permitted the establishment of the absolute stereochemistry of pinnatoxin A as **43**, the antipode of **50** originally proposed by Uemura.

8.3.3
Gymnodimine and Derivatives

Isolated from oysters collected in New Zealand [27], gymnodimine (**51**) is another spirocyclic imine marine toxin that was shown to be a powerful ligand of muscle-type nicotinic cholinergic receptors [28]. It was first described in 1995 by Yasumoto *et al.*, who extracted it from the shellfish *Tiostrea chilensis* (Ostreidae) and dinoflagellates of the *Karenia*[3)] genus (Gymnodiniaceae). They established its gross structure, which was later detailed (in terms of stereochemistry) by Blunt and Munro [29]. Gymnodimine is a macrocyclic molecule composed of a trisubstituted tetrahydrofuran, a cyclic imine spiro-fused to a cyclohexene ring and a butenolide moiety (Figure 8.5). Several years later, the group of Miles reported the isolation of gymnodimines B [30] and C [31], oxidized analogs of **51** that bear an exocyclic C17=C29 double bond and an allylic hydroxyl at C18.

With its unique structure and intriguing biological activity, gymnodimine has attracted the attention of several synthetic chemists [32]. Among them, the group of Kishi has developed a biomimetic approach [32e] headed towards the formation of the spiro-ring system by an IMDA reaction as a means to close the macrocyclic framework of the molecule. Their strategy was based on the biosynthetic hypothesis developed by Uemura for pinnatoxin A [21a] that seems reasonably applicable to gymnodimine (Scheme 8.11) considering the obvious structural similarities that exist between the two toxins.

To achieve the biomimetic construction of the macrocycle (Scheme 8.12), the authors first prepared triene **52** (obtained in 62% yield over four steps), protected amine **53** (obtained in 60% yield over two steps), and tetrahydrofuran **54** (obtained in 5% yield over 14 steps). These three precursors were then assembled to permit the construction of compound **55** in 8% yield over eight additional steps. Under classical deprotection conditions **55** was transformed into **56**, which was then treated with $(PPh_3)_4Pd$ to cleave the alloc carbamate and give α,β-unsaturated imine **57** (51% for the two steps). The key biomimetic IMDA was carried out under

3) Organisms of the *Karenia* genus where formerly included in the *Gymnodinium* genus.

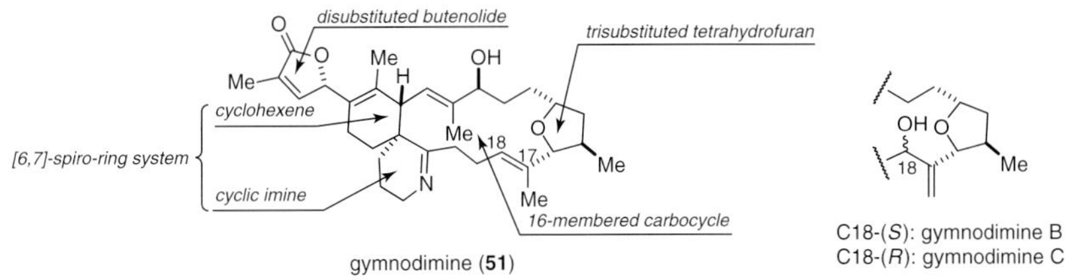

Figure 8.5 Structure overview of gymnodimine and congeners.

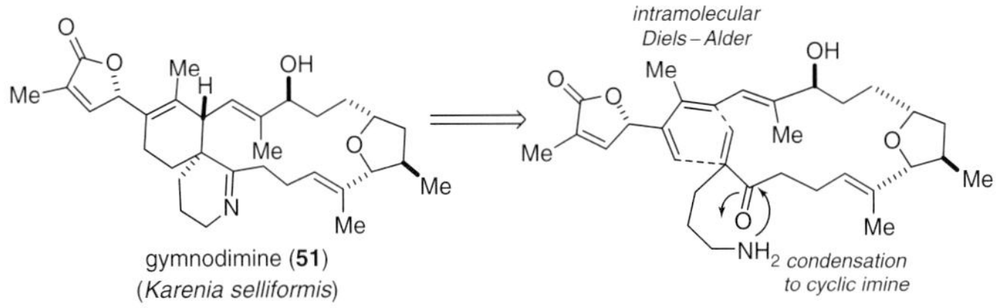

Scheme 8.11 Plausible biosynthetic origin of gymnodimine.

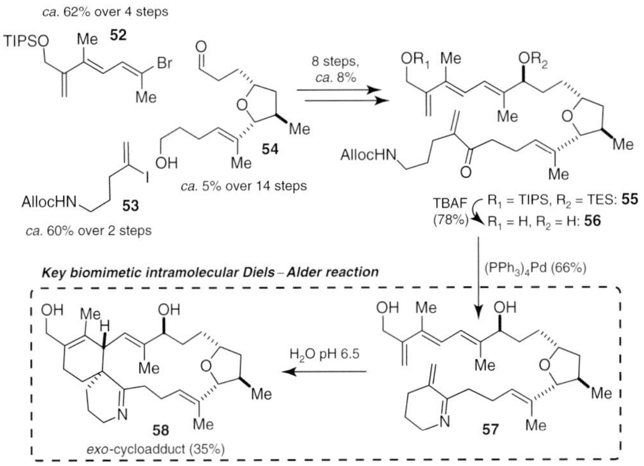

Scheme 8.12 Kishi's biomimetic approach to the macrocyclic framework of gymnodimine [32e].

mild conditions (aqueous buffer pH 6.5 at 36 °C) and afforded a mixture of three cycloadducts. One of these was the desired *exo*-cycloadduct **58** (35%) while the other two were *endo*-cycloadducts that were isolated as keto-amines, suggesting that their imine form is unstable in aqueous conditions [32e].

Notably, **58** is the only *exo*-cycloadduct that was isolated, demonstrating that the *exo*-IMDA proceeds with high facial selectivity. Also interesting is the fact that the

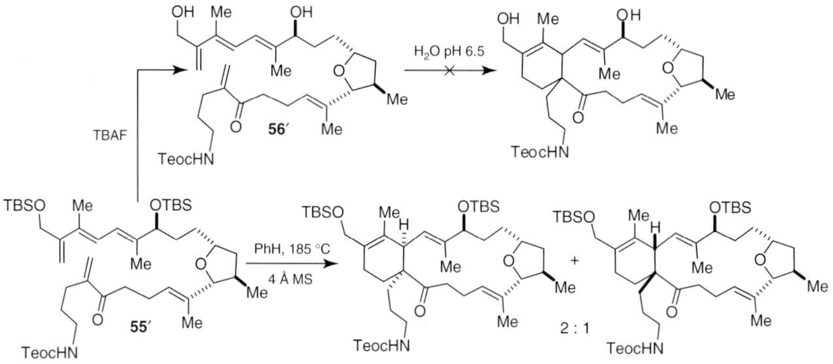

Scheme 8.13 Attempts at cycloaddition from keto-amine precursors [32e].

authors submitted deprotected keto-amine **56′** to the same aqueous conditions and observed nothing but the decomposition of the triene moiety. In addition, when they attempted to achieve cyclization of protected keto-amine **55′** they observed the formation of the two *endo*-cycloadducts with poor facial selectivity (2 : 1), but could never detect any *exo*-product (Scheme 8.13).

The results obtained in this study suggest that an IMDA reaction could account for the formation of gymnodimine without enzyme assistance but that the formation of the α,β-unsaturated imine dienophile should occur prior to the cycloaddition process.

8.4
Other Polyketide Derived Alkaloids

8.4.1
Cassiarins A and B

In 2007, the group of Morita reported the isolation of two new alkaloids [33], cassiarins A (**59**) and B (**60**), extracted from the leaves of *Cassia siamea* (Fabaceae) collected in Indonesia. The aromatic molecules were shown to exhibit promising antimalarial activities against *Plasmodium falciparum* (**59** in particular, with an IC_{50} of 0.005 $\mu g\, ml^{-1}$), and a vasodilatator effect of cassiarin A was demonstrated in a recent study [34].

Structurally, both molecules share the same tricyclic aromatic core composed of a 3-methylisoquinoline fused to a 2-methylpyran ring and mainly differ in the fact that **60** is N-substituted whereas **59** is not (Figure 8.6).

To date, only two groups have been working on the synthesis of cassiarins: the Morita group [35] and the Yao group [36]. The latter published a biomimetic total synthesis of both alkaloids based on Morita's biosynthetic proposal (Scheme 8.14) [33], which suggests that chromone precursor **61** could react with ammonia or

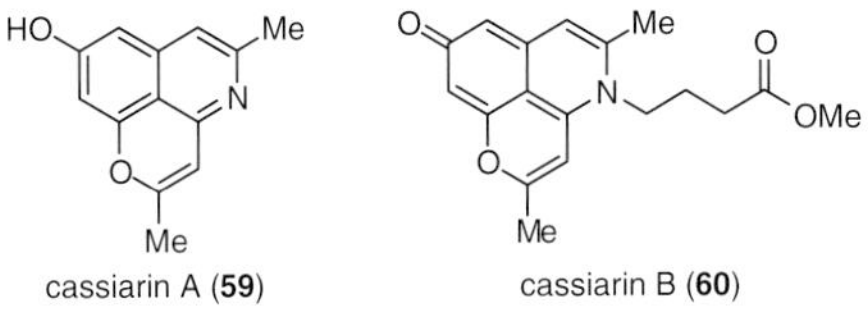

Figure 8.6 Structure of cassiarins A and B.

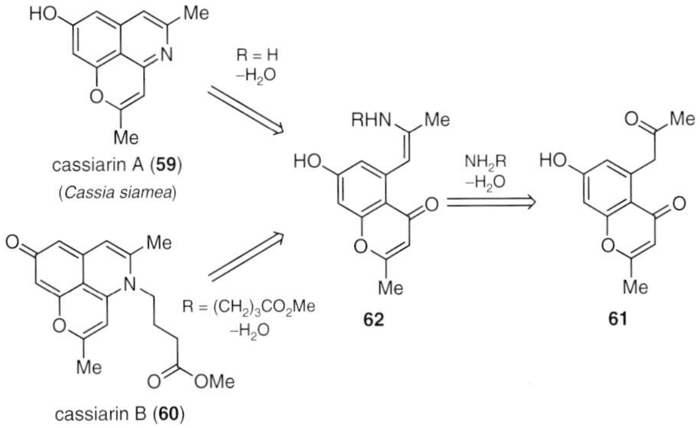

Scheme 8.14 Biosynthetic origin of cassiarins proposed by Morita.

a primary amine to give intermediate enamine **62** that would then cyclize into isoquinoline to yield cassiarins.

To conduct their biomimetic study, Yao and Yao prepared chromone precursor **63** under literature conditions [37] and obtained key intermediate **64** after three additional steps (phenol protection, triflate formation, and Negishi coupling, 65% yield overall). Treatment of **64** with trifluoroacetic acid in the presence of a catalytic amount of silver nitrate afforded **65** (synthetic equivalent of postulated biosynthetic precursor **61**), which could not be isolated. Exposing the reaction mixture containing crude **65** to either ammonia or methyl 4-aminobutyrate, followed by cleavage of the MOM ether with hydrochloric acid, permitted the obtention of cassiarin A (60% yield from **64**) and cassiarin B (52% from **64**) respectively (Scheme 8.15) [36].

8.4.2
Decahydroquinoline Alkaloids

Isolated from different terrestrial [38] and marine [39] organisms, decahydroquinoline alkaloids represent one of the major classes of the wide variety of toxic amphibian alkaloids [40] produced by tropical frogs of the Dendrobatidae [41] and Mantellidae [42] families. It is still undecided whether these molecules can be synthesized *de novo* by the amphibians and some observations suggest that they might be of dietary origin (especially since they have been found to occur in ants [38c,d]) [43].

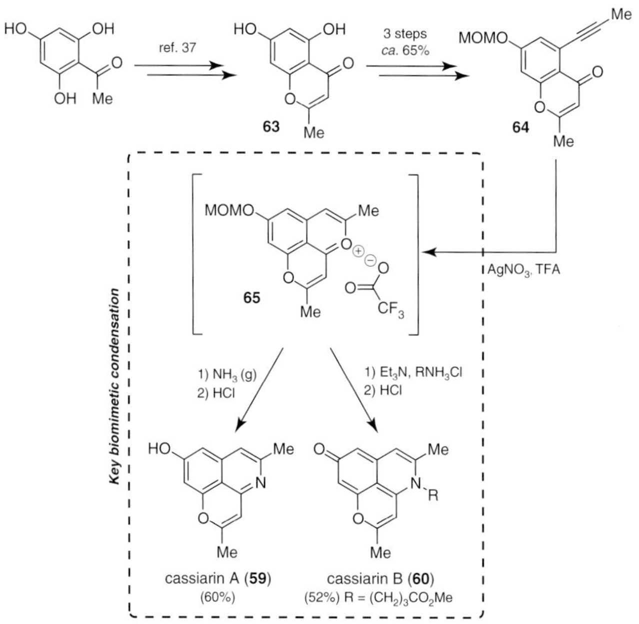

Scheme 8.15 Yao's biomimetic synthesis of cassiarins [36].

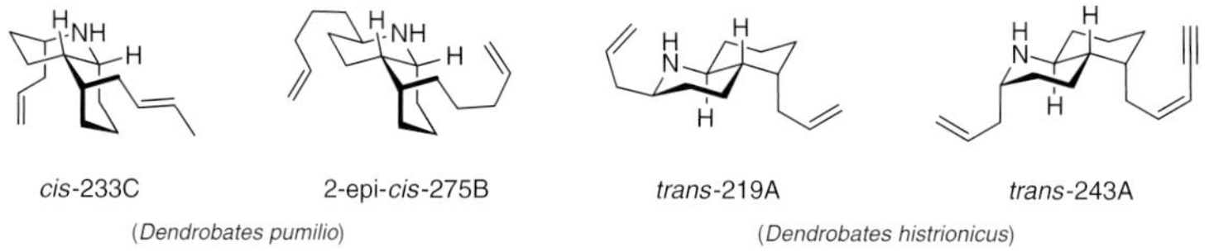

Figure 8.7 Representative examples of decahydroquinoline alkaloids.

Amphibian alkaloids have been the subject of numerous synthetic studies over the last three decades [44] and recently Amat and Bosch have been developing the biomimetic synthesis of decahydroquinoline alkaloids (Figure 8.7) [45]. These natural compounds possess a *cis*- or *trans*-fused decahydroquinoline substituted at both C2 and C5 positions. It has been suggested that they might find their biosynthetic origin in a linear polyketide precursor (such as **66**) which would undergo an intramolecular aldolization–crotonization process to form a cyclohexenone intermediate (**67**). Upon reductive amination, condensation, and reduction, the latter would lead to the 2,5-disubstituted decahydroquinoline framework (**68**) (Scheme 8.16) [46].

Based on this biosynthetic scenario, Amat *et al.* decided to undertake the synthesis of *cis*-195A (**69**) [45c], formerly known as *(−)-pumiliotoxin C* [47], a representative alkaloid of the group. As a synthetic equivalent of **66**, they prepared diketo ester

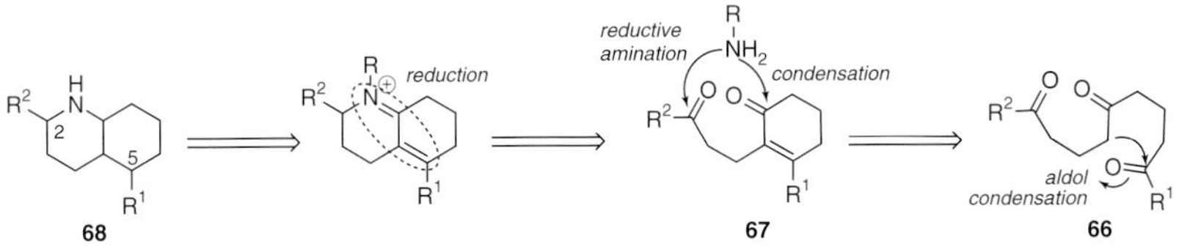

Scheme 8.16 Proposed biosynthetic origin of decahydroquinoline alkaloids.

70 (in 65% yield from 5-oxohexanoyl chloride) and submitted it to treatment with aqueous lithium hydroxide followed by trimethylsilyl chloride in methanol and obtained cyclohexenone **71** in 82% yield. Reaction of the latter with *(R)*-phenylglycinol under acidic catalysis enabled its cyclocondensation to tricyclic lactam **72** in 70% yield (Scheme 8.17). This biomimetic sequence provides experimental support for the proposed biosynthetic scenario from which it was rationalized.

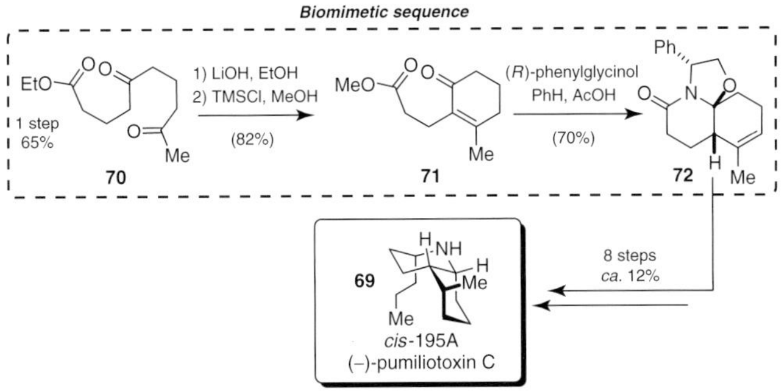

Scheme 8.17 Amat's biomimetic synthesis of *cis*-195A [45c].

At this point, eight additional synthetic steps (mostly reductions and an Eschenmoser sulfide contraction to implement the propyl side-chain at C2) were needed to complete the synthesis of the target compound **69** in 12% yield (Scheme 8.17) [45c]. Interestingly, the authors noted that, depending on the order in which the reduction of the endocyclic double bond and the Eschenmoser contraction were carried out, different series of *cis*-decahydroquinolines could be accessed.

When the tricyclic lactam intermediate **73** is first submitted to selective hydrogenation of the C5=C6 double bond, hydrogen uptake is directed to the β-face by the presence of an axial C–O bond to give **74**. After conversion of the lactone into thioamide and treatment under Eschenmoser's sulfide contraction conditions [$BrCH_2R^3$ then $P(OMe)_3$, Et_3N] intermediate **75** is obtained. Reduction of the latter begins by hydrogenation of the exocyclic double bond to give **76** on which the hydrogen uptake is directed to the less hindered α-face of the iminium, giving access to the *cis* series of decahydroquinolines (**77**), as depicted in Scheme 8.18 (Path A) [45a].

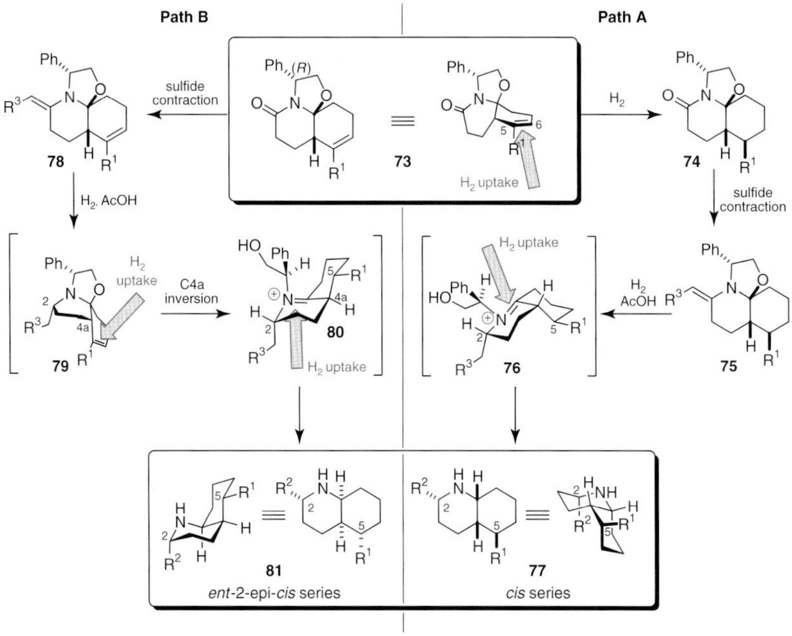

Scheme 8.18 Enantioselective synthesis of 2,5-disubstituted decahydroquinolines [45a].

In contrast, when **73** is first subjected to sulfide contraction, compound **78** is obtained and can be totally reduced in one-pot. This process begins with the hydrogenation of the exocyclic double bond to give intermediate **79**, on which the presence of a β-oriented side-chain at C2 directs the hydrogen uptake to the α-face of the C5=C6 endocyclic double bond. At that point, an inversion of the C4a center is required to put the R^1 substituent at C5 in an equatorial position and yield intermediate **80**. Hydrogen uptake on the latter is directed to the less hindered β-face of the iminium, giving access to the *ent*-2-epi-*cis* series of decahydroquinolines (**81**) (Scheme 8.18, Path B). Access to the natural 2-epi-*cis* series can be achieved in the same manner from a tricyclic lactam precursor obtained from *(S)*- instead of *(R)*-phenylglycinol [45a].

8.4.3
Zoanthamine Alkaloids

Zoanthamines are highly complex molecules extracted from polyps of the *Zoanthus* genus (Zoanthidae), collected in different regions around the world (Figure 8.8). The first members of the family, including zoanthamine (**82**), were isolated by the group of Faulkner and Rao in the 1980s from zoanthids collected off the coasts of India [48]. Later, Uemura *et al.* isolated new zoanthamine alkaloids from *Zoanthus* species collected off the coasts of southern Japan [49], Norte *et al.* reported the

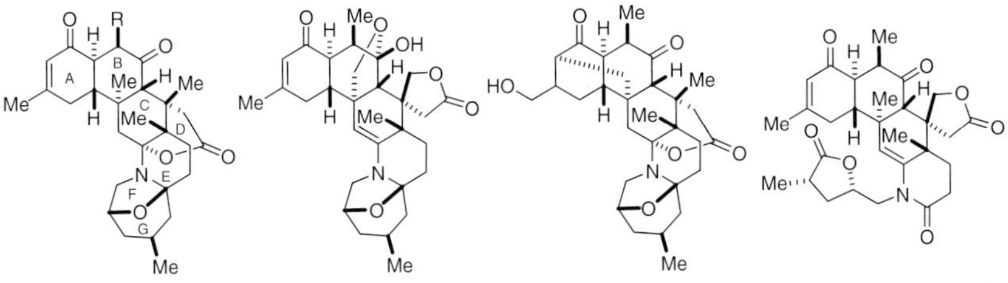

Figure 8.8 Representative examples of zoanthamine alkaloids.

isolation of additional members of the family from polyps collected in the Canary Islands [50] and other groups have reported related molecules as well [51].

Zoanthamine alkaloids display a wide range of biological activities as some of them have been shown to be antiosteoporotic [52], anti-inflammatory [48a,b], cytotoxic [49], antibacterial [53], or active against platelet aggregation [54].

The challenging structural features and interesting biological activities of zoanthamines have prompted many research groups to develop their own approach to the synthesis of these alkaloids [55]; to date, only the groups of Miyashita [56] and Kobayashi [57] have reported total syntheses.

Among the different synthetic studies toward these alkaloids, a biomimetic approach to the heterocyclic CDEFG ring system of zoanthamine has been disclosed by Kobayashi *et al.* [58]. Their strategy was based on Uemura's hypothesis [49b, 59], according to which zoanthamines could be derived from linear polyketide precursor **83** (Scheme 8.19).

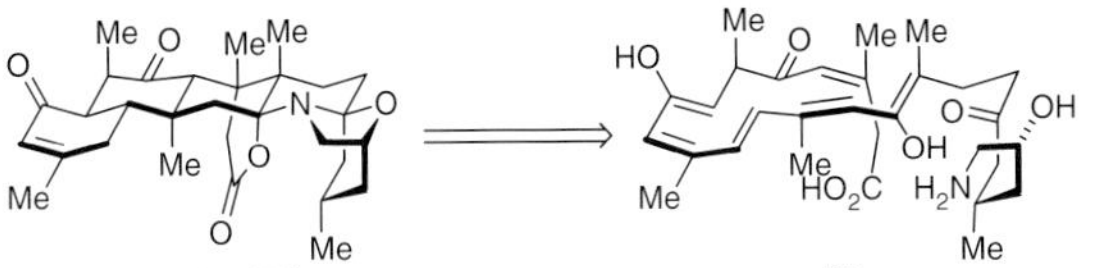

Scheme 8.19 Biosynthetic origin of zoanthamine proposed by Uemura.

Upon electrocyclization and Diels–Alder cycloaddition, enabling construction of the ABC rings, the latter could give rise to intermediate **84**, which would then be transformed into **85** after hemiaminal ether formation, providing the FG rings. The E ring would then be formed by attack of the secondary amine on the ketone of ring C followed by dehydration to yield iminium **86**. Finally, the D ring would be constructed by reduction of the cyclic iminium upon attack by the carboxylic acid to yield zoanthamine (Scheme 8.20).

To mimic this scenario, Kobayashi and coworkers decided to study the cyclization of advanced intermediate **87** under different conditions. To this end, aldehyde **88** [obtained in 24% yield over ten steps from (+)-Wieland–Miescher ketone] [58b]

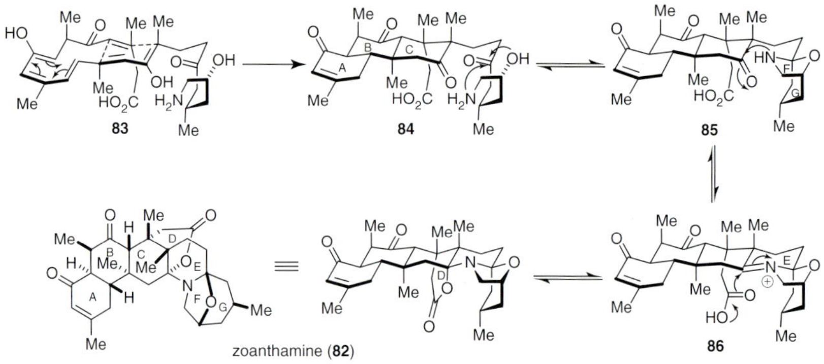

Scheme 8.20 Proposed biosynthetic route for zoanthamine [55].

and sulfone **89** (obtained in 11% yield over 17 steps from D-glutamic acid) were coupled and, after further modification, compound **87** was obtained in 38% yield over six steps (Scheme 8.21) [58a,b].

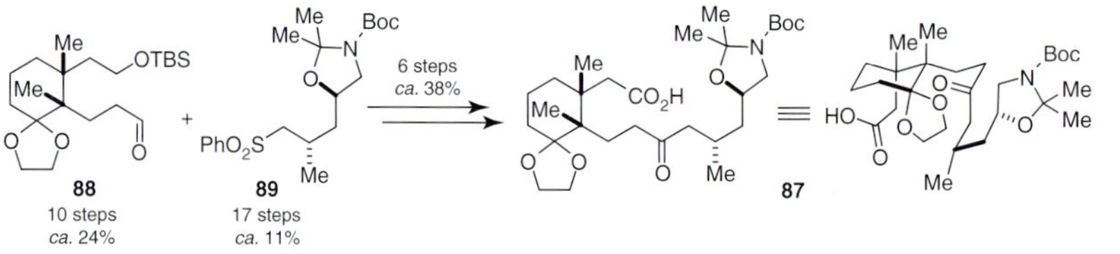

Scheme 8.21 Synthesis of cyclization substrate **87** [58a,b].

With precursor **87** in hand, the authors explored different deprotection strategies. They thus found that upon treatment with 2N hydrochloric acid in tetrahydrofuran at room temperature only partial deprotection took place, allowing the formation of **90**, which bears the bicyclohemiaminal ether FG ring system of zoanthamine and can be considered a synthetic equivalent of plausible biosynthetic intermediate **85** (Scheme 8.22). Heating of **90** in wet acetic acid for several hours and subsequent addition of a dehydrating agent permitted the deprotection of the Boc group and the formation of the D and E rings to yield compound **91** in 85% yield from **87**. When the latter conditions were applied directly to substrate **87**, pentacycle **91** was formed in 89% yield, presumably through the formation of transient intermediate **92**, synthetic equivalent of plausible biosynthetic intermediate **84** [58a]. During this simple one-pot process, the fully functionalized CDEFG ring system of zoanthamine is formed in high yield, demonstrating the efficacy of this biomimetic strategy.

Kobayashi *et al.* recently applied the stepwise version of this methodology to the total synthesis of norzoanthamine [57b].

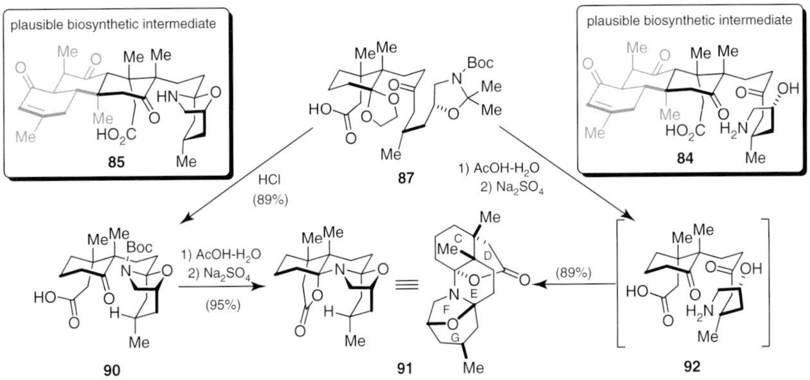

Scheme 8.22 Kobayashi's biomimetic approach of to the CDEFG rings of zoanthamine [58a].

8.4.4 Azaspiracids

Phykotoxins of the azaspiracid family were discovered in the late 1990s during an investigation into the cause of food poisonings [60] due to the consumption of Irish blue mussels (*Mytilus edulis*, Mytilidae) [61]. Subsequently, a total of eleven azaspiracids have been isolated [62] and their occurrence in different shellfish species [63] suggests that the origin of these alkaloids might be a marine dinoflagellate [64].

Azaspiracids are highly complex nonacyclic molecules with particular structural features such as a tetracyclic ring system composed of a substituted tetrahydrofuran side-fused to a trioxadispiroketal, a central highly substituted tetrahydropyran, a [6,6]-bicycloketal (dioxabicyclo[3.3.1]nonane) and a piperidine ring spiro-fused to a tetrahydrofuran, resulting in a spiro-hemiaminal ether moiety (Figure 8.9). Most azaspiracids differ from one another by the methylation (at C8 and C22) and hydroxylation (at C3 and C23) patterns and all share the same C26–C40 region. Recently, 20 new azaspiracid derivatives were identified, including dihydroxy (at C3 and C23) and carboxy-derivatives (at C22) [65].

The exceptional structural complexity of azaspiracids along with their biological activities (mainly toxic effects [66]) have motivated numerous synthetic efforts [67] that culminated in total syntheses by the groups of Nicolaou [68] (which led to the unambiguous elucidation of the structure of azaspiracid-1 (**93**) [69]) and Evans [70].

Like other polyether compounds produced by dinoflagellates, azaspiracids are thought to be of polyketide origin [71]. Based on the assumption that a linear precursor (**94**) might organize into the polycyclic framework of azaspiracids (Scheme 8.23), Forsyth *et al.* have devised two biomimetic strategies for the synthesis of the spiro-hemiaminal ether moiety of these molecules [67m].

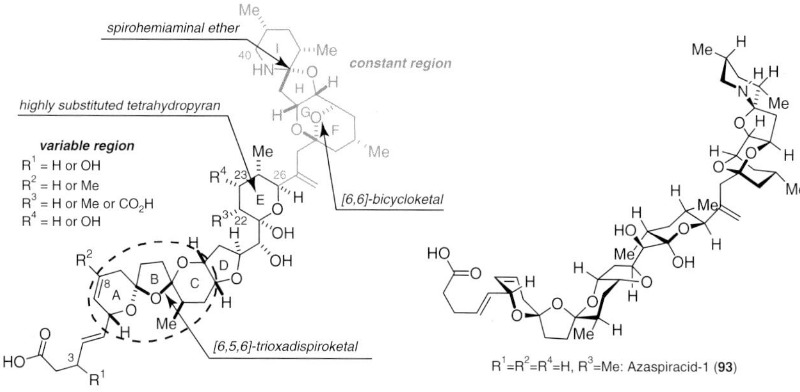

Figure 8.9 General structure of azaspiracids and 3D structure of azaspiracid-1.

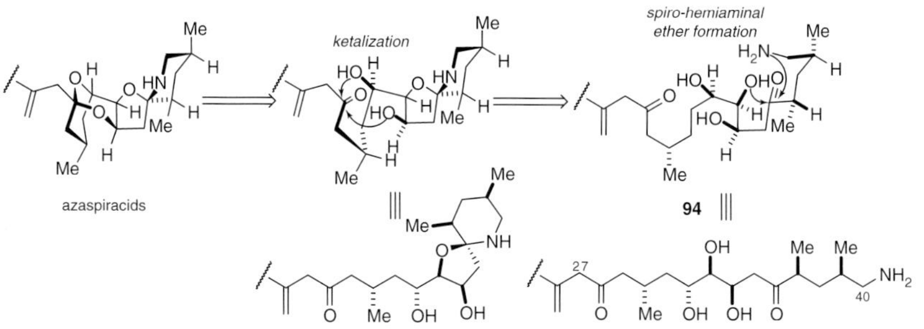

Scheme 8.23 Plausible biosynthetic origin of the FGHI rings of azaspiracids.

The first of these routes is based on a Staudinger/aza-Wittig sequence that was applied to intermediate **95** (obtained in 7% yield over 13 steps), synthetic equivalent of the C27–C40 region of postulated biosynthetic intermediate **94** (Scheme 8.24). Treatment of **95** with triethylphosphine furnished spiro-hemiaminal ether **96** in 75% yield, presumably through formation of iminophosphorane **97**, which would cyclize into **98** by an intramolecular aza-Wittig reaction and then into **96** by addition of the C33 hydroxyl to the newly formed imine. The fact that **96** is obtained as the major product along with its C36 isomer (4 : 1 ratio) reflects either a kinetic pseudo-axial attack of the C33 hydroxyl on the Si face of the imine or a post-addition thermodynamic equilibration (or both) [67m].

The second route is based on a Staudinger/hetero-Michael sequence that was applied to carbamate-ynone **99** (obtained in 6% yield over 14 steps), another synthetic equivalent of the C27–C40 end (**94**). Conversion of **99** into spiro-hemiaminal ether **100** was carried out either stepwise or in a one-pot process. Under dichloro-dicyano-quinone (DDQ) deprotection conditions, *p*-methoxybenzyl ether was cleaved but the conjugate addition of the nitrogen on the ynone was also initiated

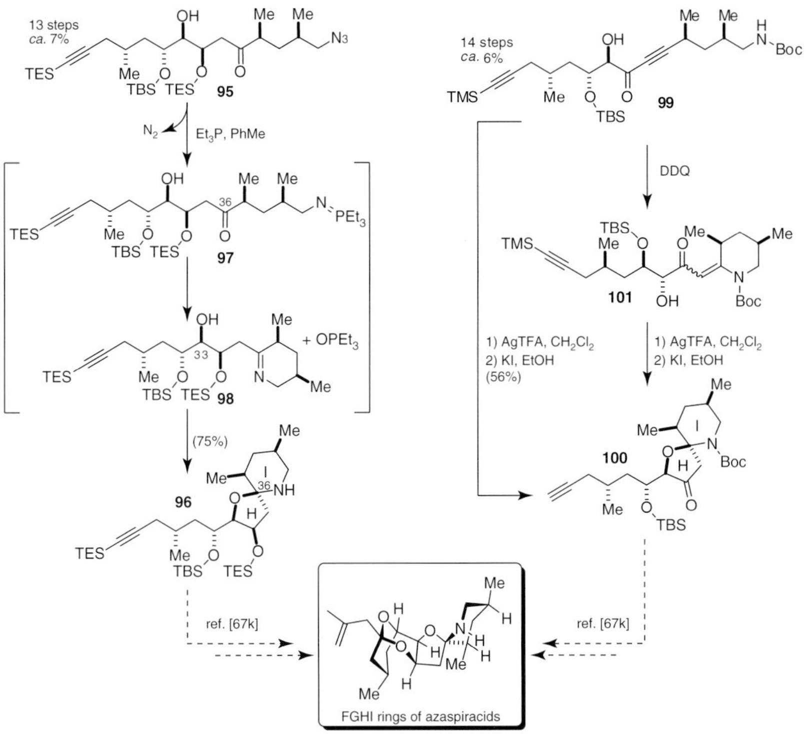

Scheme 8.24 Forsyth's biomimetic approach to the HI rings of azaspiracids [67m].

to give **101**. Treatment with silver trifluoroacetate and subsequent addition of ethanolic potassium iodine surprisingly afforded **100** (instead of the expected iodo-alkyne) by addition of the C33 hydroxyl to the enone-enamine. When the conditions initially intended to convert the terminal alkynyl silyl group into the corresponding iodo-alkyne were directly applied to **99**, the two hetero-Michael additions were initiated and the C27 and C33 protecting groups were cleaved in a one-pot process to yield **100** in 56% yield (Scheme 8.24) [67m].

Both spiro-hemiaminal ethers **96** and **100** can then be easily converted into the FGHI ring-system of azaspiracids under conventional conditions, previously described by the Forsyth group [67k].

8.5
Alkaloids Derived from Terpene Precursors

Various alkaloids derived from terpene precursors have been reported in the literature (Figure 8.10). These include several monoterpenoid alkaloids (such as gardenamide [72]), a large number of diterpenoid alkaloids [73] (e.g., barbaline),

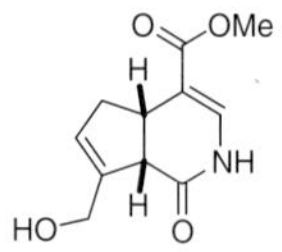

gardenamide
(*Rothmania urcelliformis*)

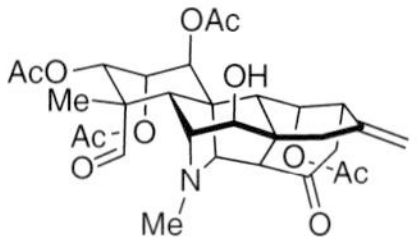

barbaline
(*Delphinium barbeyi*)

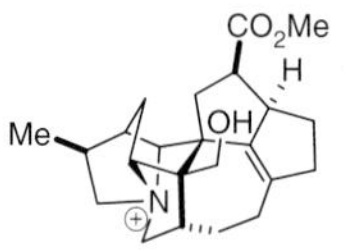

daphniglaucine A
(*Daphniphyllum glaucescens*)

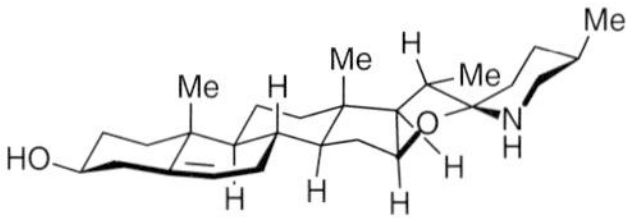

solasodine
(*Solanum* sp.)

Figure 8.10 Examples of terpenoid alkaloids.

numerous triterpenoid alkaloids, either steroidal [74] (such as solasodine) or non-steroidal (such as *Daphniphyllum* alkaloids, *vide infra* [75]), and others.

The following sections are devoted to the bis-steroidal alkaloids of the cephalostatin/ritterazine family and to the efforts developed around the biomimetic synthesis of *Daphniphyllum* alkaloids.

8.5.1
Cephalostatins and Ritterazines

Even though they were not isolated from the same marine organisms, cephalostatins and ritterazines are two structurally related groups of bis-steroidal pyrazine alkaloids [76]. Cephalostatins were extracted from the marine tubeworm *Cephalodiscus gilchristi* (Cephalodiscidae) collected off the coasts of southeastern Africa by the group of Pettit in the early 1970s. It took 15 years before the authors were able to isolate and elucidate the structure of cephalostatin 1 (**102**, Figure 8.11) [77], and in the following decade 18 more cephalostatins were reported by the same group [78]. Ritterazines, on the other hand, were extracted from the marine tunicate *Ritterella tokioka* (Polyclinidae) collected by Fusetani *et al.* off the coast of Japan. The first member of the family, ritterazine A (**103**), was reported in 1994 [79] and, in subsequent years, 25 other ritterazines were described [80].

The two groups share the same general structure consisting of two hexacyclic steroidal spiroketal units (one of which is called South and the other North)[4] attached by a central pyrazine ring (Figure 8.11). Most of the differences between cephalostatins are found in the outer CDEF rings and especially in the spiroketal

4) South and North hemispheres refer to the Fuchs nomenclature. Pettit calls these regions Left and Right, whereas Fusetani speaks of West and East regions. The numbers or letters that follow these denominations correspond to the alkaloids to which the regions belong (e.g., South 1 refers to the southern region of cephalostatin 1 and North A refers to the northern region of ritterazine A) as depicted on Figure 8.11.

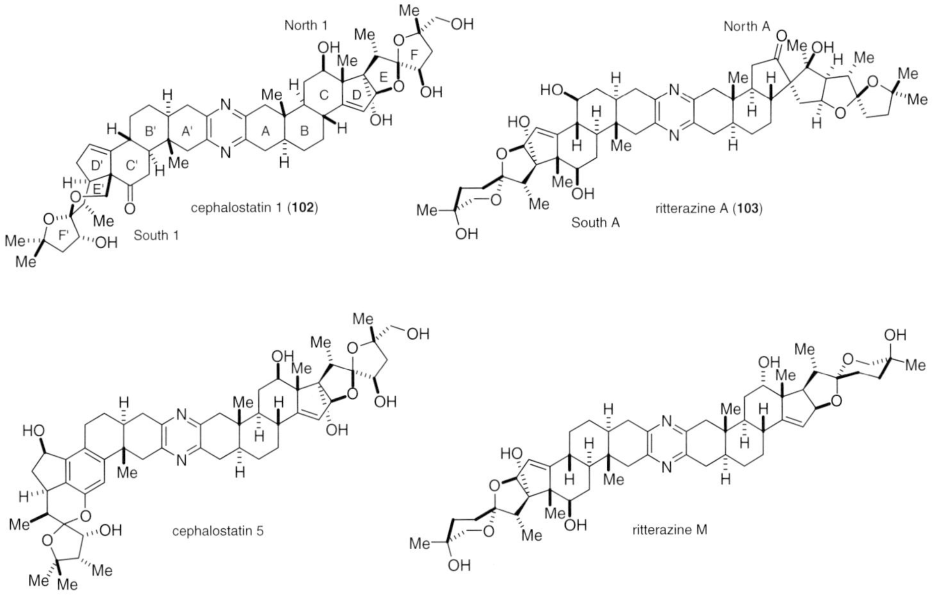

Figure 8.11 Structures representative of cephalostatins and ritterazines.

end regions. Ritterazines also show little variability in their AB rings but can have different CD junction patterns (either side-fused 6,5 or spiro-fused 5,5). The striking structural similarities along with the fact that the two groups of alkaloids were isolated from two animals of different phyla (i.e., Hemicordata for *Cephalodiscus* sp. and Chordata for *Ritterella* sp.) suggest that the biosynthesis of these compounds might actually find its origin in a marine microorganism ingested by both *C. gilchristi* and *R. tokioka* [78h, 80a].

The cytotoxic activities displayed by both cephalostatins [77, 78] and ritterazines [79, 80] are very impressive. Cephalostatin 1, for instance, is one of the most potent anticancer agents tested to this day, with femtomolar activity against the P388 cell line. Ritterazines globally seem to be a little less active than cephalostatins but some of them nonetheless show subnanomolar activities against the P388 cell line as well (e.g., ritterazine B).

Since the structure of cephalostatin 1 was disclosed in the late 1980s, the cephalostatin/ritterazine family has been the subject of numerous synthetic efforts [81]. The attractive structures and biological activities, along with the fact that only minute amounts of these alkaloids can be isolated from their natural sources, have prompted chemists to develop efficient synthetic routes to complete their evaluation as drug candidates.

The group of Fuchs has had a prominent role in the chemistry of the cephalostatin/ritterazine family and has developed biomimetic approaches to different regions of the molecules. In 2005, Lee and Fuchs proposed biosynthetic pathways to both the southern region of cephalostatin 7 and the northern region of cephalostatin 1 [82]. This hypothesis postulates that hecogenin (**104**) can be modified to give **105**, which then undergoes electrophilic spiroketal ring opening followed by elimination to produce **106**. Dihydroxylation of the C25=C26 olefin, [4+2]-cycloaddition of singlet oxygen on the cyclopentadiene moiety and endoperoxide cleavage will then convert **106** into **107**. The latter can, at that point, undergo an acid-catalyzed cyclization cascade by attack of either the C26 or the C25 hydroxyl on the C22 ketone to yield South 7 or North 1, respectively (Scheme 8.25).

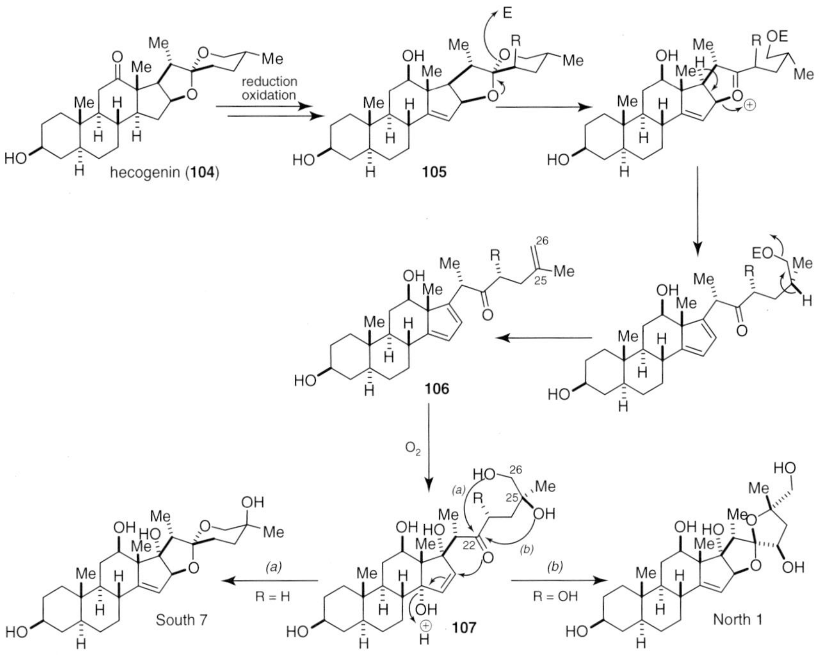

Scheme 8.25 Biosynthetic hypothesis for the formation of South 7 and North 1 (E = electrophile).

Inspired by this biosynthetic scenario, the authors designed a synthesis that would reproduce the major steps described above. They started from **108** [obtained in 75% yield from hecogenin (**104**)], which underwent reductive spiroketal opening followed by iodination and elimination to afford **109** in 81% yield. Dihydroxylation by the Sharpless protocol and subsequent selective protection of the primary C26 alcohol followed by benzoylation of the C25 hydroxyl gave **110** in 94% yield over the three steps. Tetrahydrofuran opening upon treatment with trifluoroacetyl triflate (TFAT), removal of the trifluoroacetate under mild basic conditions, oxidation of the obtained hydroxyl into the corresponding ketone, and protection of the latter with 1,3-propanediol afforded **111** in 81% yield. Cycloaddition of singlet oxygen to the cyclopentadiene moiety of the latter took place stereospecifically to give only the α-face adduct and the endoperoxide was reductively cleaved by treatment with activated zinc and acetic acid to give **112** in 83% yield. *In situ* generated hydrogen cyanide unexpectedly led to the formation of hydroxypropyl ether **114** in 86% yield, presumably via formation of transient oxonium **113**. After oxidation of the primary alcohol of **114** to the corresponding aldehyde, reaction with tetraisobutylammonium fluoride (TBAF) permitted the concurrent cleavage of the silyl ether and acrolein to directly furnish tetrahydrofuran hemiacetal **115** in 84% yield. Treatment of the latter with pyridinium *p*-toluenesulfonate (PPTS) completed the cyclization sequence and afforded a protected version of the southern hemisphere of cephalostatin 7 in 66% yield[5] (Scheme 8.26) [82].

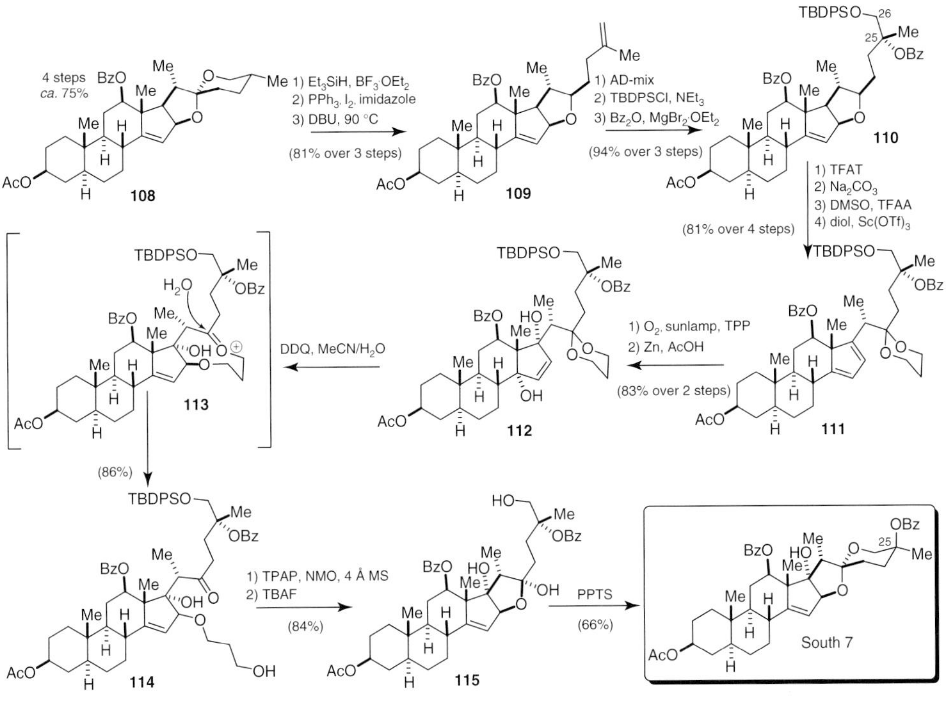

Scheme 8.26 Biomimetic synthesis of South 7 [82].

This biomimetically inspired synthesis (carried out in 16 steps and 24% overall yield from **108**) is a great improvement on the previously published method [83] (25 steps, 2% overall yield) for the preparation of South 7 (which is found in 18 of the 45 members of the cephalostatin/ritterazine family). The same strategy could be applied to the synthesis of North 1 but to date no report of progress in this direction has been disclosed.

The Fuchs group has also studied the biomimetic coupling of the northern and southern units [84], following the hypothesis that Nature might utilize the random

5) The reaction actually has a yield of 81%, but 15% corresponds to the C25 epimer of South 7 carried onward from the Sharpless dihydroxylation reaction.

coupling of different steroidal α-aminoketones to generate various bis-steroidal pyrazines [78c,h].

In this work, the authors reduced α-azidoketones **116** and **117**, to generate α-aminoketones **118** and **119** *in situ*, by treatment with ethanolic NaHTe for an hour. They then added silica gel and exposed the mixture to the air and, 18 h later, observed the formation of heterodimer **120** and homodimers **121** and **122** in 35%, 23%, and 14% yields respectively. Upon deprotection with an excess of TBAF in refluxed THF, the corresponding cephalostatin 7, ritterazine K, and cephalostatin 12 were each obtained in about 80% yield (Scheme 8.27) [84b].

This biomimetic one-pot process led to the simultaneous syntheses of cephalostatin 7, cephalostatin 12, and ritterazine K and provided the first synthetic samples of these molecules.

Considering that the most biologically active bis-steroidal pyrazines are the unsymmetrical ones, different groups have developed methods specifically designed to avoid the formation of symmetrical dimers. The Heathcock group disclosed their approach based on the reaction of 2-acetoxy-3-ketones (**123**) with 2-amino-3-methoximes (**124**) in 1992 [85]. In the following years, Guo and Fuchs proposed a similar approach based on the reaction of 2-azido-3-ketones (**125**) with 2-amino-3-methoximes [86] and the group of Winterfeldt also published a different strategy involving vinyl azides and α-aminoketones [87a], which was more recently improved to a simpler approach relying on the reaction between 2-hydroxy-3-ketones (**126**) and 2-amino-3-ketones (**127**) in the presence of ammonium acetate [87b] (Figure 8.12).

8.5.2
Daphniphyllum Alkaloids

At the beginning of the twentieth century, Yagi reported the isolation of daphnimacrine, a complex alkaloid from a common Japanese tree, *Daphniphyllum macropodum* (Daphniphyllaceae) also known as Yuzuriha [88]. At that time, the available spectroscopic techniques did not permit elucidation of the molecule's structure.[6] Almost 60 years later, Hirata and Yamamura initiated research on *Daphniphyllum* alkaloids and, since then, different groups have reported the structures of more than a 100 representatives of this growing family of natural compounds

6) The structure of daphnimacrine was elucidated by X-ray analysis and reported in 1968 [89].

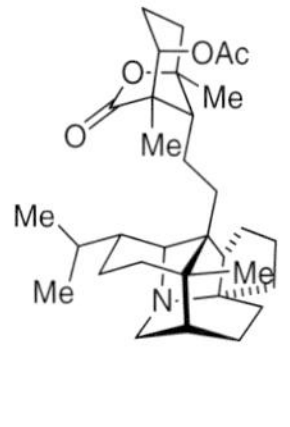

daphnimacrine

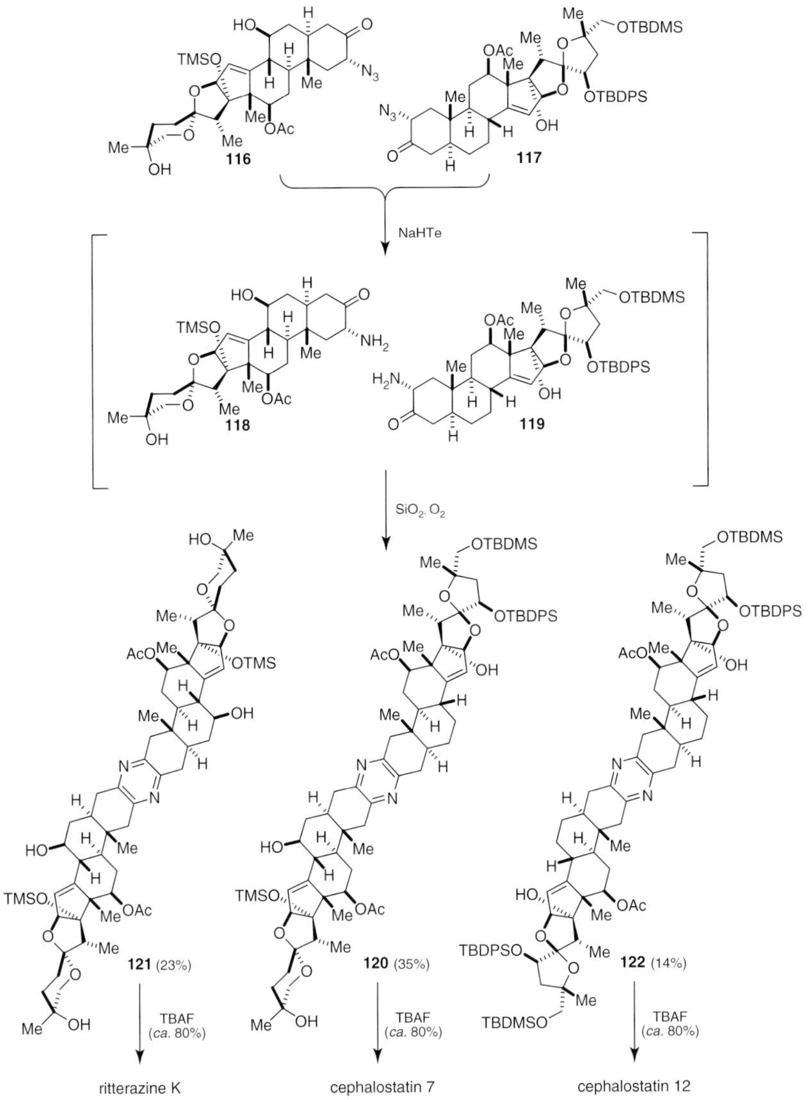

Scheme 8.27 Biomimetic pseudo-combinatorial approach to bis-steroidal pyrazines [84b].

isolated from plants of the *Daphniphyllum* genus found in Japan, China, Taiwan, and New Guinea [75].

Some of these alkaloids have been shown to possess biological activities, including platelet aggregation inhibition [90], antioxidant activity [91], modulation of the expression of nerve growth factor (NGF) [92], and cytotoxicity [93].

Upon discovery, *Daphniphyllum* alkaloids have been classified into different structural types with respect to their backbone skeleton (Figure 8.13 below presents

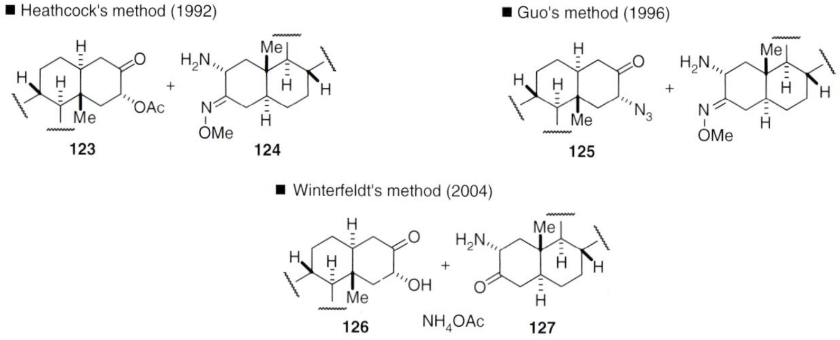

Figure 8.12 Different approaches to unsymmetrical bis-steroidal pyrazines.

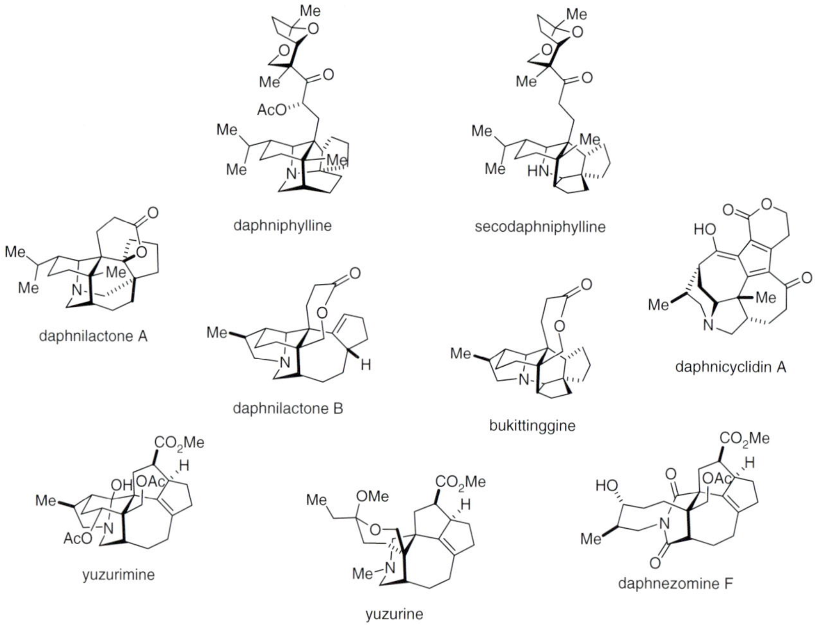

Figure 8.13 Examples of *Daphniphyllum* alkaloids, representative of their structural diversity.

major alkaloids representative of different structural types). Despite the chemical diversity of these often strikingly complex molecules, it has been proposed that they all arise from a common terpene precursor derived from squalene [94].

Since the 1980s, the Heathcock group has had a prominent role in the synthesis of *Daphniphyllum* alkaloids [95]. They have reported several biomimetic approaches based on their own biosynthetic hypothesis for the formation of a postulated pentacyclic intermediate, named *proto*-daphniphylline, that is likely to be common

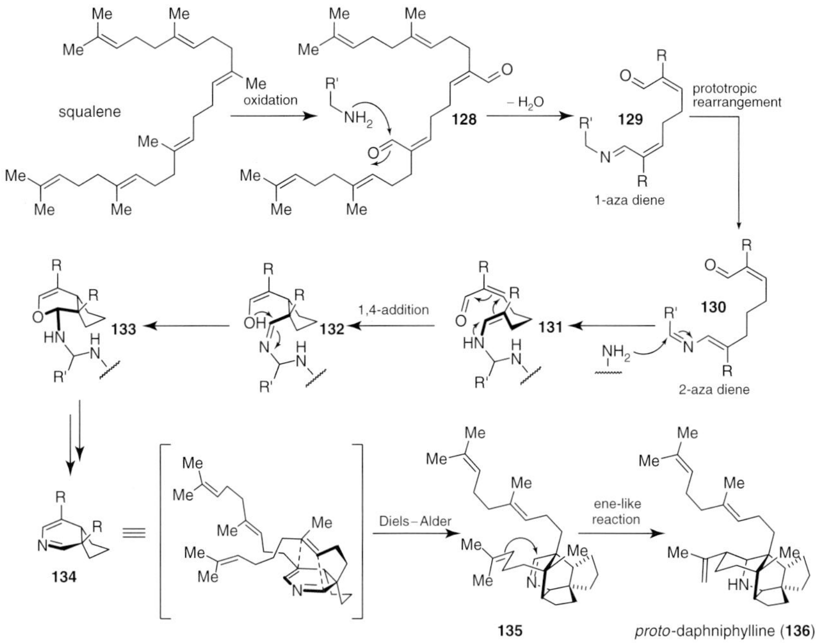

Scheme 8.28 Heathcock's hypothesis for the biosynthesis of *proto*-daphniphylline.

to many types of *Daphniphyllum* alkaloids (Scheme 8.28) [96]. According to this scenario, squalene is oxidized into dialdehyde **128**, which, upon condensation with a primary amine (e.g., amino acid or pyridoxamine), gives rise to 1-azadiene (**129**) that then isomerizes by a prototropic rearrangement to the corresponding 2-azadiene (**130**). The enimine double bond of the later is not very nucleophilic and the authors propose that, at that point, another nucleophilic species reduces the imine double bond to yield nucleophilic enamine **131**. An intramolecular Michael addition of the enamine on the conjugated aldehyde gives monocyclic **132**, which is then converted into **133** by hemiaminal ether formation. Proton-mediated addition and elimination processes allow the conversion of the bicyclic dihydropyran intermediate into a dihydropyridine derivative **134**, which can undergo an IMDA reaction and yield tetracyclic intermediate **135**. Finally, an ene-like reaction enables the formation of the fifth ring to form *proto*-daphniphylline (**136**).

This hypothesis has been supported by the successful biomimetic total syntheses of different alkaloids that Heathcock *et al.* have managed to complete over the years.

In 1988, they reported the synthesis of methyl homosecodaphniphyllate (**137**) [97, 98] via a tetracyclization reaction cascade from monocyclic diol **138** (obtained in 60% yield over eight steps). Under Swern oxidation conditions, **138** was converted into dialdehyde **139** (synthetic equivalent of **132**) and upon subsequent addition of gaseous ammonia, replacement of the solvent with acetic acid, and heating at 70 °C for 1.5 h, pentacyclic product **140** was formed in 77% overall yield. Removal of the benzyl protecting group and reduction of the terminal double bond were conducted simultaneously by hydrogenation over Pd/C. Jones oxidation of the obtained primary alcohol into the corresponding carboxylic acid followed by Fischer esterification then afforded (±)-methyl homosecodaphniphyllate (**137**) in 85% yield over the three steps (Scheme 8.29).

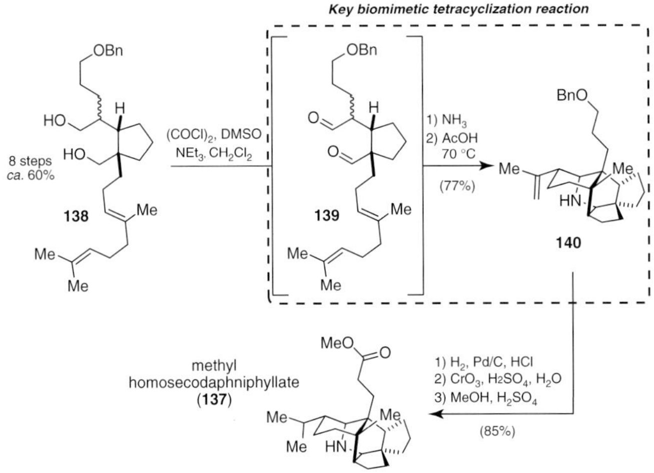

Scheme 8.29 Heathcock's synthesis of methyl homosecodaphniphyllate [98].

The key tetracyclization reaction, during which six bonds and four rings are formed, proceeds with impressive efficacy. In subsequent years, this strategy allowed the authors to achieve the total syntheses of other *Daphniphyllum* alkaloids, including secodaphniphylline [99, 100], bukittinggine[7] (**141**) [101, 102], methyl homodaphniphyllate (**142**) [99, 103], daphnilactone A (**143**) [103b–105], and codaphniphylline [106, 107].

In the case of bukittinggine [102], the tetracyclization process was carried out from precursor **144** (benzyloxy derivative of **138**, obtained in 12% yield over twelve steps) and gave pentacyclic **145** in 74–78% yield (Scheme 8.30). A palladium-assisted oxidative cyclization then permitted the formation of hexacyclic intermediate **146** in 70% yield. The latter was submitted to hydroboration-oxidation

7) Bukittinggine was isolated from *Sapium baccatum* (Euphorbiaceae) [101a] but is considered a member of the *Daphniphyllum* alkaloids because of obvious structural similarities with alkaloids such as methyl homosecodaphniphyllate and daphnilactone B. Moreover, three bukittinggine-type alkaloids have been isolated recently [101b] from *Daphniphyllum calycinum*:

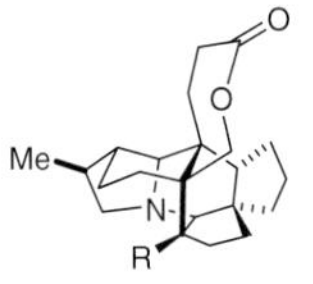

R = H: bukittiggine
R = OH: caldaphnidine N

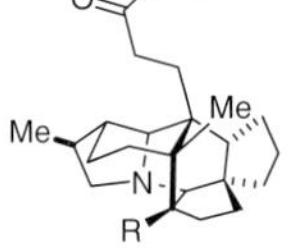

R = H: caldaphnidine O
R = OH: caldaphnidine P

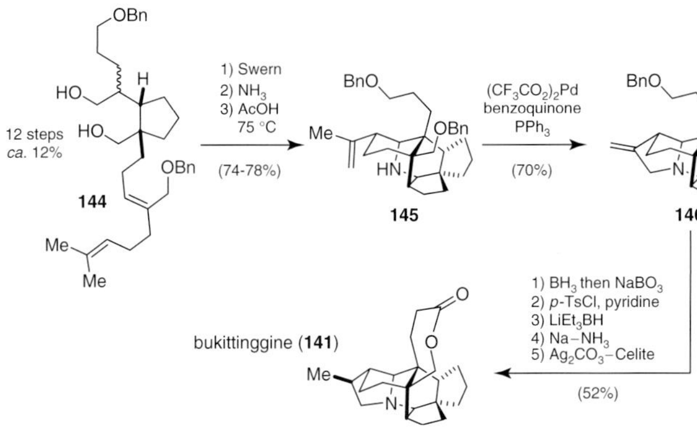

Scheme 8.30 Heathcock's synthesis of bukittinggine [102].

followed by tosylation and reduction to allow the conversion of the exocyclic double bond into the β-methyl derivative (along with negligible amounts of its α-methyl isomer). Removal of the benzyl protecting groups with sodium in condensed ammonia followed by oxidative lactonization with silver carbonate on Celite (Fétizon's reagent) completed the synthesis of (±)-bukittinggine (**141**) in 52% yield from **146**.

The syntheses of methyl homodaphniphyllate (**142**) and daphnilactone A (**143**) [103b] gave another example of biomimetic tetracyclization reaction, but also provided arguments in favor of the hypothesis according to which the secodaphnane skeleton might be the initial biosynthetic skeleton of *Daphniphyllum* alkaloids [96a, 108] (despite the scarce occurrence of the related alkaloids compared to those of the daphnane type). In a plausible scenario [109], the secodaphnane skeleton (**147**) can be converted into intermediate **148** by a fragmentation process initiated by oxidation either at C9 (**149a**, path A in Scheme 8.31) or on the nitrogen atom

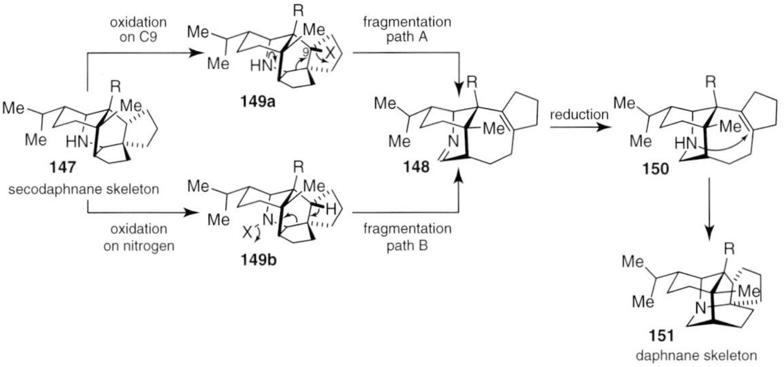

Scheme 8.31 Plausible mechanism for the conversion of the secodaphnane skeleton into the daphnane skeleton [109].

(**149b**, path B in Scheme 8.31). Reduction of the imine to secondary amine **150** and nucleophilic attack of the nitrogen on the double bond can then lead to the daphnane skeleton (**151**).

For the syntheses of **142** and **143** [103b], cyclic ether **152** (obtained in 73% yield over six steps) was submitted to the tetracyclization conditions described above and gave hexacyclic intermediate **153** (synthetic equivalent of **149a**) in 47% yield (Scheme 8.32). Selective reduction of the terminal double bond by hydrogenation over Adam's catalyst followed by treatment with diisobutylaluminum hydride (DIBAL-H) provided fragmentation product **154** (synthetic equivalent of **150**) in 70% yield for the two steps. In this process, DIBAL-H serves both as activator for the alkoxy leaving group and as reducing agent for the transient iminium ion. Intermediate **154** can then lead to two different skeletons: Jones oxidation of the primary hydroxyl into the corresponding carboxylic acid followed by treatment with aqueous formaldehyde at pH 7 yielded (±)-daphnilactone A (**143**) in 50% yield, whereas successive treatment of **154** with phenyl isothiocyanate, refluxing formic acid, and methanolic potassium hydroxide furnished the daphnane skeleton-bearing intermediate **155**. Oxidation of the hydroxyl followed by Fischer esterification completed the synthesis of (±)-methyl homodaphniphyllate (**142**) in 70% yield over the last five steps.

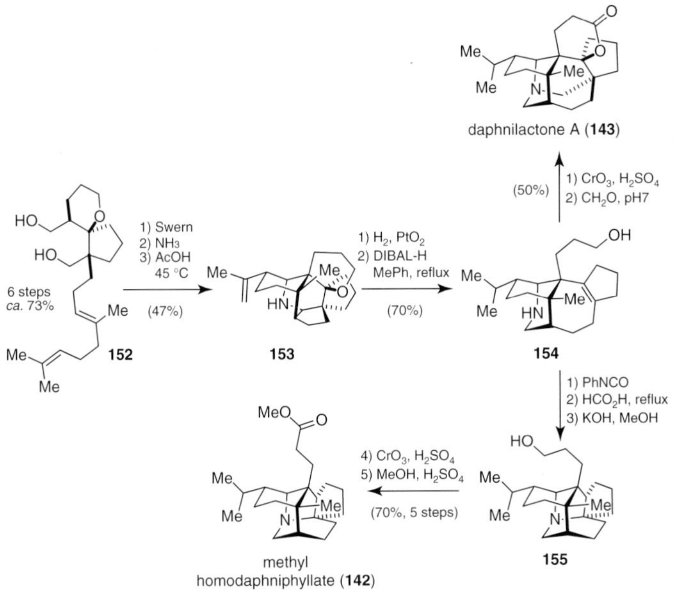

Scheme 8.32 Heathcock's syntheses of methyl homodaphniphyllate and daphnilactone A [103b].

These results are in good agreement with the postulated biosynthetic conversion of the secodaphnane skeleton into the daphnane skeleton through oxidation at C9 (path A on Scheme 8.31). The relevance of the N-oxidation-mediated fragmentation scenario (path B on Scheme 8.31) was also studied by the authors [109] but will not be detailed here.

In the early 1990s [110], the Heathcock group reported results concerning the biomimetic total synthesis of *proto*-daphniphylline (**136**) in which they took their key polycyclization reaction one step further. In this work they achieved the pentacyclization of dialdehyde **156** (partially reduced equivalent of biosynthetic intermediate **128**, see Scheme 8.28), directly into **136**, under conditions similar to those described above for the tetracyclization process (i.e., treatment with gaseous ammonia followed by warm acetic acid), but in rather low yield (15%). Surprisingly, a few years later [95a], the authors reported that under the same conditions, but with methylamine (which was used by accident) in place of ammonia, the pentacyclization took place and afforded compound **157** (reduced equivalent of **136**) in 65% overall yield (Scheme 8.33).

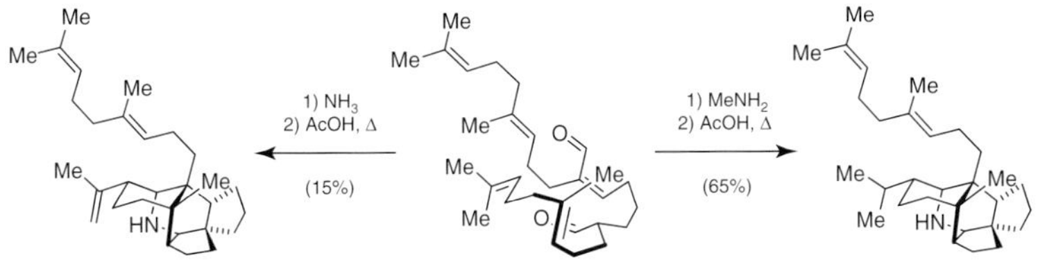

proto-daphniphylline (**136**) **156** **157**

Scheme 8.33 Pentacyclization of **156** into either **136** or **157** [95a].

The difference in efficacy between the tetracyclization and pentacyclization processes, when carried out with ammonia, points out the formation of the first cyclopentane ring as the limiting reaction. When methylamine is used instead of ammonia, the secondary enamine **158b** that is formed after condensation with **156** is certainly more nucleophilic than primary enamine **158a** (Scheme 8.34), thus resulting in much higher overall yield.

As for the spontaneous reduction of the isoprenyl group, it is probably due to the fact that after the Prins-like reaction converting **159** into **160**, the reduction of the carbocation leading to **161** occurs via a 1,5-hydride shift instead of a 1,5-proton transfer as proposed for the conversion of **162** into **136-H$^+$** (Scheme 8.34). The authors pointed out that, since no isoprenyl-bearing *Daphniphyllum* alkaloid has ever been reported, the source of the biosynthetic nitrogen is very likely to be an alkylamine such as pyridoxamine or an amino acid [95a].

During the conversion of **156** into **157**, seven bonds and five rings are formed in a fully diastereoselective way and one out of three similar double bonds is regioselectively saturated.

8.6 Conclusion

Throughout this short survey we have seen that biomimetic strategies can be applied successfully to alkaloids of non-amino acid origin. In some cases, the

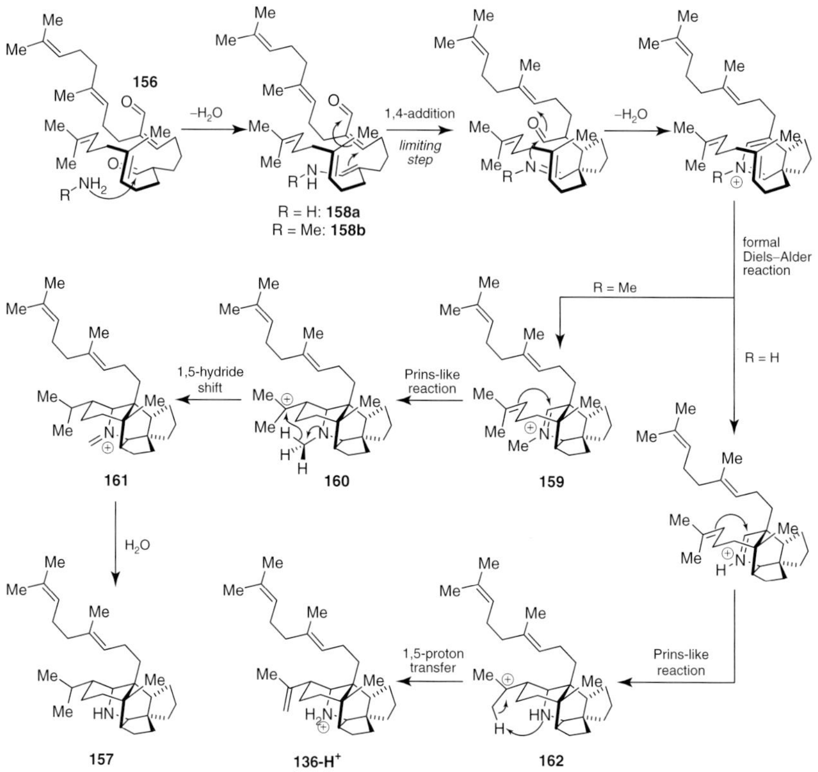

Scheme 8.34 Proposed pentacyclization mechanism.

strategies provided by biosynthetic hypotheses have been applied to total syntheses of the target molecules, sometimes drastically improving the yields of delicate transformations and, most of all, the ease of the global synthetic process. In other cases, biomimetic strategies have been shown to be useful for the formation of particular regions of interest, which often display structural complexity or originality.

Interestingly, in nearly all studies presented above, the key biomimetic steps concern the nitrogen-containing moieties of the molecules (the only exception is found in the biomimetic transformations of the cephalostatin/ritterazine spiroketal end). In these examples, the nitrogen atom is either involved directly in the chemical transformations or serves as an activator for cycloaddition processes.

Many of the efforts presented in this chapter are only models and still need to be confirmed or taken a few steps further by completing the syntheses or evaluating their relevance with appropriate substrates. Moreover, several families of alkaloids of non-amino acid origin have not (or very scarcely) been studied in terms of biomimetic synthesis yet and would definitely deserve to be. This is especially true for terpene-derived molecules, such as diterpenoid alkaloids, which can display impressive complex structures (e.g., *Delphinium* alkaloids; see the structure of barbaline, Figure 8.10).

Another issue that, in many cases, needs to be dealt with is the biosynthetic origin of the nitrogen atom(s) found in the molecules. The work of the Heathcock group on *Daphniphyllum* alkaloids regarding this question stands as an exception and the findings were, to some extent, the fruit of thorough analysis of accidental results.

References

1. Brown, R.F.C., Drummond, R., Fogerty, A.C., Hughes, G.K., Pinhey, J.T., Ritchie, E., and Taylor, W.C. (1956) *Aust. J. Chem.*, **9**, 283–287.
2. (a) Pinhey, J.T., Ritchie, E., and Taylor, W.C. (1961) *Aust. J. Chem.*, **9**, 106–134; (b) Binns, S.V., Dunstan, P.J., Guise, G.B., Holder, G.M., Hollis, A.F., McCredie, R.S., Pinhey, J.T., Prager, R.H., Rasmussen, M., Ritchie, E., and Taylor, W.C. (1965) *Aust. J. Chem.*, **18**, 569–573; (c) Mander, L.N., Prager, R.H., Rasmussen, M., Ritchie, E., and Taylor, W.C. (1967) *Aust. J. Chem.*, **20**, 1473–1491; (d) Mander, L.N., Prager, R.H., Rasmussen, M., and Ritchie, E. (1967) *Aust. J. Chem.*, **20**, 1705–1718.
3. Ritchie, E. and Taylor, W.C. (1967) in *The Alkaloids*, vol. IX (eds R.H.F. Manske and H.L. Holmes), Academic Press, New York, pp. 529–544.
4. The stereochemistry of Class II and Class II alkaloids has recently been revised. See: Willis, A.C., O'Connor, P.D., Taylor, W.C., and Mander, L.N. (2006) *Aust. J. Chem.*, **59**, 629–632.
5. Mander, L.N., Willis, A.C., Hertl, A.J., and Taylor, W.C. (2009) *Tetrahedron Lett.*, **50**, 7089–7092.
6. (a) Tchabanenko, K., Adlington, R.M., Cowley, A.R., and Baldwin, J.E. (2005) *Org. Lett.*, **7**, 585–588; (b) Tchabanenko, K., Chesworth, R., Parker, J.S., Anand, N.K., Russell, A.T., Adlington, R.M., and Baldwin, J.E. (2005) *Tetrahedron*, **61**, 11649–11656.
7. Movassaghi, M., Hunt, D.K., and Tjandra, M. (2006) *J. Am. Chem. Soc.*, **128**, 8126–8127 (see Supporting Information for full biosynthetic hypothesis).
8. Kozikowski, A.P., Fauq, A.H., Miller, J.H., and McKinney, M. (1992) *Bioorg. Med. Chem. Lett.*, **2**, 797–802.
9. (a) Hart, D.J., Wu, W.-L., and Kozikowski, A.P. (1995) *J. Am. Chem. Soc.*, **117**, 9369–9370; (b) Hart, D.J., Li, J., Wu, W., and Kozikowski, A.P. (1997) *J. Org. Chem.*, **62**, 5023–5033; (c) Chackalamannil, S., Davies, R.J., Asberom, T., Doller, D., and Leone, D. (1996) *J. Am. Chem. Soc.*, **118**, 9812–9813; (d) Chakalamannil, S., Davies, R.J., Wang, Y., Asberom, T., Doller, D., Wong, J., and Leone, D. (1999) *J. Org. Chem.*, **64**, 1932–1940; (e) Takadoi, M., Katoh, T., Ishiwata, A., and Terashima, S. (1999) *Tetrahedron Lett.*, **40**, 3399–3402; (f) Takadoi, M., Katoh, T., Ishiwata, A., and Terashima, S. (2002) *Tetrahedron*, **58**, 9903–9923; (g) De Baecke, G., and De Clercq, P.J. (1995) *Tetrahedron Lett.*, **36**, 7515–7518; (h) Hofman, S., De Baecke, G., Kenda, B., and De Clercq, P.J. (1998) *Synthesis*, 479–489; (i) Hofman, S., Gao, L.-J., Van Dingenen, H., Hosten, N.G.C., Van Haver, D., De Clercq, P.J., Milanesio, M., and Viterbo, D. (2001) *Eur. J. Org. Chem.*, 2851–2860; (j) Wong, L.S.-M. and Sherburn, M.S. (2003) *Org. Lett.*, **5**, 3603–3606.
10. Movassaghi, M., Tjandra, M., and Qi, J. (2009) *J. Am. Chem. Soc.*, **131**, 9648–9650.
11. O'Connor, P.D., Mander, L.N., and McLachlan, M.M.W. (2004) *Org. Lett.*, **6**, 703–706.
12. (a) Zi, W., Yu, S., and Ma, D. (2010) *Angew. Chem., Int. Ed. Engl.*, **49**, 5887–5890; (b) Larson, K.K. and Sarpong, R. (2009) *J. Am. Chem.

Soc., **131**, 13244–13245; (c) Evans, D.A. and Adams, D.J. (2007) *J. Am. Chem. Soc.*, **129**, 1048–1049; (d) Shah, U., Chackalamannil, S., Ganguly, A.K., Chelliah, M., Kolotuchin, S., Bulevich, A., and McPhail, A. (2006) *J. Am. Chem. Soc.*, **128**, 12654–12655; (e) Mander, L.N. and McLachlan, M.M. (2003) *J. Am. Chem. Soc.*, **125**, 2400–2401.

13. For a recent review, see: (a) Guéret, S.M. and Brimble, M.A. (2010) *Nat. Prod. Rep.*, **27**, 1350–1366; (b) Hu, T.M., Burton, I.W., Cembella, A.D., Curtis, J.M., Quilliam, M.A., Walter, J.A., and Wright, J.L.C. (2001) *J. Nat. Prod.*, **64**, 308–312; (c) Falk, M., Burton, I.W., Hu, T., Walter, J.A., and Wright, J.L.C. (2001) *Tetrahedron*, **57**, 8659–8665; (d) Hu, T.M., Curtis, J.M., Walter, J.A., and Wright, J.L.C. (1996) *Tetrahedron Lett.*, **37**, 7671–7674.
14. (a) Torigoe, K., Murata, M., Yasumoto, T., and Iwashita, T. (1988) *J. Am. Chem. Soc.*, **110**, 7876–7877; (b) Hu, T.M., deFreitas, A.S.W., Curtis, J.M., Oshima, Y., Walter, J.A., and Wright, J.L.C. (1996) *J. Nat. Prod.*, **59**, 1010–1014.
15. Lu, C.K., Lee, G.H., Huang, R., and Chou, H.N. (2001) *Tetrahedron Lett.*, **42**, 1713–1716.
16. Kita, M., Kondo, M., Koyama, T., Yamada, K., Matsumoto, T., Lee, K.-H., Woo, J.-T., and Uemura, D. (2004) *J. Am. Chem. Soc.*, **126**, 4794–4795.
17. (a) Maier, M. and Varseev, G. (2006) *Angew. Chem., Int. Ed. Engl.*, **45**, 4767–4771; (b) Sakai, E., Araki, K., Takamura, H., and Uemura, D. (2006) *Tetrahedron Lett.*, **47**, 6343–6345; (c) Snider, B.B. and Che, Q. (2006) *Angew. Chem., Int. Ed. Engl.*, **45**, 932–935; (d) Zou, Y., Che, Q., and Snider, B.B. (2006) *Org. Lett.*, **8**, 5605–5608; (e) Kim, J. and Thomson, R.J. (2007) *Angew. Chem., Int. Ed. Engl.*, **46**, 3104–3106; (f) Born, S., Bacani, G., Olson, E.E., and Kobayashi, Y. (2008) *Synlett*, 2877–2881; (g) Born, S. and Kobayashi, Y. (2008) *Synlett*, 2479–2482.
18. Kita, M., Ohishi, N., Washida, K., Kondo, M., Koyama, T., Yamada, K., and Uemura, D. (2005) *Bioorg. Med. Chem.*, **13**, 5253–5258.
19. Roush, W.R. (1992) in *Comprehensive Organic Synthesis*, vol. 5 (ed. B.M. Trost), Pergamon Press, Oxford, pp. 513–550.
20. Burke, J.P., Sabat, M., Iovan, D.A., Myers, W.H., and Chruma, J.J. (2010) *Org. Lett.*, **12**, 3192–3195.
21. (a) Uemura, D., Chou, T., Haino, T., Nagatsu, A., Fukuzawa, S., Zheng, S.-Z., and Chens, H.-S. (1995) *J. Am. Chem. Soc.*, **117**, 1155–1156; (b) Chou, T., Kamo, O., and Uemura, D. (1996) *Tetrahedron Lett.*, **37**, 4023–4026.
22. (a) Takada, N., Umemura, N., Suenaga, K., Chou, T., Nagatsu, A., Haino, T., Yamada, K., and Uemura, D. (2001) *Tetrahedron Lett.*, **42**, 3491–3494; (b) Matsuura, F., Hao, J., Reents, R., and Kishi, Y. (2006) *Org. Lett.*, **8**, 3327–3330.
23. Chou, T., Haino, T., Kuramoto, M., and Uemura, D. (1996) *Tetrahedron Lett.*, **37**, 4027–4030.
24. Selwood, A.I., Miles, C.O., Wilkins, A.L., van Ginkel, R., Munday, R., Rise, F., and McNabb, P. (2010) *J. Agric. Food Chem.*, **58**, 6532–6542.
25. (a) Takada, N., Umemura, N., Suenaga, K., and Uemura, D. (2001) *Tetrahedron Lett.*, **42**, 3495–3497; (b) Hao, J., Matsuura, F., Kishi, Y., Kita, M., Uemura, D., Asai, N., and Iwashita, T. (2006) *J. Am. Chem. Soc.*, **128**, 7742–7743.
26. See, for example: (a) McCauley, J.A., Nagasawa, K., Lander, P.A., Mischke, S.G., Semones, M.A., and Kishi, Y. (1998) *J. Am. Chem. Soc.*, **120**, 7647–7648; (b) Matsuura, F., Peters, R., Anada, M., Harried, S.S., Hao, J., and Kishi, Y. (2006) *J. Am. Chem. Soc.*, **128**, 7463–7465; (c) Sakamoto, S., Sakazaki, H., Hagiwara, K., Kamada, K., Ishii, K., Noda, T., Inoue, M., and Hirama, M. (2004) *Angew. Chem., Int. Ed.*, **43**, 6505–6510; (d) Stivala, C.E., and Zakarian, A. (2008) *J. Am. Chem. Soc.*, **130**, 3774–3776; (e) Ishihara, J., Tojo, S., Kamikawa, A., and Murai, A. (2001) *Chem. Commun.*, 1392–1393; (f) Suthers, B.D., Jacobs, M.F., and Kitching, W. (1998) *Tetrahedron Lett.*,

39, 2621–2624; (g) Nakamura, S., Kikuchi, F., and Hashimoto, S. (2008) *Angew. Chem., Int. Ed.*, **47**, 7091–7094.

27. Seki, T., Satake, M., Mackenzie, L., Kaspar, H.F., and Yasumoto, T. (1995) *Tetrahedron Lett.*, **36**, 7093–7096.

28. Kharrat, R., Servent, D., Girard, E., Ouanounou, G., Amar, M., Marrouchi, R., Benoit, E., and Molgó, J. (2008) *J. Neurochem.*, **107**, 952–963.

29. Stewart, M., Blunt, J.W., Munro, M.H.G., Robinson, W.T., and Hannah, D.J. (1997) *Tetrahedron Lett.*, **38**, 4889–4890.

30. Miles, C.O., Wilkins, A.L., Stirling, D.J., and MacKenzie, A.L. (2000) *J. Agric. Food Chem.*, **48**, 1373–1376.

31. Miles, C.O., Wilkins, A.L., Stirling, D.J., and MacKenzie, A.L. (2003) *J. Agric. Food Chem.*, **51**, 4838–4840.

32. The efforts of the Romo group have led to the only total synthesis known to date: (a) Kong, K., Romo, D., and Lee, C. (2009) *Angew. Chem., Int. Ed.*, **48**, 7402–7405; (b) Kong, K., Moussa, Z., and Romo, D. (2005) *Org. Lett.*, **7**, 5127–5130; (c) Ahn, Y., Cardenas, G.I., Yang, J., and Romo, D. (2001) *Org. Lett.*, **3**, 751–754; (d) Yang, J., Cohn, S.T., and Romo, D. (2000) *Org. Lett.*, **2**, 763–766; For synthetic studies by other research groups, see: (e) Johannes, J.W., Wenglowsky, S., and Kishi, Y. (2005) *Org. Lett.*, **7**, 3997–4000; (f) Tsujimoto, T., Ishihara, J., Horie, M., and Murai, A. (2002) *Synlett*, 399–402; (g) Ishihara, J., Miyakawa, J., Tsujimoto, T., and Murai, A. (1997) *Synlett*, 1417–1419; (h) White, J.D., Quaranta, L., and Wang, G. (2007) *J. Org. Chem.*, **72**, 1717–1728; (i) White, J.D., Wang, G., and Quaranta, L. (2003) *Org. Lett.*, **5**, 4983–4986; (J) White, J.D., Wang, G., and Quaranta, L. (2003) *Org. Lett.*, **5**, 4109–4112.

33. Morita, H., Oshimi, S., Hirasawa, Y., Koyama, K., Honda, T., Ekasari, W., Indrayanto, G., and Zaini, N.C. (2007) *Org. Lett.*, **9**, 3691–3693.

34. Matsumoto, T., Kobayashi, T., Ishida, K., Hirasawa, Y., Morita, H., Honda, T., and Kamata, K. (2010) *Biol. Pharm. Bull.*, **33**, 844–848.

35. (a) Rudyanto, M., Tomizawa, Y., Morita, H., and Honda, T. (2008) *Org. Lett.*, **10**, 1921–1922; (b) Morita, H., Tomizawa, Y., Deguchi, J., Ishikawa, T., Arai, H., Zaima, K., Hosoya, T., Hirasawa, Y., Matsumoto, T., Kamata, K., Ekasari, W., Widyawaruyanti, A., Wahyuni, T.S., Zaini, N.C., and Honda, T. (2009) *Bioorg. Med. Chem.*, **17**, 8234–8240.

36. Yao, Y.S. and Yao, Z.J. (2008) *J. Org. Chem.*, **73**, 5221–5225.

37. Murti, S. (1949) *Proc. Indian Sci. Sect. A*, **30**, 107–112.

38. (a) Garraffo, H.M., Spande, T.F., Daly, J.W., Baldessari, A., and Gros, E.G. (1993) *J. Nat. Prod.*, **56**, 357–373; (b) Spande, T.F., Jain, P., Garraffo, H.M., Pannell, L.K., Yeh, H.J.C., Daly, J.W., Fukumoto, S., Imamura, K., Tokuyama, T., Torres, J.A., Snelling, R.R., and Jones, T.H. (1999) *J. Nat. Prod.*, **62**, 5–21; (c) Jones, T.H., Gorman, J.S.T., Snelling, R.R., Delabie, J.H.C., Blum, M.S., Garraffo, H.M., Jain, P., Daly, J.W., and Spande, T.F. (1999) *J. Chem. Ecol.*, **25**, 1179–1193; (d) Daly, J.W., Garraffo, H.M., Jain, P., Spande, T.F., Snelling, R.R., Jaramillo, C., and Rand, A.S. (2000) *J. Chem. Ecol.*, **26**, 73–85.

39. (a) Steffan, B. (1991) *Tetrahedron*, **47**, 8729–8732; (b) Kubanek, J., Williams, D.E., Dilip de Silva, E., Allen, T., and Andersen, R.J. (1995) *Tetrahedron Lett.*, **36**, 6189–6192; (c) Davis, R.A., Carroll, A.R., and Quinn, R.J. (2002) *J. Nat. Prod.*, **65**, 454–457; (d) Wright, A.D., Goclik, E., König, G.M., and Kaminsky, R. (2002) *J. Med. Chem.*, **45**, 3067–3072.

40. (a) Witkop, B. and Gossinger, E. (1983) in *The Alkaloids*, vol. 21 (ed. A. Brossi), Academic Press, New York, pp. 139–253; (b) Daly, J.W. and Spande, T.F. (1986) in *Alkaloids: Chemical and Biological Perspectives*, vol. 4 (ed. S.W. Pelletier), John Wiley & Sons, Inc., New York, pp. 1–274; (c) Daly, J.W., Garraffo, H.M., and Spande, T.F. (1993) in *The Alkaloids*, vol. 43 (ed. G.A. Cordell), Academic Press, San Diego, pp. 185–288; (d) Daly, J.W. (1998) in *The Alkaloids*, vol. 50 (ed. G.A. Cordell), Academic Press, New York, pp. 141–169; (e) Daly, J.W., Garraffo, H.M., and Spande, T.F. (1999) in *Alkaloids: Chemical and Biological Perspectives*, vol. 13 (ed. S.W. Pelletier), Pergamon, New York, pp. 1–161; (f) Daly, J.W., Spande, T.F., and Garraffo, H.M. (2005) *J. Nat. Prod.*, **68**, 1556–1575.

41. (a) Daly, J.W., Ware, N., Saporito, R.A., Spande, T.F., and Garraffo, H.M. (2009) *J. Nat. Prod.*, **72**, 1110–1114; (b) Daly, J.W., Spande, T.F., and Garraffo, H.M. (2005) *J. Nat. Prod.*, **68**, 1556–1575; (c) Tokuyama, T., Tsujita, T., Shimada, A., Garraffo, H.M., Spande, T.F., and Daly, J.W. (1991) *Tetrahedron*, **47**, 5401–5414; (d) Tokuyama, T., Nishimori, N., Karle, I.L., Edwards, M.W., and Daly, J.W. (1986) *Tetrahedron*, **42**, 3453–3460.

42. Garraffo, H.M., Caceres, J., Daly, J.W., Spande, T.F., Andriamaharavo, N.R., and Andriantsiferana, M. (1993) *J. Nat. Prod.*, **56**, 1016–1038.

43. (a) Daly, J.W., Kaneko, T., Wilham, J., Garraffo, H.M., Spande, T.F., Espinosa, A., and Donnelly, M.A. (2002) *Proc. Natl. Acad. Sci. U.S.A.*, **99**, 13996–14001; (b) Daly, J.W. (1995) *Proc. Natl. Acad. Sci. U.S.A.*, **92**, 9–13.

44. Some recent achievements: (a) Voituriez, A., Ferreira, F., Pérez-Luna, A., and Chemla, F. (2007) *Org. Lett.*, **9**, 4705–4708; (b) Lee, C.-L.K. and Loh, T.-P. (2005) *Org. Lett.*, **7**, 2965–2967; (c) Holub, N., Neidhöfer, J., and Blechert, S. (2005) *Org. Lett.*, **7**, 1227–1229; (d) Davis, F.A. and Yang, B. (2005) *J. Am. Chem. Soc.*, **127**, 8398–8407; (e) Wrobleski, A., Sahasrabudhe, K., and Aubé, J. (2004) *J. Am. Chem. Soc.*, **126**, 5475–5481; (f) Wrobleski, A., Sahasrabudhe, K., and Aubé, J. (2002) *J. Am. Chem. Soc.*, **124**, 9974–9975; (g) Toyooka, N., Fukutome, A., Nemoto, H., Daly, J.W., Spande, T.F., Garraffo, H.M., and Kaneko, T. (2002) *Org. Lett.*, **4**, 1715–1717; (h) for a review on older syntheses of dendrobatid alkaloids, see: Kibayashi, C., Aoyagi, S. (1997) in *Studies in Natural Products Chemistry*, vol. 19 (ed. A.U. Rahman), Elsevier Science B.V., Amsterdam, pp. 3–88; for biomimetic approaches to dendrobatid alkaloids, see: (i) Glanzmann, M., Karalai, C., Ostersehlt, B., Schön, U., Frese, C., and Winterfeldt, E. (1982) *Tetrahedron*, **38**, 2805–2810; (J) Bonin, M., Royer, J., Grierson, D.S., and Husson, H.-P. (1986) *Tetrahedron Lett.*, **27**, 1569–1572; for reviews on biomimetic syntheses, see: (k) Scholz, U. and Winterfeldt, E. (2000) *Nat. Prod. Rep.*, **17**, 349–366; (l) Nicolaou, K.C., Montagnon, T., and Snyder, S.A. (2003) *Chem. Commun.*, 551–564; (m) De la Torre, M.C. and Sierra, M.A. (2004) *Angew. Chem., Int. Ed. Engl.*, **43**, 160–181.

45. (a) Amat, M., Fabregat, R., Griera, R., Florindo, P., Molins, E., and Bosch, J. (2010) *J. Org. Chem.*, **75**, 3797–3805; (b) Amat, M., Fabregat, R., Griera, R., and Bosch, J. (2009) *J. Org. Chem.*, **74**, 1794–1797; (c) Amat, M., Griera, R., Fabregat, R., Molins, E., and Bosch, J. (2008) *Angew. Chem., Int. Ed.*, **47**, 3348–3351; (d) Escolano, C., Amat, M., and Bosch, J. (2006) *Chem. Eur. J.*, **12**, 8198–8207.

46. Winterfeldt, E. (1979) *Heterocycles*, **12**, 1631–1650.

47. (a) Numerous enantioselective syntheses of *cis*-195A have been reported so far, see for example: (a) Oppolzer, W. and Flaskamp, E. (1977) *Helv. Chim. Acta*, **60**, 204–207; (b) Murahashi, S.-I., Sasao, S., Saito, E., and Naota, T. (1993) *Tetrahedron*, **49**, 8805–8826; (c) Comins, D.L. and Dehghani, A. (1993) *J. Chem. Soc., Chem. Commun.*, 1838–1839; (d) Naruse, M., Aoyagi, S., and Kibayashi, C. (1996) *J. Chem. Soc., Perkin Trans. 1*, 1113–1124; (e) Riechers, T., Krebs, H.C., Wartchow, R., and Habermehl, G. (1998) *Eur. J. Org. Chem.*, 2641–2646; (f) Dijk, E.W., Panella, L., Pinho, P., Naasz, R., Meetsma, A., Minaard, A.J., and Feringa, B.L. (2004) *Tetrahedron*, **60**, 9687–9693.

48. (a) Rao, C.B., Anjaneyula, A.S.R., Sarma, N.S., Venkatateswarlu, Y., Rosser, R.M., Faulkner, D.J., Chen, M.H.M., and Clardy, J. (1984) *J. Am. Chem. Soc.*, **106**, 7983–7984; (b) Rao,

C.B., Anjaneyula, A.S.R., Sarma, N.S., Venkatateswarlu, Y., Rosser, R.M., and Faulkner, D.J. (1985) *J. Org. Chem.*, **50**, 3757–3760; (c) Rao, C.B., Roa, D.V., and Raju, V.S.N. (1989) *Heterocycles*, **28**, 103–106.

49. (a) Fukuzawa, Y., Hayashi, Y., Uemura, D., Nagatsu, A., Yamada, Y., and Ijuin, Y. (1995) *Heterocycl. Commun.*, **1**, 207–214; (b) Kuramoto, M., Hayashi, K., Fujitani, Y., Yamaguchi, K., Tsuji, T., Yamada, K., Ijuin, Y., and Uemura, D. (1997) *Tetrahedron Lett.*, **38**, 5683–5686.

50. (a) Daranas, A.H., Fernández, J.J., Gavín, J.A., and Norte, M. (1998) *Tetrahedron*, **54**, 7891–7896; (b) Daranas, A.H., Fernández, J.J., Gavín, J.A., and Norte, M. (1999) *Tetrahedron*, **55**, 5539–5546.

51. See, for example: (a) Rahman, A.U., Alvi, K.A., Abbas, S.A., Choudhary, M.I., and Clardy, J. (1989) *Tetrahedron Lett.*, **30**, 6825–6828; (b) Fattorusso, E., Romano, A., Taglialatela-Scafati, O., Achmad, M.J., Bavestrello, G., and Cerrano, C. (2008) *Tetrahedron Lett.*, **49**, 2189–2192.

52. Yamaguchi, K., Yada, M., Tsuji, T., Kuramoto, M., and Uemura, D. (1999) *Biol. Pharm. Bull.*, **22**, 920–924.

53. Venkateswarlu, Y., Reddy, N.S., Ramesh, P., Reddy, P.S., and Jamil, K. (1998) *Heterocycl. Commun.*, **4**, 575–580.

54. Villar, R.M., Gil-Longo, J., Daranas, A.H., Souto, M.L., Fernández, J.J., Peixinho, S., Barral, M.A., Santafé, G., Rodríguez, J., and Jiménez, C. (2003) *Bioorg. Med. Chem.*, **11**, 2301–2306.

55. For a recent review, see: Behenna, D.C., Stockdill, J.L., and Stoltz, B.M. (2008) *Angew. Chem., Int. Ed.*, **47**, 2365–2386.

56. (a) Miyashita, M., Sasaki, M., Hattori, I., Sakai, M., and Tanino, K. (2004) *Science*, **305**, 495–499; (b) Sakai, M., Sasaki, M., Tanino, K., and Miyashita, M. (2002) *Tetrahedron Lett.*, **43**, 1705–1708; (c) Miyashita, M. (2007) *Pure Appl. Chem.*, **79**, 651–655; (d) Yoshimura, F., Sasaki, M., Hattori, I., Komatsu, K., Sakai, M., Tanino, K., and Miyashita, M. (2009) *Chem. Eur. J.*, **15**, 6626–6644.

57. (a) Murata, Y., Yamashita, D., Kitahara, K., Minasako, Y., Nakazaki, A., and Kobayashi, S. (2009) *Angew. Chem., Int. Ed.*, **48**, 1400–1403; (b) Yamashita, D., Murata, Y., Hikage, N., Takao, K., Nakazaki, A., and Kobayashi, S. (2009) *Angew. Chem., Int. Ed.*, **48**, 1404–1406.

58. (a) Hikage, N., Furukawa, H., Takao, K., and Kobayashi, S. (2000) *Chem. Pharm. Bull.*, **48**, 1370–1372; (b) Hikage, N., Furukawa, H., Takao, K., and Kobayashi, S. (1998) *Tetrahedron Lett.*, **39**, 6237–6240; (c) Hikage, N., Furukawa, H., Takao, K., and Kobayashi, S. (1998) *Tetrahedron Lett.*, **39**, 6241–6244.

59. (a) Kuramoto, M., Hayashi, K., Yamaguchi, K., Yamada, K., Tsuji, T., and Uemura, D. (1998) *Bull. Chem. Soc. Jpn.*, **71**, 771–779; (b) Uemura, D. (2006) *Chem. Rec.*, **6**, 235–248.

60. (a) McMahon, T. and Silke, J. (1996) *Harmful Algae News*, **14**, 2; (b) for a review on azaspiracid shellfish poisoning see: Twiner, M.J., Rehmann, N., Hess, P., and Doucette, G.J. (2008) *Mar. Drugs*, **6**, 39–72.

61. Satake, M., Ofuji, K., Naoki, H., James, K.J., Furey, A., McMahon, T., Silke, J., and Yasumoto, T. (1998) *J. Am. Chem. Soc.*, **120**, 9967–9968.

62. (a) Ofuji, K., Satake, M., McMahon, T., Silke, J., James, K.J., Naoki, H., Oshima, Y., and Yasumoto, T. (1999) *Nat. Toxins*, **7**, 99–102; (b) Ofuji, K., Satake, M., McMahon, T., James, K.J., Naoki, H., Oshima, Y., and Yasumoto, T. (2001) *Biosci. Biotechnol. Biochem.*, **65**, 740–742; (c) James, K.J., Sierra, M.D., Lehane, M., Magdalena, A.B., and Furey, A. (2003) *Toxicon*, **41**, 277–283.

63. Furey, A., Moroney, C., Magdalena, A.B., Saez, M.J.F., Lehane, M., and James, K.J. (2003) *Environ. Sci. Technol.*, **37**, 3078–3084.

64. (a) First thought to be *Protoperidinium crassipes* (Protoperidiniaceae), see: James, K.J., Moroney, C., Roden, C., Satake, M., Yasumoto, T., Lehane, M., and Furey, A. (2003) *Toxicon*, **41**, 145–151; (b) but later corrected to

Azadinium spinosum (Peridiniphycidae), see: Tillmann, U., Elbracchter, M., Krock, B., John, U., and Cembella, A. (2009) *Eur. J. Phycol.*, **44**, 63–79.

65. Rehmann, N., Hess, P., and Quilliam, M.A. (2008) *Rapid Commun. Mass Spectrom.*, **22**, 549–558.

66. (a) Twiner, M.J., Hess, P., Dechraoui, M.Y.B., McMahon, T., Samons, M.S., Satake, M., Yasumoto, T., Ramsdell, J.S., and Doucette, G.J. (2005) *Toxicon*, **45**, 891–900; (b) Colman, J.R., Twiner, M.J., Hess, P., McMahon, T., Satake, M., Yasumoto, T., Doucette, G.J., and Ramsdell, J.S. (2005) *Toxicon*, **45**, 881–890; (c) Ronzitti, G., Hess, P., Rehmann, N., and Rossini, G.P. (2007) *Toxicol. Sci.*, **95**, 427–435; (d) Vilarino, N., Nicolaou, K.C., Frederick, M.O., Cagide, E., Ares, I.R., Louzao, M.C., Vieytes, M.R., and Botana, L.M. (2006) *Chem. Res. Toxicol.*, **19**, 1459–1466; (e) Kulagina, N.V., Twiner, M.J., Hess, P., McMahon, T., Satake, M., Yasumoto, T., Ramsdell, J.S., Doucette, G.J., Ma, W., and O'Shaughnessy, T.J. (2006) *Toxicon*, **47**, 766–773; (f) Alfonso, A., Vieytes, M.R., Ofuji, K., Satake, M., Nicolaou, K.C., Frederick, M.O., and Botana, L.M. (2006) *Biochem. Biophys. Res. Commun.*, **346**, 1091–1099; (g) Vale, C., Nicolaou, K.C., Frederick, M.O., Gomez-Limia, B., Alfonso, A., Vieytes, M.R., and Botana, L.M. (2007) *J. Med. Chem.*, **50**, 356–363; (h) Bellocci, M., Sala, G.L., Callegari, F., and Rossini, G.P. (2010) *Toxicol. Sci.*, **117**, 109–121; (i) Cao, Z., LePage, K.T., Frederick, M.O., Nicolaou, K.C., and Murray, T.F. (2010) *Toxicol. Sci.*, **114**, 323–334; (J) Vale, C., Nicolaou, K.C., Frederick, M.O., Vieytes, M.R., and Botana, L.M. (2010) *Toxicol. Sci.*, **113**, 158–168; (k) Kellmann, R., Schaffner, C.A., Grønset, T.A., Satake, M., Ziegler, M., and Fladmark, K.E. (2009) *J. Proteomics*, **72**, 695–707.

67. (a) Nicolaou, K.C., Pihko, P.M., Diedrichs, N., Zou, N., and Bernal, F. (2001) *Angew. Chem., Int. Ed.*, **40**, 1262–1265; (b) Nicolaou, K.C., Qian, W.Y., Bernal, F., Uesaka, N., Pihko, P.M., and Hinrichs, J. (2001) *Angew. Chem., Int. Ed.*, **40**, 4068–4071; (c) Carter, R.G. and Weldon, D.J. (2000) *Org. Lett.*, **2**, 3913–3916; (d) Carter, R.G., Bourland, T.C., and Graves, D.E. (2002) *Org. Lett.*, **4**, 2177–2179; (e) Carter, R.G., Graves, D.E., Gronemeyer, M.A., and Tschumper, G.S. (2002) *Org. Lett.*, **4**, 2181–2184; (f) Zhou, X.T. and Carter, R.G. (2004) *Chem. Commun.*, 2138–2140; (g) Zhou, X.T., Lu, L., Furkert, D.P., Wells, C.E., and Carter, R.G. (2006) *Angew. Chem., Int. Ed.*, **45**, 7622–7626; (h) Zhou, X.T. and Carter, R.G. (2006) *Angew. Chem., Int. Ed.*, **45**, 1787–1790; (i) Dounay, A.B. and Forsyth, C.J. (2001) *Org. Lett.*, **3**, 975–978; (J) Aiguade, J., Hao, J.L., and Forsyth, C.J. (2001) *Org. Lett.*, **3**, 979–982; (k) Forsyth, C.J., Hao, J.L., and Aiguade, J. (2001) *Angew. Chem., Int. Ed.*, **40**, 3663–3667; (l) Geisler, L.K., Nguyen, S., and Forsyth, C.J. (2004) *Org. Lett.*, **6**, 4159–4162; (m) Nguyen, S., Xu, J.Y., and Forsyth, C.J. (2006) *Tetrahedron*, **62**, 5338–5346; (n) Forsyth, C.J., Xu, J.Y., Nguyen, S.T., Samdal, I.A., Briggs, L.R., Rundberget, T., Sandvik, M., and Miles, C.O. (2006) *J. Am. Chem. Soc.*, **128**, 15114–15116; (o) Li, Y.F., Zhou, F., and Forsyth, C.J. (2007) *Angew. Chem., Int. Ed.*, **46**, 279–282; (p) Ishikawa, Y. and Nishiyama, S. (2004) *Heterocycles*, **63**, 539–565; (q) Ishikawa, Y. and Nishiyama, S. (2004) *Heterocycles*, **63**, 885–893; (r) Oikawa, M., Uehara, T., Iwayama, T., and Sasaki, M. (2006) *Org. Lett.*, **8**, 3943–3946; (s) Yadav, J.S., Joyasawal, S., Dutta, S.K., and Kunwar, A.C. (2007) *Tetrahedron Lett.*, **48**, 5335–5340; (t) Yadav, J.S. and Venugopal, C. (2007) *Synlett*, 2262–2266; (u) Li, X., Li, J., and Mootoo, D.R. (2007) *Org. Lett.*, **9**, 4303–4306.

68. (a) Nicolaou, K.C., Li, Y.W., Uesaka, N., Koftis, T.V., Vyskocil, S., Ling, T.T., Govindasamy, M., Qian, W., Bernal, F., and Chen, D.Y.K. (2003) *Angew. Chem., Int. Ed. Engl.*, **42**, 3643–3648; (b) Nicolaou, K.C., Chen, D.Y.K., Li, Y.W., Qian, W.Y., Ling, T.T., Vyskocil, S., Koftis, T.V., Govindasamy, M., and Uesaka, N. (2003) *Angew. Chem., Int. Ed.*, **42**,

3649–3653; (c) Nicolaou, K.C., Pihko, P.M., Bernal, F., Frederick, M.O., Qian, W.Y., Uesaka, N., Diedrichs, N., Hinrichs, J., Koftis, T.V., Loizidou, E., Petrovic, G., Rodriquez, M., Sarlah, D., and Zou, N. (2006) *J. Am. Chem. Soc.*, **128**, 2244–2257; (d) Nicolaou, K.C., Chen, D.Y.K., Li, Y.W., Uesaka, N., Petrovic, G., Koftis, T.V., Bernal, F., Frederick, M.O., Govindasamy, M., Ling, T.T., Pihko, P.M., Tang, W.J., and Vyskocil, S. (2006) *J. Am. Chem. Soc.*, **128**, 2258–2267.

69. (a) Nicolaou, K.C., Vyskocil, S., Koftis, T.V., Yamada, Y.M.A., Ling, T.T., Chen, D.Y.K., Tang, W.J., Petrovic, G., Frederick, M.O., Li, Y.W., and Satake, M. (2004) *Angew. Chem., Int. Ed.*, **43**, 4312–4318; (b) Nicolaou, K.C., Koftis, T.V., Vyskocil, S., Petrovic, G., Ling, T.T., Yamada, Y.M.A., Tang, W.J., and Frederick, M.O. (2004) *Angew. Chem., Int. Ed.*, **43**, 4318–4324; (c) Nicolaou, K.C., Koftis, T.V., Vyskocil, S., Petrovic, G., Tang, W.J., Frederick, M.O., Chen, D.Y.K., Li, Y.W., Ling, T.T., and Yamada, Y.M.A. (2006) *J. Am. Chem. Soc.*, **128**, 2859–2872; (d) Nicolaou, K.C., Frederick, M.O., Loizidou, E.Z., Petrovic, G., Cole, K.P., Koftis, T.V., and Yamada, Y.M.A. (2006) *Chem. Asian J.*, **1**, 245–263; (e) Nicolaou, K.C., Frederick, M.O., Petrovic, G., Cole, K.P., and Loizidou, E.Z. (2006) *Angew. Chem., Int. Ed.*, **45**, 2609–2615.

70. (a) Evans, D.A., Kværnø, L., Mulder, J.A., Raymer, B., Dunn, T.B., Beauchemin, A., Olhava, E.J., Juhl, M., and Kagechika, K. (2007) *Angew. Chem., Int. Ed.*, **46**, 4693–4697; (b) Evans, D.A., Dunn, T.B., Kværnø, L., Beauchemin, A., Raymer, B., Olhava, E.J., Mulder, J.A., Juhl, M., Kagechika, K., and Favor, D.A. (2007) *Angew. Chem., Int. Ed.*, **46**, 4698–4703; (c) Evans, D.A., Kværnø, L., Dunn, T.B., Beauchemin, A., Raymer, B., Mulder, J.A., Olhava, E.J., Juhl, M., Kagechika, K., and Favor, D.A. (2008) *J. Am. Chem. Soc.*, **130**, 16295–16309.

71. Kalaitzis, J.A., Chau, R., Kohli, G.S., Murray, S.A., and Neilan, B.A. (2010) *Toxicon*, **56**, 244–258.

72. Bringmann, G., Hamm, A., Kraus, J., Ochse, M., Noureldeen, A., and Jumbam, D.N. (2001) *Eur. J. Org. Chem.*, 1983–1987.

73. For a recent review, see: Wang, F.-P., Chen, Q.-H., and Liu, X.-Y. (2010) *Nat. Prod. Rep.*, **27**, 529–570.

74. See, for example: (a) Ata, A. and Andersh, B.J. (2008) *Alkaloids Chem. Biol.*, **66**, 191–213; (b) Li, H.J., Jiang, Y., and Li, P. (2006) *Nat. Prod. Rep.*, **23**, 735–752; (c) Rahman, A.U. and Choudhary, M.I. (1999) *Nat. Prod. Rep.*, **16**, 619–635, and references therein.

75. For a recent review, see: Kobayashi, J. and Kubota, T. (2009) *Nat. Prod. Rep.*, **26**, 936–962.

76. For a review, see: Moser, B.R. (2008) *J. Nat. Prod.*, **71**, 487–491.

77. Pettit, G.R., Inoue, M., Kamano, Y., Herald, D.L., Arm, C., Dufresne, C., Christie, N.D., Schmidt, J.M., Doubek, D.L., and Krupa, T.S. (1988) *J. Am. Chem. Soc.*, **110**, 2006–2007.

78. (a) Pettit, G.R., Inoue, M., Kamano, Y., Dufresne, C., Christie, N.D., Niven, M.L., and Herald, D.L. (1988) *J. Chem. Soc., Chem. Commun.*, 865–867; (b) Pettit, G.R., Kamano, Y., Dufresne, C., Inoue, M., Christie, N.D., Schmidt, J.M., and Doubek, D.L. (1989) *Can. J. Chem.*, **67**, 1509–1513; (c) Pettit, G.R., Kamano, Y., Inoue, M., Dufresne, C., Boyd, M.R., Herald, C.L., Schmidt, J.M., Doubek, D.L., and Christie, N.D. (1992) *J. Org. Chem.*, **57**, 429–431; (d) Pettit, G.R., Xu, J.P., Williams, M.D., Christie, N.D., Doubek, D.L., and Schmidt, J.M. (1994) *J. Nat. Prod.*, **57**, 52–63; (e) Pettit, G.R., Ichihara, Y., Xu, J.P., Boyd, M.R., and Williams, M.D. (1994) *Bioorg. Med. Chem. Lett.*, **4**, 1507–1512; (f) Pettit, G.R., Xu, J.P., Ichihara, Y., and Williams, M.D. (1994) *Can. J. Chem.*, **72**, 2260–2267; (g) Pettit, G.R., Xu, J.P., and Schmidt, J.M. (1995) *Bioorg. Med. Chem. Lett.*, **5**, 2027–2032; (h) Pettit, G.R., Tan, R., Xu, J.P., Ichihara, Y., Williams, M.D., and Boyd, M.R. (1998) *J. Nat. Prod.*, **61**, 955–958.

79. Fukuzawa, S., Matsunaga, S., and Fusetani, N. (1994) *J. Org. Chem.*, **59**, 6164–6166.

80. (a) Fukuzawa, S., Matsunaga, S., and Fusetani, N. (1995) *J. Org. Chem.*, **60**, 608–614; (b) Fukuzawa, S., Matsunaga, S., and Fusetani, N. (1995) *Tetrahedron*, **51**, 6707–6716; (c) Fukuzawa, S., Matsunaga, S., and Fusetani, N. (1997) *J. Org. Chem.*, **62**, 4484–4491.

81. (a) For a recent review, see: Lee, S., LaCour, T.G., and Fuchs, P.L. (2009) *Chem. Rev.*, **109**, 2275–2314; for recent achievements, see: (b) Fortner, K.C., Kato, D., Tanaka, Y., and Shair, M.D. (2010) *J. Am. Chem. Soc.*, **132**, 275–280; (c) Poza, J.J., Rodríguez, J., and Jiménez, C. (2010) *Bioorg. Med. Chem.*, **18**, 58–63.

82. Lee, J.S. and Fuchs, P.L. (2005) *J. Am. Chem. Soc.*, **127**, 13122–13123.

83. (a) Jeong, J.U. and Fuchs, P.L. (1994) *Tetrahedron Lett.*, **35**, 5385–5388; (b) Jeong, J.U. and Fuchs, P.L. (1995) *Tetrahedron Lett.*, **36**, 2431–2434.

84. (a) Jeong, J.U., Sutton, S.C., Kim, S., and Fuchs, P.L. (1995) *J. Am. Chem. Soc.*, **117**, 10157–10158; (b) Jeong, J.U., Guo, C., and Fuchs, P.L. (1999) *J. Am. Chem. Soc.*, **121**, 2071–2084.

85. Smith, S.C. and Heathcock, C.H. (1992) *J. Org. Chem.*, **57**, 6379–6380.

86. Guo, C., Bhandaru, S., Fuchs, P.L., and Boyd, M.R. (1996) *J. Am. Chem. Soc.*, **118**, 10672–10673.

87. (a) Drogemuller, M., Jautelat, R., and Winterfeldt, E. (1996) *Angew. Chem., Int. Ed. Engl.*, **35**, 1572–1574; (b) Haak, E. and Winterfeldt, E. (2004) *Synlett*, 1414–1418.

88. Yagi, S. (1909) *Kyoto Igaku Zasshi*, **6**, 208–222.

89. Nakano, T., Saeki, Y., Gibbons, C.S., and Trotter, J. (1968) *J. Chem. Soc., Chem. Commun.*, 600–601.

90. Li, C.S., Di, Y.T., He, H.P., Gao, S., Wang, Y.H., Lu, Y., Zhong, J.L., and Hao, X.J. (2007) *Org. Lett.*, **9**, 2509–2512.

91. Mu, S.Z., Yang, X.W., Di, Y.T., He, H.P., Wang, Y., Wang, Y.H., Li, L., and Hao, X.J. (2007) *Chem. Biodivers.*, **4**, 129–138.

92. Saito, S., Yahata, H., Kubota, T., Obara, Y., Nakahata, N., and Kobayashi, J. (2008) *Tetrahedron*, **64**, 1901–1908.

93. (a) Morita, H. and Kobayashi, J. (2002) *Tetrahedron*, **58**, 6637–6641; (b) Kobayashi, J., Ueno, S., and Morita, H. (2002) *J. Org. Chem.*, **67**, 6546–6549; (c) Zhang, Y., He, H.P., Di, Y.T., Mu, S.Z., Wang, Y.H., Wang, J.S., Li, C.S., Kong, N.C., Gao, S., and Hao, X.J. (2007) *Tetrahedron Lett.*, **48**, 9104–9107.

94. (a) Suzuki, K.T., Okuda, S., Niwa, H., Toda, M., Hirata, Y., and Yamamura, S. (1973) *Tetrahedron Lett.*, **14**, 799–802; (b) Niwa, H., Hirata, Y., Suzuki, K.T., and Yamamura, S. (1973) *Tetrahedron Lett.*, **14**, 2129–2132.

95. (a) Heathcock, C.H. (1996) *Proc. Natl. Acad. Sci. U.S.A.*, **93**, 14323–14327; (b) Heathcock, C.H. (1992) *Angew. Chem., Int. Ed. Engl.*, **31**, 665–681.

96. (a) Ruggeri, R.B. and Heathcock, C.H. (1989) *Pure Appl. Chem.*, **61**, 289–292; (b) Heathcock, C.H., Piettre, S., Ruggeri, R.B., Ragan, J.A., and Kath, J.C. (1992) *J. Org. Chem.*, **57**, 2554–2566.

97. For the isolation of methyl homosecodaphniphyllate, see: Toda, M., Hirata, Y., and Yamamura, S. (1972) *Tetrahedron*, **14**, 1477–1484.

98. (a) Ruggeri, R.B., Hamen, M.M., and Heathcock, C.H. (1988) *J. Am. Chem. Soc.*, **110**, 8734–8736; (b) Heathcock, C.H., Hansen, M.M., Ruggeri, R.B., and Kath, J.C. (1992) *J. Org. Chem.*, **57**, 2544–2553.

99. For the isolation of secodaphniphylline, see: Irikawa, H., Toda, M., Yamamura, S., and Hirata, Y. (1969) *Tetrahedron Lett.*, **9**, 1821–1824.

100. (a) Stafford, J.A. and Heathcock, C.H. (1990) *J. Org. Chem.*, **55**, 5433–5434; (b) Heathcock, C.H. and Stafford, J.A. (1992) *J. Org. Chem.*, **57**, 2566–2574.

101. (a) For the isolation of bukittinggine, see: Arbain, D., Byrne, L.T., Cannon, J.R., Patrick, V.A., and White, A.H. (1990) *Aust. J. Chem.*, **43**, 185–190; (b) for bukittinggine-related alkaloids, see: Zhang, C.R., Yang, S.P., and Yue, J.M. (2008) *J. Nat. Prod.*, **71**, 1663–1668.

102. Heathcock, C.H., Stafford, J.A., and Clark, D.L. (1992) *J. Org. Chem.*, **57**, 2575–2584.

103. (a) Ruggeri, R.B. and Heathcock, C.H. (1990) *J. Org. Chem.*, **55**, 3714–3715; (b) Heathcock, C.H., Ruggeri, R.B., and McClure, K.F. (1992) *J. Org. Chem.*, **57**, 2585–2594.

104. For the isolation of daphnilactone A, see: (a) Sasaki, K. and Hirata, Y. (1972) *Tetrahedron Lett.*, **13**, 1275–1278; (b) Sasaki, K. and Hirata, Y. (1972) *J. Chem. Soc., Perkin Trans. 2*, 1411–1415.

105. Ruggeri, R.B., McClure, K.F., and Heathcock, C.H. (1989) *J. Am. Chem. Soc.*, **111**, 1530–1531.

106. For the isolation of codaphniphylline, see: Irakawa, H., Sakabe, N., Yamamura, S., and Hirata, Y. (1968) *Tetrahedron*, **24**, 5691–5700.

107. Heathcock, C.H., Kath, J.C., and Ruggeri, R.B. (1996) *J. Org. Chem.*, **60**, 1120–1130.

108. (a) Niwa, H., Toda, M., Ishimaru, S., Hirata, Y., and Yamamura, S. (1974) *Tetrahedron*, **30**, 3031–3036; (b) Yamamura, S. and Hirata, Y. (1976) *Chem. Lett.*, 1381–1382.

109. Heathcock, C.H. and Joe, D. (1995) *J. Org. Chem.*, **60**, 1131–1142.

110. Piettre, S. and Heathcock, C.H. (1990) *Science*, **248**, 1532–1534.

9
Biomimetic Synthesis of Azole- and Aryl-Peptide Alkaloids

Hans-Dieter Arndt, Roman Lichtenecker, Patrick Loos, and Lech-Gustav Milroy

Das Runde muss ins Eckige.
Sepp Herberger

9.1
Introduction

In general, peptides are not traditionally considered alkaloids. Therefore, this and the following chapter will have to stretch the borders of the field to identify common principles among structurally related molecules of various origins. We illustrate similarities and differences between typical peptide-derived natural products, which either are "alkaloids" in the sense of the typically assumed definition – that is, a generally alkaline (=basic), bioactive compound, preferentially of green plant origin – or are so much related to them either by biosynthetic roots or by structure that a unifying discussion may be appropriate.

Owing to the high number of molecules, synthetic works, and structure–function investigations – all triggered by highly intriguing molecular structures and many promising bioactivities of peptide alkaloids – it is impossible to give a comprehensive treatment here. We apologize for the unavoidable, arbitrary selection of topics, molecules, and syntheses, and for the omission of all the excellent studies we could not cover in detail or even reference.

9.1.1
Peptide Alkaloids: An Overview

Peptide alkaloids occupy an ill-defined subsection of the rich structural universe of alkaloid natural products. They show similarities to many other peptide natural products, for example, lantibiotics or cyclo(depsi)peptides. A somewhat arbitrary distinction has to be made between the many amino-acid-derived alkaloids, for example, indole, diketopiperazine, or the ergot alkaloids such as lysergic acid (**1**) on the one hand, and linear peptides with basic properties such as dolastatin 15 (**2**) on the other hand. Historically, only cyclopeptides like **3**, which defied the "typical" alkaloid or peptide classification and were basic and from plant origin, were termed "*peptide alkaloids.*" However, substances with various degrees of

Biomimetic Organic Synthesis, First Edition. Edited by Erwan Poupon and Bastien Nay.

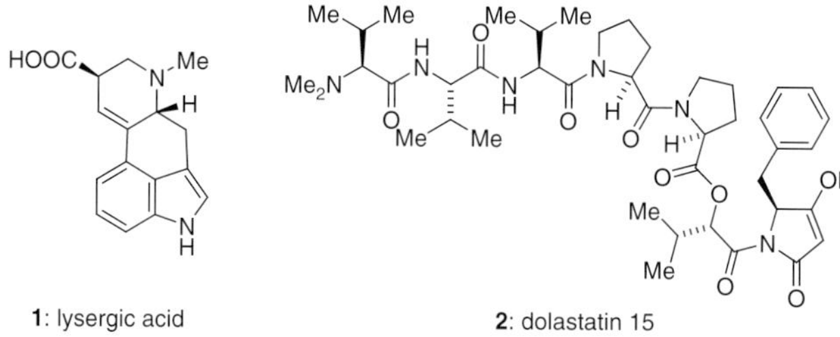

1: lysergic acid   **2**: dolastatin 15

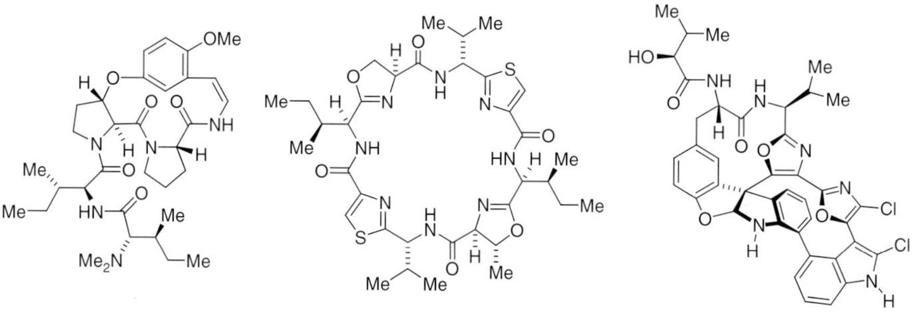

3: ziziphine A   **4**: patellamide A   **5**: diazonamide A

Figure 9.1 Common alkaloid structures (**1** and **2**) in comparison with typical peptide alkaloids (**3**–**5**).

alkaloid and peptide elements are more often found and their number is steadily growing (Figure 9.1). Nonetheless, a recent classification that includes microbial and marine alkaloids (such as **4** or **5**) remains unavailable. Therefore, this chapter and the one following are devoted to molecules that:

- are composed of at least three amino acids with peptide bonds;
- are preferentially cyclic, but go beyond ordinary cyclo(depsi)peptide structure;
- feature additional motifs such as non-trivial (i.e., non-amino acid side chain) heterocycles, biaryls, or condensation loci distinguishing them from regular peptides.

9.1.2
Sources of Peptide Alkaloids

Peptide alkaloids have been first isolated from plants, but can be found in many lower organisms as well, for example, in bacteria, cyanobacteria, fungi, and marine invertebrates such as ascidians. For many of the compounds the actual producing organism has not been finally elucidated. Some materials isolated from plants in fact resemble products from bacterial or cyanobacterial origin. Indeed, in some cases it was shown that biosynthetically complex, highly modified secondary metabolites were produced by symbiotic microorganisms within the higher organism host from which the compounds have been isolated [1].

9.1.3 Key Features of Biosynthesis

The common feedstock for all peptide alkaloids is amino acids, which get connected to canonical peptide chains and are then further modified. The backbones are assembled either by ribosomal peptide synthesis (RPS) [2] from genetically encoded templates on ribosomes or by non-ribosomal peptide synthesis (NRPS) [3] on dedicated multiprotein clusters (Figure 9.2). With RPS, a homochiral L-peptide is generated at the ribosome as a synthesis template ("structure peptide"), which typically carries a "leader peptide" as a tag for molecular recognition by the maturation enzymes. After maturation of the structure peptide, the leader peptide is removed. This process has only recently become more apparent [4]. With NRPS, more building blocks than only the gene-encoded amino acids are available for chain growth, which ensues in a conveyor-belt fashion on multiprotein complexes. In this case chemical modification can occur already during the assembly process. Both types of products then can undergo final enzymatic tailoring after assembly, such as methylation, prenylation, glycosidation, or acylation. Sometimes integration with other biosynthesis pathways (RPS→NRPS, NRPS→polyketide, NRPS→alkaloid) is observed, leading to additional building blocks or mixed-biosynthesis products.

Canonical peptide chains are rather flexible and tend to fold into stable secondary structures only if their length exceeds 50–70 residues – with some exceptions in highly disulfide-rich "miniproteins." Modification by post-assembly maturation and tailoring enzymes let the peptide mature into a structurally more rigid peptide alkaloid (Scheme 9.1a). Typical rigidifying elements are macrocycle formation, heterocycles formed from side-chain condensation reactions, or oxidative connections of side-chains, primarily at electron-rich aromatic side chains from Phe, Tyr, or Trp (Scheme 9.1b). Conceptually, this is reminiscent of terpene biosynthesis, where a flexible oligo-isoprenoid template is set up by oligomerization from isopentenyl pyrophosphate, which is followed by a distinct cyclization and tailoring phase [5]. This generates polycycles and oxidizes the chains at reactive loci (mostly double bonds) to arrive at very unique, defined structures. Just as in the case of terpenes, it can be assumed that diverse target structures can be reached from a single peptide precursor material by such a "covalent folding" – only depending on the topology of the maturation pathway. This will certainly offer benefits during evolution processes and organism adaptation, because small changes in biosynthesis can then generate entirely different structures. Of course, peptide alkaloids rarely reach the close-knit structural complexity of carbocyclic terpenes, but their richer heteroatom chemistry and the many functional groups available can lead to very specific shapes and drug-like activity from only a few amino acids as precursor material. Finally, post-synthetic tailoring might further complete the biosynthesis by adding alkyl groups or sugar residues to the core scaffold.

For many peptide alkaloids it is not yet clear how the peptide template becomes assembled. Both RPS and NRPS have been documented in several cases. Furthermore, the interconnection of assembly and maturation remains mostly speculative. Biosynthesis research will still have to show at which steps the maturation happens

Ribosomal peptide synthesis (RPS)	**Non-ribosomal peptide synthesis (NRPS)**
- Gene-encoded, mRNA templated synthesis at the ribosome	- Dedicated, MDa-protein-cluster-architecture-encoded synthesis
- Generic variation of oligonucleotide template possible	- Less generic protein-cluster template, variation difficult
- tRNA-mediated delivery of 20 proteinogenic amino acids (AA)	- Amino acyl-AMP-mediated delivery of >200 non-proteinogenic AA
- Chain growth as tRNA-ester at the ribosome	- Chain growth as phosphopantheinyl-thioester at the NRPS cluster
- No chain modifications during growth	- Chain modifications during growth typical (redox, epimerization)
- Hydrolytic release as canonical L-peptide chain tagged with a leader peptide	- Diverse release pathways (hydrolysis, reduction, macrocycle formation) generating an untagged synthesis product
- Leader-peptide-mediated enzymatic maturation - Protease-mediated liberation from leader peptide	- Maturation during growth and/or after release possible
- Post-release modifications (alkylations, building block addition)	- Post-assembly tailoring (alkylations, glycosidations)

Figure 9.2 Distinct biosynthetic aspects of RPS and NRPS pathway logics.

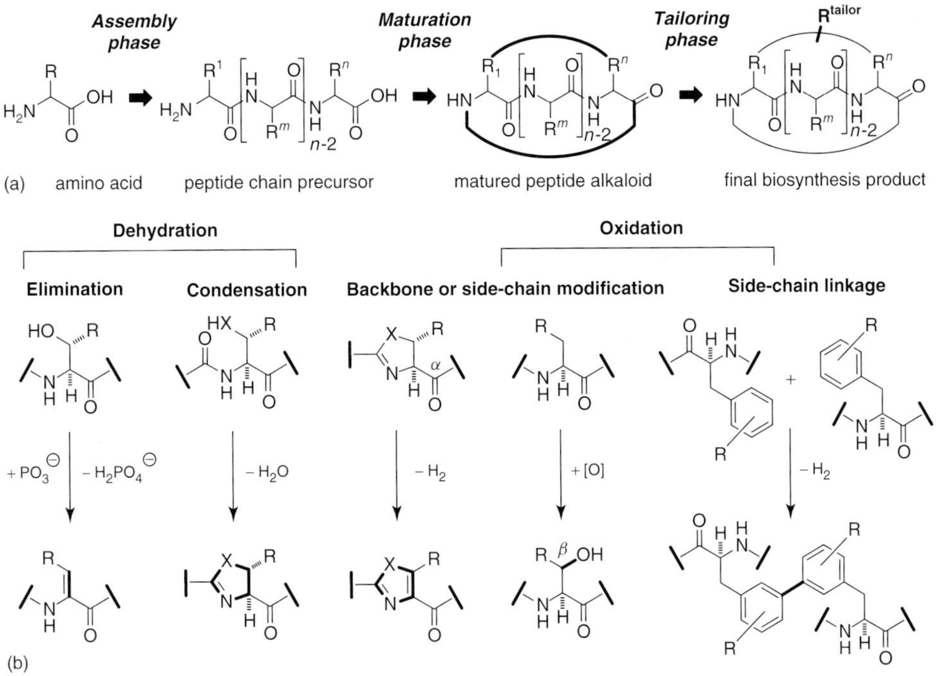

Scheme 9.1 (a) Covalent folding of peptide chains into peptide alkaloid scaffolds (schematically simplified); (b) typical maturation steps in peptide alkaloid biosyntheses.

and if it indeed follows a model with distinct phases. However, such a model will provide a guiding principle for biomimetic synthesis strategies. For the sake of chemical discussion and to highlight similar features, we will often boldly simplify the structures in question in a "retrobiosynthetic" simplification guided by the typically available biosynthetic transformations (Scheme 9.1b), and attempt to align key synthetic contributions with the characteristic features of this impressive natural product class. This chapter is devoted to azole- and aryl-peptide alkaloids, whereas the following chapter will illustrate indole peptide alkaloids and the ecteinascidines.

9.2 Azole-Containing Peptide Alkaloids

9.2.1 Structural Features

Azole-containing molecules comprise probably the most diverse family among the peptide alkaloids. Their unifying feature is the occurrence of thioazol(in)e and oxazol(in)e units embedded in a peptide chain (Figure 9.3). These modifications

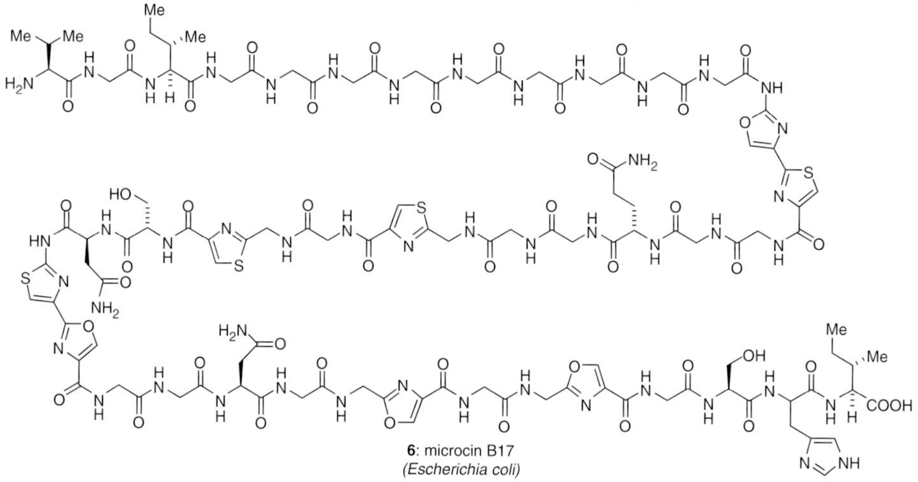

6: microcin B17
(Escherichia coli)

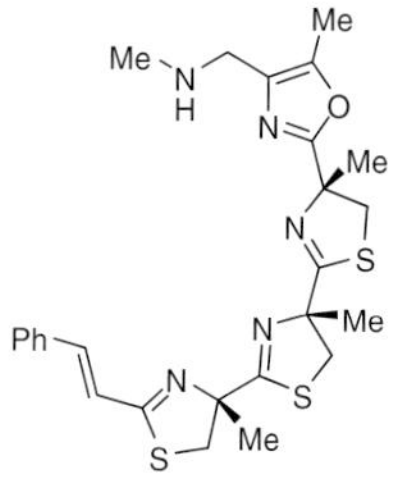

7: thiangazole
(Polyangium sp.*)*

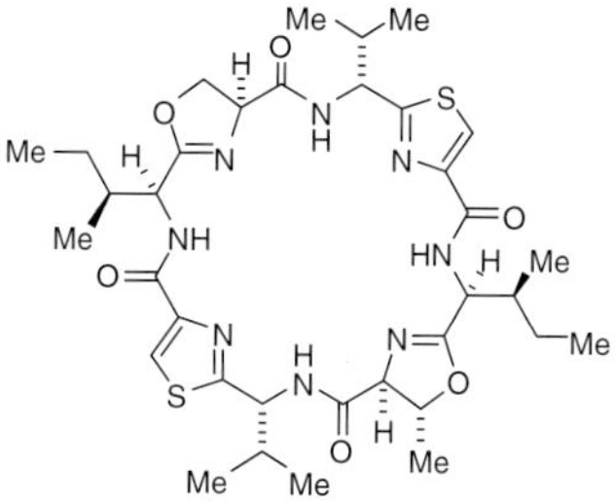

4: patellamide A
(Prochloron spp.*)*

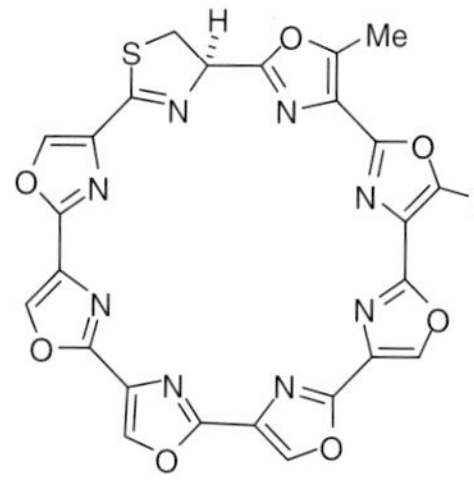

8: telomestatin
(Streptomyces annulatus)

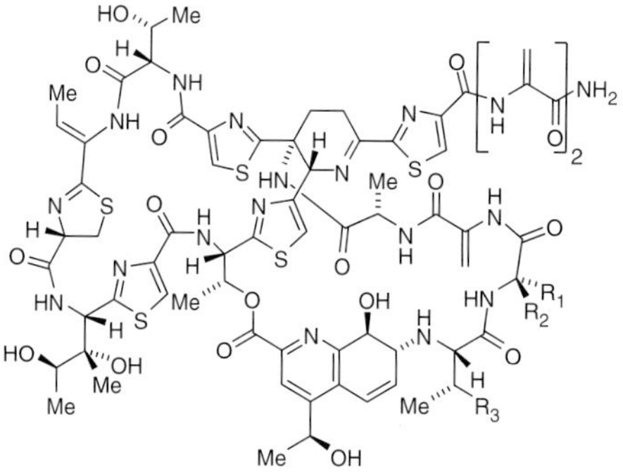

9: thiostrepton (*Streptomyces azureus*): $R_1 = CH_3$, $R_2 = H$, $R_3 = CH_2CH_3$
10: siomycin A (*Streptomyces sioyaensis*): $R_1 = R_2 = CH_2$ (dehydroalanine), $R_3 = CH_3$

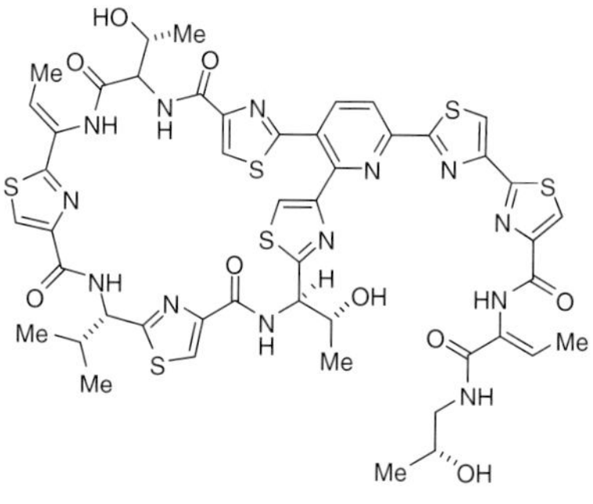

11: micrococcin P1
(*Micrococcus* spp.)

Figure 9.3 Azole-containing peptide alkaloids.

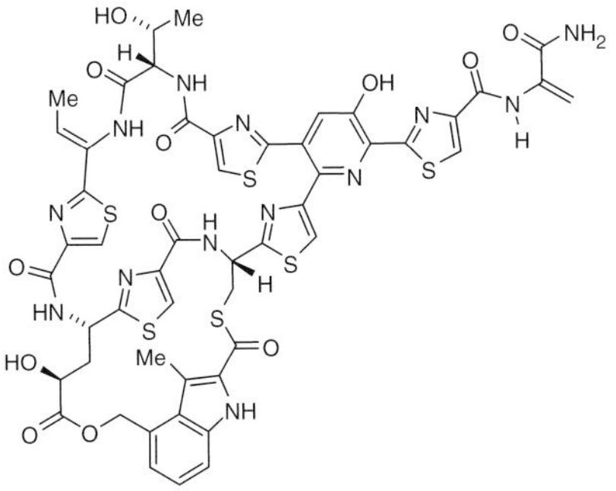

12: nosiheptide
(Streptomyces actuosus)

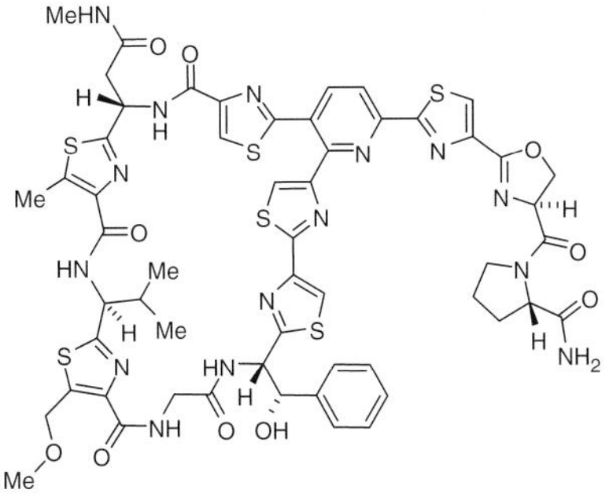

13: GE2270A
(Planobispora rosea)

Figure 9.3 *(Continued)*

can occur with a low frequency, such as exemplified in the long-chain molecule microcin B17 (**6**), or with impressive density, as can be seen with telomestatin (**8**) or thiostrepton (**9**), for instance. However, in most cases the amino acid nature of their backbone composition is still readily discerned. They can be found as linear [microcin B17, thiangazole (**7**)] and monocyclic azol(in)e-containing peptide alkaloids [lissoclinamides, patellamides (**4**), telomestatin], and as the structurally more complex thiopeptide (also called *thiazolyl peptide*) antibiotics, such as thiostrepton, micrococcin P1 (**11**), nosiheptide (**12**), or GE2270A (**13**). Most of them were retrieved from microorganisms of terrestrial or marine origin, prominently from *Actinomyces* and cyanobacteria.

9.2.2
Biomimetic Elements in Azole-Containing Peptide Alkaloids

Tailoring by hydroxylation, epoxidation, glycosidation, or alkylation/prenylation has been found for some azole-containing peptide alkaloids, but the large majority of them are composed purely from naturally occurring amino acid building blocks, also with respect to the stereochemistry of individual residues. Their key feature is the occurrence of azole heterocycles in the running peptide chain – mostly oxazol(in)es and/or thiazol(in)es. Owing to their complex molecular structures and their mostly microbial origin, it was assumed for many of them that their assembly would arise from NRPS pathways. However, Walsh *et al.* showed that microcin B17 (**6**) is formed from genetically encoded peptides by RPS [6]. A multi-enzyme complex transforms the linear peptide precursor by heterocycle formation [7].

While it is certainly too early to generalize, this trend has gained momentum in recent years, when RPS was shown to be involved with the *Lissoclinum* peptides, the patellamides (**4**), and the thiopeptide antibiotics as well (**9–13**). In all cases a serine, threonine, or cysteine peptide is dehydrated (cyclodehydratase, McbB for **6**) and then sometimes further oxidized (dehydrogenase, McbC for **6**) to generate peptide-embedded heterocycles [8].

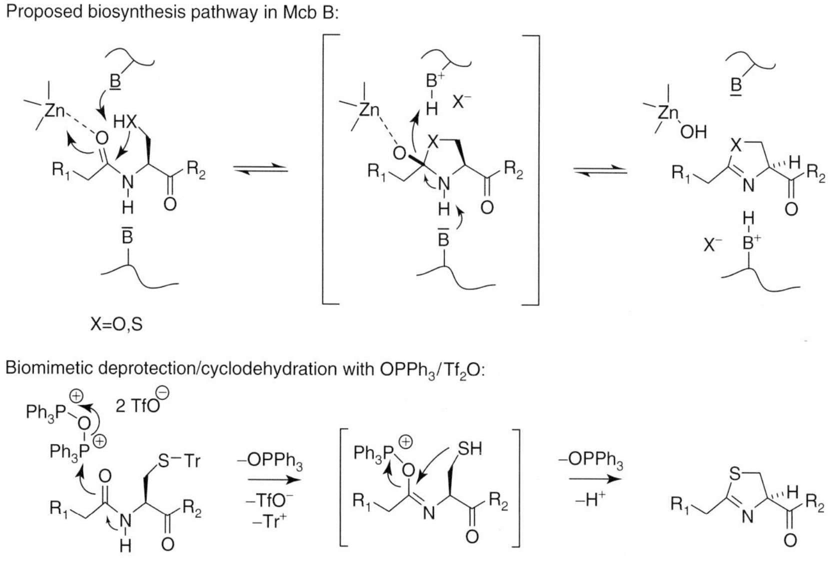

Scheme 9.2 Enzymatic heterocycle formation and biomimetic cyclodehydration.

Mechanistically, the dehydratase enables a nucleophilic side-chain attack at an amide carbon (Scheme 9.2), which is not easily induced with chemical reagents. These typically activate the side chain [9, 10]. However, for thiazolines such a condensation was achieved with oxophilic Lewis acids such as $TiCl_4$ [11] or Tf_2O-activated triphenylphosphine oxide in a biomimetic fashion [12], leading directly to peptide heterocycles (Scheme 9.2). The access to oxazoles by this method is less biomimetic, as the corresponding hydroxy group has to be oxidized to the ketone prior to ring closure and then serves as the electrophile [13]. Furthermore, at sensitive functionalities and longer peptide chains the limits of mimicking enzymes operating on peptide chain backbones with bulk reagents are currently reached rather quickly.

9.2.3
Thiangazole

The oligo-thiazoline thiangazole (**7**) was isolated from *Polyangium* spp. (strain PI 3007) and showed HIV-inhibitory properties [14, 15]. Together with the similar mirabazoles and tantazoles, its biosynthesis remains unclear to date, but is considered to involve SAM-dependent α-methylation of cysteine to account for the rare 2-methylcysteine subunits. Interestingly, when assuming a Hofmann-type deamination on a terminal phenylalanine and cyclodehydratation and dehydrogenation steps, **7** can be formally simplified to the pentapeptide **15** (Scheme 9.3). In a RPS-grounded assembly, a single methyltransferase would suffice to install all methyl groups, either on the peptide (**15**→**14**) or on the azoline level after azole ring formation. The former order of events seems more

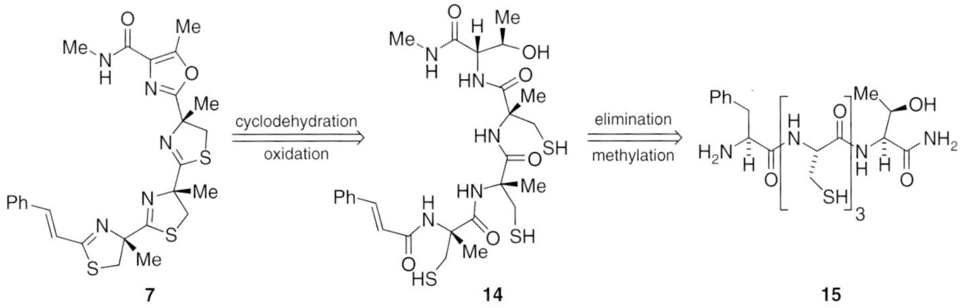

Scheme 9.3 Hypothetical biomimetic simplification of thiangazole (**7**) to pentapeptide **15**.

plausible since catenated thiazolines tend to auto-oxidize easily. Alternatively, 2-methylcysteine might be biosynthetically generated and independently integrated into the growing chain by NRPS.

Such a bold proposal has not yet been realized in synthetic practice, but elements of it inspired the synthesis design in various groups. In most successful syntheses, the C-terminal oxazole was not part of a multiple ring closure but had to be independently generated by cyclodehydratation reactions (Figure 9.4). To assemble the oligo-thiazoline, several alternatives were worked out. The group of Pattenden successfully established an iterative condensation of nitriles with 2-methylcysteine to extend the chain C-terminally [16]. In a potentially more biomimetic variant, Heathcock showed that a string of 2-methylcysteines can be condensed to an oligothiazole with considerable efficiency. This method was later used by Wipf (below) [17] and Akaji [18] in their syntheses of **7**, which featured further optimizations.

To furnish the uncommon 2-methylcysteine, Seebach's self-regeneration of chirality was instrumental. Despite the tendency of cysteine enolates to eliminate the sulfur substituent, Pattenden could prepare 2-methylcysteine via the *tert*-butyl

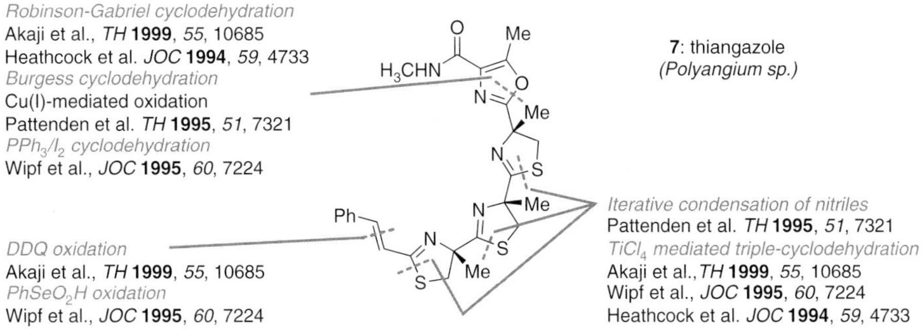

Figure 9.4 Strategic disconnections in thiangazole (**7**) total syntheses. To improve readability, the following abbreviations have been adopted throughout: ACIE = Angew. Chem. Int. Ed., BMCL = Bioorg. Med. Chem. Lett., CAJ = Chem. Asian. J., CPB = Chem. Pharm. Bull., CEJ = Chem. Eur. J., JACS = J. Am. Chem. Soc., JOC = J. Org. Chem., OL = Org. Lett., PNAS = Proc. Natl. Acad. Sci. U.S.A., TH = Tetrahedron, TL = Tetrahedron Lett.

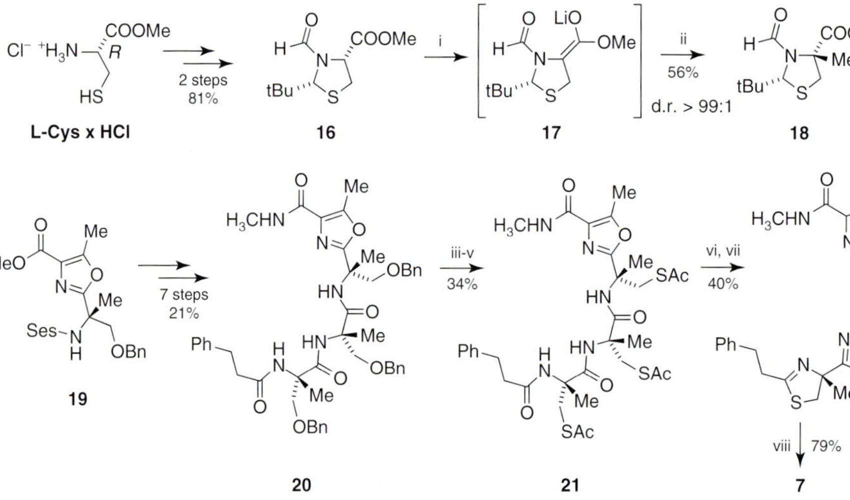

Scheme 9.4 2-Methylcysteine synthesis and oxazoline–thiazoline conversion by Wipf. *Reagents and conditions*: (i) LDA, DMPU, THF, −90 °C; (ii) MeI; (iii) $Pd(OH)_2$, H_2, MeOH; (iv) Burgess reagent, THF; (v) AcSH; (vi) NH_3, MeOH; (vii) $TiCl_4$, CH_2Cl_2; and (viii) PhSeOOH, benzene, 60 °C; Ses: (trimethylsilyl)ethylsulfonyl.

thioaminal **16**, which was alkylated to **18** via enolate **17** with a dr of >99 : 1 (Scheme 9.4) [19].

To circumvent the preparation of the cysteine building blocks and to generate variants for structure–activity relationship studies, Wipf developed an interesting oxazoline–thiazoline conversion. Trisubstituted oxazole **19** was extended with 2-methylserine building blocks to oligomer **20** (seven steps, 21%) [17]. After debenzylation, multiple oxazoline formation was cleanly induced with the Burgess reagent (60%), giving a structural analog of **7**. Treatment of the labile oxazolines with thioacetic acid then delivered the 2-methylcysteine precursor **21**, which was processed by biomimetic dehydration and oxidation of the terminal phenylethyl group (PhSeOOH) to give thiangazole (**7**). Overall, these biomimetic investigations were setting the stage for approaches to more complex peptide alkaloid syntheses.

9.2.4
Lissoclinamide 7

In contrast to the linear oligothiazolines, much more is known on the biosynthesis of cyclic azole-containing peptide alkaloids. The gene clusters for the monocyclic lissoclinamides, the ulithiacyclamides, and the patellamides have been identified and found to be quite similar [20]. These data suggested a RPS assembly with a genetically encoded peptide template guided by a leader peptide [21]. Typical enzymatic activities, dehydration, dehydrogenation, and macrolactam formation, could be identified. For lissoclinamide 4, the peptide ring closure site was located between a Cys and a Phe residue. Notably, many *Lissoclinum* peptides contain D-amino acid residues, but a formal epimerase enzyme was not found. It was then noted that the chemically synthesized L-Val lissoclinamide 7 converted into the D-isomer (**22**) upon treatment with pyridine, showing a conformation-driven preference [22]. It can hence be assumed that the D-residues at the acidic thiazoline α-carbons are formed by epimerization of **23** following biosynthetic assembly [23]. An according simplification of lissoclinamide 7 (**22**) is shown in Scheme 9.5. Notably, during the biosynthesis, azole-formation (**24**→**23**) could precede macrocyclization (**25**→**24**), but the full peptide chain **25** will be assembled first.

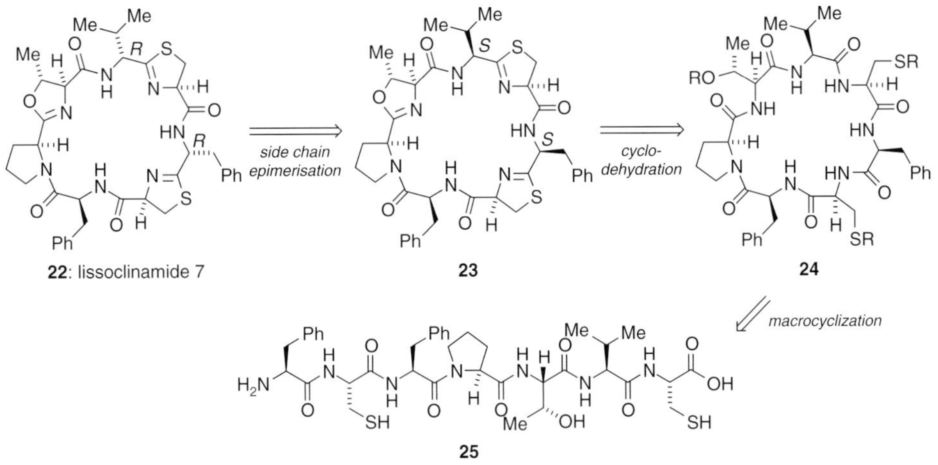

Scheme 9.5 Biosynthesis-guided simplification of lissoclinamide 7 (**22**) to heptapeptide **25**.

Many members of this peptide alkaloid class have been synthesized from oxazole- or thiazole-amino-acid building blocks, but very few approaches have been reported where the azoles are installed *after* assembly of the complete peptide chain. Additionally, for the lissoclinamides their exceptional configurational lability had to be taken into account [24]. Wipf *et al.* used multiple cyclodehydrations, oxazoline–thiazoline interconversions, and an pentafluorophenyl diphenylphosphinate (FDPP)-mediated macrocyclization (Figure 9.5), giving access to the full configurational assignment of lissoclinamide **7** and to a library of variants for biological testing.

The suitably protected heptapeptide **26** was assembled (Scheme 9.6) and converted into macrocycle **27** by deprotection, macrolactam formation, and TIPS (triisopropylsilyl)-protection of the secondary *allo*-Thr-OH while unraveling the primary Ser-OH groups. These were used to initiate a double cyclocondensation to **28** followed by thiolysis to install thioamides in the peptide backbone (**29**). At this stage, deprotection of the Thr side chain enabled a triple azoline ring closure to furnish lissoclinamide 7 (**22**) in a stereochemically pure form with exceptional efficiency (90% yield).

Using this strategy with late-stage introduction of the thiazoline rings, the stereochemistry of **22** could be firmly assigned. Notably, the intermediate oxazoline

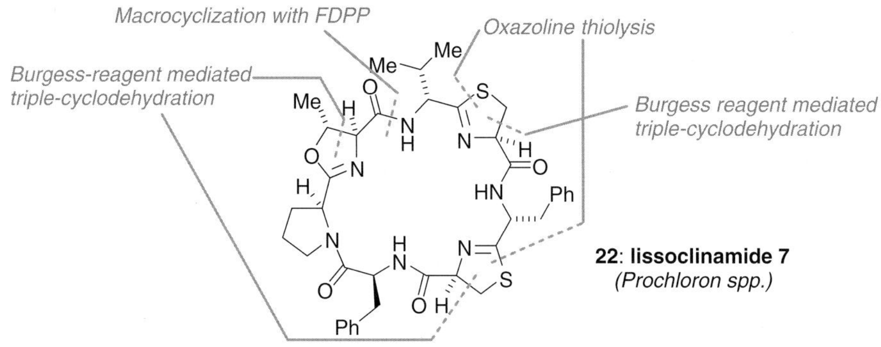

Figure 9.5 Strategic disconnections of lissoclinamide 7 (**22**) used by Wipf [22, 24].

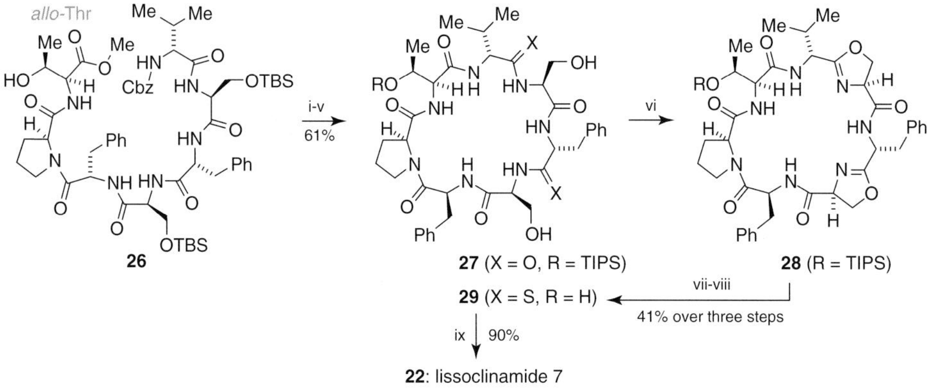

Scheme 9.6 Completion of the total synthesis of lissoclinamide 7 (**22**). *Reagents and conditions*: (i) H_2, Pd/C, MeOH; (ii) NaOH, THF–MeOH–H_2O; (iii) FDPP, $NaHCO_3$, DMF-CH_2Cl_2 (3 : 1), 40 °C; (iv) TIPS-OTf, 2,6-lutidine, CH_2Cl_2; (v) TsOH, THF–H_2O; (vi) Burgess reagent, THF, 65 °C; (vii) H_2S, MeOH–NEt_3, 22 °C, 4 d; and (viii) TBAF, THF, 37 °C; (ix) Burgess reagent, THF, 40–70 °C.

formation located the introduction of thioamides in the backbone. Owing to activation of all the OH-groups for S_N2-type substitution by the Burgess reagent in the last step [25] *allo*-Thr had to be protected during chain assembly. But, overall, without the use of enzymes this synthesis could hardly come closer to a biomimetic strategy – before the biosynthesis was elucidated.

9.2.5
Thiostrepton

Thiopeptide antibiotics are highly modified, sulfur-rich macrocyclic peptide alkaloids that share another common motif beyond the azoles in the peptide chain: a tri- or tetrasubstituted pyridine core [26]. Early studies on their biosynthesis were made by feeding labeled amino acids and it was shown that the whole framework of thiostrepton (**9**) and nosiheptide (**12**) is exclusively derived from proteinogenic amino acids [27–30]. Surprisingly, genetics studies conclusively established that all important thiopeptide antibiotics (thiostrepton, thiocillin, siomycin, thiomuracin, nosiheptide) [31–33] are derived from RPS [2,33–35]. One of the most complex members, thiostrepton (**9**), was isolated in the 1950s from *Streptomyces azureus*. It was identified as a very potent antibiotic with strong activity against Gram-positive bacteria (Figure 9.3) [7].

The X-ray structure of **9** revealed a globular structure reminiscent of folded protein domains [9]. Thiostrepton inhibits bacterial protein biosynthesis by binding tightly to the GTPase-associated region (GAR) on the 70S ribosome. Scheme 9.7 summarizes key features of its biosynthesis. The genetically encoded, Cys-rich structural peptide is N-terminally tagged with a leader peptide sequence (**30**) [4]. Dehydrations lead to the development of five thiazole rings and seven dehydroamino acids (**31**). In a highly unprecedented step – proposed early on by Bycroft and Gowland [36] – two dehydroamino acids engage in a formal 2-azadiene Diels–Alder cycloaddition (**32**) to generate a macrocycle and the central piperidine core. This core is reduced in the case of thiostrepton (**33**), but in other thiopeptides a pyridine ring results from elimination reactions. The second macrocycle is completed by insertion of an activated quinaldic acid (**34**), which is apparently incorporated in a more NRPS-like fashion. After protease-mediated cleavage of the leader peptide, nucleophilic attack of the N-terminus at the epoxidized aromatic ring is assumed to insert the unconventional final linkage between the N-terminus and a threonine OH-group in the A-ring.

Notably, this series of events suggests that the A-ring closes first. It remains to be clarified if the crucial cycloaddition during the biosynthesis is stepwise or concerted (Scheme 9.8), but the anticipated hydroxypiperidine intermediate certainly becomes reduced by an NADPH-dependent enzyme. The unconventional quinaldic acid is biosynthesized from tryptophan by a sequence involving C2 methylation, hydrolysis of the indole subunit, oxidative deamination to a putative diketone, condensation with the more electrophilic α-keto acid, and enantioselective tautomerization to a pyridyl alcohol. P450-mediated oxidation could generate a quinoline epoxide, to serve as an electrophile for the terminal Ile residue's amino group.

Two total syntheses of thiostrepton and the very similar siomycin have been published, the former by the Nicolaou group [37–40] and the latter by Nakata *et al.* [41, 42]. Figure 9.6 summarizes the key disconnections. Not surprisingly, the strategies realized do not slavishly follow the biosynthetic track – which was only clarified five years after the total synthesis was completed! Nonetheless, several structural elements are accessed in ingenious anticipation of potential biosynthetic possibilities, and the final ring closure of the B-ring is achieved at the same site where it probably occurs in biosynthesis.

In setting up the peptide chains, effective but mostly conventional technologies are used in both syntheses, which we will not discuss in detail here. However, to set up the crucial central dehydropiperidine ring two distinct approaches were developed. The Nakata team developed an interesting auxiliary-controlled imine addition to a dehydropyrrolidine-derived azomethine ylide followed by a smart,

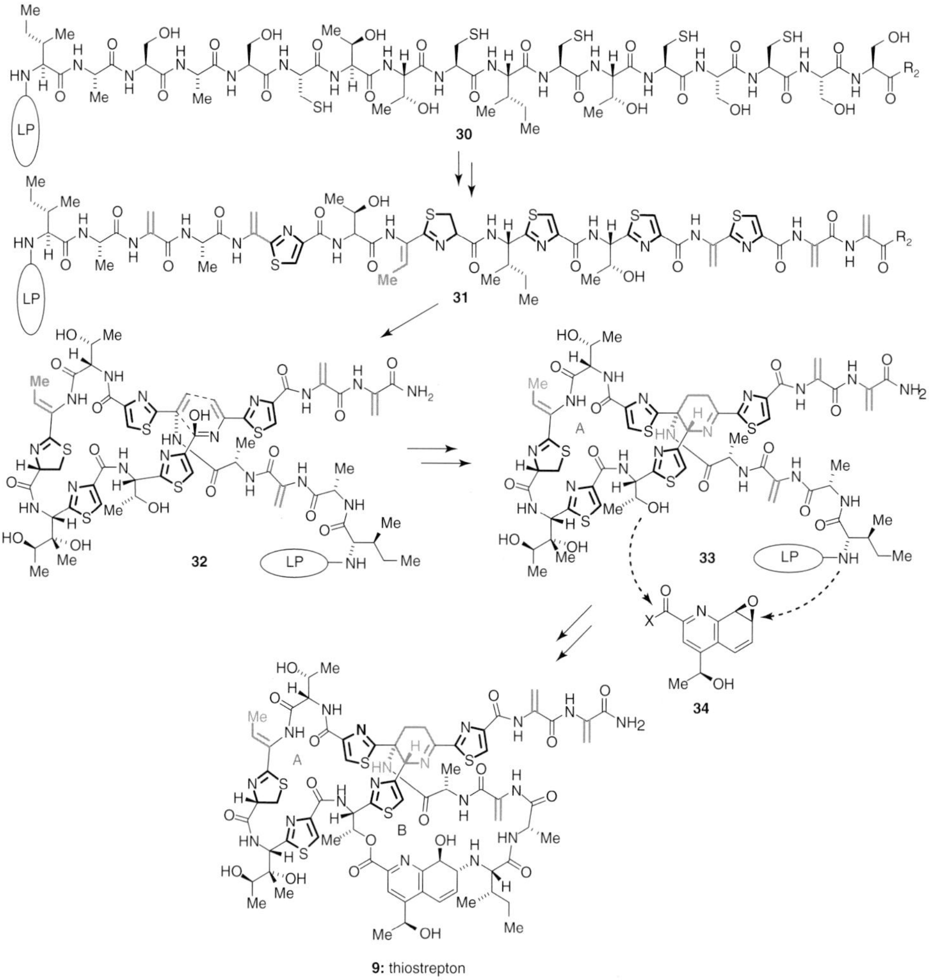

Scheme 9.7 Thiostrepton biosynthesis as deduced mostly from genetic information (LP: leader peptide).

directed five-ring-six-ring interconversion. Nicolaou's group on the other hand realized the piperidine ring formation in alignment with the biosynthetic hypothesis of Bycroft and Gowland (Scheme 9.9a) by a 2-azadiene hetero-Diels–Alder reaction. Using Ag^{+} salts, 2-azadiene **36** was liberated from a thiazolidine precursor **35**. Homodimerization of **36** was found to take place in Diels–Alder type fashion (**37**) even at low temperatures. In accordance with earlier work [43–45] the initial dehydropiperidine product **38** was highly prone to imine–enamine tautomerization, followed by an intramolecular aza-Mannich reaction. This unintended sequence could be suppressed using a scavenging amine ($BnNH_2$), which removed the side-chain imine and liberated the labile amino-dehydropiperidine product **39** as a 1 : 1 mixture of diastereomers (separable later in the synthesis).

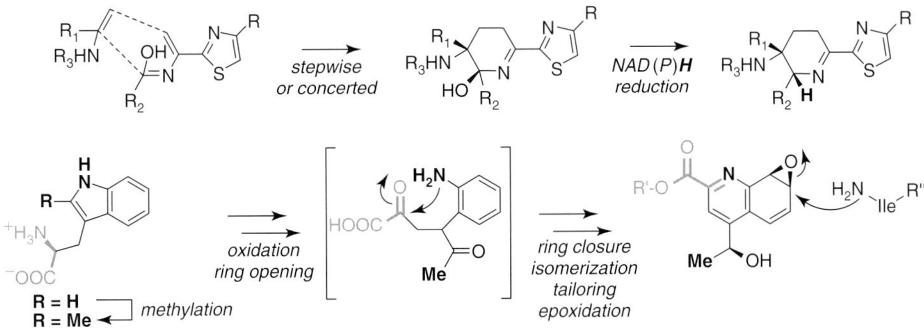

Scheme 9.8 Biosynthesis of the dehydropiperidine core and the quinoline unit of thiostrepton.

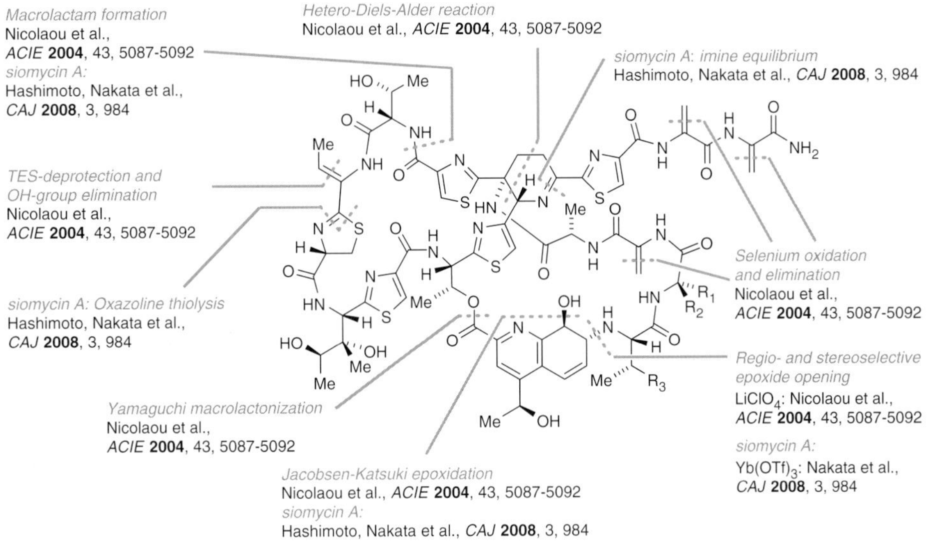

9: thiostrepton (*Streptomyces azureus*, R_1 = Me, R_2 = H, R_3 = Et)
10: siomycin A (*Streptomyces sioyaensis*, $R_1 = R_2 = CH_2$ (dehydroalanine), R_3 = Me)

Figure 9.6 Key synthetic disconnections in thiostrepton (**9**) and siomycin A (**10**) syntheses.

This sequence features both a breaking of symmetry and recycling of the accessory aldehyde **40** and is biomimetically inspired. The high reactivity and tendency to homodimerize can be attributed to the lower oxidation state of the 2-azadiene carbon-1 (imine instead of amide as in a peptide chain): studies by Moody with amide-derived 2-azadienes indicated much lower reactivity [46]. Whether the putative biosynthesis enzymes really promote a similar reaction between two crossing peptide strands remains to be elucidated. Nonetheless, it was shown that such transformations are chemically possible, and could be directed and/or catalyzed by suitable enzymes during biosynthesis.

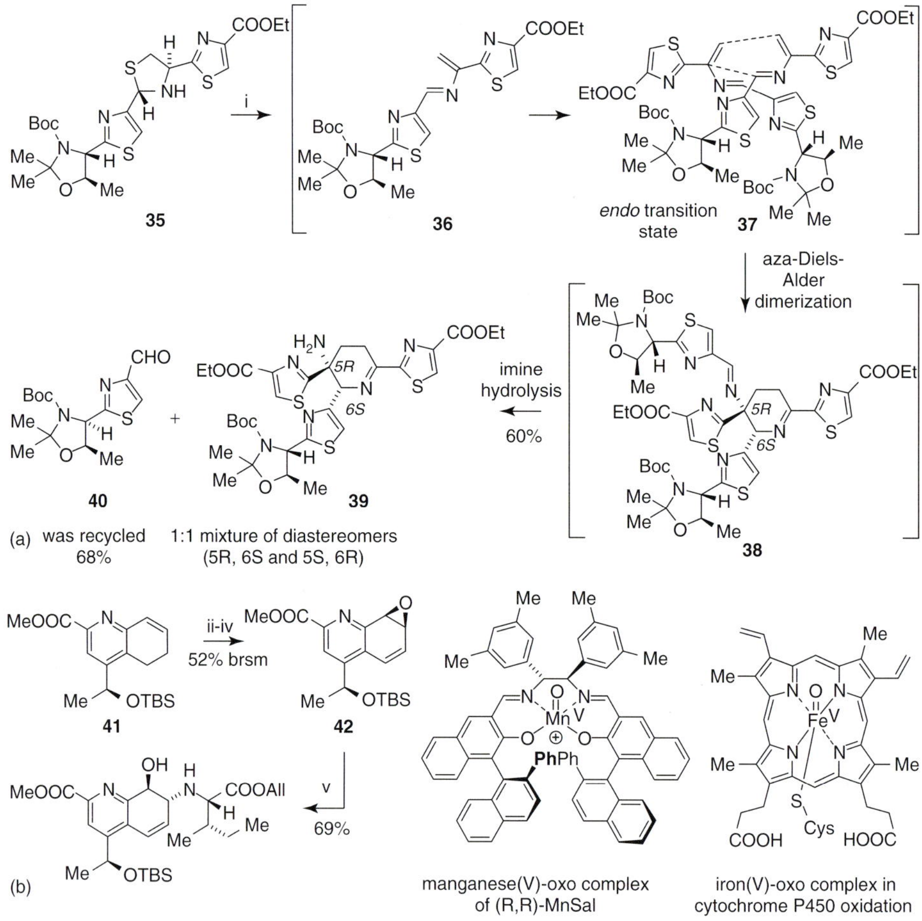

Scheme 9.9 (a) Biomimetic dehydropiperidine synthesis; (b) biomimetic quinoline epoxide opening. *Reagents and conditions*: (i) Ag_2CO_3, pyridine, DBU, $BnNH_2$, $-12\,^\circ C$; (ii) NaOCl, 4-Ph-py-*N*-oxide, *(R,R)*-MnSal, phosphate buffer, CH_2Cl_2; (iii) NBS, AIBN, CCl_4; (iv) DBU, THF; and (v) H-L-Ile-OAll, $LiClO_4$, MeCN; bvsm = based on recovered starting material.

During biosynthesis of the quinoline unit, a heme-catalyzed epoxidation is anticipated, which was followed in building block synthesis as well. Stereoselective epoxidation of the dihydroquinoline **41** was successful with Katsuki's catalyst [47] (Scheme 9.9b) in moderate dr (87 : 13). Radical bromination and elimination installed the double bond (**42**), and regioselective opening with isoleucine esters under Li^+ (Nicolaou) or Yb^{3+}-catalysis (Nakata) completed a biomimetic synthesis using the intrinsic reactivity [48, 49] of the biosynthetic intermediates.

Successful completion of the thiostrepton synthesis again featured a biomimetic strategy (Scheme 9.10). The B-ring chain **44** was attached to the A-ring scaffold

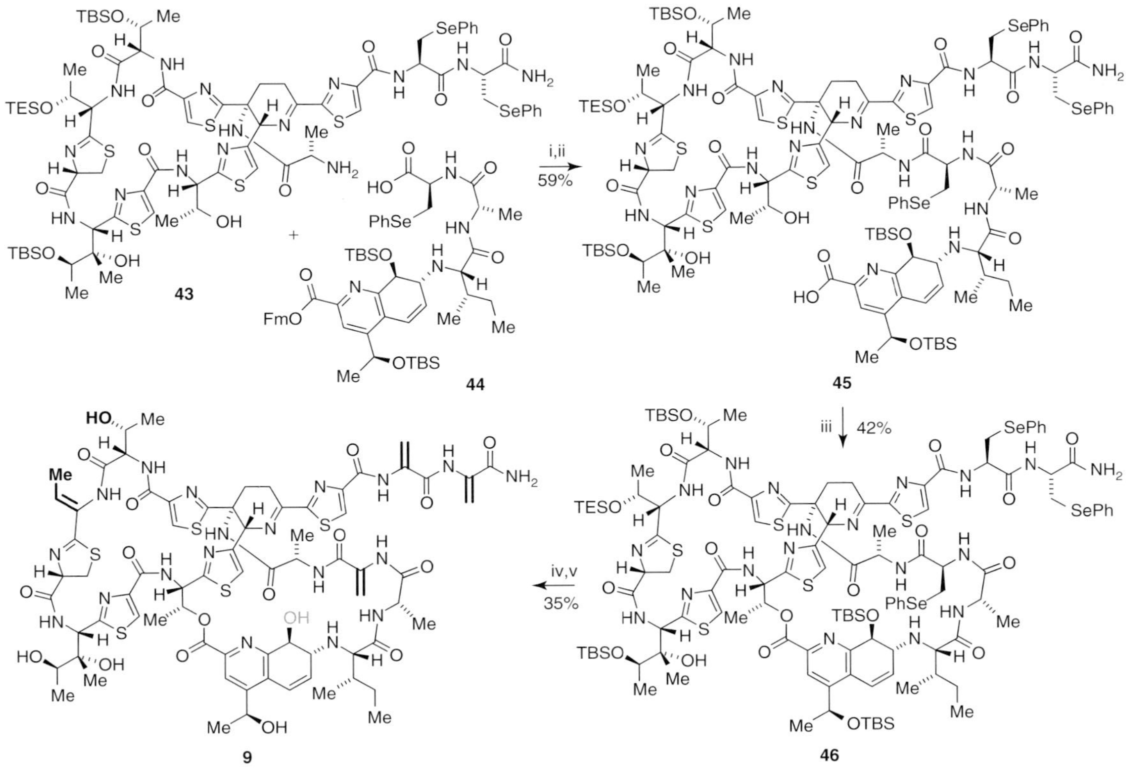

Scheme 9.10 Completion of the synthesis of thiostrepton (**9**). *Reagents and conditions*: (i) HATU, HOAt, i-Pr_2NEt, DMF; (ii) Et_2NH, CH_2Cl_2; (iii) 2,4,6-trichlorobenzoyl chloride, Et_3N, DMAP; (iv) t-BuOOH; and (v) HF–pyridine. Fm: 9-fluorenylmethyl.

43 by a HATU-mediated coupling. After C-terminal deprotection, hydroxy-acid **45** was ring-closed to the bicyclic macrolactam **46** at the putative ring closure site in the biosynthesis. The unmasking of the final product **9** was initiated with the oxidation-triggered elimination of the phenylselenyl alanines to the dehydro amino acids, and then HF treatment. During this process the Thr-OH group adjacent to the thiazoline ring was eliminated to give the *(Z)*-configured enamide side chain. This indicates that elimination of this residue may also be extremely facile during biosynthesis. Overall, the synthetic works toward thiostrepton (**9**) and siomycin (**10**) define milestones in peptide alkaloid total synthesis, and show how biomimetic considerations can successfully guide a complex total synthesis project toward crucial and productive bond formation chemistry.

9.2.6
GE2270A

In comparison it is worth taking a look at the thiopeptide GE2270A (**13**). This compound, isolated from *Planobispora rosea*, has been identified as an antibiotic as well, but its mode of action entails an allosteric blockade of elongation factor TU, a ribosome-associated factor. Structurally, some different features are noteworthy. Only one macrocycle is present, but it is larger in size than the A-ring of thiostrepton. The connecting core is now an aromatized pyridine ring, and some post-assembly tailoring is apparent (methylations, hydroxylations). Its biosynthesis has only been investigated by labeling experiments [50] but is believed to follow a similar pathway to thiostrepton, closely related to its relative thiomuracin [51].

Synthetically, GE2270 allows us to compare biomimetic and non-biomimetic strategies. Two distinct syntheses for this molecule have been worked out: one with biomimetic key steps [52, 53] and the second making extensive use of non-biological Pd-mediated sp^2-couplings [54, 55], even for the macrocyclization (Figure 9.7).

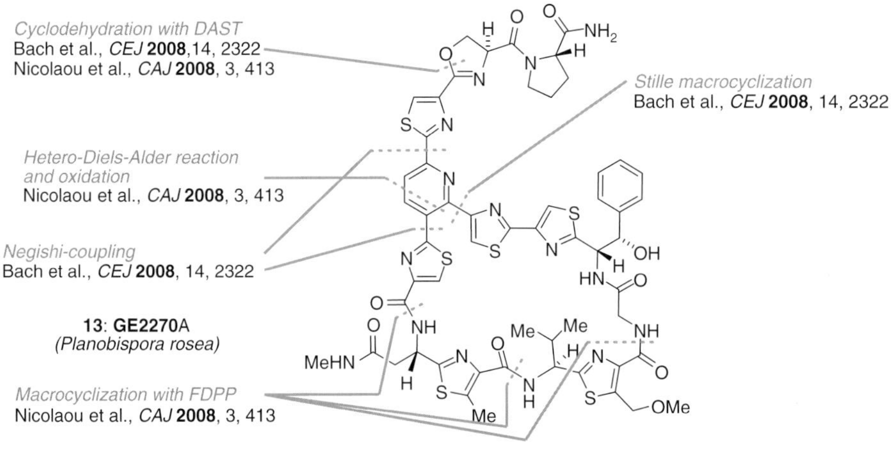

Figure 9.7 Key disconnections for GE2270A (**13**).

Both approaches installed the side chain at the end stage of the synthesis. In Nicolaou's biomimetic synthesis [52, 53], GE2270A and some of its variants were prepared. The strategy used to obtain the pyridine core was similar to that employed in their earlier synthesis of thiostrepton. Starting from the dehydropiperidine cycloadduct, NH_3 was eliminated using DBU (1,8-diazabicyclo[5.4.0]undec-7-ene), and oxidation of the dihydropyridine delivered the aromatized pyridine **47** (Scheme 9.11). Similar steps are anticipated for the biosynthesis of pyridine-containing thiopeptides. However, the building block resulting from this sequence contained two methyl esters in highly similar chemical environments due to its homodimeric origin.

The methyl esters in **47** (Scheme 9.11) were challenging to differentiate downstream. Furthermore, chain extension by peptide couplings worked, but the

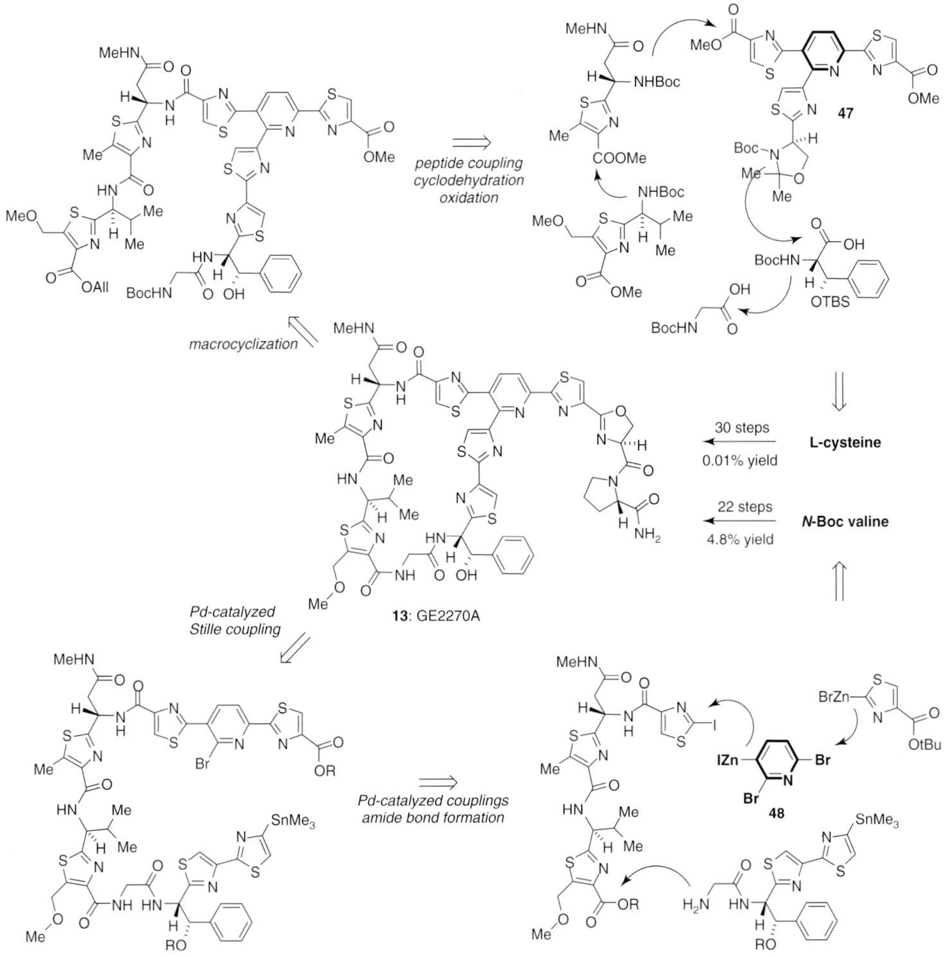

Scheme 9.11 Building blocks and key intermediates used by Nicolaou (top) and Bach (bottom).

macrocycle formation by macrolactamization turned out to be rather difficult. The macrocycle could be closed with FDPP, but in yields ranging from only 15 to 25%, depending on the site of ring closure. Overall, this led to a synthesis in 30 steps in the longest linear sequence in modest overall yield.

In the second synthesis by Bach [54, 55], a less biomimetic strategy was followed. The plan was to assemble the scaffold by consecutive Pd-mediated couplings around the pyridine core (Scheme 9.11, bottom). In prior work, methods for the coupling chemistry were developed from the zinc reagent **48**, which allowed access to all of the building blocks. Similar to Nicolaou, macrolactam formations were found to be less effective in ring closure (30%). However, when a Stille coupling was used to close the macrocycle, the ring closing efficiency strongly improved (75%). In this way, more complex building blocks could be connected, leading to a shorter linear sequence (22 steps). Macrocycle formation included, the efficiency of this synthesis was fairly high, with an overall yield of 4.8%.

These data show that biomimetic synthesis is not necessarily superior to a non-biomimetic approach, especially when crucial synthesis tools are not available or cannot be easily transferred. It critically shows that further method developments for the biomimetic synthesis of azole-containing peptide alkaloids are needed, with respect to chain assembly, maturation, and interconnection chemistry. Such efforts should be guided by close interaction with biosynthesis research to provide maximal mutual benefits.

9.3 Peptide Alkaloids Cyclized by Oxidation of Aryl Side Chains

9.3.1 Cyclic Peptides Containing Aryl-Alkyl Ethers

Numerous cyclopeptide alkaloids containing aryl-alkyl ethers have been isolated from a host of different plant families, such as Rhamnaceae, Rubiaceae, Olacaceae, and others [56, 57]. The structure of these 13- to 15-membered macrocycles features a phenyl alkyl ether with an enamide in *ortho-para* or *ortho-meta* orientation (Figure 9.8). This styrylamine moiety bridges two or three α-amino acids. *N*-Methyl or *N,N*-dimethyl substitutions are often found in the side-chains of these cyclic alkaloids.

Although various biological activities such as antibacterial, antifungal, sedative, and immunostimulant properties have been identified for these compounds, detailed understanding about their biological function and their biosynthesis still needs to be developed [58]. However, their likely precursors are oligopeptides. Feeding experiments with labeled amino acids in the plant *Ceanothus americanus* resulted in the formation of the linear precursors of the corresponding cyclopeptides, suggesting that the precursor molecules are more likely to be generated by a specific enzyme system than by cleavage from genetically encoded peptides [59]. Three possible biosynthetic pathways leading to aryl-alkyl linked cyclopeptides can

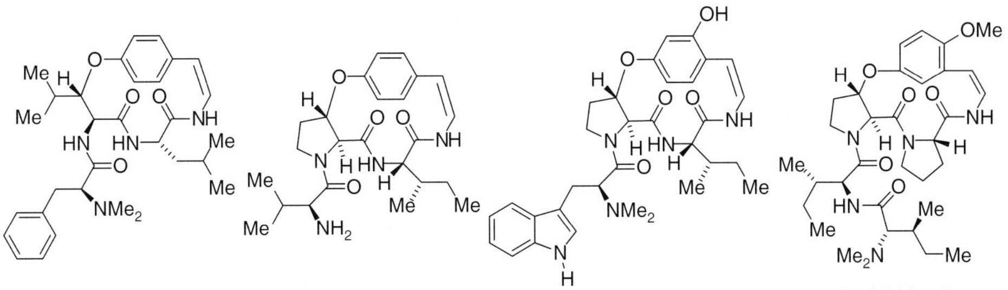

49: sanjoinine A (frangufoline) *Ziziphus vulgaris* | **50**: lotusine G *Ziziphus lotus* | **51**: hemsine A *Paliurus hemleyanus* | **3**: ziziphine A *Ziziphus oenoplia*

Figure 9.8 Selected examples of aryl-alkyl ether peptide alkaloids.

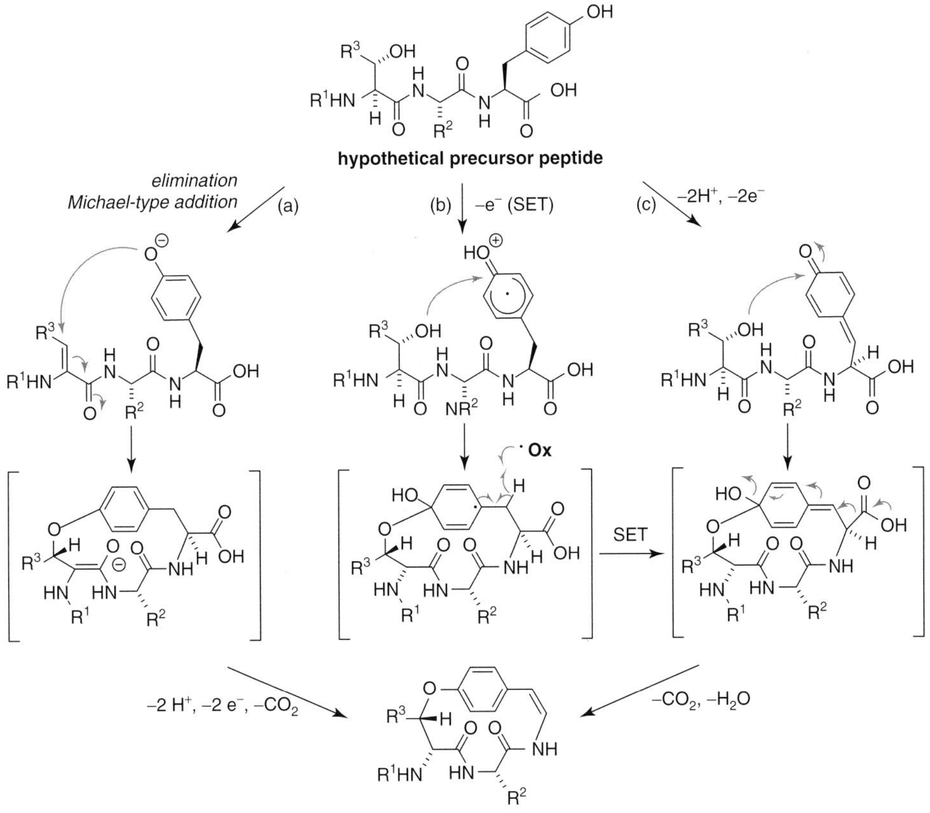

Scheme 9.12 Possible ring formations in the biosynthesis of aryl-alkyl ether peptide alkaloids, SET = single electron transfer.

be proposed, which differ only in the oxidation state of the intermediate aryl-ring (Scheme 9.12). Isolation of the dehydroamino-acid containing linear Rhamnaceae alkaloid lasiodine A led to speculation about a possible ring closure via *endo*-attack of a phenolic group to this double bond via Michael addition (Scheme 9.12, path a). This mechanism, however, is not fully suited to explain the *ortho-meta* orientation found in some other cyclopeptide alkaloids [60].

Oxidative pathways could more generally explain these patterns. Enzymatic oxidation of a tyrosine side chain could produce a radical cation that may be intercepted by a β-hydroxylated amino acid (Scheme 9.12, path b) and then further oxidized. Alternatively, a direct two-electron oxidation could be envisioned, forming an electrophilic *para*-quinone intermediate, which re-aromatizes after loss of water (Scheme 9.12, path c). In both cases a concomitant loss of the carboxyl group could ensue (cf. TMC-95A–D described in Chapter 10), potentially explaining the prevalence of an endocyclic (*Z*)-configured enamide found frequently among these peptide alkaloids.

Owing to the limited availability of these alkaloids, several strategies for their total syntheses have been developed. No genuine biomimetic approach to the macrocyclization according to those already proposed (Scheme 9.12) has been established thus far. Ring closure as the key step in each of these synthetic routes has been accomplished at different positions of the molecular scaffold (Figure 9.9) [61]. Schmidt *et al.* have introduced a macrolactamization approach based on the activation of a carboxyl-group as a pentafluorophenyl ester, which was further applied to the synthesis of several cyclopeptide alkaloids (Schmidt *et al.* [62–69], Joullié *et al.* [70–76], Han *et al.* [77–79]).

More biomimetic ring closures at the aromatic core have less frequently been realized. In their total synthesis of sanjonine G1 (**52**), Zhu *et al.* used an intramolecular S_NAr reaction of a 4-fluoro-3-nitroaryl-derivative with a β-hydroxy amino acid residue (**53**) and subsequent reductive removal of the activating nitro-group from **54** via the diazonium intermediate to obtain the desired cyclopeptide scaffold **55** (Scheme 9.13) [80–83].

A different method for macrocyclization was established by Evano, using a copper-catalyzed Ullmann-type vinyl amidation. This was applied to the synthesis of paliurine F (**56**, Scheme 9.14) [84–86]. The hydroxypyrrolidine fragment **57** was

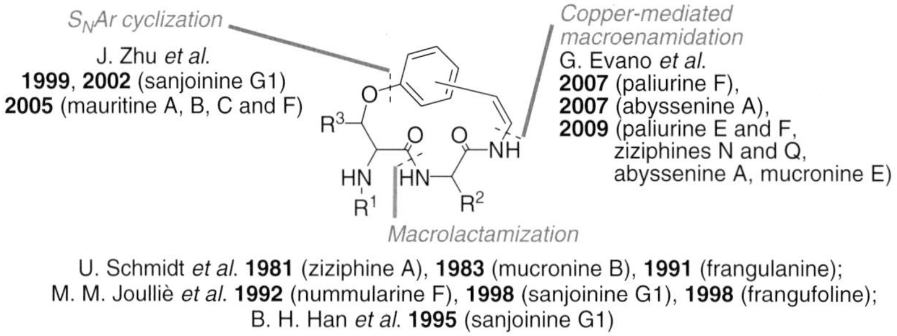

Figure 9.9 Ring-closing strategies in the synthesis of aryl-alkyl ether peptide alkaloids.

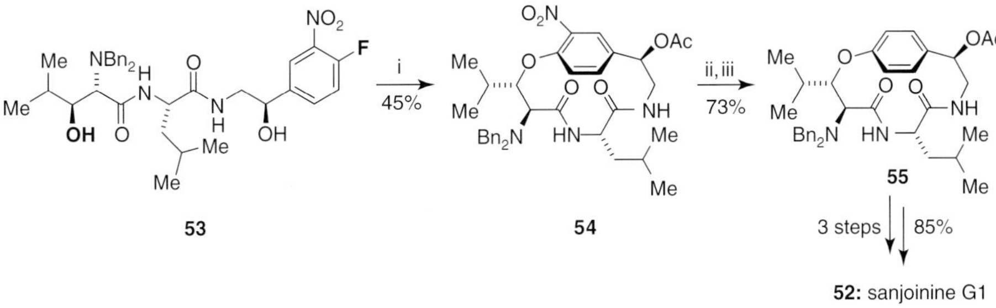

Scheme 9.13 Synthesis of sanjoinine G1 (**52**) [79]. *Reagents and conditions*: (i) TBAF, DMSO, 85 °C, then Ac_2O, Et_3N, DMAP, CH_2Cl_2; (ii) $SnCl_2$, DMF, 60 °C; and (iii) $NaNO_2$, H_3PO_2, Cu_2O, THF-H_2O.

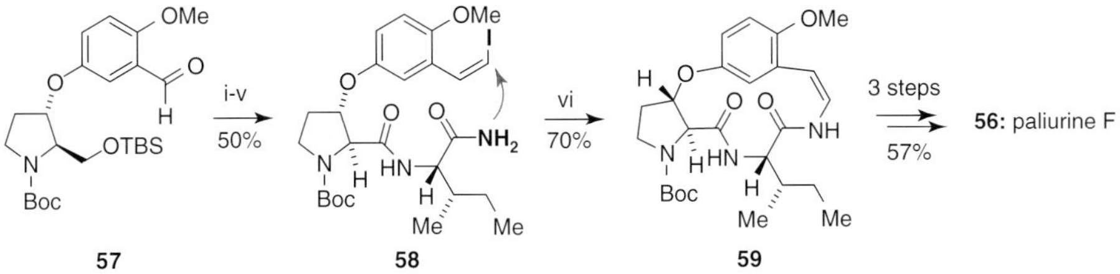

Scheme 9.14 Synthesis of paliurine F (**56**) [85]. *Reagents and conditions*: (i) $[Ph_3PCH_2I]^+ I^-$, NaHMDS, THF-HMPA, −78 °C; (ii) TBAF, THF, −10 °C to room temp. (rt); (iii) $(COCl)_2$, DMSO, Et_3N, CH_2Cl_2, −78 °C; (iv) $NaClO_2$, NaH_2PO_4, 2-methylbut-2-ene, *t*BuOH–THF–H_2O, rt; (v) L-isoleucinamide, AcOH, EDC, HOBt, NMM, DMF, rt; and (vi) CuI (10%), *N*, *N*′-dimethylethylenediamine, Cs_2CO_3, THF, 60 °C.

constructed from *N*-Boc-D-serine and a chelation-controlled borohydride reduction. The (*Z*)-vinyl iodide in **58** was introduced by Stork–Zhao olefination, while isoleucinamide was coupled to the C-terminus of the linear precursor. For the macrocyclization step to **59**, CuI/*N*,*N*′-dimethylethylenediamine gave the best results, which was followed by appendage of L-isoleucine and *N*,*N*-dimethyl-L-leucine to afford paliurine F in 6.5% overall yield over 16 steps.

9.3.2
Cyclic Peptides Containing Biaryl Ethers

The biaryl ether is a common motif found in several constrained, biologically active cyclic peptides. The simplest ones are cyclic tripeptides containing isodityrosine **60** as a common element (Figure 9.10). Biosynthetically, they are derived from oxidative cyclization of a terminal tyrosine, either at the C-terminus (**61**) or at the N-terminus (**62–70**). These different biphenyl-ether linkages lead to diverse biological activities of otherwise related structures. While K-13 (**61**), isolated from *Micromonospora halophytica* [87] acts as a potent angiotensin I converting enzyme (ACE) inhibitor, its inhibition of aminopeptidase B is very weak [88]. The OF4949

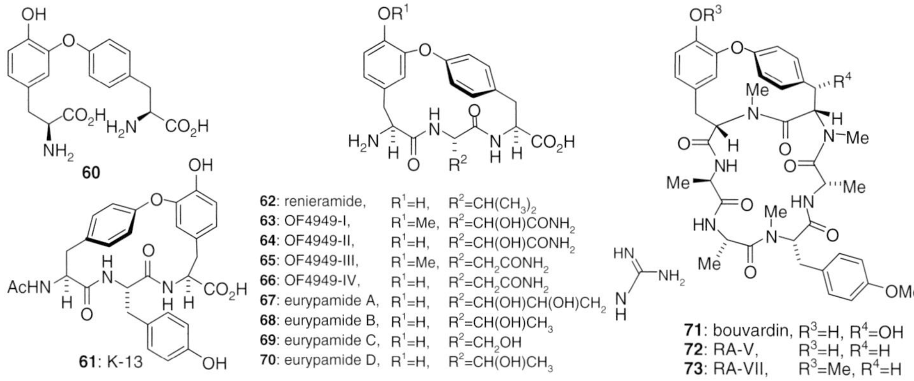

Figure 9.10 Peptides containing the cyclic isodityrosine subunit.

compound class on the other hand (**63–66**), produced by *Penicillium rugulosum*, shows potent inhibition of aminopeptidase B [89].

Bouvardin (**71**) was first isolated from *Bouvardia ternifolia* (Rubiaceae) and features an 18-membered peptide ring next to the cycloisodityrosine unit (Figure 9.10) [90]. Several members of this peptide family have been found (deoxybouvardin, 6-*O*-methylbouvardin, RA-I to RA-XVII) and were investigated as potent antitumor leads [91, 92]. Bouvardin and RA-VII (**73**) bind to the eukaryotic ribosome and inhibit protein biosynthesis.

In principle, two different approaches have been used to construct the cycloisodityrosine ring. Either the biaryl ether linkage has been formed prior to ring closure by macrolactamization or biaryl-ether formation is used to cyclize a linear peptide precursor (Figure 9.11) [93–98].

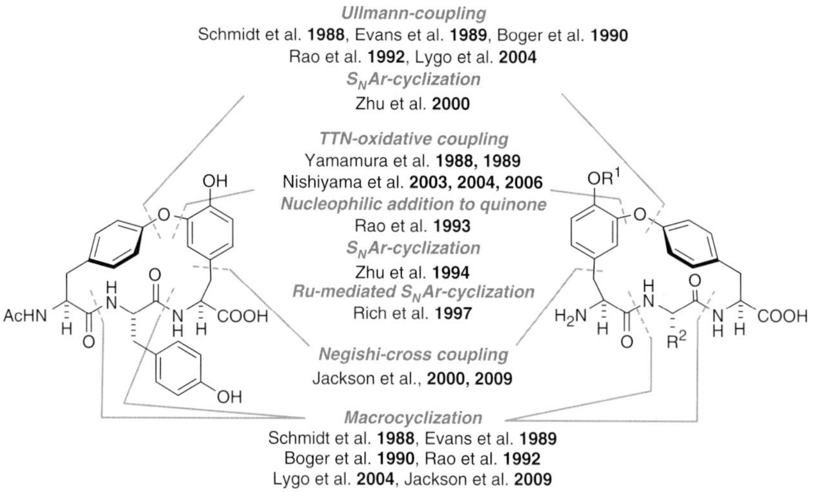

Figure 9.11 Strategies for cyclic isodityrosine synthesis.

Alternatively, Negishi couplings [99] or Horner–Wadsworth–Emmons type olefinations [100] were used to cyclize the open-chain peptide. Nishiyama and Yamamura applied thallium trinitrate (TTN) mediated cyclization of halogenated bis-phenols as a biomimetic way to access cycloisodityrosine structures [101, 102]. The synthesis of eurypamide B (**68**, first isolated from *Microciona eurypa*) [103] began with the formation of a linear tripeptide (**74**) containing one diiodo- and one dibromotyrosine derivative, which served to direct a regioselective oxidation to **75** with TTN (Scheme 9.15). Dehalogenation using hydrogenolysis and deprotection yielded the target compound **68** without any observable racemization [104].

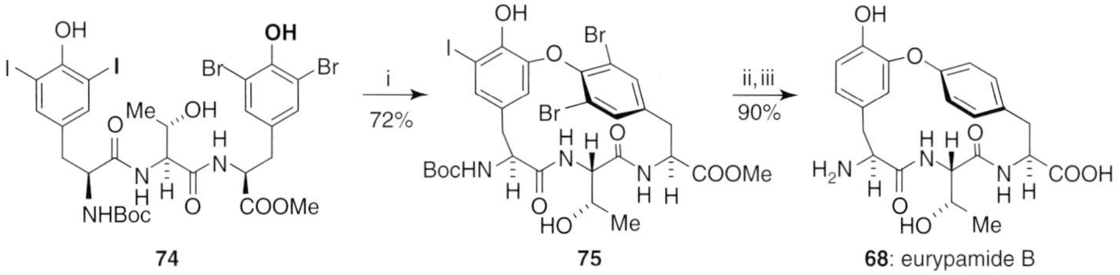

Scheme 9.15 Synthesis of eurypamide B (**68**) [103]. *Reagents and conditions*: (i) thallium trinitrate (TTN), THF–MeOH; (ii) H_2, NaOAc, 10% Pd/C, MeOH; and (iii) 1 M NaOH, MeOH, then TFA, CH_2Cl_2.

Biaryl ethers can also be formed by an Ullmann reaction, where a phenolic hydroxyl group is made to couple with an aryl iodide upon addition of $CuBr \cdot SMe_2$ [105–107]. Other approaches have used S_NAr coupling to an electron-deficient aromatic ring, which could be achieved using electron-withdrawing substituents [108] or through the intermediate formation of a metal π-arene complex. Rich *et al.* applied an intramolecular S_NAr reaction of a preformed cationic ruthenium complex as the key step in their synthesis of OF4949-III and K-13 (Scheme 9.16) [109]. 3-Chlorotyrosine (**76**) was converted into the cationic cyclopentadienylruthenium complex **77** and coupled to the protected Tyr-Tyr dipeptide **78**. The resultant cyclization precursor **79** smoothly formed the biaryl ether upon exposure to weak, non-nucleophilic base. Liberation of the auxiliary Ru-fragment by UV-irradiation and demethylation gave K-13 (**61**) in eight steps.

Figure 9.12 summarizes the main approaches to the bouvardin scaffold. Again, macrolactamization, S_NAr-[110] and Ullmann macrocyclization [107] have been used to furnish this peptide alkaloid. The first total synthesis of RA-VII was achieved via a potentially biomimetic route using TTN-mediated coupling of halogenated aryl-phenols. However, low yields hampered the critical cyclization step [111]. More recent reports describe the synthesis of the cycloisodityrosine by copper(II)-mediated *O*-arylation of phenylboronic acids [112].

A class of compounds that is biosynthesized by oxidation of bromotyrosine derivatives is the bastadins [113]. More than 20 members of this family have been isolated from marine *Ianthella* sponges. Bastadins occur in linear or cyclic form, possessing one biaryl or one or two biaryl-ether bonds (Scheme 9.17) as well as *(E)*-configured oximes. Several bastadins exhibit antimicrobial, cytotoxic,

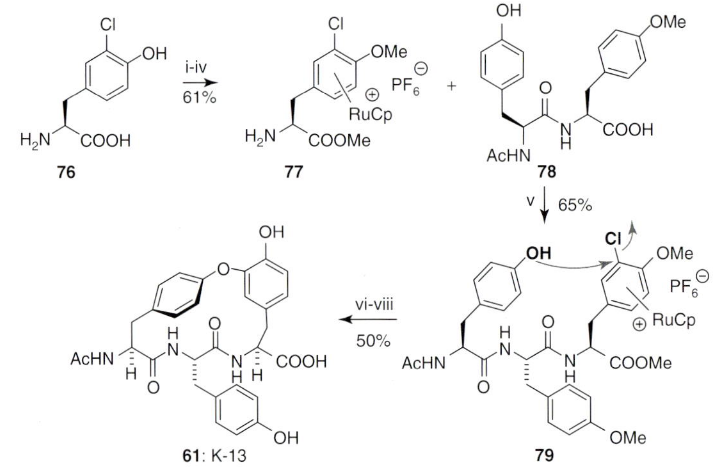

Scheme 9.16 Synthesis of K-13 (**61**) via transition-metal activated S_NAr [108]. *Reagents and conditions*: (i) Boc_2O, NaOH–dioxane; (ii) TMS-CHN_2 (4.0 equiv), THF–MeOH; (iii) 4 N HCl, dioxane, then $NaHCO_3$; (iv) $RuCp(CH_3CN)_3PF_6$, $ClCH_2CH_2Cl$, reflux; (v) EDCI, HOBt, DMF, 0 °C; (vi) sodium 2,6-di-*t*Bu-phenoxide, THF; (vii) $h\nu$, CH_3CN; and (viii) $AlBr_3$, EtSH.

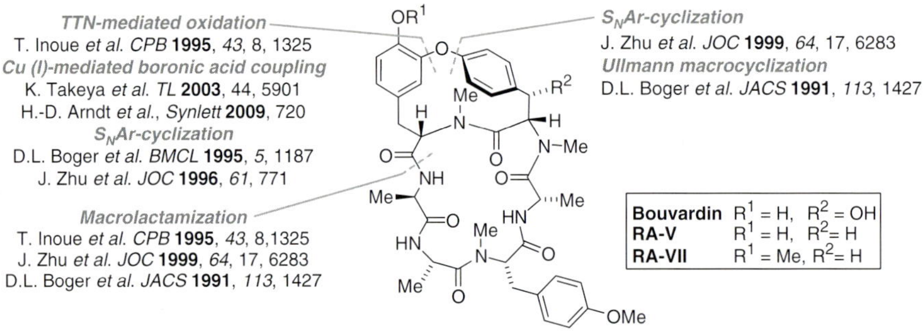

Figure 9.12 Different strategies to synthesize the bouvardin (**71**) and RA-family scaffolds (**72,73**).

and anti-inflammatory activity, as well as interesting modulation of intracellular calcium channels [114–116].

The bastadin 6 (**83**) biosynthesis is presumed to proceed via condensation of two brominated tyrosines followed by double oxidation of the primary amine yielding an *N,N'*-dihydroxylated intermediate, which forms the oxime upon dehydration (Scheme 9.17). It is then assumed that the resulting hemibastadin is dimerized by oxidation-catalyzed biaryl ether formation. A second oxidative coupling reaction completes the bastadin macrocycle [117].

Analogous to the biosynthetic pathway three main protocols have been established for bastadin synthesis that apply an oxidative coupling (Scheme 9.18). Yamamura *et al.* used TTN for aryl ether formation from **84** (Scheme 9.18, A)

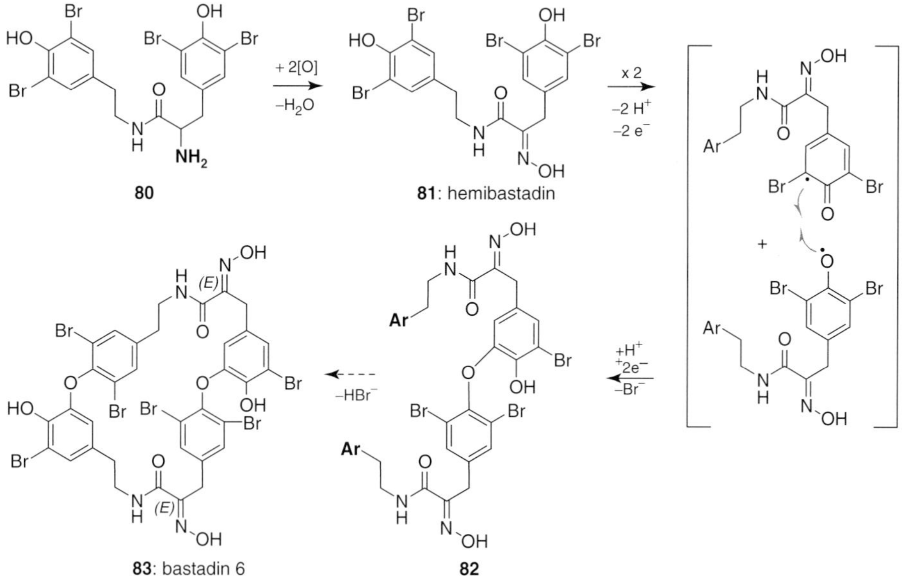

Scheme 9.17 Bastadin 6 (**83**) and its presumed biosynthesis.

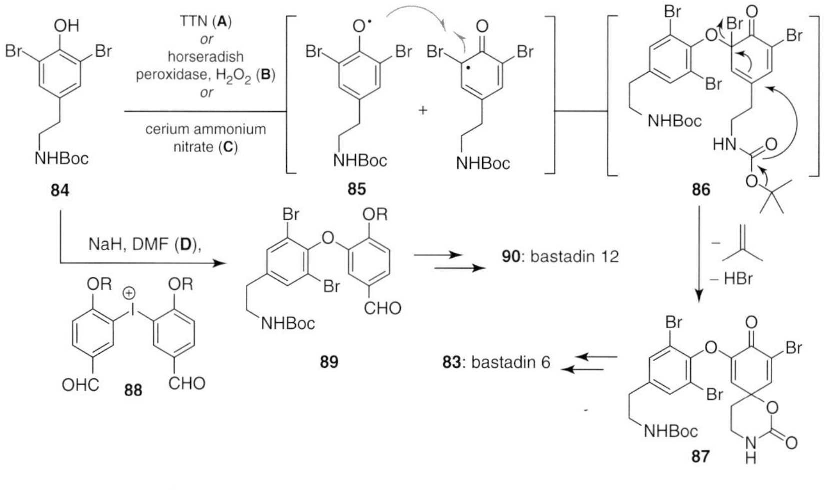

Scheme 9.18 Biomimetic synthesis approaches to bastadin aryl ethers.

[118, 119]. The second protocol from Shi *et al.* (Scheme 9.18, B) employed a horseradish-peroxidase-mediated coupling, which enabled the synthesis of bastadins 2, 3, and 6 [120]. More recently, Kobayashi *et al.* reported the synthesis of bastadin 6 via a Ce(IV)-mediated oxidative coupling (Scheme 9.18, C) [121]. All these methods proceed through radical intermediate **85**. Alternatively, different members of the bastadin family have been accessed via phenol coupling of aryl iodonium salts (**84**→**89**) [122].

9.3.3
Cyclopeptides Containing Biaryls

Biphenomycins **91–93** are cyclopeptides containing a *meta-meta* biaryl unit which bridges a tripeptide by forming a 15-membered ring (Figure 9.13). The compounds have been isolated from broths of *Streptomyces filipinensis* and *Streptomyces griseorubuginosus* and display potent antibacterial activities against various β-lactam resistant bacteria *in vivo* [123, 124].

Although very little is known about the biosynthesis of these biaryl peptide alkaloids, one could assume that a cytochrome-mediated oxidation of one aryl group forms either a radical or cationic intermediate, which is intercepted by the second aryl group to forge the biaryl linkage (Scheme 9.19). However, the generation of the uncommon *ortho*-hydroxylation remains unclear and might involve more complex biosynthetic events.

A biomimetic analogon of this reaction was realized by applying an intramolecular Suzuki–Miyaura coupling to cyclize a linear precursor tripeptide by forming the

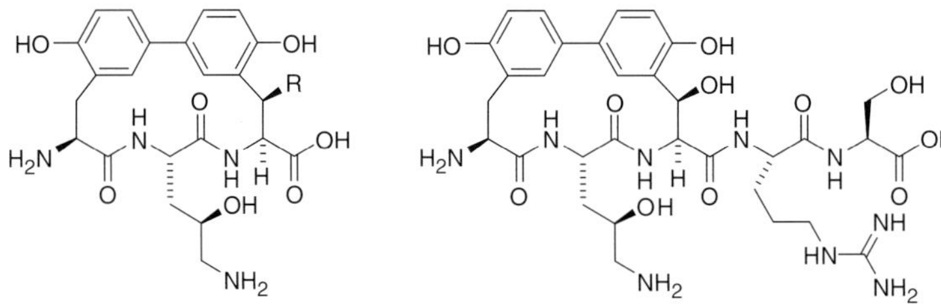

91: biphenomycin A (R = OH)
92: biphenomycin B (R = H)

93: biphenomycin C

Figure 9.13 Structures of the biphenomycins A–C (**91–93**).

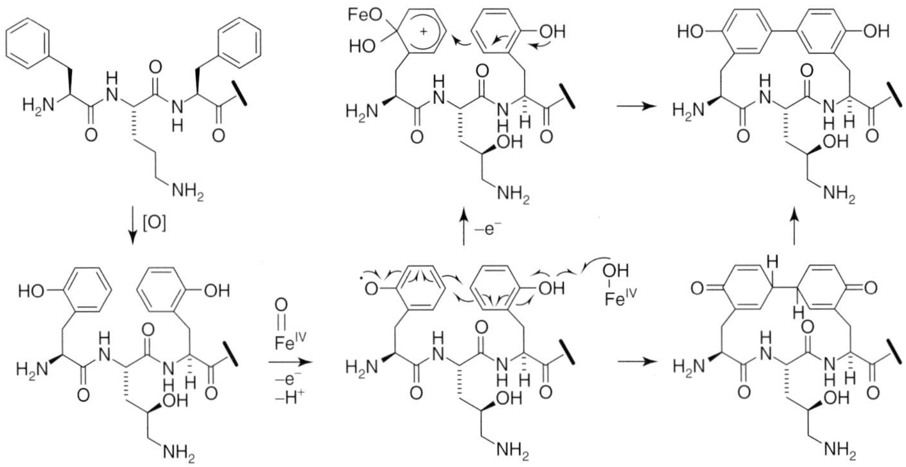

Scheme 9.19 Possible biosynthetic routes to biphenomycin biaryls.

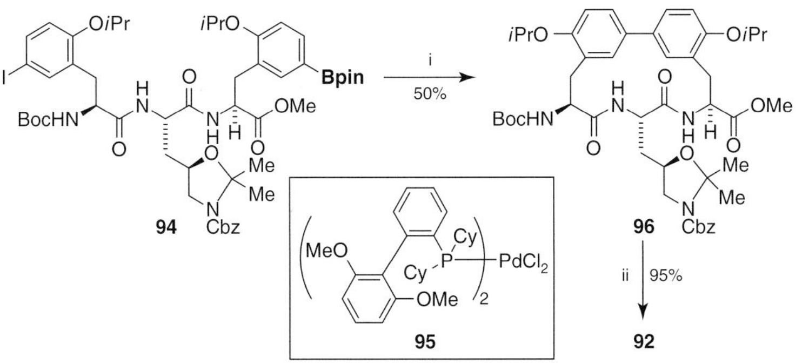

Scheme 9.20 Synthesis of biphenomycin B (**92**) by intramolecular Suzuki coupling [125]. *Reagents and conditions*: (i) **95** (0.06 equiv), K_2CO_3, ΔT(μW); and (ii) BCl_3, CH_2Cl_2, 0 °C, then 2 M LiOH, rt.

biaryl linkage (Scheme 9.20) [125]. Not unexpectedly, most synthetic approaches featured macrolactamization at the critical ring closing step [126, 127]. In this synthetic route, coupling of three non-proteinogenic amino acids yields the linear tripeptide **94** containing an arylboronate and aryl iodide at the C- and N-terminus, respectively. The aryl–aryl bond in **96** was formed by microwave-assisted Suzuki–Miyaura cross-coupling triggered by Buchwald's palladium catalyst (**95**), which proceeded in 50% yield after extensive optimization.

An intramolecular Suzuki–Miyaura cross coupling reaction was also employed for the key ring-forming step in the total synthesis of arylomycin A_2 (**97**, Scheme 9.21). This compound is a biaryl-bridged lipohexapeptide isolated from *Streptomyces* Tü 6075 [128], displaying antibacterial activity against various Gram-positive bacteria by inhibiting the bacterial type I signal peptidase [129]. Romesberg *et al.* compared a route with final macrolactamization to a more biomimetic route (Scheme 9.21) featuring a final cyclization by biaryl bond formation [130].

In fact, poor yields were observed on attempted macrolactam formation from **98** to **99**, owing to the 14-membered ring in the final product as well as the strained biphenyl linkage topology. Biaryl coupling from **100** was more successful, likely assisted by peptide backbone pre-organization and the templating effect of the palladium catalyst.

9.3.4
Vancomycin

Probably the most complex and most well studied example of an aryl-oxidized peptide alkaloid is vancomycin (**101**, Figure 9.14). This prominent glycopeptide is one of the most important agents against multidrug-resistant Gram-positive bacterial pathogens [131]. It acts as an inhibitor of peptidoglycan crosslinking during bacterial cell wall biosynthesis.

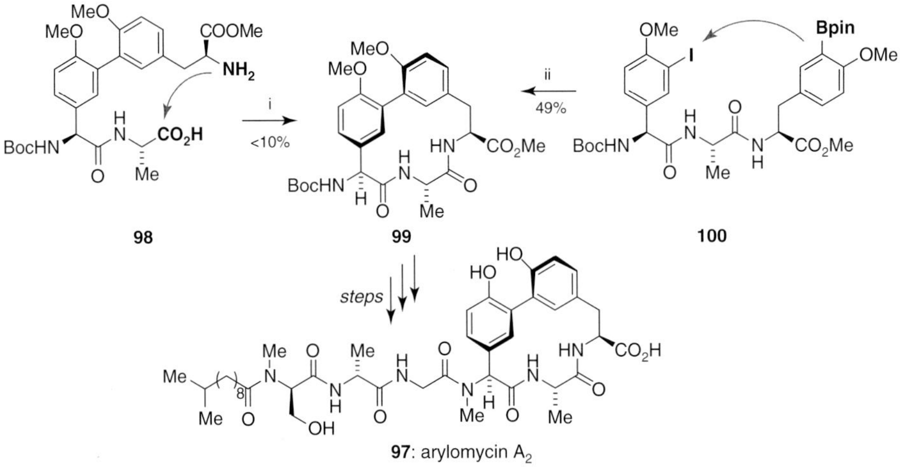

Scheme 9.21 Comparative syntheses of the arylomycin A_2 core [130]. *Reagents and conditions*: (i) EDC, HOBt, DMF; and (ii) $PdCl_2(PPh_3)_2$, K_2CO_3, CH_3CN.

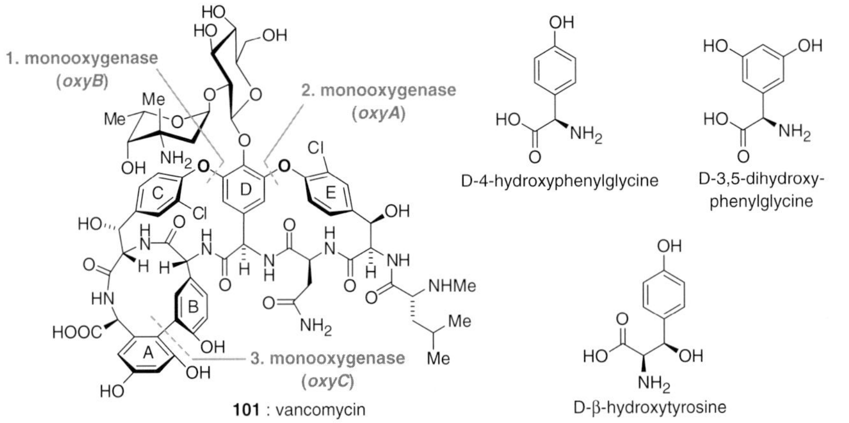

Figure 9.14 Structure of vancomycin and some of its uncommon amino acids.

The biosynthesis of vancomycin was studied in quite some detail. A multi-domain non-ribosomal peptide synthetase (NRPS) builds up the hexapeptide core of this complex molecule via a thioester-templated mechanism [132, 133]. This hexapeptide contains several non-proteinogenic amino acids: a 4-hydroxyphenylglycine, 3,5-dihydroxyphenylglycine, and β-hydroxytyrosine (Figure 9.15) [134–136]. After chlorination at the β-hydroxytyrosine aryl side-chains, the biaryl- and biaryl ether oxidative coupling steps are accomplished while the precursor peptide is still attached to the NRPS assembly line [137, 138]. This process is catalyzed by three homologous P450 oxygenases (oxyABC) [139]. Gene knockout experiments

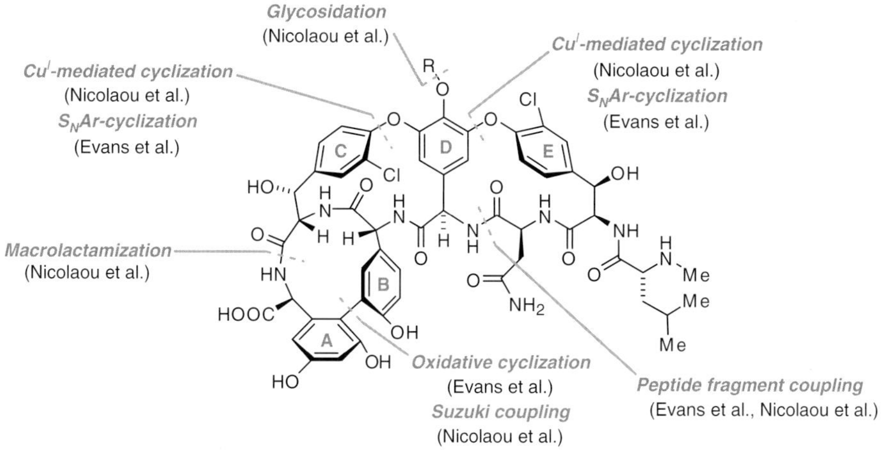

Figure 9.15 Overview of the key disconnections in vancomycin (**101**) synthesis (R = disaccharide).

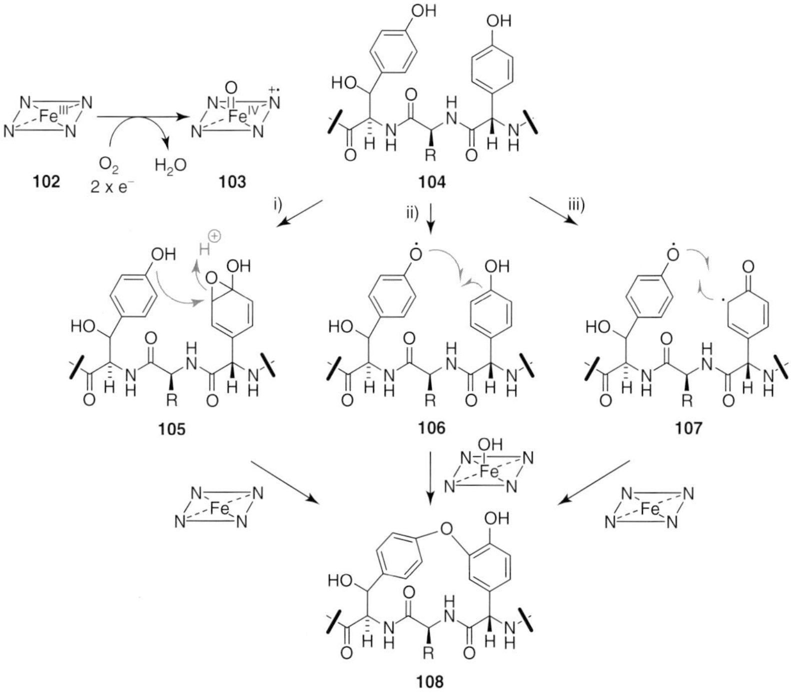

Scheme 9.22 Possible mechanisms of biaryl-ether formation during the biosynthesis of vancomycin.

specified the order of the oxidative coupling steps [140, 141]. A heme-containing oxygenase (oxyB) couples ring C and D, followed by the oxidative coupling of ring D and E (oxyA), while the final biaryl linkage is furnished by oxyC.

X-Ray structures of OxyB and C show a cytochrome P450-type fold featuring a cysteine-bound heme iron [142, 143]. Labeling experiments with OxyB-catalyzed reactions in the presence of $^{18}O_2$ showed no ^{18}O incorporation into the product [144, 145]. By analogy to other P450 enzymes, this indicates a reaction mechanism that proceeds via dioxygen addition to a high-spin Fe(III) species (**102**→**103**, Scheme 9.22). The remaining transformations (**104**→**108**) are still under investigation [145]. Possible pathways include the formation of an arene epoxide (**105**), which is opened by an intramolecular attack of a phenol (Scheme 9.22, path i).

Alternatively, a radical could be formed at one of the aromatic rings (**106**), which adds to a second aryl ring and then abstracts a hydroxyl radical from the Fe-heme complex (path ii). Finally, enzyme-mediated formation of diradical **107** could ensue, followed by coupling (path iii).

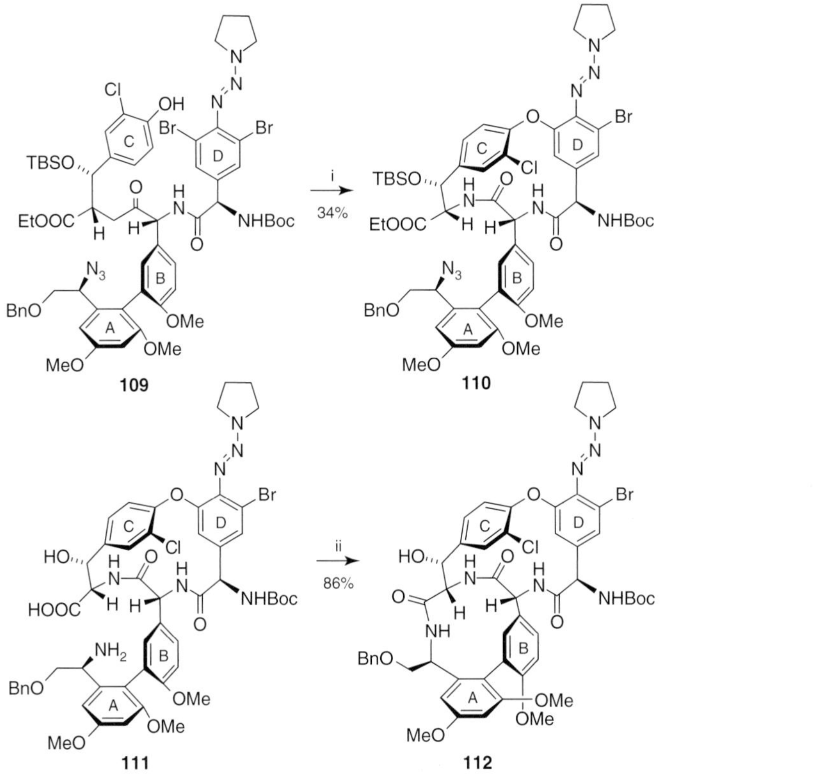

Scheme 9.23 Ring closures in Nicolaou's synthesis of vancomycin [146–151]. *Reagents and conditions*: (i) CuBr · Me_2S, K_2CO_3, pyridine, MeCN; and (ii) FDPP, iPr_2NEt, DMF.

The complex structure of vancomycin (**101**) poses an enormous challenge for synthesis (Figure 9.15). The main demands of a successful synthetic route include the asymmetric synthesis of the amino acid building blocks, their assembly, and most importantly the control over the atropisomerism ensuing from hindered rotation about the aryl–aryl and aryl–*O*-aryl axes.

Two main strategies have been developed for the synthesis of the vancomycin aglycon. The first, developed by the group of K. C. Nicolaou [146–148], is based on a triazene driven copper-mediated coupling reaction of **109** to form the C-O-D ring **110** (Scheme 9.23) [149–151]. This coupling was not atropisomer-selective, but the undesired isomer could be separated. The A-B ring (**112**) was then closed starting from **111** by macrolactamization in an excellent yield and with high atropselectivity,

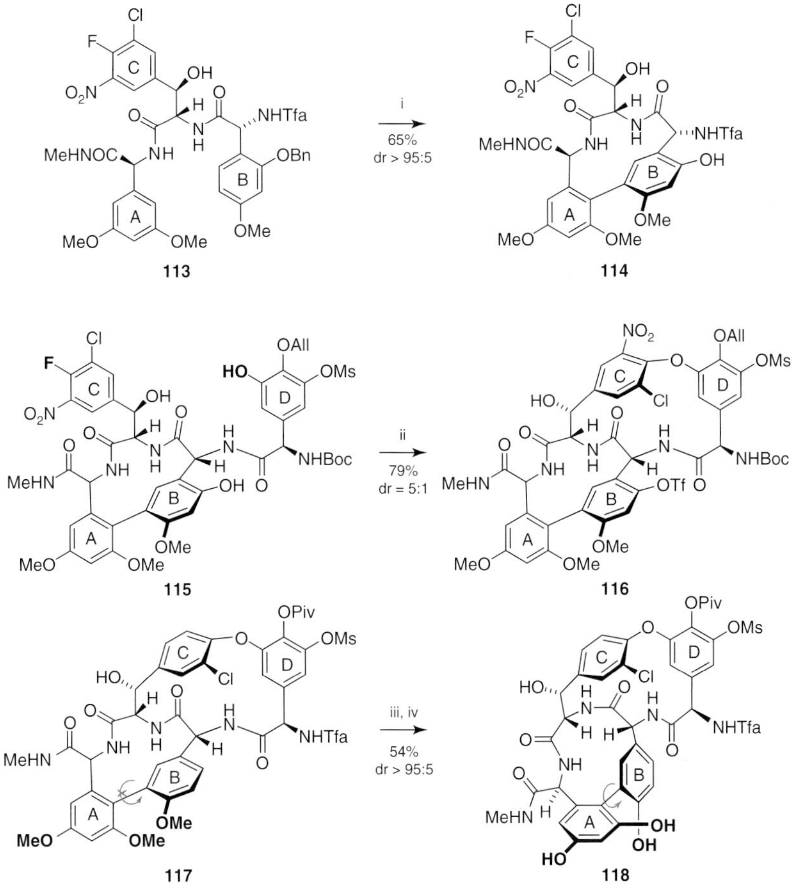

Scheme 9.24 Ring closures in Evans' synthesis of vancomycin [152, 153]. *Reagents and conditions*: (i) VOF_3, $BF_3 \cdot Et_2O$, $AgBF_4$, TFA-CH_2Cl_2, then $NaHB(OAc)_3$; (ii) Na_2CO_3, DMSO, 1.5 h, then Tf_2NPh; (iii) $AlBr_3$, then EtSH; and (iv) MeOH, 60 °C.

probably benefiting from very favorable conformational preorganization. Subsequently, the preformed tripeptide containing a phenolic hydroxyl was coupled to the C-O-D ring to set the stage for the final D-O-E ring closure, which was realized using a copper-mediated reaction of a phenolic hydroxyl and the triazene-activated aromatic ring D.

The second synthetic route to the vancomycin aglycon was established by the group of D. A. Evans (Scheme 9.24) [152, 153]. In this approach, the AB ring **114** was formed first by a biomimetic oxidative biaryl coupling of **113** using VOF_3. This interesting transformation was atropselective (19 : 1), but unfortunately in the wrong stereochemical sense! A building block containing the aryl-ring D was then coupled (**115**) and the electron-withdrawing nitro-group on ring C then allowed for a S_NAr cyclization, which preferentially formed the desired atropisomer of the C-O-D macrocycle **116**. In the biosynthetic pathway, the global structure of the molecule directs the geometry of rings A and B in the last oxidation step catalyzed by OxyC. During synthesis, gentle warming gave the correct AB ring atropisomer **118** in a ratio of 95 : 5 – after the methyl ethers had been cleaved from **117**, which prohibited atropisomer interconversion. The strong directing character of the entire molecular scaffold was illustrated by thermal equilibration of the isolated biaryl, which favors the "non-natural" atropisomer in a ratio of 2 : 1. Dedicated atropisomer equilibrations were then extensively used in Boger's total synthesis of vancomycin [154].

References

1. (a) Piel, J. (2004) *Nat. Prod. Rep.*, **21**, 519–538; (b) Schmidt, E.W. (2007) *Nat. Chem. Biol.*, **4**, 466–473; (c) Piel, J. (2009) *Nat. Prod. Rep.*, **26**, 338–362.
2. Nolan, E.M. and Walsh, C.T. (2009) *ChemBioChem*, **10**, 34–53.
3. Sieber, S.A. and Marahiel, M.A. (2005) *Chem. Rev.*, **105**, 715–738.
4. Oman, T.J. and van der Donk, W.A. (2010) *Nat. Chem. Biol.*, **6**, 9–18.
5. Yoder, R.A. and Johnston, J.N. (2005) *Chem. Rev.*, **105**, 4730–4756.
6. Milne, J.C., Roy, R.S., Eliot, A.C., Kelleher, N.L., Wokhlu, A., Nickels, B., and Walsh, C.T. (1999) *Biochemistry*, **38**, 4768–4781.
7. Li, Y.-M., Milne, J.C., Madison, L.L., Kolter, R., and Walsh, C.T. (1996) *Science*, **274**, 1188–1193.
8. Roy, R.S., Gehring, A.M., Milne, J.C., Belshaw, P.J., and Walsh, C.T. (1999) *Nat. Prod. Rep.*, **16**, 249–263.
9. Hughes, R.A. and Moody, C.J. (2007) *Angew. Chem.*, **119**, 8076–8101; *Angew. Chem. Int. Ed.*, **46**, 7930–7954.
10. Wipf, P. (1995) *Chem. Rev.*, **95**, 2115–2134.
11. Parsons, R.L. and Heathcock, C.H. Jr. (1994) *J. Org. Chem.*, **59**, 4733–4734.
12. You, S.-L., Razavi, H., Kelly, and J.W. (2003) *Angew. Chem.*, **115**, 87–89; *Angew. Chem. Int. Ed.*, **42**, 83–85.
13. You, S.-L. and Kelly, J.W. (2003) *J. Org. Chem.*, **68**, 9506–9509.
14. Jansen, R., Kunze, B., Reichenbach, H., Jurkiewicz, E., Hunsmann, G., and Höfle, G. (1992) *Liebigs Ann. Chem.*, 357–359.
15. Jansen, R., Schomburg, D., and Höfle, G. (1993) *Liebigs Ann. Chem.*, 701–704.
16. Boyce, R.J., Mulqueen, G.C., and Pattenden, G. (1995) *Tetrahedron*, **51**, 7321–7330.
17. Wipf, P. and Venkatraman, S. (1995) *J. Org. Chem.*, **60**, 7224–7229.
18. Akaji, K. and Kiso, Y. (1999) *Tetrahedron*, **55**, 10685–10694.
19. Pattenden, G., Thom, S.M., and Jones, M.F. (1993) *Tetrahedron*, **49**, 2131–2138.

20. Donia, M.S., Ravel, J., and Schmidt, E.W. (2008) *Nat. Chem. Biol.*, **4**, 341–343.

21. Donia, M.S., Hathaway, B.J., Sudek, S., Haygood, M.G., Rosovitz, M.J., Ravel, J., and Schmidt, E.W. (2006) *Nat. Chem. Biol.*, **2**, 729–735.

22. Wipf, P., Fritch, P.C., Geib, S.J., and Sefler, A.M. (1998) *J. Am. Chem. Soc.*, **120**, 4105–4112.

23. Milne, B.F., Long, P.F., Starcevic, A., Hranueli, D., and Jaspars, M. (2006) *Org. Biomol. Chem.*, **4**, 631–638.

24. Wipf, P. and Fritch, P.C. (1996) *J. Am. Chem. Soc.*, **118**, 12358–12367.

25. Wipf, P. and Miller, C.P. (1993) *J. Org. Chem.*, **58**, 3604–3606.

26. Bagley, M.C., Dale, J.W., Merritt, E.A., and Xiong, X. (2005) *Chem. Rev.*, **105**, 685–714.

27. Priestley, N.D., Smith, T.M., Shipley, P.R., and Floss, H.G. (1996) *Bioorg. Med. Chem.*, **4**, 1135–1147.

28. Frenzel, T., Zhou, P., and Floss, H.G. (1990) *Arch. Biochem. Biophys.*, **278**, 35–40.

29. Mocek, U., Zeng, Z., O'Hagan, D., Zhou, P., Fan, L.D.G., Beale, J.M., and Floss, H.G. (1993) *J. Am. Chem. Soc.*, **115**, 7992–8001.

30. Smith, T.M., Priestley, N.D., Knaggs, A.R., Nguyen, T., and Floss, H.G. (1993) *J. Chem. Soc., Chem. Commun.*, 1612–1614.

31. Kelly, W.L., Pan, L., and Li, C. (2009) *J. Am. Chem. Soc.*, **131**, 4327–4334.

32. Yu, Y., Duan, L., Zhang, Q., Liao, R., Ding, Y., Pan, H., Wendt-Pienkowski, E., Tang, G., Shen, B., and Liu, W. (2009) *ACS Chem. Biol.*, **4**, 855–864.

33. Li, C. and Kelly, W.L. (2010) *Nat. Prod. Rep.*, **27**, 153–164.

34. Arndt, H.-D., Schoof, S., and Lu, J.-Y. (2009) *Angew. Chem.*, **121**, 6900–6904; *Angew. Chem. Int. Ed.*, **48**, 6770–6773.

35. McIntosh, J.A., Donia, M.S., and Schmidt, E.W. (2009) *Nat. Prod. Rep.*, **26**, 537–559.

36. Bycroft, B.W. and Gowland, M.S. (1978) *J. Chem. Soc., Chem. Commun.*, 256–258.

37. Nicolaou, K.C., Safina, B.S., Zak, M., Estrada, A.A., and Lee, S.H. (2004) *Angew. Chem.*, **116**, 5197–5202; *Angew. Chem. Int. Ed.*, **43**, 5087–5092.

38. Nicolaou, K.C., Zak, M., Safina, B.S., Lee, S.H., and Estrada, A.A. (2004) *Angew. Chem.*, **116**, 5202–5207; *Angew. Chem. Int. Ed.*, **43**, 5092–5097.

39. Nicolaou, K.C., Safina, B.S., Zak, M., Lee, S.H., Nevalainen, M., Bella, M., Estrada, A.A., Funke, C., Zecri, F.J., and Bulat, S. (2005) *J. Am. Chem. Soc.*, **127**, 11159–11175.

40. Nicolaou, K.C., Zak, M., Safina, B.S., Estrada, A.A., Lee, S.H., and Nevalainen, M. (2005) *J. Am. Chem. Soc.*, **127**, 11176–11183.

41. Mori, T., Higashibayashi, S., Goto, T., Kohno, M., Satouchi, Y., Shinko, K., Suzuki, K., Suzuki, S., Tohmiya, H., Hashimoto, K., and Nakata, M. (2008) *Chem. Asian J.*, **3**, 984–1012.

42. Mori, T., Higashibayashi, S., Goto, T., Kohno, M., Satouchi, Y., Shinko, K., Suzuki, S., Tohmiya, H., Hashimoto, K., and Nakata, M. (2008) *Chem. Asian J.*, **3**, 1013–1025.

43. Wulff, G. and Klinken, H.T. (1992) *Tetrahedron*, **48**, 5985–5990.

44. Wulff, G. and Böhnke, H. (1986) *Angew. Chem.*, **98**, 101–102; *Angew. Chem. Int. Ed. Engl.*, **25**, 90–92.

45. Wulff, G., Lindner, H.-J., Böhnke, H., Steigel, A., and Klinken, H.-T. (1989) *Liebigs Ann. Chem.*, 527–531.

46. Moody, C.J., Hughes, R.A., Thompson, S.P., and Alcaraz, L. (2002) *Chem. Commun.*, 1760–1761.

47. Sasaki, H., Irie, R., Hamada, T., Suzuki, K., and Katsuki, T. (1994) *Tetrahedron*, **50**, 11827–11838.

48. Boyd, D.R., Davies, R.J.H., Hamilton, L., McCullough, J.J., Malone, J.F., Porter, H.P., Smith, A., Carl, J.M., Sayer, J.M., and Jerina, D.M. (1994) *J. Org. Chem.*, **59**, 984–990.

49. Bushman, D.R., Sayer, J.M., Boyd, D.R., and Jerina, D.M. (1989) *J. Am. Chem. Soc.*, **111**, 2688–2691.

50. De Pietro, M.T., Marazzi, A., Sosio, M., Donadio, S., and Lancini, G. (2001) *J. Antibiot.*, **54**, 1066–1071.

51. Morris, R.P., Leeds, J.A., Nägeli, H.-U., Oberer, L., Memmert, K., Weber, E., LaMarche, M.J., Parker, C.N., Burrer, N., Esterow, S., Hein, A.E.,

Schmitt, E.K., and Krastel, P. (2009) *J. Am. Chem. Soc.*, **131**, 5946–5955.

52. Nicolaou, K.C., Zou, B., Dethe, D.H., Li, D.B., and Chen, D.Y.-K. (2006) *Angew. Chem.*, **118**, 7950–7956; (2006) *Angew. Chem. Int. Ed.*, **45**, 7786–7792.
53. Nicolaou, K.C., Dethe, D.H., Leung, G.Y.C., Zou, B., and Chen, D.Y.-K. (2008) *Chem. Asian J.*, **3**, 413–429.
54. Müller, H.-M., Delgado, O., and Bach, T. (2007) *Angew. Chem.*, **119**, 4855–4858; *Angew. Chem. Int. Ed.*, **46**, 4771–4774.
55. Delgado, O., Müller, H.-M., and Bach, T. (2008) *Chem. Eur. J.*, **14**, 2322–2339.
56. Tan, N.-H. and Zhou, J. (2006) *Chem. Rev.*, **106**, 840–895.
57. El-Seedi, H.R., Zahra, M.H., Goransson, U., and Verpoorte, R. (2007) *Phytochem. Rev.*, **6**, 143–165.
58. Gournelis, D.C., Laskaris, G.G., and Verpoorte, R. (1997) *Nat. Prod. Rep.*, **14**, 75–82.
59. Baig, M.A., Banthorpe, D.V., Coleman, A.A., Tampion, M.D., Tampion, J., and White, J.J. (1993) *Phytochemistry*, **34**, 171–174.
60. Marchand, J., Païs, M., Moseur, X., and Jarreau, F.-X. (1969) *Tetrahedron*, **25**, 937–954.
61. Joullié, M.M. and Richard, D.J. (2004) *Chem. Commun.*, 2011–2015.
62. Schmidt, U., Lieberknecht, A., Bökens, H., and Griesser, H. (1981) *Angew. Chem.*, **93**, 1121–1122; *Angew. Chem. Int. Ed. Engl.*, **20**, 1026–1027.
63. Schmidt, U., Griesser, H., Lieberknecht, A., and Talbiersky, J. (1981) *Angew. Chem.*, **93**, 271–272; *Angew. Chem. Int. Ed Engl.*, **20**, 280–281.
64. Schmidt, U., Lieberknecht, A., Griesser, H., and Häusler, J. (1981) *Angew. Chem.*, **93**, 272–273; *Angew. Chem. Int. Ed. Engl.*, **20**, 281–282.
65. Schmidt, U., Lieberknecht, A., Grisser, H., and Talbiersky, J. (1982) *J. Org. Chem.*, **47**, 3261–3264.
66. Schmidt, U. and Schanbacher, U. (1983) *Angew. Chem.*, **93**, 150–151; *Angew. Chem. Int. Ed. Engl.*, **22**, 152–153.
67. Schmidt, U., Lieberknecht, A., Bökens, H., and Griesser, H. (1983) *J. Org. Chem.*, **48**, 2680–2685.
68. Schmidt, U. and Schanbacher, U. (1984) *Liebigs Ann. Chem.*, 1205–1215.
69. Schmidt, U., Zäh, M., and Lieberknecht, A. (1991) *J. Chem. Soc., Chem. Commun.*, 1002–1004.
70. Nutt, R.F., Chen, K.M., and Joullié, M.M. (1984) *J. Org. Chem.*, **49**, 1013–1021.
71. Heffner, J. and Joullié, M.M. (1989) *Tetrahedron Lett.*, **30**, 7021–7024.
72. Heffner, J., Jiang, J., and Joullié, M.M. (1992) *J. Am. Chem. Soc.*, **114**, 10181–10189.
73. Jiang, J., Li, W.R., Przeslawski, R.M., and Joullié, M.M. (1993) *Tetrahedron Lett.*, **34**, 6705–6708.
74. Williams, L., Zhang, Z., Shao, F., Carroll, P.J., and Joullié, M.M. (1996) *Tetrahedron*, **52**, 11673–11694.
75. East, S.P. and Joullié, M.M. (1998) *Tetrahedron Lett.*, **39**, 7211–7214.
76. East, S.P., Shao, F., Williams, L., and Joullié, M.M. (1998) *Tetrahedron*, **54**, 13371–13390.
77. Han, B.H., Kim, Y.C., Park, M.K., Park, J.H., Go, H.J., Yang, H.O., Suh, D.Y., and Kang, Y.C. (1995) *Heterocycles*, **41**, 1909–1914.
78. Suh, D.Y., Kim, Y.C., Han, Y.M., and Han, B.H. (1996) *Heterocycles*, **43**, 2347–2351.
79. Kim, Y.-A., Shin, H.-N., Park, M.-S., Cho, S.-H., and Han, S.-Y. (2003) *Tetrahedron Lett.*, **44**, 2557–2560.
80. Temal-Laïb, T., Chanstanet, J., and Zhu, J. (2002) *J. Am. Chem. Soc.*, **124**, 583–590.
81. Temal-Laïb, T. and Zhu, J. (1999) *Tetrahedron Lett.*, **40**, 83–86.
82. Temal-Laïb, T., Bois-Choussy, M., and Zhu, J. (2000) *Tetrahedron Lett.*, **41**, 7645–7648.
83. Cristau, P., Temal-Laïb, T., Bois-Choussy, M., Martin, M.-T., Vors, J.-P., and Zhu, J. (2005) *Chem. Eur. J.*, **11**, 2668–2679.
84. Toumi, M., Couty, F., and Evano, G. (2007) *Angew. Chem.*, **119**, 578–581; *Angew. Chem. Int. Ed.*, **46**, 572–575.
85. Evano, G., Toumi, M., and Coste, A. (2009) *Chem. Commun.*, 4166–4175.

86. Toumi, M., Rincheval, V., Young, A., Gergeres, D., Turos, E., Couty, F., Mignotte, B., and Evano, G. (2009) *Eur. J. Org. Chem.*, 3368–3386.
87. Kase, H., Kaneko, M., and Yamada, K. (1987) *J. Antibiot.*, **40**, 450–454.
88. Yasuzawa, T., Shirahata, K., and Sano, H. (1987) *J. Antibiot.*, **40**, 455–458.
89. Sano, S., Kuroda, H., Ueno, M., Yoshikawa, Y., Nakamura, T., and Obayashi, A. (1987) *J. Antibiot.*, **40**, 519–525.
90. Jolad, S.D., Hoffman, J.J., Torrance, S.J., Wiedhopf, R.M., Cole, J.R., Arora, S.K., Bates, R.B., Gargiulo, R.L., and Kriek, G.R. (1977) *J. Am. Chem. Soc.*, **99**, 8040–8044.
91. Itokawa, H., Takeya, K., Mihara, K., Mori, N., Hamanaka, T., and Sonobe, T. (1983) *Chem. Pharm. Bull.*, **31**, 1424–1427.
92. Itokawa, H., Takeya, K., Mori, N., Hamanaka, T., Sonobe, T., and Mihara, K. (1984) *Chem. Pharm. Bull.*, **32**, 284–290.
93. Burgess, K., Lim, D., Bois-Choussy, M., and Zhu, J. (1997) *Tetrahedron Lett.*, **38**, 3345–3348.
94. Nolasco, L., Pérez-Gonzälez, M., Caggiano, L., and Jackson, R.F.W. (2009) *J. Org. Chem.*, **74**, 8280–8289.
95. Boger, D.L. and Yohannes, D. (1990) *J. Org. Chem.*, **55**, 6000–6017.
96. Evans, D.A. and Ellman, J.A. (1989) *J. Am. Chem. Soc.*, **111**, 1063–1072.
97. Schmidt, U., Weller, D., Holder, A., and Lieberknecht, A. (1988) *Tetrahedron Lett.*, **29**, 3227–3230.
98. Paerson, A.J., Zhang, P.L., and Lee, K.J. (1996) *J. Org. Chem.*, **61**, 6581–6586.
99. Pérez-Gonzälez, M. and Jackson, R.F.W. (2000) *Chem. Commun.*, 2423–2424.
100. Chattopadhyay, S.K., Bandyopadhyay, A., and Pal, B.K. (2007) *Tetrahedron Lett.*, **48**, 3655–3659.
101. Nishiyama, S., Suzuki, Y., and Yamamura, S. (1988) *Tetrahedron Lett.*, **29**, 559–562.
102. Nishiyama, S., Suzuki, Y., and Yamamura, S. (1989) *Tetrahedron Lett.*, **30**, 379–382.
103. Reddy, M.V.R., Harper, R.M., and Faulkner, D.J. (1998) *Tetrahedron*, **54**, 10649–10656.
104. Ito, M., Yamanaka, M., Kutsumura, N., and Nishiyama, S. (2004) *Tetrahedron*, **60**, 5623–5634.
105. Rama Rao, A.V., Chakraborty, T.K., Laxma Reddy, K., and Rao, A.S. (1992) *Tetrahedron Lett.*, **33**, 4799–4802.
106. Rama Rao, A.V., Gurjar, M.K., Reddy, A.B., and Khare, V.B. (1994) *Tetrahedron Lett.*, **34**, 1657–1660.
107. Boger, D.L. and Yohannes, D. (1991) *J. Am. Chem. Soc.*, **113**, 1427–1429.
108. Bigot, A., Bois-Choussy, M., and Zhu, J. (2000) *Tetrahedron Lett.*, **41**, 4573–4577.
109. Janetka, J.W. and Rich, D.H. (1997) *J. Am. Chem. Soc.*, **119**, 6488–6495.
110. Bigot, A., Tran Huu Dau, M.E., and Zhu, J. (1999) *J. Org. Chem.*, **64**, 6283–6296.
111. Inoue, T., Inaba, T., Umezawa, I., Yuasa, M., Itokawa, H., Ogura, K., Komatsu, K., Hara, H., and Hoshino, O. (1995) *Chem. Pharm. Bull.*, **43**, 1325–1335.
112. (a) Hitotsuyanagi, Y., Ishikawa, H., Naito, S., and Takeya, K. (2003) *Tetrahedron Lett.*, **44**, 5901–5903; (b) Kilitoglu, B., Arndt, H.-D. (2009) *Synlett*, 720–723.
113. Kazlauskas, R., Lidgard, R.O., Murphy, P.T., Wells, R.J., and Blount, J.F. (1998) *Aust. J. Chem.*, **34**, 765–786.
114. Carney, J.R., Scheuer, P.J., and Kelly-Borges, M. (1993) *J. Nat. Prod.*, **56**, 153–157.
115. Pordesimo, E.O. and Schmitz, F.J. (1990) *J. Org. Chem.*, **55**, 4704–4709.
116. Mack, M.M., Molinski, T.F., Buck, E.D., and Pessah, I.N. (1994) *J. Biol. Chem.*, **269**, 23236–23249.
117. Jaspars, M., Rali, T., Laney, M., Schatzman, R.C., Diaz, M.C., Schmitz, F.J., Pordesimo, E.O., and Crews, P. (1994) *Tetrahedron*, **50**, 7367–7374.
118. Nishiyama, S. and Yamamura, S. (1984) *Tetrahedron Lett.*, **23**, 1281–1284.
119. Nishiyama, S., Suzuki, T., and Yamamura, S. (1982) *Tetrahedron Lett.*, **23**, 3699–3702.

120. Guo, Z.-W., Machiya, K., Salamonczyk, G.M., and Sih, C.J. (1998) *J. Org. Chem.*, **63**, 4269–4276.
121. Kotoku, N., Tsujita, H., Hiramatsu, A., Mori, C., Koizumi, N., and Kobayashi, M. (2005) *Tetrahedron*, **61**, 7211–7218.
122. Couladouros, E.A., Pitsinos, E.N., Moutsos, V.I., and Sarakinos, G. (2005) *Chem. Eur. J.*, **11**, 406–421.
123. Chang, C.C., Morton, G.O., James, J.C., Siegel, M.M., Kuck, N.A., Testa, R.T., and Borders, D.B. (1991) *J. Antibiot.*, **44**, 674–677.
124. Ezaki, M., Shigematsu, N., Yamashita, M., Komori, T., Umehara, K., and Imanaka, H. (1993) *J. Antibiot.*, **46**, 135–140.
125. Lépine, R. and Zhu, J. (2005) *Org. Lett.*, **7**, 2981–2984.
126. Schmidt, U., Meyer, R., Leitenberger, V., Lieberknecht, A., and Griesser, H. (1991) *J. Chem. Soc., Chem. Commun.*, 275–277.
127. Waldmann, H., He, Y.-P., Tan, H., Arve, L., and Arndt, H.-D. (2008) *Chem. Commun.*, 5562–5564.
128. Höltzel, A., Schmid, D.G., Nicholson, G.J., Stevanovic, S., Schimana, J., Gebhardt, K., Fiedler, H.-P., and Jung, G. (2002) *J. Antibiot.*, **55**, 571–577.
129. Paetzel, M., Karla, A., Strynadka, N.C.J., and Dalbey, R.E. (2002) *Chem. Rev.*, **102**, 4549–4580.
130. Roberts, T.C., Smith, P.A., Cirz, R.T., and Romesberg, F.E. (2007) *J. Am. Chem. Soc.*, **129**, 15830–15838.
131. Williams, D.H. (1996) *Nat. Prod. Rep.*, **13**, 469–477.
132. Fischbach, M.A. and Walsh, C.T. (2006) *Chem. Rev.*, **106**, 3468–3496.
133. Finking, R. and Marahiel, M.A. (2004) *Annu. Rev. Microbiol.*, **58**, 453–488.
134. Widboom, P.F. and Bruner, S.D. (2009) *ChemBioChem*, **10**, 1757–1764.
135. Süssmuth, R.D. and Wohlleben, W. (2004) *Appl. Microbiol. Biotechnol.*, **63**, 344–350.
136. Hubbard, B.K. and Walsh, C.T. (2003) *Angew. Chem.*, **115**, 752–789; *Angew. Chem. Int. Ed.*, **42**, 730–765.
137. Zerbe, K., Woithe, K., Li, D.B., Vitali, F., Bigler, L., and Robinson, J.A. (2004) *Angew. Chem.*, **116**, 6877–6881; *Angew Chem. Int. Ed.*, **43**, 6709–6713.
138. Woithe, K., Geib, N., Zerbe, K., Li, D.B., Heck, M., Fournier-Rousset, S., Meyer, O., Vitali, F., Matoba, N., Abdou-Hadeed, K., and Robinson, J.A. (2007) *J. Am. Chem. Soc.*, **129**, 6887–6895.
139. Bischoff, D., Bister, B., Bertazzo, M., Pfeifer, V., Stegmann, E., Nicholson, G.J., Keller, S., Pelzer, S., Wohlleben, W., and Süssmuth, R.D. (2005) *ChemBioChem*, **6**, 267–272.
140. Bischoff, D., Pelzer, S., Höltzel, A., Nicholson, G.J., Stockert, S., Wohlleben, W., Jung, G., and Süssmuth, R.D. (2001) *Angew. Chem.*, **113**, 1736–1739; *Angew. Chem. Int. Ed.*, **40**, 1693–1696.
141. Bischoff, D., Pelzer, S., Bister, B., Nicholson, G.J., Stockert, S., Schirle, W., Wohlleben, W., Jung, G., and Süssmuth, R.D. (2001) *Angew. Chem.*, **113**, 4824–4827; *Angew. Chem. Int. Ed.*, **40**, 4688–4691.
142. Zerbe, K., Pylypenko, O., Vitali, F., Zhang, W., Rouset, S., Heck, M., Vrijbloed, J.W., Bischoff, D., Bister, B., Süssmuth, R.D., Pelzer, S., Wohlleben, W., Robinson, J.A., and Schlichting, I. (2002) *J. Biol. Chem.*, **277**, 47476–47485.
143. Pylypenko, O., Vitali, F., Zerbe, K., Robinson, J.A., and Schlichting, I. (2003) *J. Biol. Chem.*, **278**, 46727–46733.
144. Geib, N., Woithe, K., Zerbe, K., Li, D.B., and Robinson, J.A. (2008) *Bioorg. Med. Chem. Lett.*, **18**, 3081–3084.
145. Holding, A.N. and Spencer, J.B. (2008) *ChemBioChem*, **9**, 2209–2214.
146. Nicolaou, K.C., Natarajan, S., Li, H., Jain, N.F., Hughes, R., Solomon, M.E., Ramanjulu, J.M., Boddy, C.N.C., and Takayanagi, M. (1998) *Angew. Chem.*, **110**, 2872–2878; *Angew. Chem. Int. Ed.*, **37**, 2708–2714.
147. Nicolaou, K.C., Jain, N.F., Natarajan, S., Hughes, R.R., Solomon, R., Li, H., Ramanjulu, J.M., Takayanagi, M., Koumbis, A.E., and Bando, T. (1998) *Angew. Chem.*, **110**, 2879–2881; *Angew. Chem. Int. Ed.*, **37**, 2714–2716.
148. Nicolaou, K.C., Takayanagi, M., Jain, N.F., Natarajan, S., Koumbis, A.E., Bando, T., and Ramanjulu, J.M.,

(1998) *Angew. Chem.*, **110**, 2881–2883; *Angew. Chem. Int. Ed.*, **37**, 2717–2719.

149. Nicolaou, K.C., Li, H., Boddy, C.N.C., Ramanjulu, J.M., Yue, T.-Y., Natarajan, S., Chu, X.-J., and Bräse, S. (1999) *Chem. Eur. J.*, **5**, 2584–2601.

150. Nicolaou, K.C., Boddy, C.N.C., Li, H., Koumbis, A.E., Hughes, R., Natarajan, S., Jain, N.F., Ramanjulu, J.M., Bräse, S., and Solomon, M.E. (1999) *Chem. Eur. J.*, **5**, 2602–2621.

151. Nicolaou, K.C., Koumbis, A.E., Takayanagi, M., Natarajan, S., Jain, N.F., Bando, T., Li, H., and Hughes, R. (1999) *Chem. Eur. J.*, **5**, 2622–2647.

152. Evans, D.A., Wood, M.R., Trotter, B.W., Richardson, T.I., Barrow, J.C., and Katz, J.L. (1998) *Angew. Chem.*, **110**, 2864–2868; *Angew. Chem. Int. Ed.*, **37**, 2700–2704.

153. Evans, D.A., Dinsmore, C.I., Watson, P.S., Wood, M.R., Richardson, T.I., Trotter, B.W., and Katz, J.L. (1998) *Angew. Chem.*, **110**, 2868–2872; *Angew. Chem. Int. Ed.*, **37**, 2704–2708.

154. Boger, D.L., Miyazaki, S., Kim, S.H., Wu, J.H., Loiseleur, O., and Castle, S.L. (1999) *J. Am. Chem. Soc.*, **121**, 3226–3227.

10 Biomimetic Synthesis of Indole-Oxidized and Complex Peptide Alkaloids

Hans-Dieter Arndt, Lech-Gustav Milroy, and Stefano Rizzo

10.1 Indole-Oxidized Cyclopeptides

10.1.1 Introduction

In the previous chapter, biomimetic aspects of the synthesis of azole- and aryl-containing peptide alkaloids were discussed. Macrocyclic peptide alkaloids in which the ring is formed through covalent attachment to a Trp-based indole are less common in Nature than phenol-oxidized cyclopeptides, but are even more intriguing (Figure 10.1).

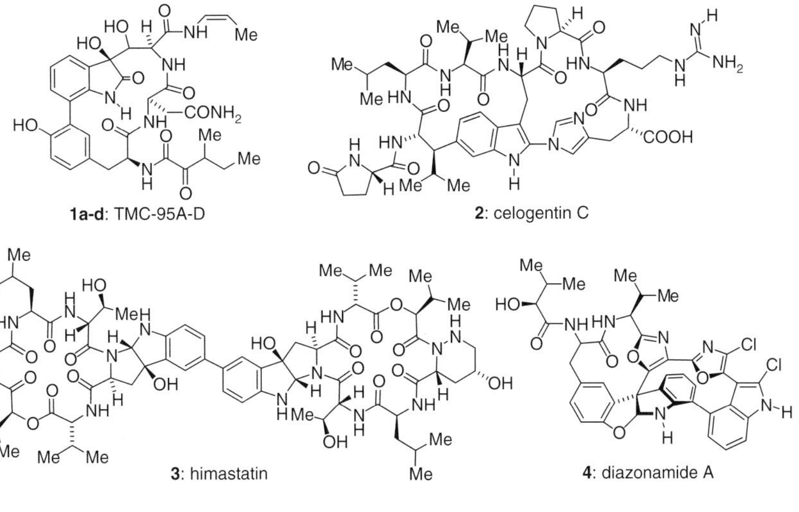

Figure 10.1 Selected indole-oxidized peptide alkaloids (**1**–**4**).

In the broader sense, electron-rich indoles are highly susceptible to oxidation reactions, much like phenol derivatives, and prone to undergo coupling reactions

Biomimetic Organic Synthesis, First Edition. Edited by Erwan Poupon and Bastien Nay.

with latent electrophiles. In the context of a densely functionalized peptide chain, this leads to a range of unusual substitution patterns, sometimes macrocycle forming, which is evidenced by the structural diversity of the molecules presented in the following section.

10.1.2 TMC-95A-D

TMC-95A-D (Figure 10.2) are selective proteasome inhibitors isolated from the fermentation broth of *Apiospora montagnei* [1]. They consist most notably of a highly oxidized Trp-Tyr biaryl linkage with an all L-configured tripeptide-derived monocycle appended by unusual *cis*-propenyl- and 3-methyl-2-oxopentanyl amide residues. Compounds **1a–d** inhibit the chymotrypsin-like, trypsin-like, and caspase-like peptidase activities of the proteasome at nanomolar concentrations [1].

Little is known about the biosynthesis of **1a–d**, although knowledge about the biosynthesis of biphenyl cyclopeptides (see Chapter 9) may be indicative. Macrocyclization is thought to proceed via an intramolecular biaryl coupling resulting from oxidative tailoring of side chains. Oxidation of the Trp indole ring at the β, C2, and C3 positions may be a pre-requisite in this case (Scheme 10.1). Some experimental evidence suggests that the oxindole more readily undergoes *intra*-molecular biaryl coupling when sp^3-hybridized at the C3 position (**6** → **1**) [2], even more than the corresponding indole (**7** → **5**) [3].

The oxindole moiety is likely to result from multiple rounds of oxidation of Trp by heme peroxidase or cytochrome P450-type enzymes at the β, C2, and C3 positions (Scheme 10.1). The indolyl–phenyl coupling (Scheme 10.2) might either occur via ring opening of an arene epoxide (Path A, **8** → **9**) or via a cytochrome P450-mediated phenoxy radical mechanism (Path B, **8** → **10**), analogous to biphenyl-couplings.

The 3-methyl-2-oxopentanoic side chain (Scheme 10.3) is predicted to result from an oxidative transamination of isoleucine followed by hydrolysis (**11** → **12**) [4]. A

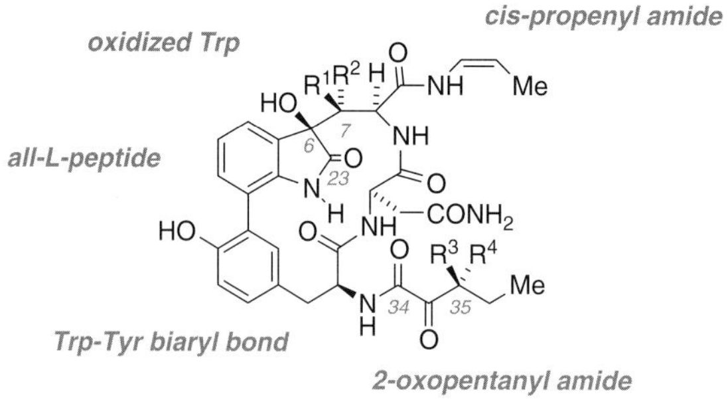

TMC-95A-D

	R^1	R^2	R^3	R^4	TMC-95
1a:	H	OH	Me	H	A
1b:	H	OH	H	Me	B
1c:	OH	H	Me	H	C
1d:	OH	H	H	Me	D

(*Apiospora montagnei* Sacc. TC 1093)

- proteasome inhibitor -

Koguchi et al., *JAB* **2000**, *53*, 105
Kohno et al., *JOC* **2000**, *65*, 990

Figure 10.2 Key features of TMC-95A-D (**1a–d**). To improve scheme readability, the following abbreviations are used throughout: *ACIE* = *Angew. Chem. Int. Ed.*, *CEJ* = *Chem. Eur. J.*, *JAB* = *J. Antibiot. JACS* = *J. Am. Chem. Soc.*, *JOC* = *J. Org. Chem.*, *OL* = *Org. Lett.*, *PNAS* = *Proc. Natl. Acad. Sci. U.S.A.*, *TH* = *Tetrahedron*, *TL* = *Tetrahedron Lett.*

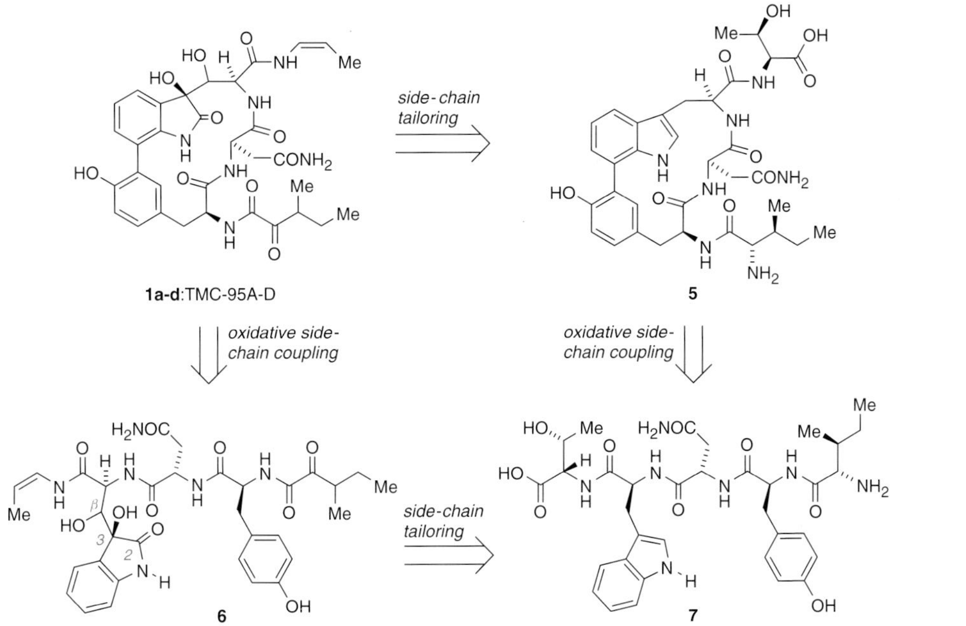

Scheme 10.1 Simplified retrobiosynthetic analysis of TMC-95A-D.

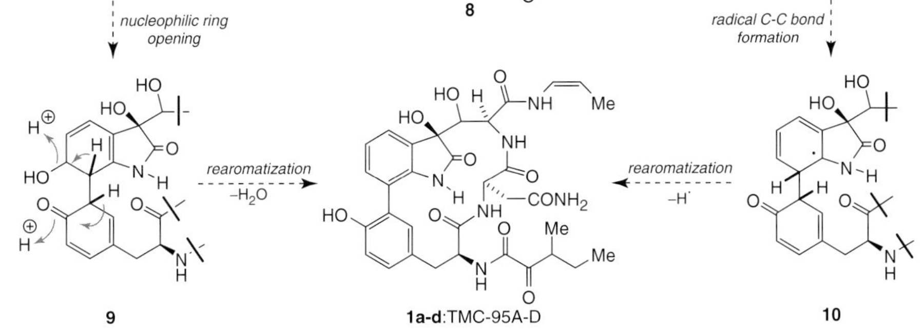

Scheme 10.2 Two alternative mechanisms for indolyl–phenyl coupling.

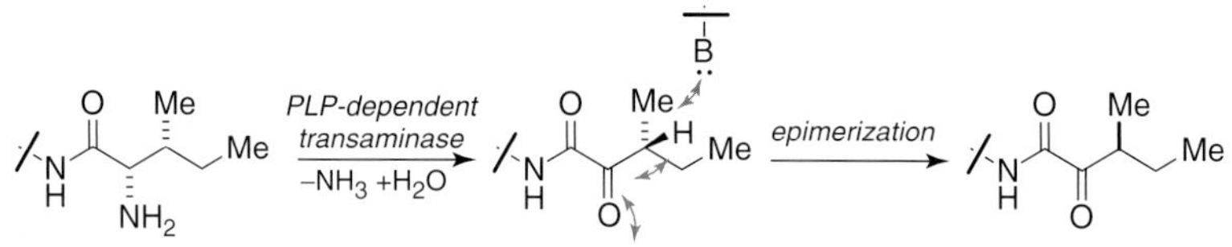

11 12 13

Scheme 10.3 Biosynthetic explanation for the origin of the 3-methyl-2-oxopentanoic side-chain [4]. PLP = pyridoxal phosphate

base-mediated epimerization of the labile β-stereogenic center (**12** → **13**) – *in vivo* or during isolation – would then offer a convenient explanation for the occurrence of two sets of TMC-95 diastereomers (Figure 10.2, TMC-95 A with C, and B with D).

The biosynthesis of the rare terminal enamide (Scheme 10.4) could possibly involve base-catalyzed *anti*-elimination of *C*-terminal *allo*-Thr **14** in a suitably *O*-activated form (e.g., **14** → **15**). This might then deliver the enamide with correct *(Z)*-geometry on concomitant decarboxylation (**15** → **16**). While biochemical evidence is limited, chemical reactivity suggests this to be a viable process (see below).

Successful total syntheses of natural TMC-95 compounds have been achieved using a combination of new and established chemical techniques – some in the end more biomimetic than others. To form the critical biaryl bond, mostly Pd-mediated coupling has been used, whereas biomimetic oxidative coupling processes have yet to be explored (Figure 10.3). Additionally, considerable synthetic work has

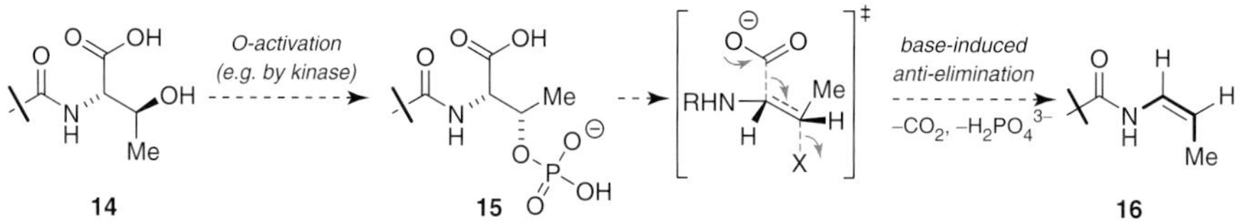

14 15 16

Scheme 10.4 Potential biosynthetic explanation for the peculiar *(Z)*-enamide side-chain.

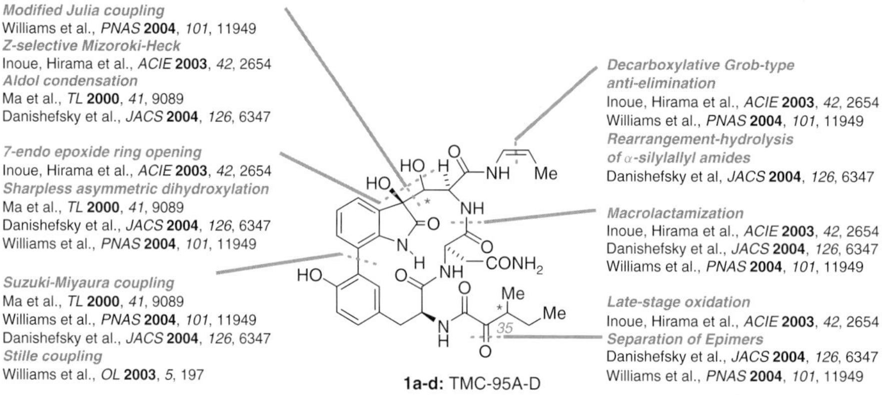

Figure 10.3 Summary of synthetic approaches to TMC-95A-D (**1a**–**d**).

been dedicated to the synthesis of unnatural analogs [5], which in some cases has produced more selective proteasomal inhibitors.

10.1.2.1 **Formation of the Trp-Tyr Biaryl Bond by Metal-Catalyzed Cross Coupling**

Palladium catalysis has been used to mediate the intermolecular coupling between appropriately activated oxindole (Scheme 10.5, **17** and **18**) and phenol fragments (**19** and **20**) to form the biaryl (**21**) in a relatively mild and selective manner. Although less biomimetic, this highly efficient and flexible approach was well suited for analog studies. The more biomimetic intramolecular cross coupling strategy gave some success during the synthesis of simplified TMC-95 analogs [3, 4]. Interestingly, ring closure was found in this case to be most effective for substrates bearing oxindoles that were sp^3-hybridized at the C3 position (Scheme 10.5, **21**). However, yields were typically lower compared with intermolecular cross-couplings.

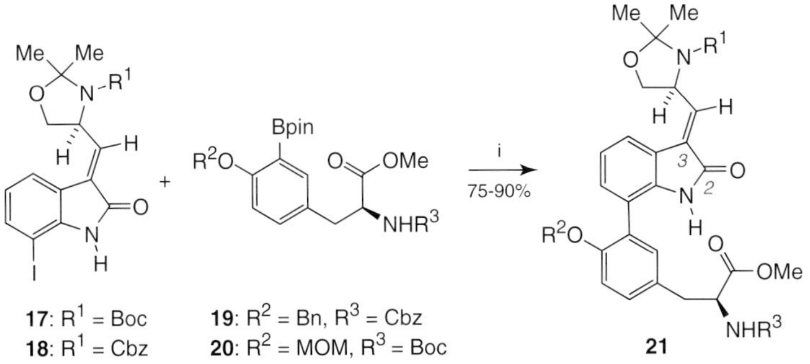

Scheme 10.5 Conventional biaryl coupling. *Reagents and conditions*: (i) **17** + **19**, cat. $PdCl_2$(dppf), K_2CO_3, DME, 80 °C [6]; or **18** + **20**, cat. $PdCl_2$(dppf), K_2CO_3, aq. DME [7]. Bpin = 4,4,5,5-tetramethyl-1,3,2-dioxaborolane-2-yl; DME = 1,2-dimethoxyethane; dppf = 1,1′-bis(diphenylphasphino)ferrocene.

10.1.2.2 **Stereocontrolled Oxidation of the Oxindole Fragment**

Metal catalysts are frequently used to perform selective oxidation reactions in a biomimetic fashion and often rival enzymatic processes in terms of their mildness and stereocontrol. The electron-poor oxindole-based vinylogous amides **22** and **24** were dihydroxylated using catalytic osmium(IV) tetraoxide in two structurally similar cases in excellent yield by inherent substrate control (Scheme 10.6).

Inoue, Hirama and coworkers employed substrate-controlled oxidation to stereoselectively introduce the requisite diol, but with a twist to install the correct stereochemistry (Scheme 10.7). After substrate-controlled epoxidation of enamide **26** with dimethyldioxirane (DMDO), treatment of epoxide **27** with a strong Lewis acid initiates an inspired transcarbamoylation via a 7-*endo* epoxide ring opening by the pendant Boc-carbonyl to form oxazinanone **28**.

Despite the elegance and efficiency of these oxidations, they may not fully reflect biosynthesis. The Trp β-hydroxyl group is isolated from Nature as a 1 : 1

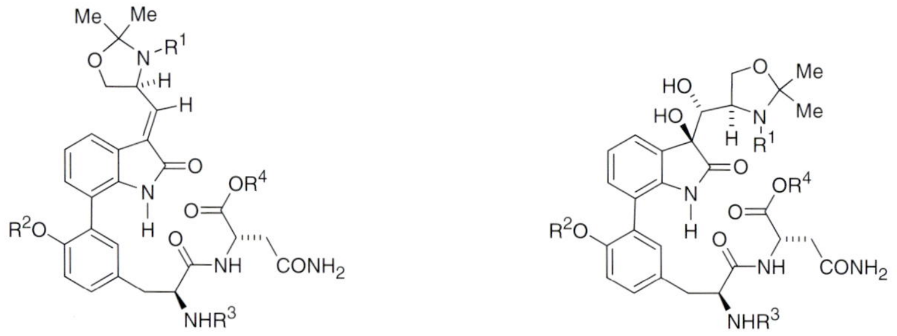

22 (R^1=Boc, R^2=Bn, R^3=Cbz, R^4=tBu) —i, 88% (d.r. = 5:1)→ **23** R^1=Boc, R^2=Bn, R^3=Cbz, R^4=tBu)

24 (R^1=Cbz, R^2=MOM, R^3=Boc, R^4=Bn) —ii, 87% (d.r. = 20:1)→ **25** (R^1=Cbz, R^2=MOM, R^3=Boc, R^4=Bn)

Scheme 10.6 *Reagents and conditions*: (i) OsO_4, NMO, $(DHQD)_2PHAL$, *t*-BuOH–H_2O, room temp. (rt) [6]; (ii) OsO_4, py, 0 °C, then sat. $NaHSO_3$ [7].

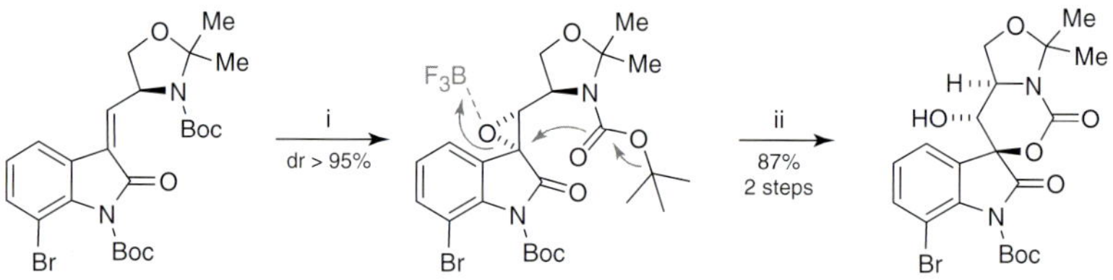

26 **27** *single diastereomer* **28**

Scheme 10.7 *Reagents and conditions*: (i) DMDO, CH_2Cl_2, rt; (ii) $BF_3{\cdot}OEt_2$, CH_2Cl_2, −78 → 0 °C [8].

epimeric mixture, whereas the C3 hydroxyl group exists as a single epimer. The two alcohol functional groups might therefore be introduced in two separate enzymatic oxidation steps.

10.1.2.3 Late-Stage Stereoselective *(Z)*-Enamide Formation

Two contrasting synthetic methods have been described for the installation of the *(Z)*-enamide: one distinctly biomimetic, the other arguably more synthetic (Scheme 10.8). The first method (Route A) [8] started from L-*allo*-threonine and operates under mild reaction conditions (0 °C). After hydrogenolysis of benzyl ester **29**, a Grob-type *anti*-elimination (**29** → **30**) of carbon dioxide was induced under Mitsunobu conditions in a good yield over the two steps. A more engineered approach (Route B) [6] – a stepwise dyotropic rearrangement involving sequential 1,4-silyl and 1,4-hydrogen shifts – necessitated a higher temperature to drive the reaction. Nevertheless, these forcing conditions were well tolerated; the intermediate silyl-enimidate (**32** → **33**) readily hydrolyzed to form the target *(Z)*-enamide (**34**).

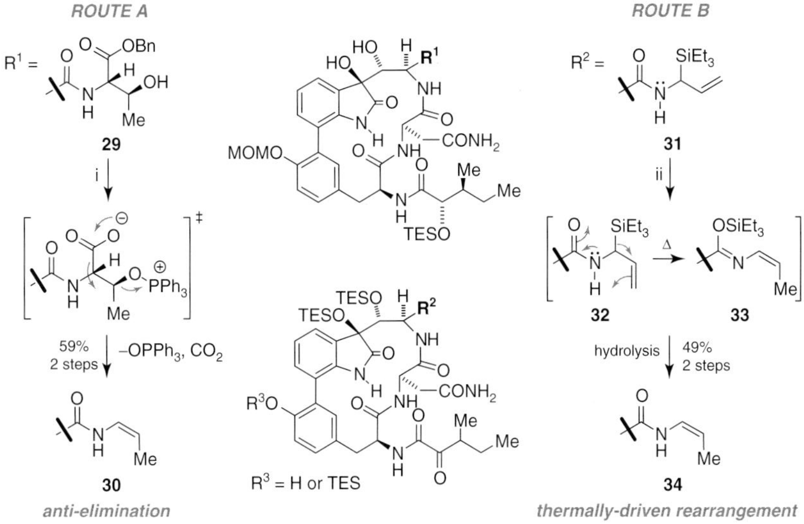

Scheme 10.8 *Reagents and conditions*: (i) (1) H_2, $Pd(OH)_2/C$, THF–H_2O (1 : 2) and (2) DEAD, PPh_3, mol. sieves, 0 °C [8]; (ii) (1) xylene, 140 °C and (2) HF–py, THF–py, then Me_3SiOMe [6].

10.1.3 Celogentin C

The bicyclic peptides celogentins A–H, J, and K [Figure 10.4 shows celogentin C (**2**) and K (**35**)] were isolated together with the analogous moroidin [9] from the seeds of *Celosia argentea* [10]. These structurally unique secondary metabolites are inhibitors of tubulin polymerization, with **2** being the most active congener.

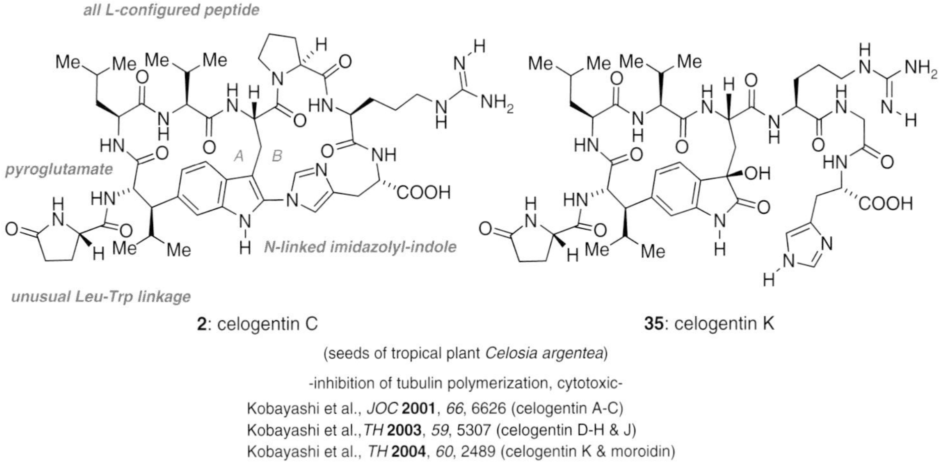

Figure 10.4 Isolation, bioactivity, and structures of celogentin C (**2**) and K (**35**).

Common to all isolated celogentins is the 17-membered A-ring (Figure 10.4), whereas the B-ring is found in different sizes and with a varying peptide sequence. The most distinguishing features of the celogentins though are the highly unusual sp^2–sp^3 Leu-Trp and sp^2–sp^2 His-Trp linkages.

Whereas the peptide origin of the celogentins is rather obvious, (Scheme 10.9) the biosynthetic logic behind the highly unusual side chain modifications is not. In this case, structural differences evident between naturally occurring variants may provide some useful insights. For example, epoxidation across the C2–C3-bond would generate iminium ion **37** and as such activate the indole nucleus for nucleophilic attack (Scheme 10.10). An intramolecular attack by His (**37** → **38**) may then be facilitated by the presence of a turn-inducing Pro in the peptide chain, for example, to deliver the B-ring of celogentin C (**2**) after desiccation. In the absence of such conformational preorganization, hydrolysis, followed by further enzymatic oxidation (**37** → **35**) could in principle become more prominent. Indeed, celogentin K (**35**) was isolated together with the corresponding ring-closed product moroidin [9].

The biosynthesis of the unprecedented Leu-Trp linkage is currently unclear, but we can propose a likely pathway. In peptide alkaloids, oxidases are known to frequently introduce hydroxyl functionality at the β-position of amino acids

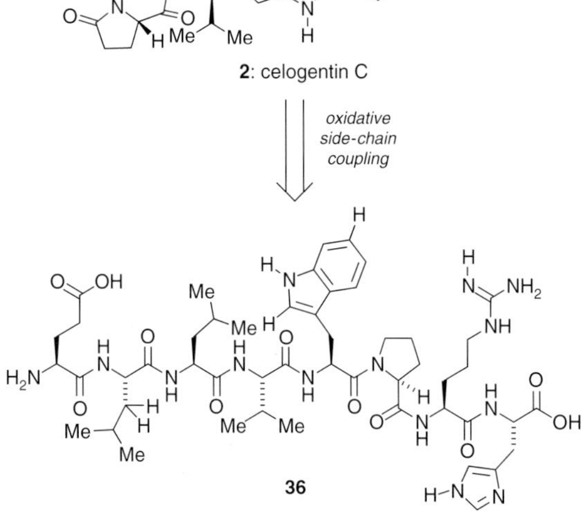

Scheme 10.9 Simplified retrobiosynthetic proposal for celogentin.

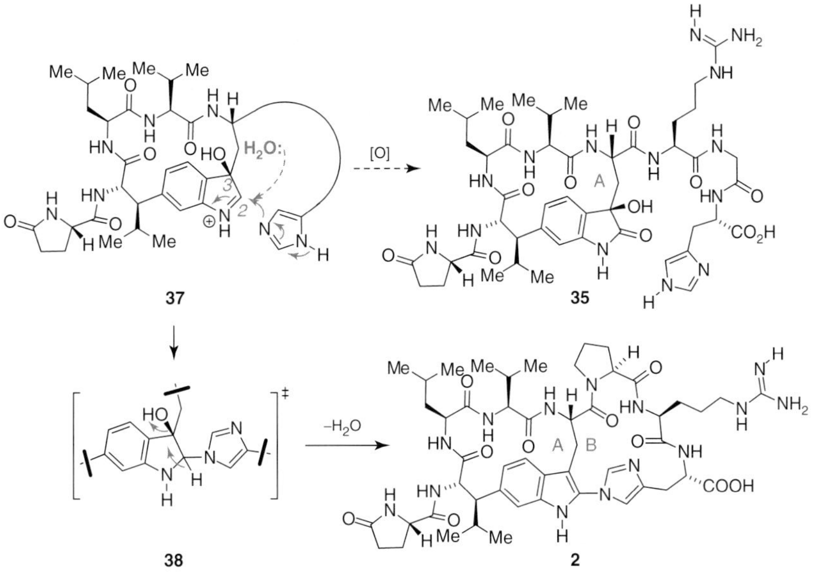

Scheme 10.10 Biosynthetic proposal for the formation of the B-ring Trp-His linkage.

(Scheme 10.11, **39** → **40**) [11]. Enzymatic activation of the alcohol (**40** → **41**) might then induce *N*-selective attack by the vicinal α-amide group (**41** → **42**) and lead to an *N*-acyl aziridine intermediate [12]. This could suffer a nucleophic attack by the Trp indole ring to close the A-ring (**42** → **43** → **44**), potentially guided by a Lewis-acidic enzyme. β-Selective ring-openings of aziridines with arenes are known [13], even in an intramolecular fashion [14].

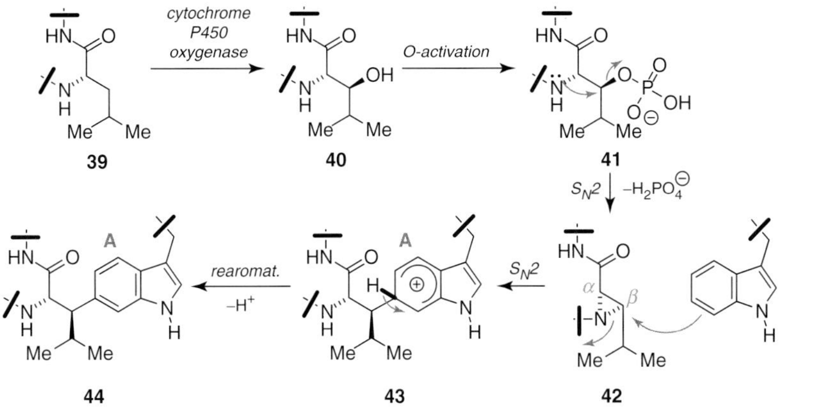

Scheme 10.11 Proposal for the biosynthetic origin of the A-ring Trp-Leu linkage.

Biosynthetic reasoning aside, celogentin C (**2**) with its most promising bioactivity has received much synthetic attention, culminating in several elegant total syntheses. In Nature, the closure of the A and B rings is thought to result from

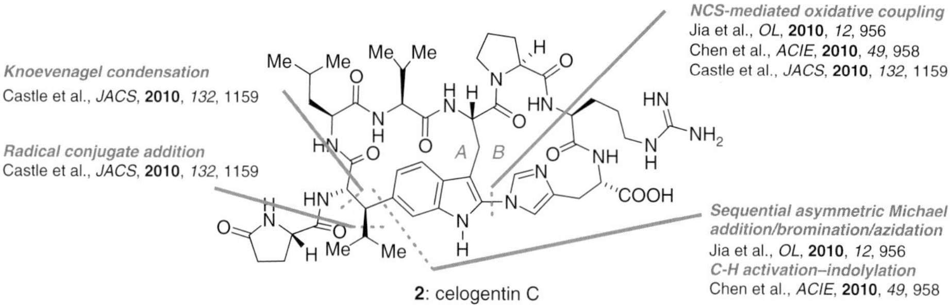

Figure 10.5 Synthetic approaches to celogentin C.

intramolecular oxidative tailoring of the peptide side chains. Naturally, the unusual Leu-Trp and His-Trp coupling arrangements provided a stiff synthetic challenge, and were preferentially formed in an *inter*-molecular fashion prior to ring closure by conventional macrolactamization (Figure 10.5).

10.1.3.1 Intramolecular Knoevenagel Condensation/Radical Conjugate Addition

The difficult Leu-Trp linkage was first addressed by Castle and coworkers in their total synthesis of celogentin C (**2**) [15]. A Knoevenagel condensation between titanated α-nitro amide tautomer **45** and Trp aldehyde **46** formed the acyclic chain and gave nitro olefin **47** in good yield and as a single isomer (Scheme 10.12). At the critical bond-forming step, substrate-controlled 1,4-addition of an isopropyl radical was executed, but gave only slight preference for the desired diastereomer. Subsequent SmI_2-mediated NO_2-group reduction then gave amine **48** in excellent yield [15].

A more stereoselective synthesis inspired by Evans's auxiliary chemistry was used by Jia to form the key Leu-Trp arrangement as a single diastereomer (Scheme 10.13)

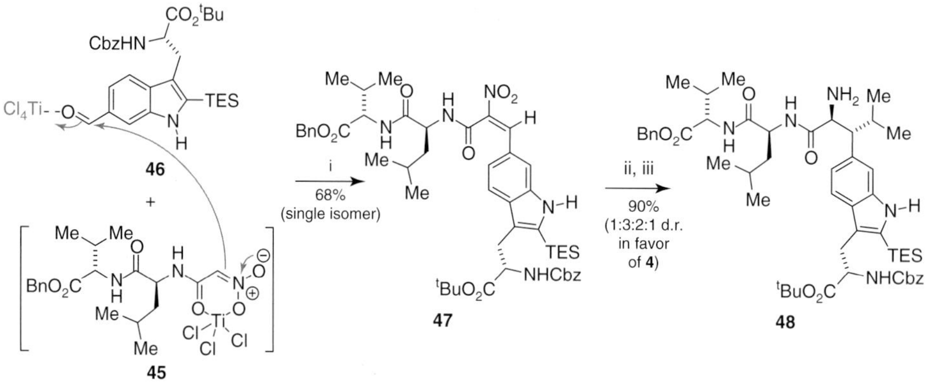

Scheme 10.12 *Reagents and conditions*: (i) $TiCl_4$, NMM, THF–Et_2O (2 : 1); (ii) Et_3B/O_2, $Zn(OTf)_2$ (2.0 equiv.), *i*PrI, Bu_3SnH; and (iii) SmI_2, THF–MeOH [15]. NMM = *N*-methyl morpholine.

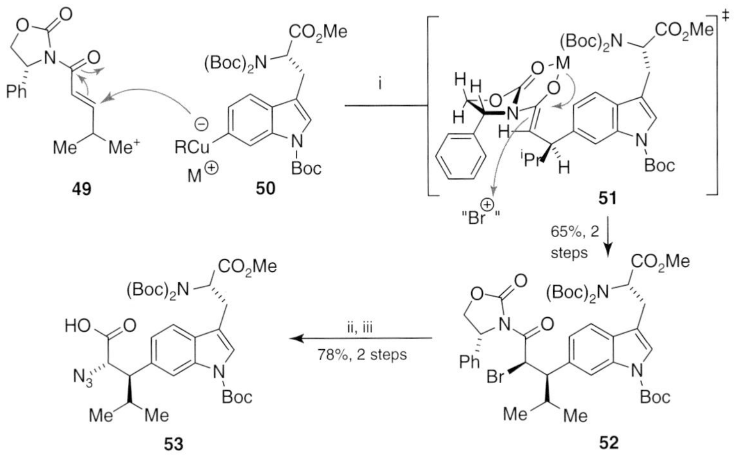

Scheme 10.13 *Reagents and conditions*: (i) (1) **50** from CuBr-Me_2S (*in situ*), THF, −30 °C, then **49**; (2) NBS; (ii) (1) NaN_3; (iii) LiOH, H_2O_2 [16].

[16]. Here, the chiral oxazolidinone controls the initial Michael-type 1,4-cuprate addition (**49** + **50** → **51**) and the subsequent enolate capture by *N*-bromosuccinimide (NBS) (**51** → **52**). The α-amino group is introduced in a latent form as an azide – via S_N2 reaction of the intermediate bromide prior to auxiliary cleavage (**52** → **53**) – to be unveiled at a later stage for the coupling chemistry.

10.1.3.2 C−H Activation–Indolylation

Using methodology developed first by Daugulis [18] and then Corey [19], Chen demonstrated that *N*-phthaloyl protected 8-aminoquinoline amide **54** couples stereoselectively to iodotryptophane **55** under Pd(II) catalysis (Scheme 10.14).

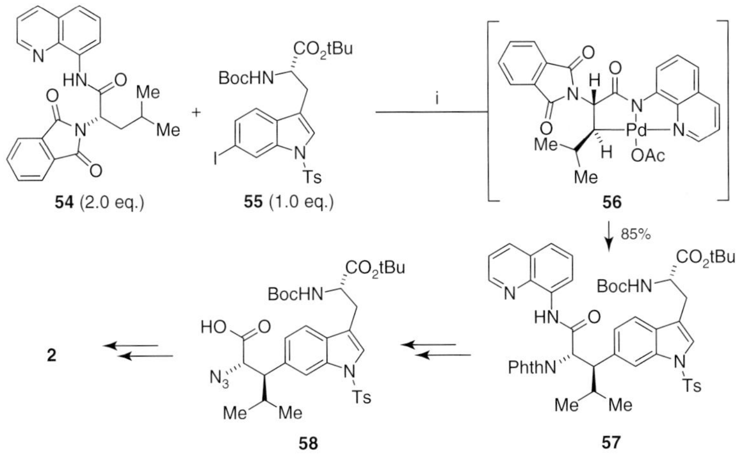

Scheme 10.14 *Reagents and conditions*: (i) $Pd(OAc)_2$ (cat.), AgOAc, *t*-BuOH, 110 °C, 36 h [17].

The quinoline serves as a chelating auxiliary for palladium coordination, and promotes the formation of *trans*-palladacycle intermediate **56**. This palladium(II) intermediate then undergoes a cross-coupling with **55** to deliver the desired product **57** as a single diastereomer. The phthaloyl group is critically involved in the arylation step, providing bis-protection of the α-amino group and a steric bias. Just like in the previous example (Scheme 10.13), an azide **58** is used as masked amine functionality in the later stages of the synthesis.

10.1.3.3 NCS-Mediated Oxidative Coupling

A serendipitous discovery by Castle and coworkers was crucial for the intermolecular side chain coupling between Trp and His [15]. They initially found that treating **59** with *N*-chlorosuccinimide (NCS) afforded an undesirable dichlorinated product of unknown structure, which was unreactive toward **60** (Scheme 10.15). When, however, Pro-OBn was preincubated with **59** and NCS, monochlorination to a presumed α-chloro-imminium ion was achieved, which led to the desired adduct **61** after hydrogenolysis. It is noteworthy that Castle's unique conditions were also used in subsequent total synthesis efforts [16, 17].

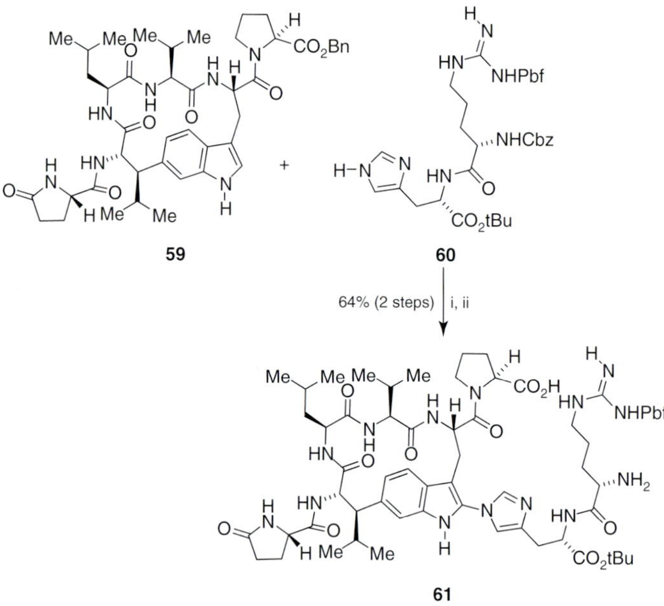

Scheme 10.15 *Reagents and conditions*: (i) Pro-OBn (2.0 equiv), NCS (3.0 equiv.), 1,4-dimethyl piperazine, CH_2Cl_2 then **60** (5.0 equiv.); (ii) 10% Pd/C, HCO_2NH_4, MeOH–H_2O (5 : 1) [15]. Pbf = 2,2,4,6,7-pentamethyl-2,3-dihydrobenzofuran-5-sulfonyl.

10.1.4
Himastatin and Chloptosin

The antibiotic and antitumor bacterial metabolite himastatin (Figure 10.6, **3**) is a rather exceptional example of a peptide alkaloid given that it is a fully symmetric dimer coupled via a pyrroloindole [20]. The more recently isolated compounds chloptosin (**62**) [21] and kutzneride 2 (**63**) – one of nine structurally related kutznerides [22] – are structurally similar to **3** in many respects, hinting at a common biosynthetic pathway. What distinguishes these metabolites from other indole-oxidized peptide alkaloids is that the enzymatic oxidative processes apparently do not mediate formation of the macrocycles, which appear to be typical NRPS (non-ribosomal peptide synthesis) cyclo(depsi)peptide products. However, oxidative tailoring of Trp and Orn side chains forms five- and six-membered rings at the periphery of the macrocycle, and seems to cause dimerization of the pyrroloindoles.

Taking chloptosin (**62**) as an example, the dimeric structure can be severed across the biaryl bond (Scheme 10.16, **62** → **64**) and at the peripheral pyrroloindole

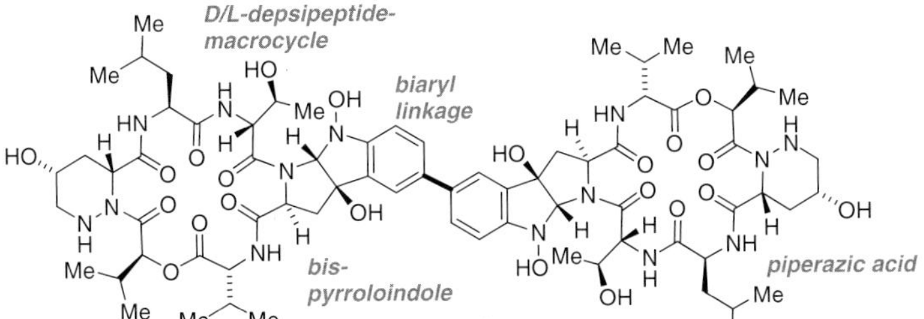

3: himastatin
(Actinomycete ATCC 53653 strain)

-antibiotic, anticancer-

Lam et al., *J. Antibiot.* **1990**, *43*, 956
Leet et al., *J. Antibiot.* **1990**, *43*, 961
Leet et al., *J. Antibiot.* **1996**, *49*, 299

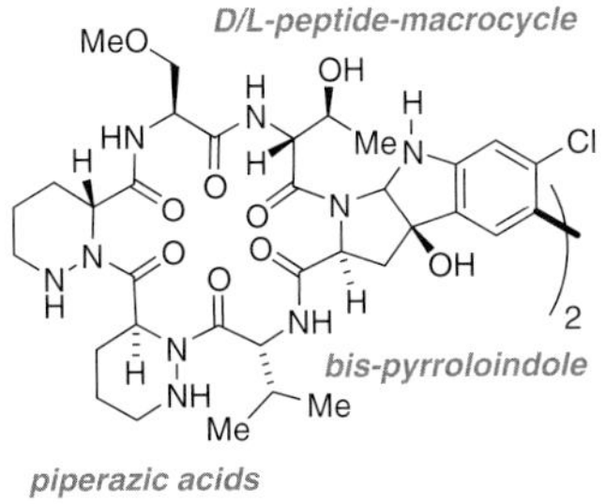

62: chloptosin
(Streptomyces MK498-98F14 strain)

-antibiotic, anticancer-

Umezawa et al., *JOC* **2000**, *65*, 459

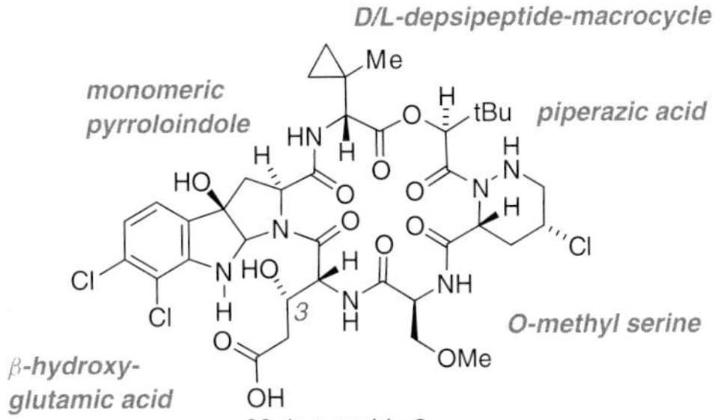

63: kutzneride 2
(Actinomycete Kutzneria sp. 744)

-antimicrobial-

Broberg et al., *J. Nat. Prod.* **2006**, *69*, 97
Pohanka et al., *J. Nat. Prod.* **2006**, *69*, 1776

Figure 10.6 Structural features and biological activities of pyrroloindole-based peptide alkaloids.

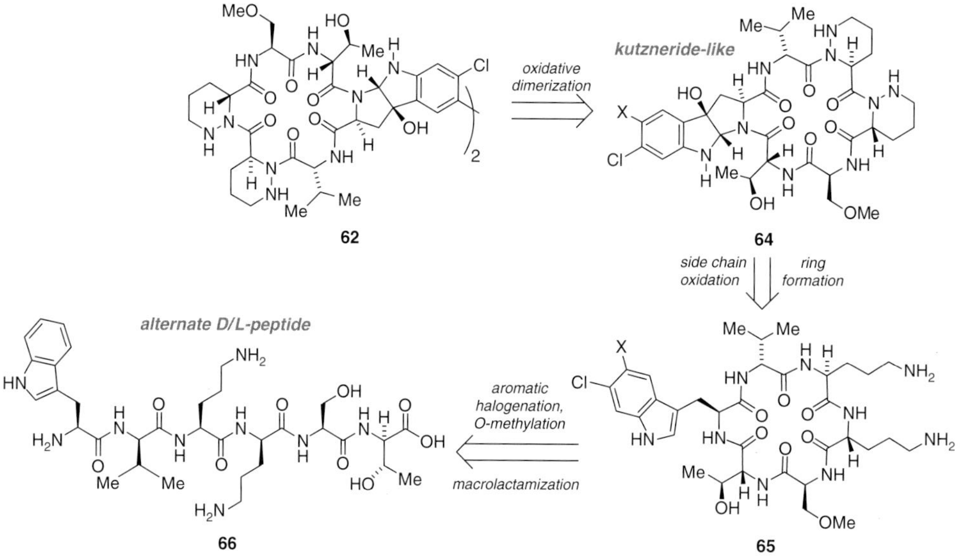

Scheme 10.16 Retrobiosynthetic simplification of chloptosin (**62**).

and piperazic acid rings (**64** → **65**), which quickly exposes a modified alternating D/L cyclic peptide (**66**). The identification of and subsequent studies into the gene cluster of the kutznerides [23] has helped to elucidate the order of steps and the function of some constituent enzymes.

While the precise order of biosynthetic events remains unclear, oxidative dimerization and pyrroloindole formation are presumed to occur *after* assembly of the cyclic (depsi)peptide monomer (Scheme 10.17a). Little is also known

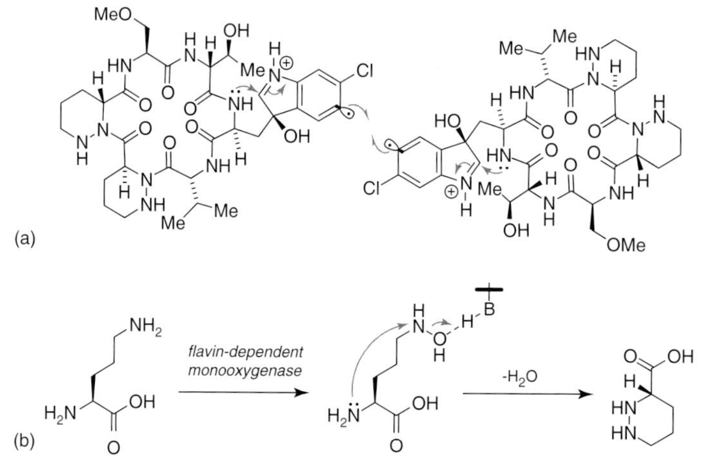

Scheme 10.17 Simplified proposal for dimerization (a) and piperazic acid formation (b) during the biosynthesis of dimeric pyrroloindole peptide alkaloids [23].

about the biosynthesis of the unusual (though not uncommon) piperazic acids (Scheme 10.17b). Feeding experiments [24, 25] suggest that their synthesis occurs prior to incorporation into the peptide chain. Similar to the hydrazo linkage in valanimycin [26], the biosynthesis of piperazate might be initiated with *N*-hydroxylation of the α-amino group of ornithine by a flavin-dependent monooxygenase (KtzI in the case of the kutznerides), followed by nucleophilic attack of the α-amino group to close the ring. The *O*-methylation of serine, by contrast, most likely occurs during assembly.

Characterization of the flavin-dependent halogenase enzymes KtzQ, KtzR, and the NADH-utilizing FAD reductase KtzS has indicated that regioselective double halogenation in kutzneride biosynthesis occurs on the free L-Trp (Scheme 10.18, **67** → **68** → **69**) through the sequential action of the KtzQR enzyme pair. It is likely, therefore, that dichlorinated Trp is the monomer incorporated into the growing kutzneride assembly line by the KtzH adenylation domain (**69** → **70**) [27]. The final conversion into the pyrroloindole moiety in the mature kutzneride is postulated to occur via epoxidation of the indole 2,3-double bond by the heme protein KtzM, followed by intramolecular capture of the epoxide by the amide nitrogen (**71** → **63**) [23]. The biosynthetic origin of the other non-proteinogenic non-ribosomal amino acids found in the kutzneride family is the subject of ongoing studies.

To date, three successful bi-directional syntheses have been reported in the literature, one for **3** and two for **62** (Scheme 10.19). Bi-directional synthesis was pursued earlier for polyketides, which are inherently more symmetric [28]. In light of the brief discussion on the biosynthesis however, this innovative strategy eventually

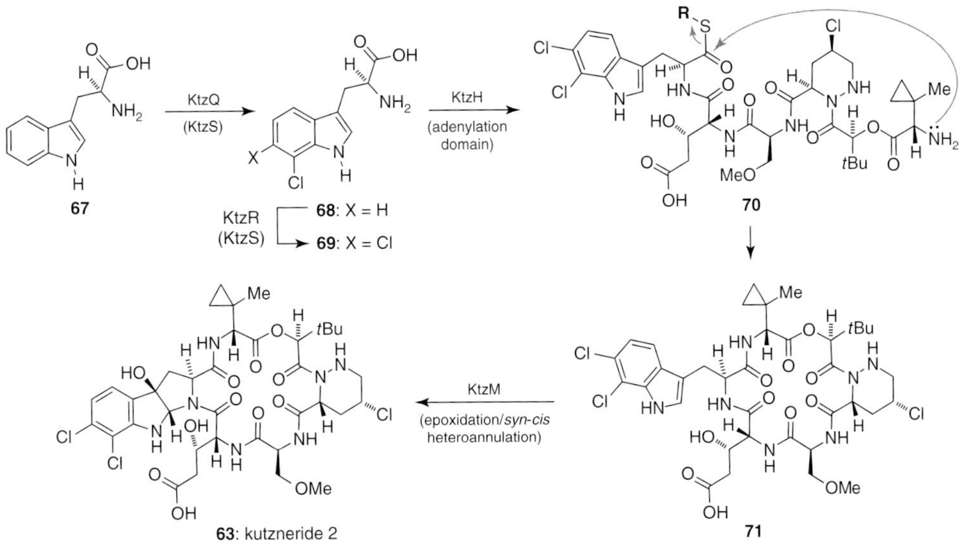

Scheme 10.18 Proposed biosynthesis of 6,7-dichlorotryptophan and its incorporation into the kutzneride 2 (**63**) assembly according to Walsh *et al.* [23, 27]; R = NRPS-thiolation domain.

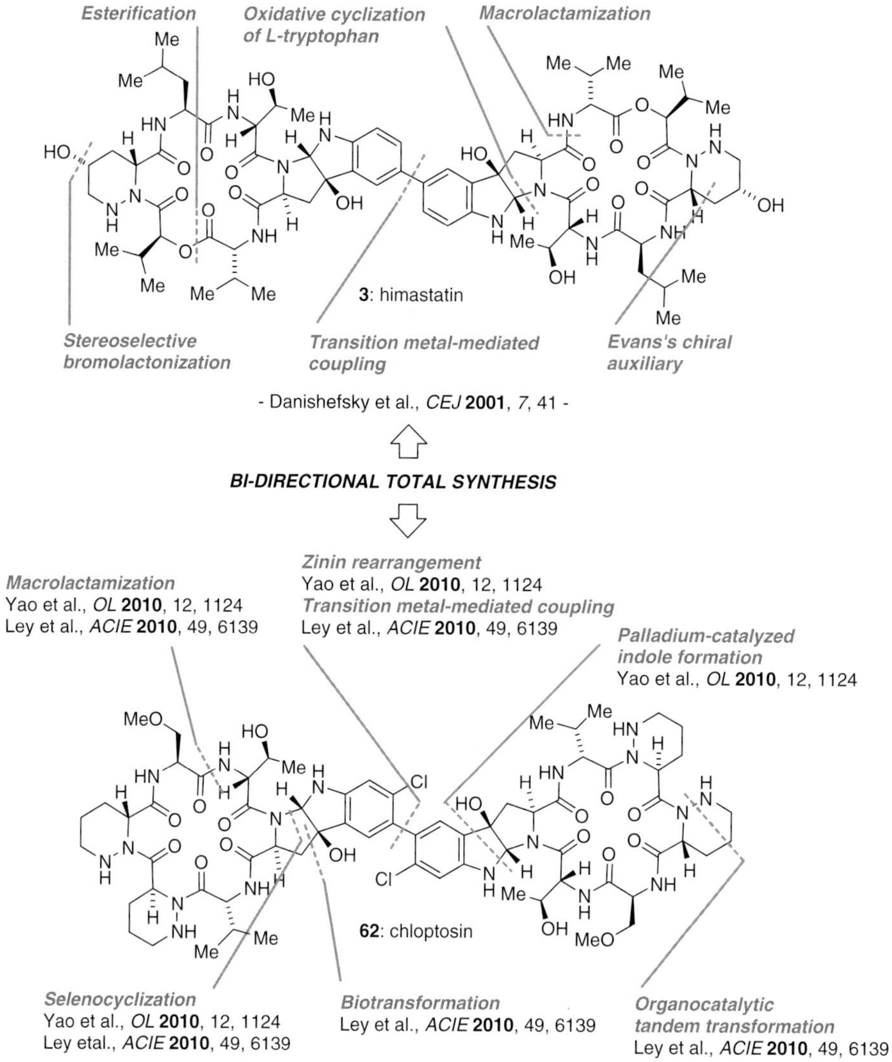

Scheme 10.19 Summary of the synthetic approaches to himastatin (**3**) and chloptosin (**62**).

adopted by three groups in order to complete their respective syntheses cannot be considered truly biomimetic. For peptide alkaloid biosynthesis, it is difficult to see how Nature might use bi-directional synthesis to her own advantage. Even so, the success stories described below indicate what is currently feasible in the laboratory.

10.1.4.1 Synthesis of the Himastatin Pyrroloindole Core

Danishefsky and coworkers were successful in installing the biaryl group after pyrroloindole cyclization [29]. Starting from protected L-tryptophan **72** (Scheme 10.20), the authors prepared a fully-protected pyrroloindole core (**74**)

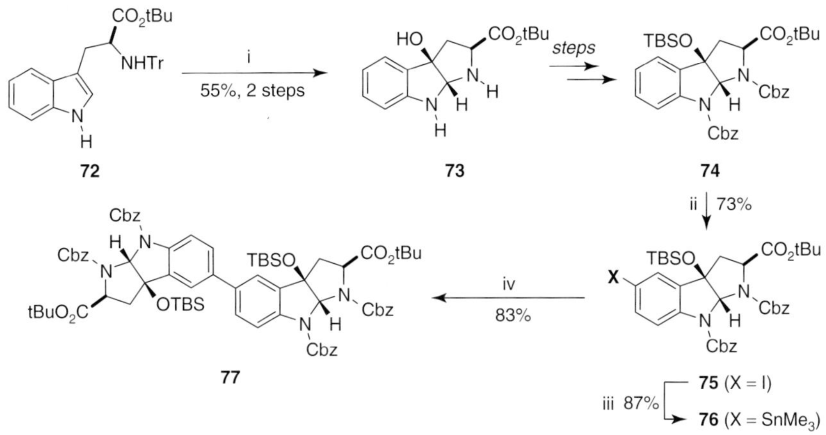

72 73 74 77

iii 87% 75 (X = I) → 76 (X = $SnMe_3$)

Scheme 10.20 Himastatin synthesis [29]. *Reagents and conditions*: (i) (1) DMDO, CH_2Cl_2, −78 °C, then (2) HOAc, MeOH, CH_2Cl_2; (ii) ICl, 2,6-di-*tert*-butylpyridine, CH_2Cl_2; (iii) Me_6Sn_2, $Pd(PPh_3)_4$, THF, 60 °C; and (iv) Pd_2dba_3, **76**, Ph_3As, DMF.

via a DMDO-mediated *syn-cis* oxidative annulation (**72** → **73**). The trityl- and *tert*-butyl protecting groups were critical for achieving the *syn-cis* selectivity. A regioselective aromatic iodination was then followed by stannylation and Pd-mediated dimerization under Stille coupling conditions (**74** → **77**), to create the biaryl bond in excellent yield and clear a path for the bi-directional synthesis.

10.1.4.2 **Synthesis of the Chloptosin Pyrroloindole Core**

The first synthesis of chloptosin (**62**) was performed by Yao and built on Danishefsky's seminal himastatin work. However, the added complexity of two 6-chloro-substituents on the biaryl pyrroloindole core had to be mastered [30]. This was achieved through formation of the biaryl bond via a Zinin rearrangement – predetermining the bi-directional synthesis – followed by a regioselective aromatic iodination to set up metal-catalyzed indole formation (Scheme 10.21, **78** → **79**). After condensation of **79** with two equivalents of L-pyroglutamic acid derivative **80**, treatment of the intermediate bis-imine with palladium(II) diacetate effected a Heck-type heteroannulation (**80** → **82**) which left the two aryl chlorides untouched.

Finally, a highly stereoselective selenocyclization mediated the formation of the pyrroloindole into the correct *syn-cis* configuration (**83**), followed by an obligatory oxidative work-up to give **85**. Ley's very recent synthesis of the dimeric pyrroloindole core *en route* to **2** is even more direct [31]. In this case, enzyme catalysis was used to prepare 6-chloro-L-Trp as a key building block.

10.1.4.3 **Macrolactamization**

All three highlighted total syntheses successfully close the macrocycle using bi-directional macrolactamization, only at different positions along the 18-membered ring (Scheme 10.19). Yao elaborated the pyrroloindole dimer (Scheme 10.21, **85**) to the macrocyclization precursor **86** (Scheme 10.22) using

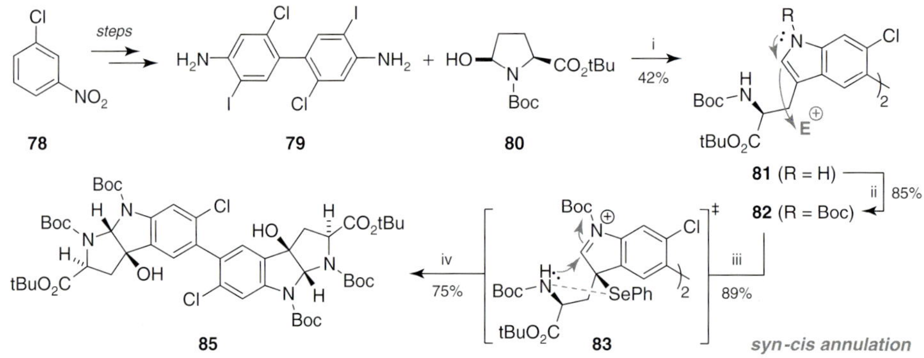

Scheme 10.21 Chloptosin core synthesis [30]. *Reagents and conditions*: (i) $Pd(OAc)_2$ (10 mol.%), DABCO (6.0 equiv), DMF (0.2 M), 120 °C, 48 h; (ii) Boc_2O, DMAP; (iii) *N*-(phenylseleno)phthalimide (**84**), PPTS, Na_2SO_4, CH_2Cl_2, rt; and (iv) *m*-CPBA, aq. $NaHCO_3$, *i*-PrOH, rt.

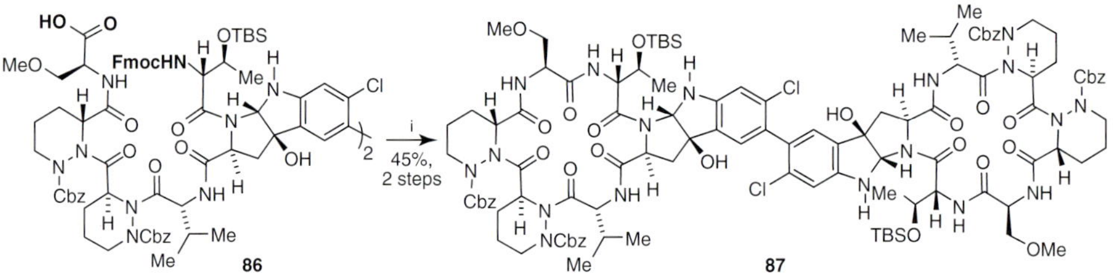

Scheme 10.22 Chloptosin synthesis [30]. *Reagents and conditions*: (i) (1) iPr_2NH, CH_3CN, rt and (2) HOAt, PyBOP, iPr_2NEt, DMF, CH_2Cl_2, −5 °C to rt.

a sequence of advanced solution-phase peptide couplings, which was then cyclized to afford **87** in an excellent yield. This complex example exceptionally showcases bi-directional macrocyclization for total synthesis (>2500 Da intermediate heavily-laden with heteroatom-rich coordinating protecting groups) [30]. Danishefsky was the first to demonstrate this in his total synthesis of **3** [29]. The Ley group has since accomplished the same feat in their total synthesis of **62** [31].

These synthetic efforts define the limits of biomimetic approaches in peptide alkaloid synthesis, and highlight the methodological needs for more selective coupling between large peptide subunits and/or individual cyclopeptide monomers. Apparently, attempts by Danishefsky and coworkers to couple two cyclopeptides to form the biaryl were unsuccessful [26]. However, in their synthesis of chloptosin (**62**), Ley and coworkers reported the successful dimerization of a fully formed macrocycle monomer using Stille- and Suzuki-cross-couplings [31]. Unfortunately, deleterious over-oxidation could not be suppressed. Nevertheless, this insightful chemical "hint" suggests that dimerization of the fully-formed monomer by oxidative coupling might indeed be biosynthetically feasible within a favorable enzymatic environment.

10.1.5 Diazonamide

Diazonamides A and B (Figure 10.7, **4** and **89**) are secondary metabolites isolated from the colonial ascidian *Diazona angulata* [32, 33]. Diazonamide A (**4**) features a unique, densely functionalized quaternary carbon (C10) located at the center of a bicyclic ring system, which is surrounded by bis-indole, 2,4-bisoxazole, and α-hydroxyvaline structural motifs. The minor isolate, diazonamide A (**4**) was initially assigned the structure **88** by analogy to a misassigned X-ray crystal structure of the *p*-bromobenzamide derivative of the major isolate diazonamide B (not shown), [32]. Recently, three additional diazonamide family members (C–E, Figure 10.7, **90–92**, respectively) were isolated alongside **4** and **89** [33]. Biological data for these new analogs has shown the importance of the side chain valine residue and the need for an α-hydroxy in place of an α-amino group for compound potency [33]. Compound **4** is endowed with high *in vitro* cytotoxicity toward human tumor cell lines, where it blocks cell division during mitosis through a unique interaction with the mitochondrial enzyme ornithine δ-amino transferase [34].

The synthesis of the originally proposed structure (Figure 10.7, **88**) by Harran and coworkers [35] prompted a structural reassignment of both diazonamide A from **88** to **4** and diazonamide B to **89** based on mismatching analytical and biological data between the natural product and the synthetic material. Despite this turn of events, several elegant approaches to the revised structure (**4**) of this structurally impressive and biologically significant secondary metabolite were quickly produced, notwithstanding the diverse and noteworthy routes to the originally-proposed diazonamide A structure (**88**) [36].

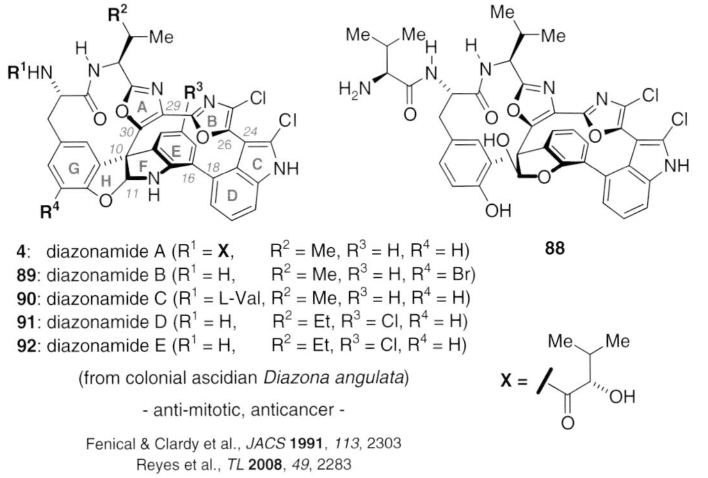

Figure 10.7 Diazonamide A (**4**), the initially proposed structure (**88**), and natural analogs (**89–92**).

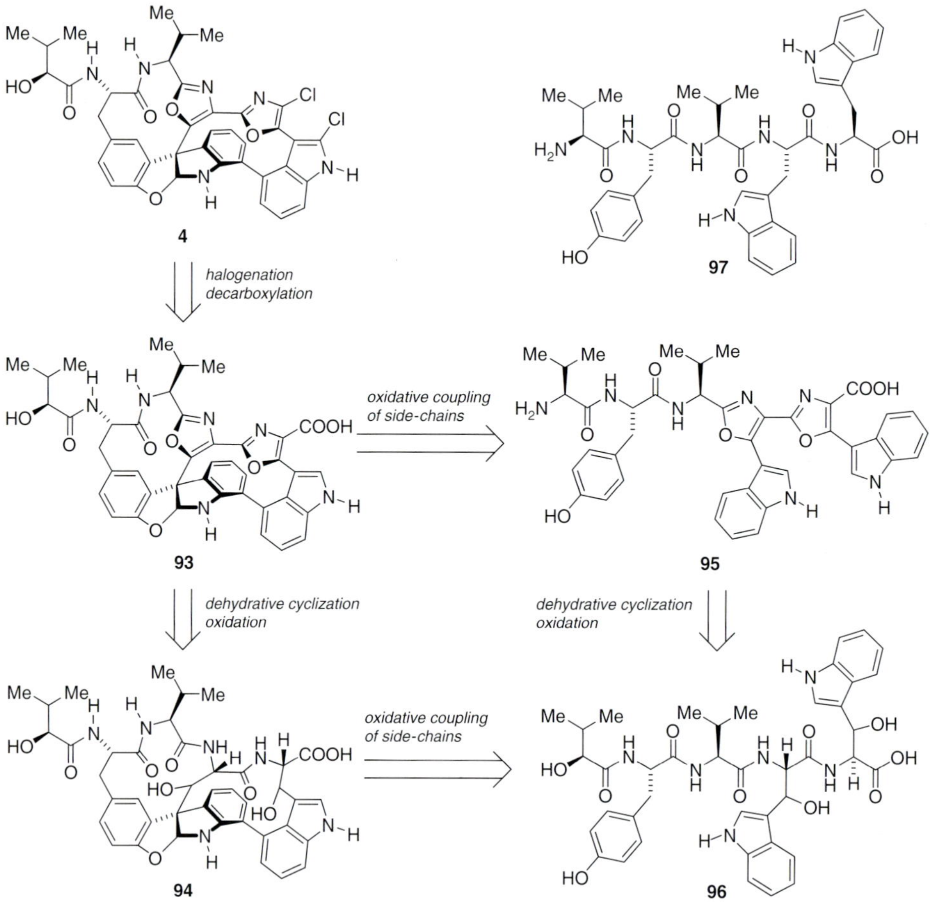

Scheme 10.23 Simplified retrobiosynthetic analysis of diazonamide A (**4**) reveals its pentapeptide origin.

At first sight, the peptide origin of **4** appears hidden (Scheme 10.23). In a retrobiosynthetic sense, pruning the aromatic chlorides is a logical first step (**4** → **93**). Regioselective chlorination may be achieved on the mature polycyclic core in the later stages of a synthesis. At this juncture, disconnection across the central C10-quaternary amino acetal–biaryl bonds (**93** → **95** or **94** → **96**) and hydrolysis of the 2,4-bisoxazole ring system (**93** → **94** or **95** → **96**) reverts the complex core back to a linear chain (**96**). Removal of the apparent oxidative tailoring eventually reveals the anticipated all-L-Val-Tyr-Val-Trp-Trp peptide **97**. How all the reactions are really executed in Nature remains unclear. Synthesis could at least make a contribution in this regard, to execute potentially feasible biomimetic strategies, and help to substantiate hypotheses.

The peculiar central aminal moiety is certainly the most striking structural element of diazonamide A. Its biosynthesis is anything but clear, although it could result from S_N1-type interception of an advanced, highly-oxidized indolyl-oxazole by the tyrosine phenol group (Scheme 10.24, **98** → **100**) followed by desiccation. For this, the "natural" reactivity of indole must be reversed, probably by oxidizing past the indolinone stage. On the other hand, the Tyr residue might, after oxidation, serve as an electrophile for nucleophilic attack by the indole C3 (**98** → **99**). Here, steric hindrance at C3 and the susceptibility of Trp to undergo oxidation might complicate matters. Synthetic approaches based on both of these biomimetic postulates have been explored [37, 38].

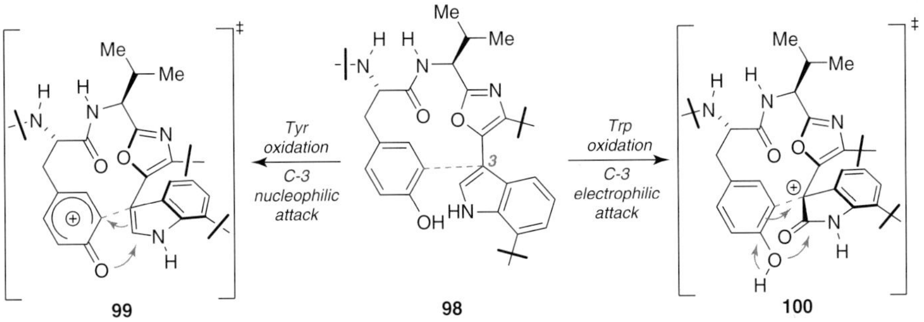

Scheme 10.24 Biosynthetic proposal for the formation of the aminal core.

The constrained bis-indole biaryl might be formed via a biradical coupling mechanism (Scheme 10.25, **101** → **102**) that is thought to take place once the aminal has been established, or at least when the central Trp is in a higher oxidized state (e.g., as an oxindole like in TMC-95A-D-*vide supra*) [37, 38]. The α-hydroxyvaline is believed to result from oxidative deamination of valine (Scheme 10.26, **103** → **104**), followed by reduction of the subsequent α-oxo functionality by a reductase (**104** → **105**) [39].

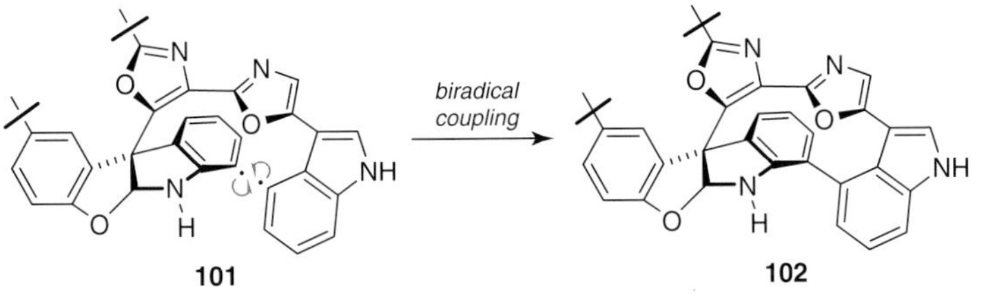

Scheme 10.25 Simplified biosynthetic proposal for biaryl coupling.

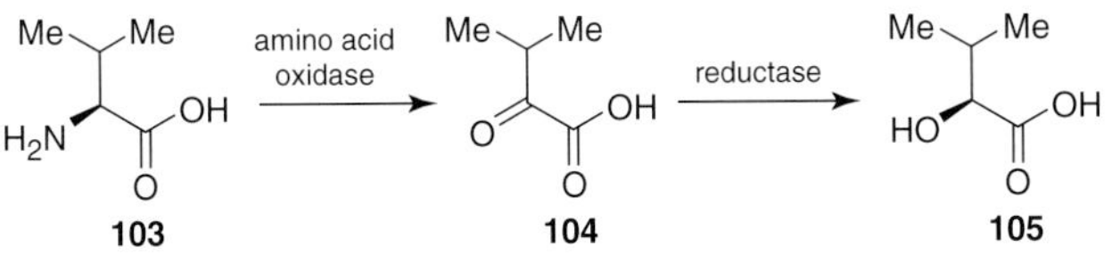

Scheme 10.26 Biosynthetic proposal for the formation of the α-hydroxyvaline.

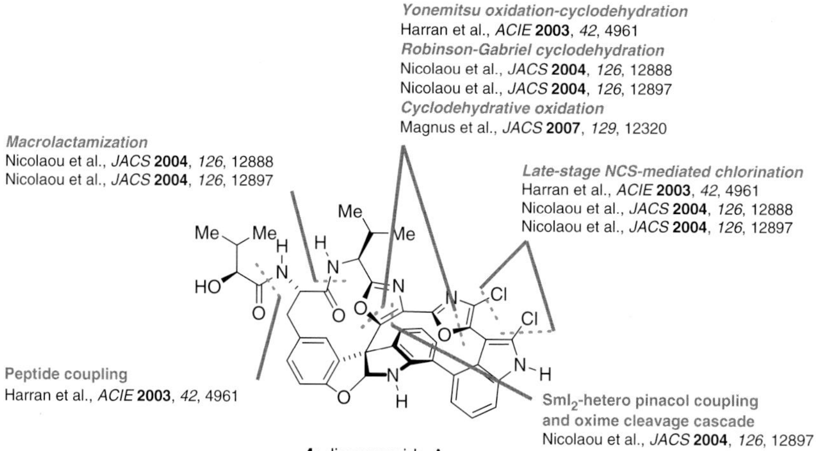

4: diazonamide A

Figure 10.8 Highlights of synthetic approaches to **4** in the peripheral regions.

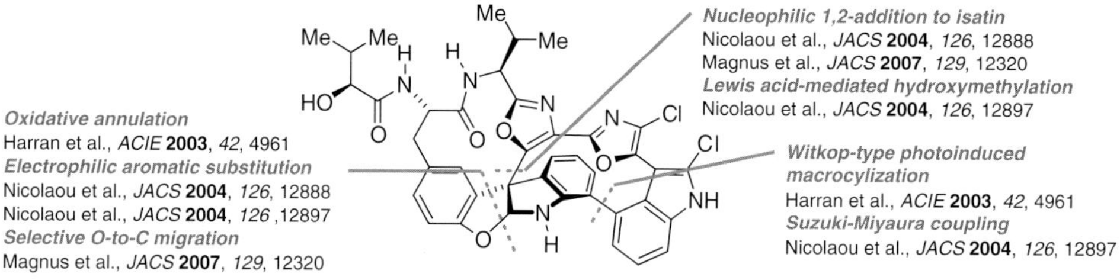

4: diazonamide A

Figure 10.9 Highlights of synthetic approaches to **4** at the polycyclic core of **4**.

Diazonamide A (**4**) is by far the most complex indole-oxidized peptide alkaloid discussed in this subsection. It has generated considerable attention in the synthetic community, not least for the significant challenge posed by the dense polycyclic core, which has elicited several unique approaches to its construction. Unfortunately, due to space restrictions, not all of disconnections (Figures 10.8 and 10.9) can be discussed in detail here. Instead, greater priority is given to more biomimetic aspects.

10.1.5.1 Late-Stage Aromatic Chlorination

Early model studies on simplified substrates highlighted the potential of installing the chlorine atoms at the later stages of the synthesis (Scheme 10.27a) [40]. This strategy could be transferred to more complex intermediates and was used in all successful total syntheses of **4**. The highlighted chlorination step used toward the first total synthesis of the correct structure of diazonamide A (**4**) (Scheme 10.27b) [38] was employed with a similar outcome later as well [37].

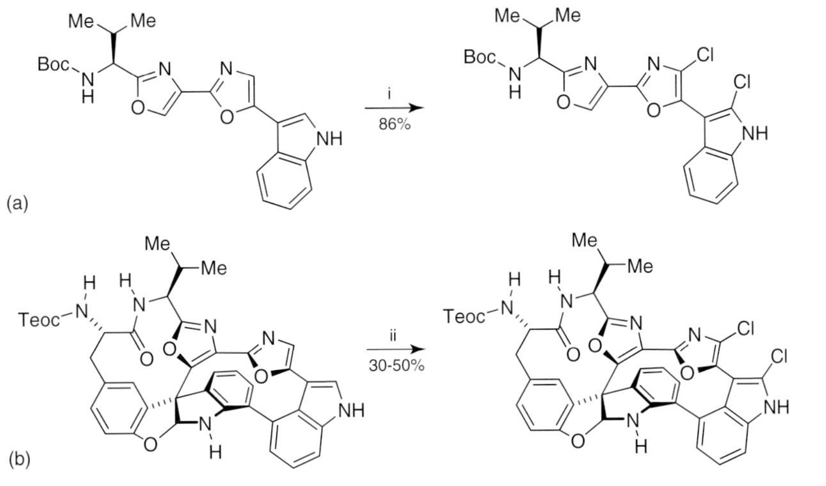

Scheme 10.27 Late stage chlorination reactions: (a) model studies [40]; (b) total synthesis of **4** [38]. *Reagents and conditions*: (i) *N*-chlorosuccinimide (4.0 equiv.), CCl_4, 40 °C, 18 h; and (ii) 2,3,4,5,6,6-hexachloro-2,4-cyclohexadien-1-one (2.5 equiv.), *N*, *N*′-dimethylformamide, rt, 24 h.

10.1.5.2 **Bisoxazole Ring System via Oxidative Dehydrative Cyclization**

Moody showed that a diazonamide precursor peptide could be directly oxidized to a putative intermediate (Scheme 10.28) [41]. Though low yielding, this approach nonetheless enables direct access to the key 2,4-bisoxazole core **107** in a single step starting from the readily accessible peptide **106**. This interesting result hints at a biosynthetic route whereby the polycyclic core may be formed in a single enzyme-mediated oxidation step.

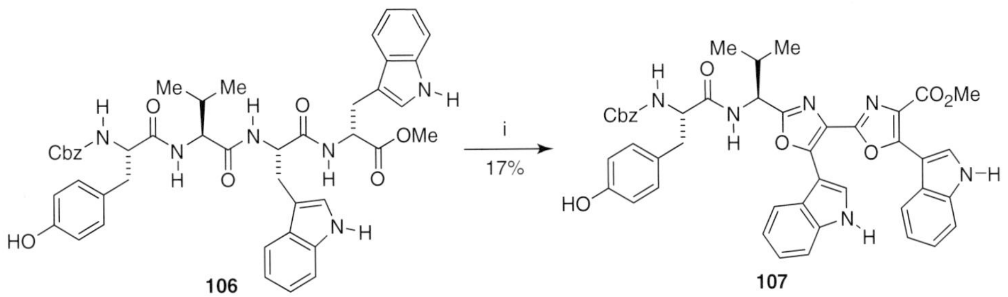

106 **107**

Scheme 10.28 Biomimetic synthesis of the 2,4-bisoxazole core [41]. *Reagents and conditions*: (i) 2,3-dichloro-5,6-dicyano-1,4-benzoquinone (DDQ), THF, reflux.

10.1.5.3 **Oxidative Annulation**

Hypervalent iodine reagents were used by Harran to form the critical aminal core by oxidative activation of the Tyr residue (Scheme 10.29) [38]. Indolyl-oxazole **108** was treated with a slight excess of $PhI(OAc)_2$ in the presence of LiOAc, which presumably activates at the phenol OH group (**109**) for nucleophilic attack by the

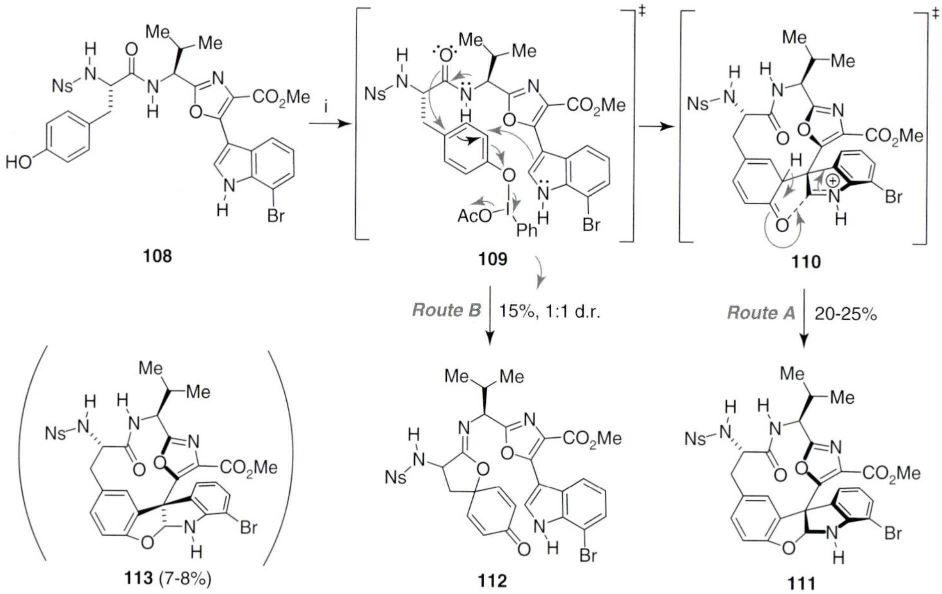

Scheme 10.29 Formation of the aminal core by oxidative annulation [38]. *Reagents and conditions*: (i) $PhI(OAc)_2$ (1.1 equiv.), LiOAc (2.0 equiv.), TFE, inverse addition, $-20\,^{\circ}C$, 10 min; Ns = 4-nitrotoluenesulfonyl (nosyl), TFE = 2,2,2-trifluoroethanol.

electron-rich indole (Route A, **109** → **111**). Formation of the "shunt" product **112** (Route B) and the undesired diastereomer **113** could easily be suppressed were the oxidation to occur in an enzymatic setting.

10.1.5.4 **Sequential Nucleophilic 1,2-Addition, Electrophilic Aromatic Substitution**

In a distinct approach by Nicolaou and coworkers, organolithium reagent **114** was reacted with isatin **115** to produce tertiary alcohol **116** as a mixture of diastereomers (Scheme 10.30) [37]. Treatment with acid then enabled an S_N1 reaction between **116** and the phenol ring of tyrosine **118**, to form the critical quaternary carbon bond, presumably via intermediate **117**. In this case, the Boc protecting group was lost during the aromatic substitution step and had to be reinstalled to facilitate the separation of the two diastereomers. Interestingly, attempts to form the ring *inter*-molecularly under various acidic conditions failed (**120** → **121**). This result could suggest that strong enzymatic assistance might be needed, were the cationic intermediate to be formed during biosynthesis.

10.1.5.5 **Reductive Aminal Formation**

Despite promising studies on simpler substrates (typically 4.0 equiv. $BH_3{\cdot}SMe_2$), aminal formation from the sterically encumbered indolinone **122** (Scheme 10.31)

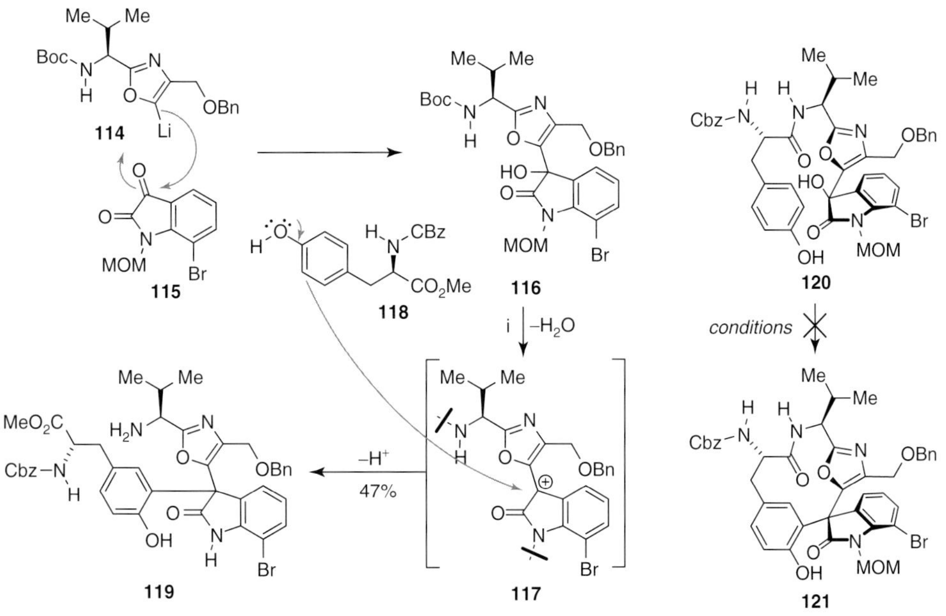

Scheme 10.30 Nucleophilic 1,2-addition followed by intramolecular electrophilic aromatic substitution and an attempted intramolecular transformation [37]. *Reagents and conditions*: (i) **118** (4.0 equiv.), *p*-TsOH (4.0 equiv.), $ClCH_2CH_2Cl$, 83 °C, 25 min.

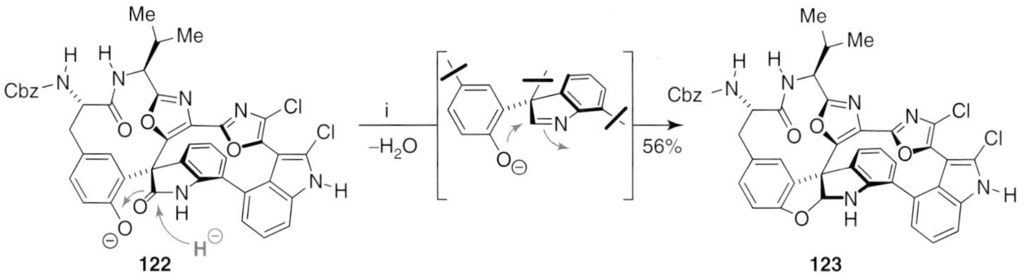

Scheme 10.31 Reductive aminal formation [37]. *Reagents and conditions*: (i) DIBAL-H (1.0 M in $PhCH_3$, 100.0 equiv), THF, −78 → 25 °C, 3 h.

proved to be highly challenging [**122** → **123**, 100.0 equiv. DIBAL-H (diisobutylaluminum hydride)!] [37]. Forged only under strong laboratory conditions, this bond construction is perhaps less likely to occur in Nature.

10.1.5.6 **Indole–Indole Coupling**

The coupling of the Trp residues is thought to proceed via a biradical mechanism. In the flask, these radicals were photochemically generated from indolyl bromide **124** (Scheme 10.32), leading to **126** via intermediate **125** in an excellent yield and as a single atropisomer. Apparently, conformational constraints control the

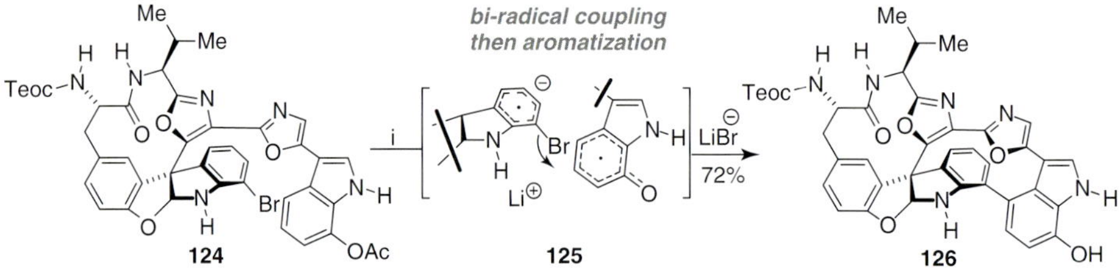

Scheme 10.32 Witkop-type photo-induced macrocyclization [38]. *Reagents and conditions*: (i) *hν* (300 nm) argon-purged CH_3CN-H_2O (3 : 1, 3.0 mM), LiOH (2.0 equiv.) rt, 3 h; Teoc = 2-(trimethylsilyl)ethoxycarbonyl.

stereochemical outcome of this reaction and the difference in electron density between the two aryl rings facilitates the biaryl coupling. The conditions for this transformation were originally developed by Harran [38]. Nicolaou applied a similar strategy *before aminal formation* after slight modification to the reaction conditions [37].

All the synthesis data indicate that in Nature an enzymatically derived radical would react in a similar fashion. Both Nicolaou and Harran have charted the intrinsic reactivity of many putative biosynthesis intermediates. By inference, the radical coupling path way could be the more accessible one for biosynthesis. Future research on the producer organism and biosynthesis genes will hopefully clarify this point, and provide insight into the enzymatic machinery that enables the formation of this highly peculiar molecule in Nature.

10.2
A Complex Peptide Alkaloid: Ecteinascidin 743 (ET 743)

The ecteinascidins (Figure 10.10, **127–132**) are a family of alkaloids derived from the sea squirt *Ecteinascidia turbinata*, which quickly raised considerable interest due to their potent antiproliferative properties [42–44]. Their activity was uncovered in 1969 during a natural product screening program of the American National Cancer Institute (NCI) [45]. Follow-up studies led to their isolation and structural identification [42], including ecteinascidin 743 (ET 743, trabectidin, **128**) as the most abundant marine alkaloid of the family – still at a very low content of 0.5–4 ppm in the biomass. The remarkable three-dimensional architecture, combined with its meager availability from natural sources rendered ET 743 a very attractive synthetic target. This led to key examples of biomimetic synthesis, which were instrumental in advancing therapeutic studies. Semi-synthetic ET 743 was recently approved in Europe for the treatment of advanced soft tissue sarcoma and is marketed as Yondelis® [46].

The ecteinascidins feature a piperazine-bridged bis(tetrahydroisoquinoline) framework and a pentacyclic A–E ring system, which they share with the bacterial safracin and saframycin alkaloids. Linked to this platform is a bridged ten-membered lactone ring that incorporates a benzylic thioether and a

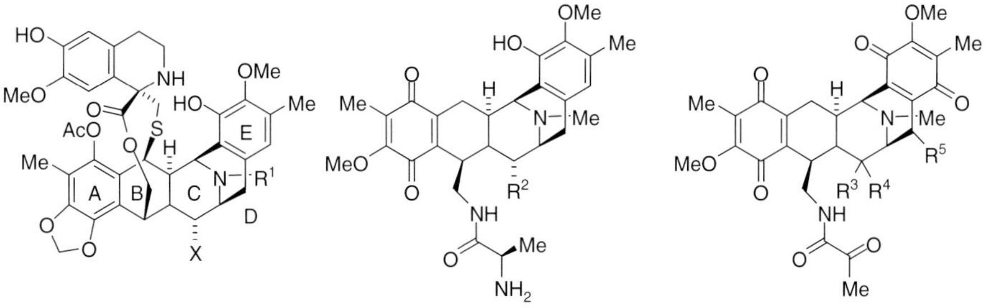

127: ET 729 (R¹ = H, X = OH)
128: ET 743 (R¹ = Me, X = OH)
129: ET 745 (R¹ = Me, X = H)
130: ET 759A (R¹ = Me, X = = O, lactam)
131: ET 759B (R¹ = Me, X = OH, S-oxide)
132: ET 770 (R¹ = Me, X = CN)

133: Safracin A (R² = H)
134: Safracin B (R² = OH)
135: Cyano-safracin B (R² = CN)

136: Saframycin A (R³ = CN, R⁴ = H, R⁵ = H)
137: Saframycin B (R³ = R⁴ = R⁵ = H)
138: Saframycin C (R³ = R⁴ = H, R⁵ = OMe)
139: Saframycin G (R³ = CN, R⁴ = H, R⁵ = OH)
140: Ranieramycin C (R³ = R⁴ = O, R⁵ = OH)
141: Ranieramycin D (R³ = R⁴ = O, R⁵ = OEt)

Figure 10.10 Some of the most representative ecteinascidins and related compounds.

spiro-tetrahydroisoquinoline unit, rendering the ecteinascidins significantly more complex than the safracins (**133–135**) and saframycins (**136–141**). The absolute stereochemistry was determined by X-ray crystal structure analysis of the natural N^{12}-oxide of ET 743 [47] and a derivative of ET 729 [48].

10.2.1 Biosynthesis and Biomimetic Strategy

It may come as a surprise that ET 743 is derived from a peptide precursor. However, feeding experiments on both enzymatic preparations and living organisms using radiolabeled amino acids showed that tyrosine and cysteine are employed in the biosynthesis of ecteinascidins [49, 50]. The two-carbon unit C1–C22 is presumably derived from glyoxylate, which probably originates from glycine [48, 51]. The gene cluster of the structurally related saframycin A was characterized and gives good precedence for ET 743 (Scheme 10.33).

A core tetrapeptide **142** (Ala-Gly-Tyr-Tyr) is thought to be assembled first by an NRPS machinery. The conversion of Tyr into the observed 3-hydroxy-5-methyl-*O*-methyltyrosine could take place either before its incorporation into the NRPS

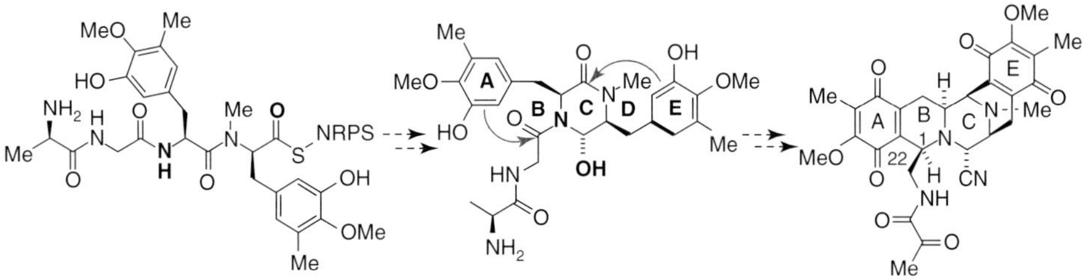

142 **143** **136**: saframycin A

Scheme 10.33 Gene cluster-based proposal for saframycin A biosynthesis.

machinery or during assembly [52, 53]. Reductive cleavage likely generates keto-piperazinol intermediate **143**, which then becomes the subject of an intricate maturation phase that ultimately delivers saframycin A (**136**) via several oxidoreduction, cyclization, *N*-methylation, and nitrile introduction steps. The order of ring closures remains to be clarified, but as the peptide chain is apparently released as an aldehyde it seems likely that ring C is closed first, followed by annelating E through D and then ring A through B later (arrows on Scheme 10.33).

The structural similarity between ET 743 (**128**) and saframycin A (**136**) (Figure 10.10) strongly suggests a common biosynthetic pathway for the pentacyclic core, which helps to guide a speculative retrobiosynthetic simplification of ET 743 (Scheme 10.34). Logical disconnections removing L-dopa (**128** → **144**) and cysteine (**144** → **145**) reveal a similar core structure **145**. However, only a complex sequence of post-assembly modifications can transform a Tyr-Tyr-Gly-based tripeptide **147** into **146**: reductive cleavage, carbinolamine formation, ring cyclization, *N*-methylation, oxidative deamination ($NH_2 \rightarrow C=O$), amide reduction ($C=O \rightarrow OH$), esterification (with Cys), benzyl cation formation, intramolecular cysteine addition, oxidative deamination, and a stereoselective Pictet–Spengler condensation!

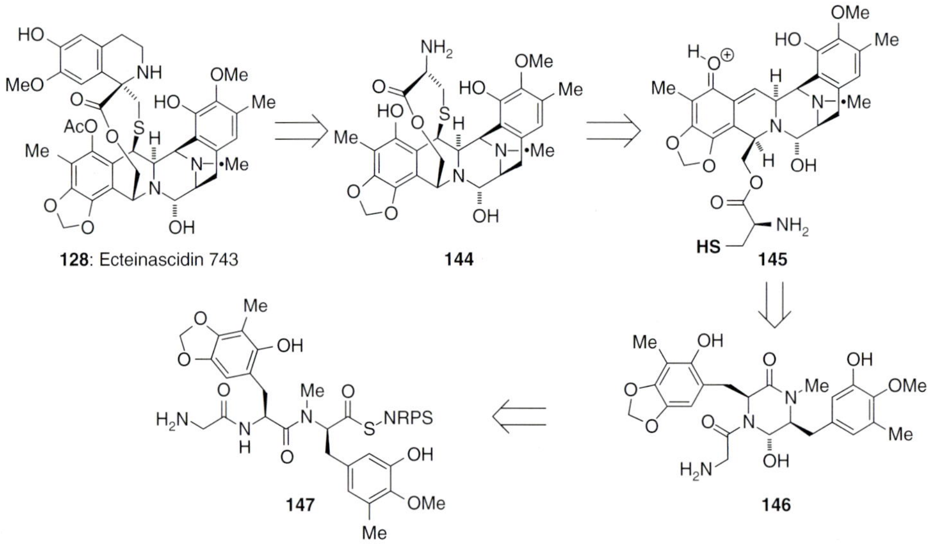

Scheme 10.34 Proposed biosynthesis for ET 743.

While the individual steps remain to be further illuminated, their formal necessity has led to the development of an array of elegant methods and total syntheses, all of which feature strongly biomimetic elements. Total syntheses were accomplished first by the group of Corey *et al.* [54], then by Fukuyama [55] and Zhu [56]. Williams [57] and Danishefsky [58] have contributed syntheses of the A–E pentacycle (Figure 10.11). Based on this precedent, a semi-synthesis starting from cyanosafracin B has been developed, which is used to produce

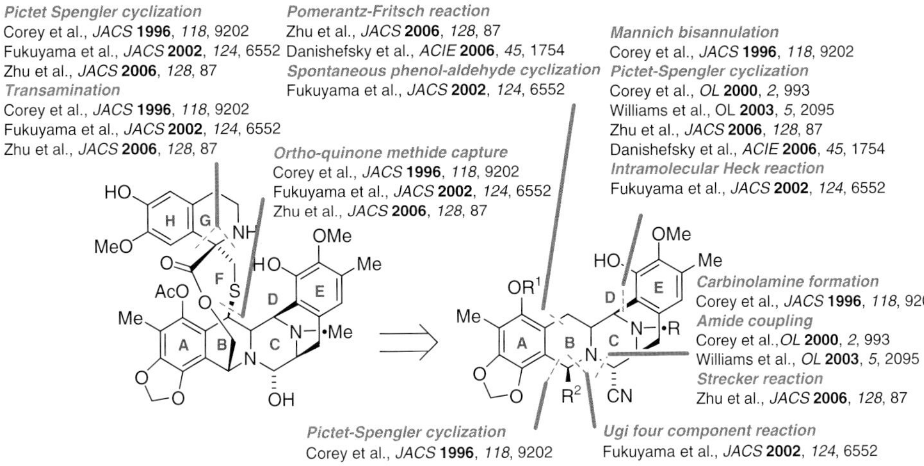

Figure 10.11 Summary of synthetic approaches to ET 743.

ET 743 on scale for drug use [59]. All the syntheses can be divided into three key phases: ABCDE-pentacycle formation, bridge installation, and the endgame. For the pentacycle several different access pathways have been developed, while the later stages of the synthesis have been realized in a fairly comparable manner by Corey, Fukuyama, and Zhu – all very close to anticipated biosynthetic events.

10.2.2
Pentacycle Formation

The seminal and optimized studies [54, 60] are based on biomimetic reactivity. Carbinolamine formation from amino acid ester **148** and a Mannich bis-annulation were used to effectively form the ABC ring system **149** (Scheme 10.35). An amino acid was appended to yield lactone **150**, which was reduced and the phenol protecting groups cleaved to yield lactol **151**. This was engaged in a biomimetic, diastereoselective Pictet–Spengler reaction via iminium-ion **152** to install the CD rings (→ **153**).

Williams *et al.* (Scheme 10.36) [57] employed an auxiliary β-lactam to control the stereochemical course of the first of two planned Pictet–Spengler reactions (**154** → **155**). After coupling of an unnatural *N*-methyltyrosine (**156**), the β-lactam was skillfully brought into play as a masked aldehyde: after chemoselective reduction with $LiEt_3BH$, the intermediate metalated aminal opened and eliminated $BnNH_2$. The resulting α,β-unsaturated aldehyde condensed with the free secondary amine to form iminium ion **157**, which underwent an intramolecular Pictet–Spengler cyclization cascade to furnish the pentacycle **158**.

In Danishefsky's approach (Scheme 10.37) [58], a Pomeranz–Fritsch isoquinoline synthesis was used to prepare AB-ring precursor **159**, which carried an aldehyde on ring E. Acid-mediated Boc-group cleavage initiated the formation of

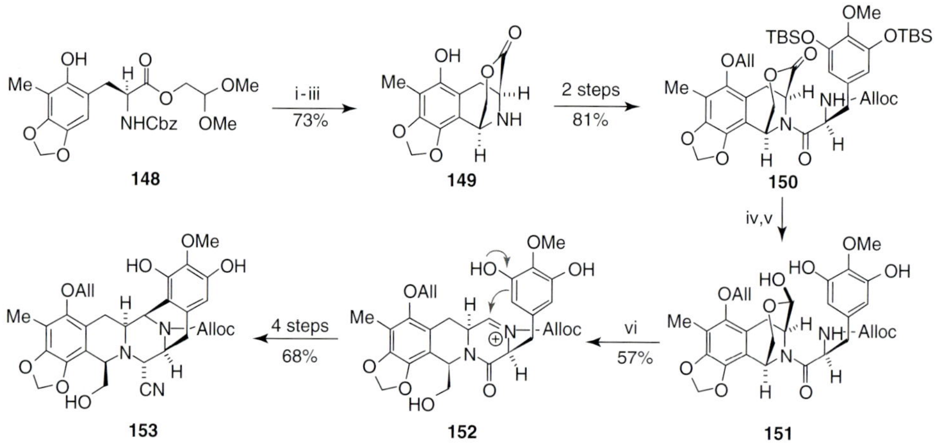

Scheme 10.35 Corey's approach to the pentacycle core **153** of ET 743. *Reagents and conditions*: (i) $BF_3{\cdot}OEt_2$, H_2O; (ii) $BF_3{\cdot}OEt_2$, 4 Å molecular sieves; (iii) H_2, Pd/C; (iv) $LiAlH_2(OEt)_2$, Et_2O, −78 °C; (v) KF, MeOH; and (vi) 0.6 M TfOH, $H_2O{-}CF_3CH_2OH$ (3 : 2), BHT, 45 °C. BHT = 2,6-di*fert* butyl-4-methyl phenol.

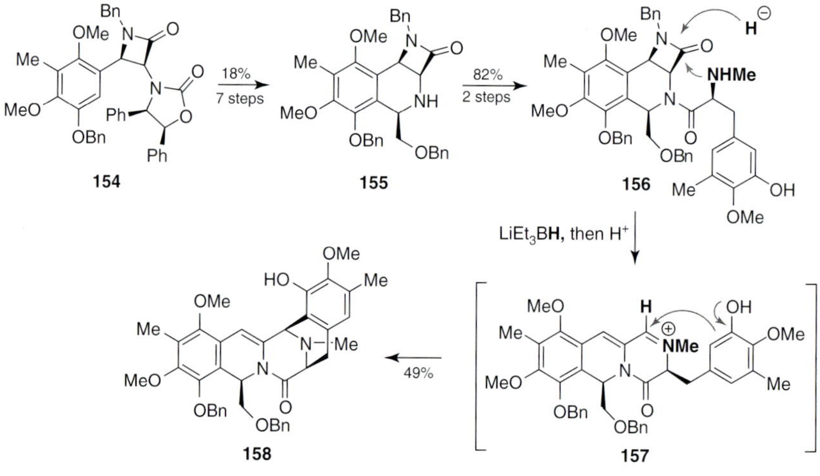

Scheme 10.36 Williams' synthesis of the pentacyclic core **158** of ET 743.

the D-ring imminium ion **160**, which engaged in a diastereoselective vinylogous Pictet–Spengler cyclization to form the CD-rings of pentacycle **161**.

Fukuyama's synthesis (Scheme 10.38) [55] featured an atom-economic Ugi reaction to assemble starting material **162**, which was transformed into diketopiperazine **163** in four steps. Chemoselective reduction of the secondary amide allowed transformation to enamine **164**, which was subjected to a (non-biomimetic) Heck cyclization to efficiently generate the CDE-ring system **165**. Multiple functional group

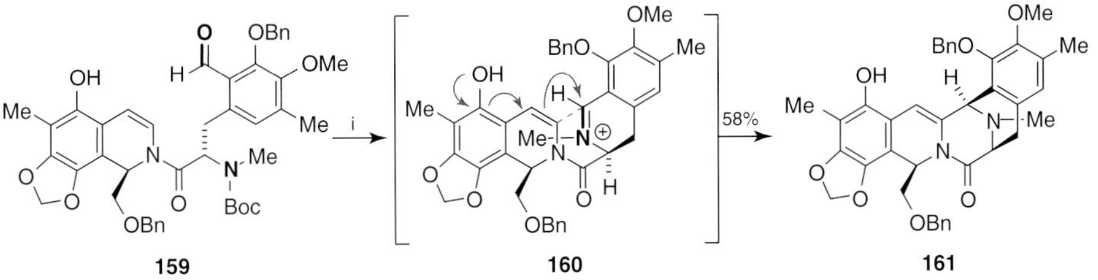

Scheme 10.37 Danishefsky's synthesis of ET 743's pentacyclic core (**161**): *Reagents and conditions*: (i) CHF_2CO_2H, $MgSO_4$, benzene.

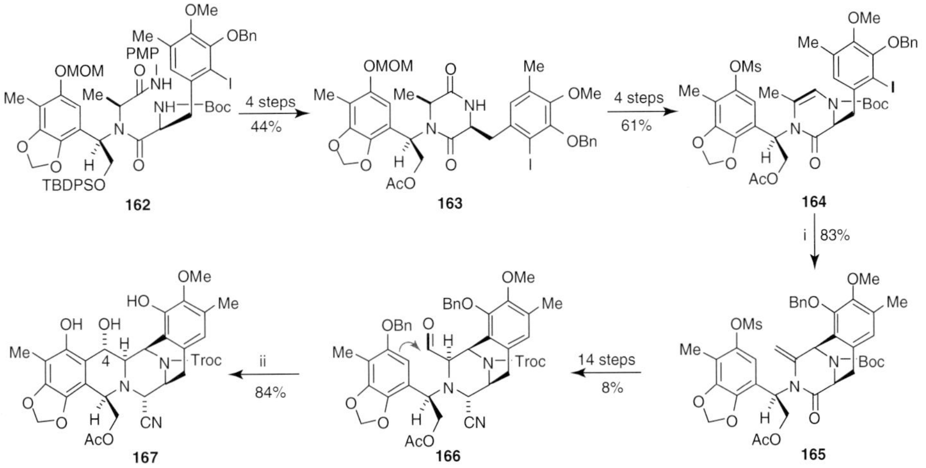

Scheme 10.38 Synthesis of the pentacyclic core (**167**) of ET 743 developed by Fukuyama. *Reagents and conditions*: (i) $Pd_2(dba)_3$ (5 mol.%), P(*o*-tol)$_3$, (20 mol.%), TEA, CH_3CN, reflux and (ii) Pd/C, H_2, THF.

interconversions gave the advanced aldehyde **166**, which smoothly ring-closed in a phenol–aldehyde condensation to the advanced ABCDE pentacycle **167** after hydrogenolytic debenzylation. This effort was based in part on previous work on saframycin A [61]. Notably, the intramolecular electrophilic substitution of the phenol by the aldehyde yielded the requisite oxidation state at C4 for the construction of the ten-membered bridge (*vide infra*).

In Zhu's approach (Scheme 10.39) [56], the synthesis of the pentacyclic core starts with a rapid construction of D–E fragment **170** using a highly diastereoselective Pictet–Spengler condensation of Garner's aldehyde [62] with substituted phenylalanol **168** (prepared from 3-methylcatechol in eight steps) [63]. After masking the secondary amine of **169** as the *N*-allyloxycarbamate and releasing the amino alcohol, the subsequent D–E fragment is coupled with **171** (prepared from sesamol in six steps) [56] to afford intermediate **172** as the major diastereomer. Ring C was

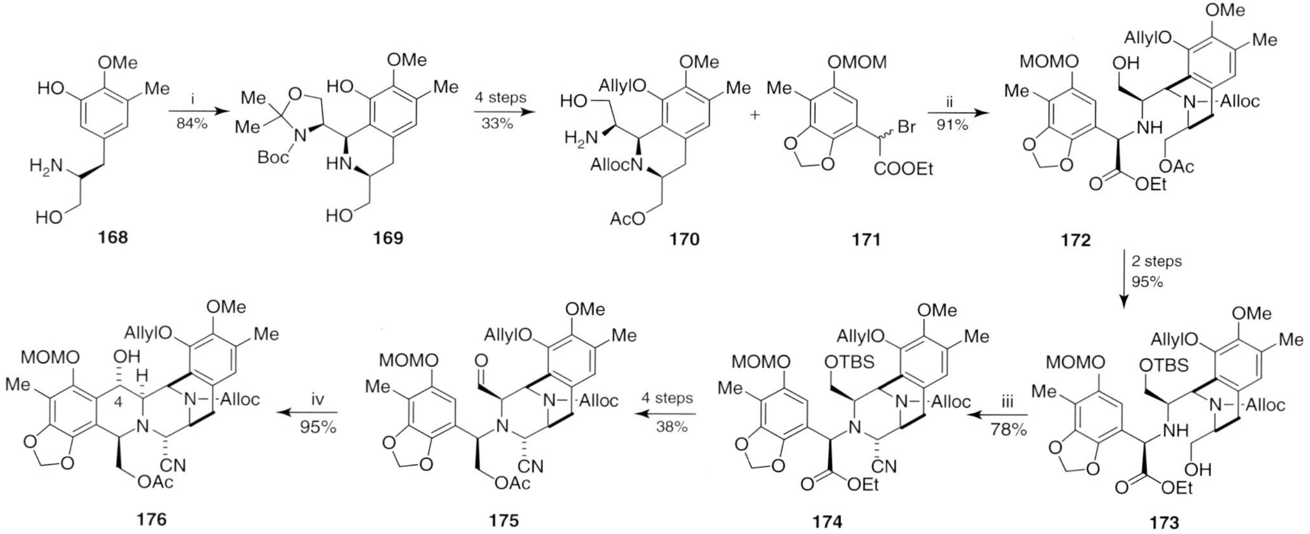

Scheme 10.39 Synthesis of the pentacyclic core (**176**) of ET 743 developed by Zhu. *Reagents and conditions*: (i) *N*-Boc-L-serinal-N,O-dimethylacetal, AcOH, CH_2Cl_2/CF_3CH_2OH (7 : 1), 3 Å molecular sieve; (ii) TEA, MeCN, 0 °C; (iii) Dess–Martin reagent then TMSCN, $ZnCl_2$; and (iv) TFA, CH_2Cl_2.

constructed into the D–E segment using a decisive zinc chloride-catalyzed Strecker reaction to give intermediate **174**. A Pomeranz–Fritsch-type cyclization installed ring B, with the corrected oxidation state at C4 affording the pentacycle **176**.

Overall, many of these key transformations were conceived at oxidation stage levels similar to the natural substrates and hence are considerably biomimetic, but their strategic placement within the scaffold deviates. Compared with their final strategic ring closures, Corey's and Williams' syntheses used the D-ring for a biomimetic annelation of a nucleophilic aromatic ring, which may be formed in biosynthesis during earlier steps. Danishefsky employed an imminium ion cyclization to close the C-ring, which in biosynthesis will be formed first. Fukuyama and Zhu close the B-ring late in synthesis, comparable to the anticipated biosynthesis, but by forming a C–C bond that in biosynthesis is part of an amino acid.

10.2.3 Bridge Formation

The centerpiece of the ecteinascidin scaffold synthesis is certainly the construction of the ten-membered-ring thioether bridge. Saliently, retrobiosynthetic analysis strongly suggested a cysteine-thiol capture, which was put into synthetic practice in two different forms (Scheme 10.40). Corey managed to selectively oxidize the A-ring arene to **177**, which was eliminated to give a highly reactive quinone methide **178**. Base-induced Fm-deprotection then unleashed the nucleophilic thiolate **179**

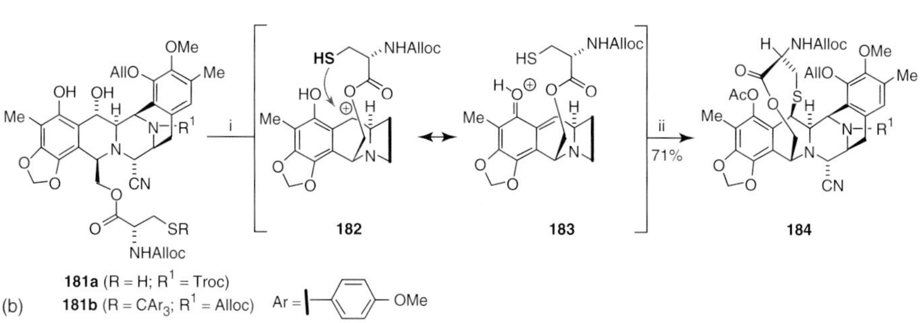

Scheme 10.40 (a) Corey's strategy for bridge formation via an ortho-quinone methide. *Reagents and conditions*: (i) DMSO, Tf_2O, −40 °C; (ii) *i*-Pr_2NEt, 0 °C; (iii) *t*-BuOH, 0 °C; (iv) $(Me_2N)_2$ C=N-*t*-Bu, 23 °C; and (v) Ac_2O, 23 °C. Fm = fluorenylmethyl. (b) Fukuyama's and Zhu's approaches for bridge formation via a benzyl cation. *Reagents and conditions*: (i) TFA, CF_3CH_2OH; and (ii) Ac_2O, pyridine, DMAP.

in situ, which added to the quinine methide and formed the sought after thioether **180** in very good yield (Scheme 10.40a). On the other hand, Fukuyama accessed the D-ring in the correct oxidation stage at the benzylic carbon (Scheme 10.40b). Hence, simple acid treatment of **181a** led to the highly stabilized benzyl cation **182** (isoelectronic to protonated quinine methide **183**), which captured the free thiol chain and delivered the thioether bridge in **184** [55]. Zhu [56] employed an efficient one-pot deprotection/cyclization of the *S*-trityl protected precursor **181b** to convert it into **184**. In all cases the product was isolated as the more stable phenol acetate. The efficient biomimetic synthesis of this bridge both by nucleophilic and by electrophilic chemistry indicates that this ring closure will readily occurs in Nature, where – according to typical oxidative biosynthesis chemistry – a cationic pathway may be in operation.

10.2.4
Endgame

To finish the synthesis of ET 743, an L-dopa derivative must be incorporated. Here, Fukuyama and Zhu essentially follow Corey's precedent (Scheme 10.41) [56, 61]. Allyloxycarbamate **185** is deprotected to the free amine, and a pyridinium salt closely related to the amino-transferase cofactor pyridoxal pyrophosphate is then used to execute an oxidative deamination to ketone **186**. After this astonishingly biomimetic move, the ketone was condensed with 5-(2-aminoethyl)-2-methoxyphenol in a Pictet–Spengler-like tetrahydroisoquinoline synthesis to afford **187**. Notably, this condensation was exquisitely facile and completely stereoselective even without the aid of an enzyme. Apparently, the steric bulk of the E-ring pointing in the same direction leads to an *exo*-attack of the methoxyphenol ring on the intermediate imminium ion. Final protecting group manipulations delivered ET 743 (**128**).

Overall, among peptide alkaloids the ecteinascidines have arguably stimulated some exquisite biomimetic synthesis plans, and the methods developed for this particularly intriguing scaffold will continue to stimulate the field of biomimetic synthesis. Notably, a large set of these transformations was adapted to the semi-synthesis of ET 743 for drug use. This illustrates that biomimetic

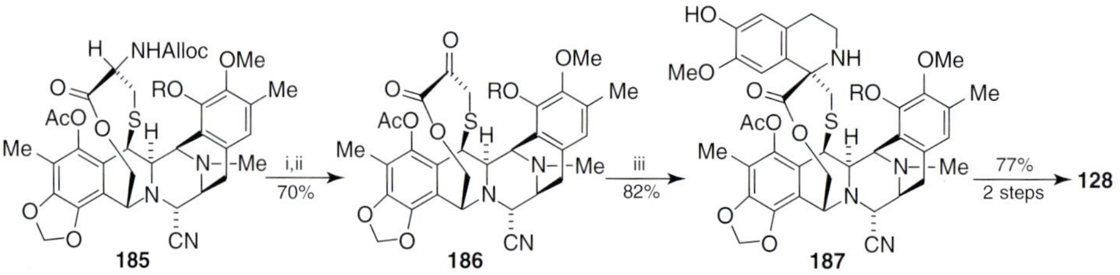

Scheme 10.41 Biomimetic completion of ET 743 total synthesis. *Reagents and conditions*: (i) *n*-Bu_3SnH, $PdCl_2(PPh_3)_2$, AcOH; (ii) *N*-methylpyridinium-4-carboxyaldehyde iodide, DBU, $(CO_2H)_2$; and (iii) 5-(2-aminoethyl)-2-methoxyphenol, silica gel. R = Mom.

syntheses can be competitive on a larger scale, and can even positively impact process research.

10.3 Outlook

In this and the preceding chapter we could neither cover all peptide alkaloid structures nor possibly refer to all the excellent work executed beyond the material we were able to discuss. The world of peptide alkaloids is certainly much larger. For example, we had to omit other glycopeptides antibiotics such as teicoplanin, complestatin (**188**/**189**, Figure 10.12), or kistamycin, the indole-crosslinked alkaloids of the kapakahine family (**190**), or the instructive phallotoxins (**191**) and amatoxins from the death cap mushroom family – to mention just a few.

However, from all the knowledge available and the selection of chemistry illustrated above, it becomes increasingly apparent that peptide alkaloid structures make use of a general toolbox of enzymatic manipulations that show similarities even between very distinguished RPS and NRPS biosynthetic pathways and different species in the kingdoms of life (bacteria, fungi, plants). Especially, the specific control Nature exerts over assembly, maturation, and tailoring of a peptide

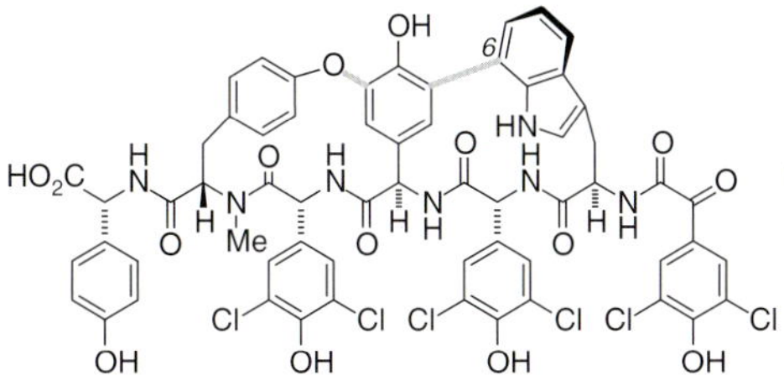

188: chloropeptin I

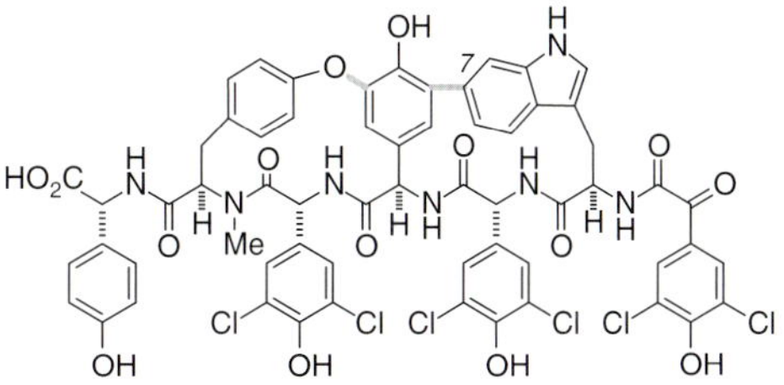

189: chloropeptin II/complestatin

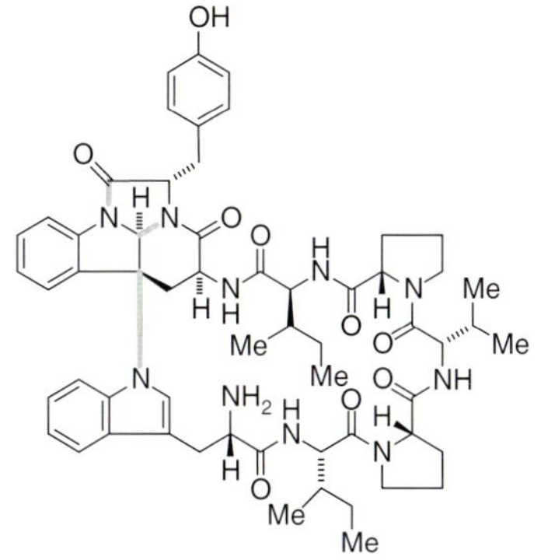

190: kapakahine A

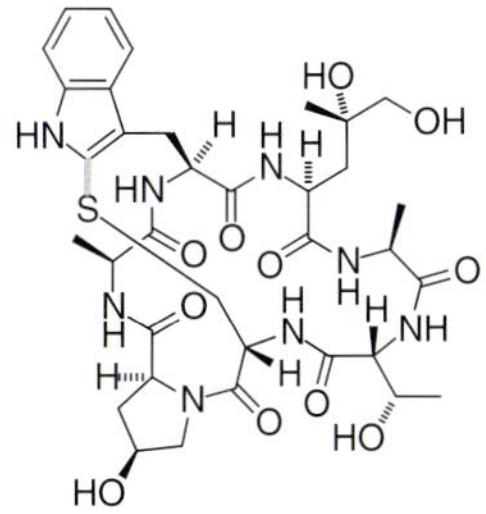

191: phalloidin

Figure 10.12 Other peptide alkaloid structures. Oxidation-derived connector bonds are highlighted.

alkaloid is highly stimulating. For more precise control of reactions on peptide precursors in the chemical sense and for targeted planning, more specific oxidation and crosslinking chemistry needs to be developed. In such synthesis endeavors, chemo- and regioselectivity will be much more necessary (and may be more difficult to achieve) than the central stereoselectivity that dominates the assembly of carbon chains. These efforts may eventually contribute to a different approach to synthesis – namely to more manipulating of oligomers, and to less assembling of building blocks.

References

1. (a) Koguchi, Y., Kohno, J., Nishio, M., Takahashi, K., Okuda, T., Ohnuki, T., and Komatsubara, S. (2000) *J. Antibiot.*, **53**, 105–109; (b) Kohno, J., Koguchi, Y., Nishio, M., Nakao, K., Kuroda, M., Shimizu, R., Ohnuki, T., and Komatsubara, S. (2000) *J. Org. Chem.*, **65**, 990–995.
2. Kaiser, M., Siciliano, C., Assfalg-Machleidt, I., Groll, M., Milbradt, A.G., and Moroder, L. (2003) *Org. Lett.*, **5**, 3435–3437.
3. Berthelot, A., Piguel, S., Le Dour, G., and Vidal, J. (2003) *J. Org. Chem.*, **68**, 9835–9838.
4. Sattely, E.S., Fischbach, M.A., and Walsh, C.T. (2008) *Nat. Prod. Rep.*, **25**, 757–793.
5. Coste, A., Couty, F., and Coste, G. (2008) *C. R. Chim.*, **11**, 1544–1573.
6. (a) Lin, S. and Danishefsky, S.J. (2001) *Angew. Chem.*, **113**, 2021–2024; *Angew. Chem. Int. Ed.*, **40**, 1967–1970; (b) Lin, S. and Danishefsky, S.J. (2002) *Angew. Chem.*, **114**, 530–533; *Angew. Chem. Int. Ed.*, **41**, 512–515; (c) Lin, S., Yang, Z.-Q., Kwok, B.H.B., Koldobskiy, M., Crews, C.M., and Danishefsky, S.J. (2004) *J. Am. Chem. Soc.*, **126**, 6347–6355.
7. (a) Albrecht, B.K. and Williams, R.M. (2003) *Org. Lett.*, **5**, 197–2000; (b) Albrecht, B.K. and Williams, R.M. (2004) *Proc. Natl. Acad. Sci. U.S.A.*, **101**, 11949–11954.
8. (a) Inoue, M., Furuyama, H., Sakazaki, H., and Hirama, M. (2001) *Org. Lett.*, **3**, 2863–2865; (b) Inoue, M., Furuyama, H., Sakazaki, and H., Hirama, M. (2003) *Angew. Chem.*, **115**, 2758–2761; *Angew. Chem. Int. Ed.*, **42**, 2654–2657.
9. (a) Leung, T.-W.C., Williams, D.H., Barna, J.C.J., Foti, S., and Oelrichs, P.B. (1986) *Tetrahedron*, **42**, 3333–3348; (b) Kahn, S.D., Booth, P.M., Waltho, J.P., and Williams, D.H. (1989) *J. Org. Chem.*, **54**, 1901–1904; correction: Kahn, S.D., Booth, P.M., Waltho, J.P., Williams, D.H. (2000) *J. Org. Chem.*, **65**, 8406.
10. (a) Kobayashi, J., Suzuki, H., Shimbo, K., Takeya, K., and Morita, H. (2001) *J. Org. Chem.*, **66**, 6626–6633; (b) Suzuki, H., Morita, H., Iwasaki, S., and Kobayashi, J. (2003) *Tetrahedron*, **59**, 5307–5315; (c) Suzuki, H., Morita, H., Shiro, M., and Kobayashi, J. (2004) *Tetrahedron*, **60**, 2489–2495.
11. Puk, O., Bischoff, D., Kittel, C., Pelzer, S., Weist, S., Stegmann, E., Süssmuth, R.D., and Wohlleben, W. (2004) *J. Bacteriol.*, **186**, 6093–6100.
12. Example for aziridine biosynthesis: Ogasawara, Y. and Lu, H.-W. (2009) *J. Am. Chem. Soc.*, **131**, 18066–18068.
13. Reviews: (a) Hu, X.E. (2004) *Tetrahedron*, **60**, 2701–2743; (b) Lu, P. (2010) *Tetrahedron*, **66**, 2549–2560.
14. Bergmeier, S.C., Katz, S.J., Huang, J., McPherson, H., Donoghue, P.J., and Reed, D.D. (2004) *Tetrahedron Lett.*, **45**, 5011–5014.
15. (a) Ma, B., Litvinov, D.N., He, L., Banerjee, B., and Castle, S.L. (2009) *Angew. Chem.*, **121**, 6220–6223; *Angew. Chem. Int. Ed.*, **48**, 6104–6104; (b) Ma, B., Banerjee, B., Litvinov, D.N., He, L., and Castle, S.L. (2010) *J. Am. Chem. Soc.*, **132**, 1159–1171.

16. Hu, W., Zhang, F., Xu, Z., Liu, Q., Cui, Y., and Jia, Y. (2010) *Org. Lett.*, **12**, 956–959.

17. Feng, Y. and Chen, G. (2010) *Angew. Chem.*, **122**, 970–973; *Angew. Chem. Int. Ed.*, **49**, 958–961.

18. (a) Zaitsev, V.G., Shabashov, D., and Daugulis, O. (2005) *J. Am. Chem. Soc.*, **127**, 13154–13155; (b) Shabashov, D. and Daugulis, O. (2005) *Org. Lett.*, **7**, 3657–3659.

19. Reddy, B.V.S., Reddy, L.R., and Corey, E.J. (2006) *Org. Lett.*, **8**, 3391–3394.

20. (a) Lam, K.S., Hesler, G.A., Mattei, J.M., Mamber, S.W., Forenza, S., and Tomita, K. (1990) *J. Antibiot.*, **43**, 956–960; (b) Leet, J.E., Figroeder, D.R., Krishnan, B.S., and Matson, J.A. (1990) *J. Antibiot.*, **43**, 961–966; (c) Leet, J.E., Figroeder, D.R., Golik, J., Matson, J.A., Doyle, T.W., Lam, K.S., Hill, S.E., Lee, M.S., Whitney, J.L., and Krishnan, B.S. (1996) *J. Antibiot.*, **49**, 299–311.

21. Umezawa, K., Ikeda, Y., Uchihata, Y., Naganawa, H., and Kondo, S. (2000) *J. Org. Chem.*, **65**, 459–463.

22. (a) Broberg, A., Menkis, A., and Vasiliauskas, R. (2006) *J. Nat. Prod.*, **69**, 97–102; (b) Pohanka, A., Menkis, A., Levenfors, J., and Broberg, A. (2006) *J. Nat. Prod.*, **69**, 1776–1781.

23. Fujimori, D.G., Hrvatin, S., Neumann, C.S., Strieker, M., Marahiel, M.A., and Walsh, C.T. (2007) *Proc. Natl. Acad. Sci. U.S.A.*, **104**, 16498–16503.

24. Arroyo, V., Hall, M.J., Hassall, C.H., and Yamasaki, K. (1976) *J. Chem. Soc., Chem. Commun.*, 845–846.

25. Umezawa, K., Ikeda, Y., Kawase, O., Naganawa, H., and Kondo, S. (2001) *J. Chem. Soc., Perkin Trans. 1*, 1550–1553.

26. (a) Parry, R.J. (1994) *J. Chem. Soc., Chem. Commun.*, 995–996; (b) Tao, T., Alemany, L.B., and Parry, R.J. (2003) *Org. Lett.*, **5**, 1213–1215.

27. Heemstra, J.R. and Walsh, C.T. Jr. (2008) *J. Am. Chem. Soc.*, **130**, 14024–14025.

28. Poss, C.S. and Figreiber, S.L. (1994) *Acc. Chem. Res.*, **27**, 9–17.

29. (a) Kamenecka, T.M. and Danishefsky, S.J. (2001) *Chem. Eur. J.*, **7**, 41–63; (b) Kamenecka, T.M. and Danishefsky, S.J. (1998) *Angew. Chem.*, **110**, 3164–3166; *Angew. Chem. Int. Ed.*, **37**, 2993–2995; (c) Kamenecka, T.M. and Danishefsky, S.J. (1998) *Angew. Chem.*, **110**, 3166–3168; *Angew. Chem. Int. Ed.*, **37**, 2995–2998

30. (a) Yu, S.-M., Hong, W.-X., Wu, Y., Zhong, C.-L., and Yao, Z.-J. (2010) *Org. Lett.*, **12**, 1124–1127; (b) Hong, W.-X., Chen, L.-J., Zhong, C.-L., and Yao, Z.-J. (2006) *Org. Lett.*, **8**, 4919–4922.

31. Oelke, A.J., France, D.J., Hofmann, T., Wuitschik, G., and Ley, S.V. (2010) *Angew. Chem.*, **122**, 6275–6278; *Angew. Chem. Int. Ed.*, **49**, 6139–6142.

32. Lindquist, N., Fenical, W., Vanduyne, G.D., and Clardy, J. (1991) *J. Am. Chem. Soc.*, **113**, 2303–2304.

33. Fernandez, R., Martin, M.J., Rodriguez-Acebes, R., Reyes, F., Francesch, A., and Cuevas, C. (2008) *Tetrahedron Lett.*, **49**, 2283–2285.

34. Wang, G., Shang, L., Burgett, A.W.G., Haran, P.G., and Wang, X. (2007) *Proc. Natl. Acad. Sci. U.S.A.*, **104**, 2068–2073.

35. (a) Li, J., Chen, X., Burgett, A.W.G., and Harran, P.G. (2001) *Angew. Chem.*, **113**, 2754–2757; *Angew. Chem. Int. Ed.*, **40**, 2682–2685; (b) Li, J., Jeong, S., Esser, L., and Harran, P.G. (2001) *Angew. Chem.*, **113**, 4901–4906; *Angew. Chem. Int. Ed.*, **40**, 4765–4770; (c) Li, J., Burgett, A.W.G., Esser, L., Amezcua, C., and Harran, P.G. (2001) *Angew. Chem.*, **113**, 4906–4909; *Angew. Chem. Int. Ed.*, **40**, 4770–4773.

36. Lachia, M. and Moody, C.J. (2008) *Nat. Prod. Rep.*, **25**, 227–253.

37. (a) Nicolaou, K.C., Chen, D.Y.K., Huang, X.H., Ling, T.T., Bella, M., and Snyder, S.A. (2004) *J. Am. Chem. Soc.*, **126**, 12888–12896; (b) Nicolaou, K.C., Hao, J.L., Reddy, M.V., Rao, P.B., Rassias, G., Snyder, S.A., Huang, X.H., Chen, D.Y.K., Brenzovich, W.E., Giuseppone, N., Giannakakou, P., and O'Brate, A. (2004) *J. Am. Chem. Soc.*, **126**, 12897–12906; (c) Nicolaou, K.C., Hao, J.L., Reddy, M.V., Rao, P.B., Rassias, G., Snyder, S.A., Huang, X.H., Chen, D.Y.K., Brenzovich, W.E., Giuseppone, N., Giannakakou, P., and O'Brate, A. (2004) *J. Am. Chem. Soc.*, **126**, 15316–15316.

38. Burgett, A.W.G., Li, Q.Y., Wei, Q., and Harran, P.G. (2003) *Angew. Chem.*, **115**, 5111–5116; *Angew. Chem. Int. Ed.*, **42**, 4961–4966.
39. Magarvey, N.A., Ehling-Schulz, M., and Walsh, C.T. (2006) *J. Am. Chem. Soc.*, **128**, 10698–10699.
40. Magnus, P. and McIver, E.G. (2000) *Tetrahedron Lett.*, **41**, 831–834.
41. (a) Sperry, J. and Moody, C.J. (2006) *Chem. Commun.*, **22**, 2397–2399; (b) Sperry, J. and Moody, C.J. (2010) *Tetrahedron*, **66**, 6483–6495.
42. Rinehart, K.L., Holt, T.G., Fregeau, N.L., Keifer, P.A., Wilson, G.R., Perun, T.J., Sakai, R., Thompson, A.G., Stroh, J.G., Shield, L.S., Seigler, D.S., Li, L.H., Martin, D.G., Grimmelikhuijzen, C.J.P., and Gäde, G. (1990) *J. Nat. Prod.*, **53**, 771–792.
43. Rinehart, K.L., Holt, T.G., Fregeau, N.L., Stroh, J.G., Keifer, P.A., Sun, F., Li, L.H., and Martin, D.G. (1990) *J. Org. Chem.*, **55**, 4512–4515.
44. Wright, A.E., Forleo, D.A., Gunawardana, G.P., Gunasekera, S.P., Koehn, F.E., and McConnel, O.J. (1990) *J. Org. Chem.*, **55**, 4508–4512.
45. Cuevas, C. and Francesch, A. (2009) *Nat. Prod. Rep.*, **26**, 322–337.
46. European Medicines Agency, Meeting highlights from the Committee for Medicinal Products for Human Use, 16-19 July 2007, Press release 19 July 2007; *http://www.ema.europa.eu/pdfs/human/press/pr/43140707en.pdf* (accessed 19 July 2007)
47. Sakai, R., Rinehart, K.L., Guan, Y., and Wang, A.H.-J. (1992) *Proc. Natl. Acad. Sci. U.S.A.*, **89**, 11456–11460.
48. Guan, Y., Sakai, R., Rinehart, K.L., and Wang, A.H.-J. (1993) *J. Biomol. Struct. Dyn.*, **10**, 793–818.
49. Russel, G.K. and Miranda, N.F. (1995) *J. Nat. Prod.*, **58**, 1618–1621.
50. Jeedigunta, S., Krenisky, J.M., and Kerr, R.G. (2000) *Tetrahedron*, **56**, 3303–3307.
51. Kapadia, G.J., Rao, G.S., Leete, E., Fayez, M.B.E., Vaishnav, Y.N., and Fales, H.M. (1970) *J. Am. Chem. Soc.*, **92**, 6943–6951.
52. Li, L., Deng, W., Song, J., Ding, W., Zhao, Q.-F., Peng, C., Song, W.-W., Tang, G.-L., and Liu, W. (2008) *J. Bacteriol.*, **190**, 251–263.
53. Velasco, A., Acebo, P., Gomez, A., Schleissner, C., Rodriguez, P., Aparicio, T., Conde, S., Munoz, R., de la Calle, F., Garcia, J.L., and Sanchez-Puelles, J.M. (2005) *Mol. Microbiol.*, **56**, 144–154.
54. Corey, E.J., Gin, D.Y., and Kania, R.S. (1996) *J. Am. Chem. Soc.*, **118**, 9202–9203.
55. Endo, A., Yanagisawa, A., Abe, M., Tohma, S., Kan, T., and Fukuyama, T. (2002) *J. Am. Chem. Soc.*, **124**, 6552–6554.
56. Chen, J., Chen, X., Bois-Choussy, M., and Zhu, J. (2006) *J. Am. Chem. Soc.*, **128**, 87–89.
57. Jin, W., Metobo, S., and Williams, R.M. (2003) *Org. Lett.*, **5**, 2095–2098.
58. Zheng, S., Chan, C., Furuuchi, T., Wright, B.J.D., Zhou, B., Guo, J., Danishefsky, S.J. (2006) *Angew. Chem.*, **118**, 1786–1791; *Angew. Chem. Int. Ed.*, **45**, 1754–1759.
59. Cuevas, C., Pérez, M., Martin, M.J., Chicharro, J.L., Fernandez-Rivas, C., Flores, M., Francesch, A., Gallego, P., Zarzuelo, M., de la Calle, F., Garcia, J., Polanco, C., Rodriguez, I., and Manzanares, I. (2000) *Org. Lett.*, **2**, 2545–2548.
60. Martinez, E.J. and Corey, E.J. (2000) *Org. Lett.*, **2**, 993–996.
61. Fukuyama, T., Yang, L., Ajeck, K.L., and Sachleben, R.A. (1990) *J. Am. Chem. Soc.*, **112**, 3712–3713.
62. (a) Garner, P. and Park, J.M. (1987) *J. Org. Chem.*, **52**, 2361–2364; (b) Garner, P. and Park, J.M. (1988) *J. Org. Chem.*, **53**, 2979–2984; (c) Garner, P., Park, J.M., and Malecki, E. (1988) *J. Org. Chem.*, **53**, 4395–4398.
63. De Paolis, M., Chen, X., and Zhu, J. (2004) *Synlett*, 729–731.